Regional Geology Reviews

The Geology of series seeks to systematically present the geology of each country, region and continent on Earth. Each book aims to provide the reader with the state-of-the-art understanding of a regions geology with subsequent updated editions appearing every 5 to 10 years and accompanied by an online "must read" reference list, which will be updated each year. The books should form the basis of understanding that students, researchers and professional geologists require when beginning investigations in a particular area and are encouraged to include as much information as possible such as: Maps and Cross-sections, Past and current models, Geophysical investigations, Geochemical Datasets, Economic Geology, Geotourism (Geoparks etc), Geo-environmental/ecological concerns, etc.

Sami Khomsi • Francois Roure
Editors

Geology of North Africa and the Mediterranean: Sedimentary Basins and Georesources

Springer

Editors
Sami Khomsi
Faculty of Earth Sciences, Geoexploration
King Abdulaziz University
Jeddah, Saudi Arabia

Francois Roure
Direction Geologie-Geochimie-Geophysique
Institut Francais du Petrole
Rueil Malmaison Cedex, France

ISSN 2364-6438 ISSN 2364-6446 (electronic)
Regional Geology Reviews
ISBN 978-3-031-18746-9 ISBN 978-3-031-18747-6 (eBook)
https://doi.org/10.1007/978-3-031-18747-6

This Springer imprint is published by the registered company Springer Nature Switzerland AG
The registered company address is: Gewerbestrasse 11, 6330 Cham, Switzerland

We dedicate this book to our dear colleague and friend Professor Mourad Bedir who passed away in December 2022.

Mourad is a highly recognized Tunisian geoscientist in the domain of subsurface geology and seismic interpretations. He was a pioneer in subsurface basin analysis in Tunisia, especially in the southern Gafsa and Sahel domains. He was also a very active member of the geologic maps project under the direction of Professor Fouad Zargouni.

He presented in 1988 a highly viewed PhD on the subsurface geology of the Sahel of Mahdia in the Atlasic Foreland Basin. Since then, he also presented an important "These d'Etat" on sedimentary basins dynamics and reservoir occurrences in the foreland domains of the Sahel and Southern Atlas. He was one of the first Tunisian geologists involved in seismic interpretation. He introduced also seismic stratigraphy in Tunisia.

Mourad published more than 70 peer-reviewed papers in international journals and conference papers. He managed and contributed to some major national and international works and research projects. He is one of the main founders, together with professors Fakher Jamoussi, Ammar Melayeh, Abdelkrim Charef, Mohamed ben Youssef, and Sami Khomsi, of the

Georesources Laboratory since 2000 . The laboratory is one of his main scientific/administrative achievements. He was also General Manager (Directeur General) of the CERTE for 6 years from 2006 to 2011.

Professor Mourad Bedir supervised more than 10 PhD on the subsurface geology of Tunisia, involving both structural and stratigraphic interpretations of geophysical records.

In such sad circumstances, we wish to present our sincere condolences to the family of Professor Mourad Bedir; his wife Wafa; and his daughters, Myriam, Malek, and Marwa.

We also present our condolences to the members of the Georesources Laboratory, the CERTE, University of Carthage and all the colleagues who worked or shared their research and time in the field or in the Lab with Mourad.

We lost a great scientist and a highly beloved father and friend.

Sami Khomsi

Francois Roure

Preface

This book constitutes an overview of the sedimentary basins of North Africa and the Mediterranean Sea. It comprises 11 original chapters covering a wide range of major regional structures and geo-resources focusing on the tectonic and geodynamic evolution of several regional Foreland and Fold-and-Thrust Belts (FFTB) such as the Rif-Tell and the Atlas. It is subdivided into 4 main topics as follows:

First Thematic: Geodynamic Evolution of the Mediterranean and Peri-Mediterranean Sedimentary Basins and Regional Geology

This theme is covered by two chapters:

1. The Mediterranean Sea: a laboratory to characterize micro-continental drift and oceanic basin formation processes. By D. Aslanian et al,
2. Three inverted Tethyan margins in the Rif belt (Morocco) Review and discussion. By A. Michard et al.

In the Chap. 1, Aslanian et al. discuss and present a synthesis of the sequence of deep geodynamic processes involving the lithosphere that lead to the genesis of passive margins and associated oceanic basins in the Mediterranean and Peri-Mediterranean realm. These processes have major consequences on the thermal and tectonic evolution of the margin, as well as on geodynamic reconstructions and tectonic evolution. They present different interpretations in the Peri-Mediterranean basins, i.e., in the Eastern Mediterranean and Levant-Northern Arabia, as well as in the Western Mediterranean, i.e., the Alboran, Algerian and Adria basins, the north African paleo-Tethyan margin and the Sicily Strait. They use multiple approaches involving interpretations of deep seismic experiments, regional geology and paleoteconic reconstructions. They also propose to conduct a wide-angle and multi-trace reflection seismic survey along 5 regional profiles (171 deployments of ocean bottom seismometers -OBS-). These profiles will allow determining the geometry, structure, and acoustic velocity of the different segments of the Algerian Basin, the base of the crust—Moho—and underlying lithospheric mantle.

In the Chap. 2, Michard et al. present and discuss a review of the westernmost Tethyan margins in the Rif belt of Morrocco in the Western Mediterranean. They describe new results on the remnants of the margins of the former basin as exposed in the Rif belt. In fact, the external zones of the belt expose remnants of the southern Jurassic Ocean-Continent Transition (OCT) of the Maghrebian Tethys and a Triassic volcanic-rich segment of the NW-African passive margin showing serpentine and gabbro slivers included in the accretionary prism derived from the inversion of the African passive margin. The northern margin of the Maghrebian Ocean is classically represented by the Dorsale Calcaire and pre-dorsalian

Triassic-Paleogene units at the external border of the Internal Zones (Alboran Domain). Their chapter based on important field works and regional reconstructions involving paleo-tectonic margin evolution provides major results concerning the tectonic slivers sampled in the former southern margin of the Maghrebian Tethys currently exposed in the External Rif accretionary prism. From NE to SW, outcrops comprise serpentinites from the Late Jurassic OCT, and gabbro bodies intruded during the Early Jurassic in the distal African margin, respectively. These authors present also new results focusing on the Rif belt formed synchronously with the Betic Cordillera (northern branch of the Gibraltar Arc) during the Cenozoic tectonic events related to the Africa-Eurasia convergence associated with the subduction of the westernmost Tethyan lithosphere of the Ligurian-Maghrebian basin. In this work, they describe the remnants of the margins of the latter basin as exposed in the Rif belt. They integrate their observations/interpretations into the regional geology, thus providing a brilliant overview of the Western Tethyan margin.

Second Thematic: Stratigraphy and Sedimentary Record

This theme is covered by two chapters:

3. Stratigraphy and structure of the Tellian thrust sheets of northeastern Algeria and its relationship with the Sellaoua and High Madjerda foreland: Souk Ahras case. By A. Chabbi et al.,
4. Facies analyses and basin evolution of the Cretaceous-Tertiary rift-related sedimentary succession of Haddat Ash Sham area, west central Arabian Shield, Saudi Arabia. By A. A Messaed et al.

In Chap. 3, Chabbi et al. present new results on their works in the High Mejerda foreland basin of the Algerian Tell. Their work is based on precise biostratigraphy and a set of new structural transects which allow them to present a very large view of the structural architecture of Souk Ahras Tellian thrust sheets and their relationship with the Sellaoua and High Medjerda Valley. They recognize and present two contractional systems in the Souk Ahras region. The northern thrust system displays an imbricate fan of thrusts involving local and intra-formational décollement driven by the specific mechanical stratigraphy. Large Triassic bodies cannot be geometrically linked to this thrust system and should rather record former salt tectonics. Their structural transects allow visualizing the overall geometry of the thrust imbricates and the role of Triassic salt involved in the deformations. Their work presents also new constraints not only on the structural style but also on the biostratigraphy by new attributions.

In Chap. 4, Mesaed et al. describe the facies architecture and sedimentary basin evolution of the rift-related Cretaceous-Tertiary succession of Haddat Ash Sham area, throughout the Arabian Shield in Saoudi Arabia. They present fieldwork and on-site paleo-stress measurements together with thin sections. They provide also lithostratigraphic logs with detailed facies descriptions and field observations, thus allowing paleo-environmental reconstitutions within the Paleogene marginal deposits of the Arabian Shield. Their interpretations are presented as block diagrams showing the depositional models of the studied sedimentary successions of Haddat Ash Sham area with the synsedimentary faulting events controlling the synsedimentary depositional configurations.

Third Thematic: Tectonic Evolution, Structural Styles, and Oil/Gas Traps of the North African Foreland Basins

This theme is covered by four chapters:

5. Petroleum system evaluation of Upper Cretaceous and Eocene plays, offshore and onshore southern Pelagian Basin, Tunisia. By K. Gruber et al.,
6. Mesozoic and Cenozoic tectono-sedimentary evolution and subsidence history of south-eastern Tunisia: Jeffara Basin petroleum prospectivity and hydrocarbon provinces. By R. Khouni et al.
7. Evaluation of Ordovician and Triassic reservoirs in the Chotts area and implied burial and thermal maturity history of their impregnating source rock. By Kraouia S. et al.
8. An overview of the Eastern Atlas folds and thrust belt and its foreland basin along the North-South Axis and Chorbane-Ktitir platform: Surface /Subsurface major structures and tectonic evolution (North Africa). By Mezni R. and Khomsi. S.

In Chap. 5, Gruber et al. present an overview of the petroleum system evaluation of Upper Cretaceous and Eocene plays, offshore and onshore southern Pelagian Basin, Tunisia. Their works give a state-of-the-art view of the oil traps potentialities in this important North African oil and gas province. Their results allow new evaluation and further understanding of the risks on key elements for hydrocarbon prospectivity in the basin in an attempt to highlight areas of low risk for individual plays. Their work is based on a regional synthesis involving subsurface maps with paleogeography and facies maps as well as a play sketch for the Pelagian Basin, illustrating the individual reservoir distribution, seal and source occurrence and trapping styles for the different plays present in the Pelagian Basin. Their works in the area constitute an updated synthesis of the Cretaceous-Paleogene HC potential of the Pelagian province. Their chapter is very rich in subsurface interpretations in terms of well-logging correlations and seismic interpretations and subsurface maturity maps of the source rocks.

In Chap. 6, Khouni et al. give an overview of the Meso-Cenozoic sedimentary basin of the Jeffara. Khouni et al. use seismic sections to describe and interpret the main structures. They focus on the magnificent salt wall affecting the Meso-Cenozoic sedimentary sequences. They also present subsidence curves and relate the subsidence changes to the major tectonic pulses operating within their study area. They also integrate the Jeffara tectonic evolution in the more general regional geology of the Pelagian-Sirt offshore basin. Their structural interpretations are used to describe the main Meso-Cenozoic oil plays and propose some ways for future oil exploration in this major petroleum province of the North African margin.

In Chap. 7, S. Karoui et al. present a complete study and assessment of the Ordovician-Upper Silurian- Triassic petroleum system in the Chotts area, an important oil and gas province limiting the Atlas belt and the Northern Saharan platform. Karaoui et al. present the Upper Silurian Fegaguira Formation which is thought to be an active source rock in the Chotts Basin of southern Tunisia. This major source rock horizon has probably contributed to the HC charge of both underlying Ordovician and overlying Triassic clastic reservoirs in the area. However, various debates are still held regarding either the source rock distribution and thermal maturity but also the reservoir's viability and extension. Their interpretations are based upon subsurface data and play evaluation focusing on the characterization of Ordovician and Triassic reservoirs through the integration of logging data of twenty wells drilled in the southern Chotts Basin aiming to better delineate prolific levels. They also applied 1D Basin Mod modeling to reconstruct the Fegaguira source rock burial and thermal histories and also to estimate its hydrocarbon generation and expulsion potential versus time. Their work constitutes a nice piece of work on the Pre-Triassic emergent petroleum systems at the scale of North Africa, and opens a new exploration perspective for oil and gas exploration in the southern part of the Atlas belt, at the northern edge of the Sahara platform.

In Chap. 8, Mezni and Khomsi present a regional overview of the Eastern Atlas fold and thrust belt and its foreland basin along the North-South Axis and Chorbane-Ktitir platform: major surface /subsurface structures and tectonic evolution. This overview is based on surface observations through the N-S Axis, northern branch of the South Atlas Front together with presentations and interpretations of seismic sections with a focus on some regional key structures such as the Chorbane anticline. The interpretations allow us to visualize and characterize important diapiric structures related to the Triassic salt, thrusting and positive inversions of the former inherited structures of the Jurassic-Cretaceous Tethyan rifting.

Fourth Thematic: Geology and Water Resources in Northern Africa: Case Studies

This theme is covered by three chapters:

9. Aquifer structuring and hydrogeological investigation in North African regions using geophysical methods: a case study of the aquifer system in the Kairouan plain (Central Tunisia). By F. Lachaal et al.
10. Use of geochemical tracers for the characterization and quantification of water leakage at the Joumine dam site, Tunisia. By F. Ben Hamouda et al.
11. Determination of suspended sediments using a nuclear probe in the Medjerda River, Tunisia. By F. Ben Hamouda et al.

In Chap. 9, F. Lachaal et al. apply the subsurface geology methods to study the aquifer structuring and hydrogeological investigation in North African regions using geophysical methods: a case study of the aquifer system in the Kairouan plain (Central Tunisia). They present a set of seismic sections to characterize the geometry of some subsurface reservoirs focusing on Nasrallah-Chorbane-Sidi El Hani and Southern Cherichira, east of the Cherahil mountain belt. They identify potential Upper Miocene-Quaternary reservoirs through seismic interpretations. They also present a piezometric map of the western plain of Kairouan and they characterize the piezometric fluctuations of Kairouan aquifer, thus evidencing a decrease in the water level, in an area suffering constant aridity.

In Chap. 10, Ben Hamouda et al. present a study on the use of geochemical tracers for the characterization and quantification of water leakage at the Joumine dam site, Northern Tunisian Atlas. In fact, the Joumine reservoir, located in the northwest of Tunisia, has an upstream watershed area of 418 km^2 and the reservoir capacity is 130 Mm^3. Shortly after the first filling of the reservoir in 1987, an important water leak was detected at the dam toe immediately after its construction. The emerging flow rate at the maximum level in the reservoir was close to 500 L/s. An important sinkhole was detected in a limestone block outcropping in the left abutment. The emergency work aimed at impermeabilizing the sinkhole by the injection of mine tailings, resulting in a decrease lof the flow rate of the leakage to about 120 L/s. The flow rate of the leakage was monitored in two drains D1 and D2, located at the dam toe. The current study by ben Hammouda constitutes a good example illustrating the usefulness of geochemical tracer methods for obtaining more precise and rapid information on the main features of water leakages, which, if ignored, would have resulted in greater repair costs or could even have affected the stability of a dam.

In Chap. 11, Ben Hamouda et al. present a study on suspended sediments using nuclear probe in the Medjerda River, which is the largest river in the northern Tunisian Atlas. Their contribution focuses on the suspended sediment concentration in Medjerda River as detected by the attenuation of radioactivity emitted from a radioactive source of Americium (241Am) dipped in the water. This attenuation of Gamma rays emitted from the 241Am, due to the presence of the sediments transported in great amounts during flood events, is measured using a scintillation detector made up of a crystal of NaI. The suspended particles are measured in

Slouguia hydrometric station (NW) located on the Medjerda River. Their study allowed measuring the number of particles that were transported by the river during the period of observation in 2005, which is estimated at 778.103 tons, i.e., equivalent to a volume of 294. 103 m^3.

This book was initiated in the fall of 2019, starting 2020 i.e., shortly before the first official cases of the Pandemia. Despite numerous delays due to COVID impacts on both reviewers and authors, we want to thank Springer for its continuous support, as well as the patience of the authors.

The Editors

Jeddah, Saudi Arabia Pr. Sami Khomsi
Rueil Malmaison Cedex, France Pr. Francois Roure

Contents

Part I

First Thematic: Geodynamic Evolution of the Mediterranean and Peri-Mediterranean Sedimentary Basins and Regional Geology

1 The Mediterranean Sea: A Laboratory to Characterize Micro-Continental Drift and Oceanic Basin Formation Processes

Daniel Aslanian, Philippe Schnürle, Maryline Moulin, Mikael Evain, Romain Pellen, Marina Rabineau, Alexandra Afilhado, Nuno Dias, and Camille Noûs

Abstract

The sequence of processes that lead to the genesis of passive margins and their associated oceanic basins remains debated. Following the continental rifting and prior to the establishment of a homogeneous and stable oceanic crust that characterize oceanic drifting, crustal thinning and the exhumation of lower continental crust together with astenospheric mantle appear to be omnipresent in the formation of an intermediate continent-ocean transitional domain. These processes have major consequences on the thermal and tectonic evolution of the margin, as well as on the geodynamic paleo-reconstructions. In the Mediterranean Sea, numerous sub-basins of very diverse ages, crustal nature, formation mechanisms, and geodynamic evolution are observed. In two areas, the Levantine Basin and the Algerian Basin, on either side of the Strait of Sicily, respectively in the old Eastern Mediterranean sea and in the younger Western Mediterranean sea, the nature and geometry of the different segments of the Earth's crust, their boundaries and links, their geodynamical evolution remain controversial and are connected with the presence and evolution of micro-continental blocks, respectively the Eratosthenes block and the AlKaPeCa group. Wide-angle seismic imaging of continental margins and adjacent oceanic basins has proven as a powerful mean to elucidate these controversies. Hence, we propose to conduct wide-angle and streamer seismic surveys along 7 profiles at the Levant and Cyprus margins (onshore-offshore with 106 deployments of ocean bottom seismometers, 15 short-period seismometers in Lebanon and 6 in Cyprus), and along 5 profiles (offshore with 171 deployments of ocean bottom seismometers) in Algerian Basin. The first experiment aims to establish whether the Levant margin is a transform margin in the direct connection with Neo-Tethys, or a conjugate normal margin of the micro-continental block of Eratosthenes. The second surveys aims to constrain the kinematic evolution (North–South versus East–West extension), the crustal nature (oceanic versus thinned continental) and the geodynamical process (slab rollback- versus continental delimination-driven) of the Algerian Basin. Both surveys are targeted to characterize, in two basins of very different age, the active processes during rifting, the formation of the ocean-continent transition, the nature and geometry of the different segments, and the geodynamic active forces, in areas where few wide-angle seismic has been acquired in the past, and then, by placing this new information in a geodynamic context, to better constrain the kinematic history of the Mediterranean Sea.

Keywords

Mediterranean geodynamics • Passive margins • Crustal architecture • Proto-oceanic • Levantine basin • Algerian basin • Wide-angle • Deep multi-channel seismic

1.1 Introduction

Mantle evolution and tectonic forces constitute the engine of the ocean floor, creating and destroying ocean basins, and driving complex structural and sedimentological evolution at

D. Aslanian (✉) · P. Schnürle · M. Moulin · M. Evain · R. Pellen · M. Rabineau
Geo-Ocean, University of Brest, CNRS, Ifremer, UMR6538, F-29280 Plouzane, France
e-mail: Daniel.Aslanian@ifremer.fr

A. Afilhado · N. Dias
Instituto Dom Luis, Lisboa, Faculdade das Ciencias da Universidade de Lisboa, Lisboa, Portugal

Instituto Superior de Engenharia de Lisboa, Lisboa, Portugal

C. Noûs
Laboratoire Cogitamus, IUEM, Place Nicolas Copernic, 29280 Plouzané, France

S. Khomsi and F. Roure (eds.), *Geology of North Africa and the Mediterranean: Sedimentary Basins and Georesources*, Regional Geology Reviews, https://doi.org/10.1007/978-3-031-18747-6_1

ocean margins: Earth's surface features, on land and at the ocean floor, are intimately related to the dynamics of the Earth's interior. One of the still debated questions in Earth Science concerns the nature of the ocean-continent transition and its links to the first "true" oceanic crust: the sequence of processes that lead from continental rifting to oceanic drifting during the genesis of passive margins remains debated. This question is intrinsically connected to vertical (subsidence/uplift) and lateral crustal movements, and therefore to palinspastic reconstructions. This link is so strong that the initial position of the plates in these reconstructions highly depends on the information we have on the Passive Margins structure and segmentation. As a consequence, the vertical, thermal (??) and tectonic motions of these domains may have a drastic different evolution.

This has been demonstrated in the Equatorial and South Atlantic Ocean by joined approaches to geodynamic studies and specifically designed wide-angle seismic experiments (Contrucci et al. 2004; Moulin et al. 2005; Aslanian et al. 2009; Aslanian and Moulin 2010; Aslanian and Moulin 2012; Moulin et al. 2012; Klingelhoefer et al. 2015; Evain et al. 2015; Loureiro et al. 2018; Pinheiro et al. 2018; Aslanian et al. 2021a; Moulin et al. 2021) and more recently thanks to the PAMELA project, an industrial and academic collaboration, in the Indian Ocean (Thompson et al. 2019; Moulin et al. 2020; Lepretre et al. 2021; Evain et al.; Aslanian et al. 2021b; Watremez et al. 2021; Schnürle et al. in press).

While in North-Western Mediteranean Sea, similar combined works have been conducted for Valencia and the Liguro-Provencal basins (Pellen et al. 2016, 2019; Moulin et al. 2015; Watremez et al. 2021; Afilhado et al. 2015; Leroux et al. 2015a, b; Bellucci et al. 2021), unsolved geodynamic and crustal questions remain in the Western Mediterranean Sea (in the Alboran-Algerian-Adria Basins, see Leroux et al. 2018) and in the Eastern Mediterranean Sea, in the Levantine Basin. In both cases, the incertitude of the crustal nature allows opposite geodynamic scenarios (and therefore subsidence evolution).

This paper presents a rapid overview of the different hypotheses in both cases and argues for indispensable wide-angle experiments in both regions. Wide-angle seismic has, over the past decades, proven an efficient tool to image the architecture and quantify the acoustic velocity of continental margins. These data, combined with structural interpretation and stratigraphic/kinematic/mechanical numerical modeling, will lead to better characterization of the crustal and mantellic processes active from rifting to drifting and constrain the global geodynamics.

1.2 The Geodynamic Problem

The evolution of the Mediterranean is closely linked to the creation of the Alpine chains (Doglioni et al. 1997; Gelabert et al. 2002) and more broadly to the relative movements of the African and European plates that are largely derived from magnetic anomalies of the Atlantic Ocean (Olivet et al. 1982; Dewey et al. 1989; Ricou 1994; Rosenbaum et al. 2002a).

The opening of the Alpine Tethys (one of the several branches of the Neotethyan Ocean) resulted from the break-up of Pangaea in the late Paleozoic (Frizon De Lamotte et al. 2011). This area opened in response to lateral sinistral movement between Iberia, Laurussia, and Africa, as a consequence of the Central Atlantic Ocean expansion (Cavazza et al. 2004) starting in the Sinemurian (Sahabi et al. 2004). Until the Cretaceous, this EW-oriented domain divided the European and African plates, forming an oceanic corridor characterized by a left-lateral transtension. From the Lower Cretaceous ($\sim$125 Ma), the trajectory of the African plate with respect to Europe changed drastically. This led to the start of the convergence between Europe and Africa (Olivet et al. 1982, Olivet 1996; Dewey et al. 1989; Ricou 1994; Rosenbaum et al. 2002a; Cavazza et al. 2004; Handy et al. 2010; Schettino and Turco 2011). This new stress regime initiated the gradual closure of the Alpine Tethys with its northward subduction beneath the southern European paleo-margin, eventually ending with the collision of the African plate with the European plate during Eocene and the development of the Alpine Chain.

This very specific context produced a very segmented domain where continental micro-blocks are numerous. The presence of such microblocks are common in "buffer" areas, local links of different kinematic systems like triple junctions (for instance, the Danakil block, Sichler 1980; McClusky et al. 2010; the Jan Mayen microcontinent, Talwani and Eldholm 1977; Gaina et al. 2009; the Iberian sub-plate, Olivet 1996).

In the case of the Santos continental sub-plate (Evain et al. 2015) at the boundary between the Central Segment and the Austral Segment of the South Atlantic Ocean (Moulin et al. 2010), interpreted as a kinematic buffer which allows an oblique, Sud-East movement in the general East–West setting (Moulin et al. 2012). In the Indian Ocean, offshore Mozambique, the Beira High, at the corner between a strike-slip and a pure divergent margin, is recognized as of continental nature (Mueller et al. 2016), shifts the Continent-Ocean Transition (COT) at that area about

150 km shoreward and has drastic consequences on the initial assemblage of the Gondwana pieces (Thompson et al. 2019; Moulin et al. 2020; Lepretre et al. 2021; Evain et al. 2021; He et al. 2021). These features play therefore an important role in the reconstruction and the nature of the surrounding areas. While in the Western Mediterraean sea, the role and the evolution of the Sardinia-Corsica micro-block (see Leroux et al. 2018 for a review) and of the Minorca block (Pellen et al. 2016) are well defined, the original position and the kinematic motion AlKaPeCa block (for Alboran, Kabylies, Peloritan, and Calabria; Bouillin 1986) are still debated as, in the eastern Mediterraean sea, the original position and the kinematic motion of the micro-continental blocks of Eratosthenes and Cyprus (Stampfli et al. 1991; Keeley 1994; Stampfli and Borel 2004; Garfunkel 1998; Barrier and Vrielynck 2008; Gardosh et al. 2010).

Consequently, the complete understanding of the geodynamic evolution of the Mediterranean Sea is intimately connected to the understanding of the features.

1.2.1 Eastern Mediterranean (Levant Margin)

The Eastern Mediterranean, located to the west of the Arabian Peninsula, is bordered by the Saharan Platform of Egypt and Libya to the south and the Cyprus Arc and the Mediterranean Ridge to the north (Fig. 1.1). The Eastern Mediterranean domain includes the Herodotus Basin, the adjacent continental margins (Levant and Tubruk basins) and their onshore continuation (Garfunkel 1998). It is an active zone of plate convergence between the African Plate and the Aegean-Anatolian microplate, currently undergoing a gradual and diachronic shift from subduction of oceanic crust to continental collision (Jolivet and Brun 2010): the last remnants of Neo-Tethyan oceanic crust on the northern edge of the African Plate are believed to be consumed by northward subduction below the Hellenic and Cyprus Arcs (Robertson 1998). This subduction has almost ceased around Cyprus, due to the collision in the subduction of isolated micro-continental blocks, in particular the Eratosthenes seamount in the trough south of Cyprus (Robertson 1998).

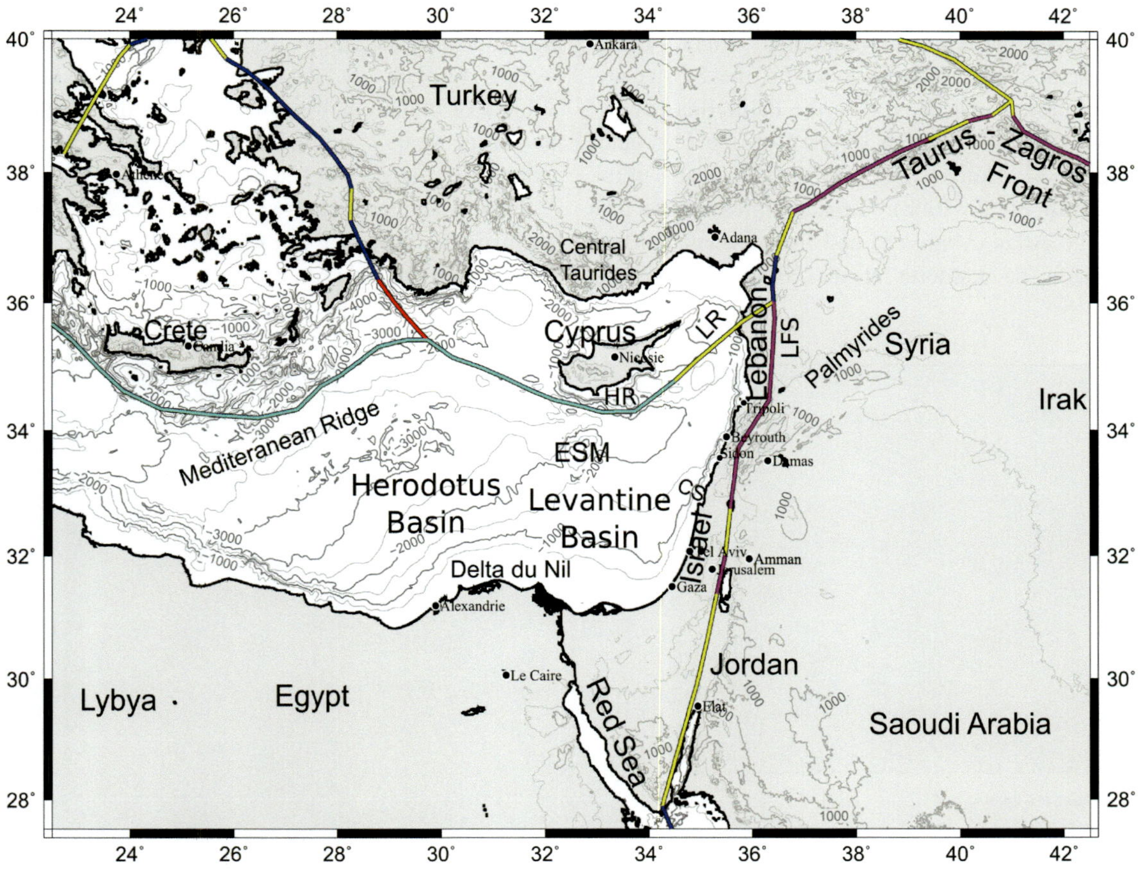

Fig. 1.1 Regional map of the Eastern Mediterranean and the Near East. Plate boundaries are indicated by double lines (from P. Bird PB2002: cyan convergent, yellow transpressive, blue transtensive, magenta strike-slip). ESM = Eratosthenes seamount, HR = Hecataeus Ridge, LR = Latakia Ridge, CS = Carmel Structure, LFS = Levant Fault System

Finally, the transform fault of the Dead Sea forms the contemporary boundary to the east of this domain.

At the end of the Palaeozoic, the Tethysian oceanic opening leads to the dislocation of Pangaea along several branches; thus, the Alpine Tethys separates southeastern Europe from north-western Africa and is connected towards the south-west to the Central Atlantic (Frizon de Lamotte et al. 2011 - Fig. 1.2); to the east, the Neo-Tethys separates Arabia from the Cimmeride microblock (most of Iran). The Permian age of the rifting that precedes the opening of the Neo-Tethys was established in Oman by analysis of the sediments of the distal passive margin (Stampfli et al. 1991; Pillevuit et al. 1997; Chauvet et al. 2009). In the Late Jurassic, Alpine Tethys and Neo-Tethys were probably linked to each other in eastern Sicily; the Gavrovo-Tripoliza and Taurus platforms were detached from Africa to form the Eastern Mediterranean domain (Fig. 1.2).

During the Lower Jurassic and Cretaceous, the movement between Africa and Eurasia is sinistral transtensible (Frizon de Lamotte et al. 2011). In the Santonian, the convergence between Africa and Eurasia begins (Rosenbaum et al. 2002a, b) leading to the gradual closure of the Alpine Tethys and Neo-Tethys, and the emplacement of ophiolites around the Arabian promontory which are almost synchronous from Crete, Cyprus to the Omani Mountains (Ricou 1971). Inversion structures are well expressed in the Syrian arc, which extends from Sinai (Egypt) to the Palmyrids (Syria), in north-central Egypt (Western Desert), on the Cyrenaica Platform (Jebel Al Akhdar rise) and the adjacent Tubruk Basin (Bosworth et al. 2008). This compression is also observed in the Atlas (Maghreb) and the adjacent Saharan platform (Frizon de Lamotte et al. 2009). Toward the Middle-lower Eocene the African-Eurasian convergence slows down, accompanied by the initiation of the collision between northern Arabia and Eurasia. The Dead Sea transform fault was initiated in the Miocene in response to the sinistral strike-slip between the Arab and African plates in probable relation to the opening of the Red Sea (Garfunkel 1981): this fault extends over 1,000 km from the Gulf of Aqaba to southern Turkey and has a thrown in the Arava Valley (Israel) of 107 km, including 40 km from the Plio-Pleistocene. Finally, during the Pliocene, the Cyprus Arc collided with the Eratosthenes Seamount (and the Hecataeus Ridge); the subducting oceanic crustal slab is continuously imaged under Cyprus (Mackenzie et al. 2006; Feld et al. 2017) but would be completely detached further east, below the Bitlis-Zagros suture (Faccenna et al. 2006). Hardy et al. (2010) propose a structural scheme of the evolution of the Levantine margin since the Mesozoic (Fig. 1.3).

Major gas discoveries and to a lesser extent hydrocarbon discoveries have revived subsea studies in the Levantine Basin region: These include the Aphrodite (Noble Energy 2011) field in Cypriot waters, Leviathan, Tamar, Karlsh, and Dalit in Israeli waters (2009), and more recently Zohr (ENI 2015) and Calypso (ENI-Total SA 2016) in the Eratosthenes Seamount carbonate platforms in Egyptian and Cypriot waters, respectively. As a result, this area has been the subject of numerous borehole and seismic imaging surveys in recent years, complemented by sedimentological and stratigraphic studies at sea and on land.

Recent studies have focused on the stratigraphy of the sedimentary series of the Cyprus, Herodotus, and Levant basins with thicknesses exceeding 10 km (Gardosh et al.

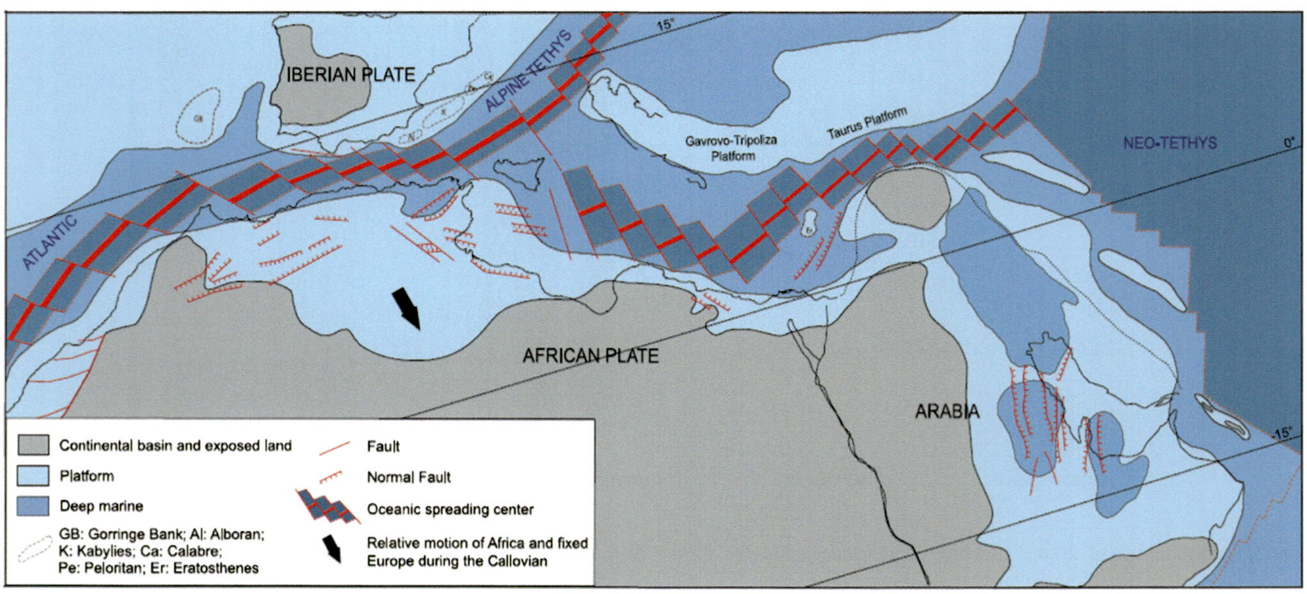

Fig. 1.2 Paleo-tectonic map of North Africa and Arabia at Callovian (from Frizon de Lamotte et al. 2011)

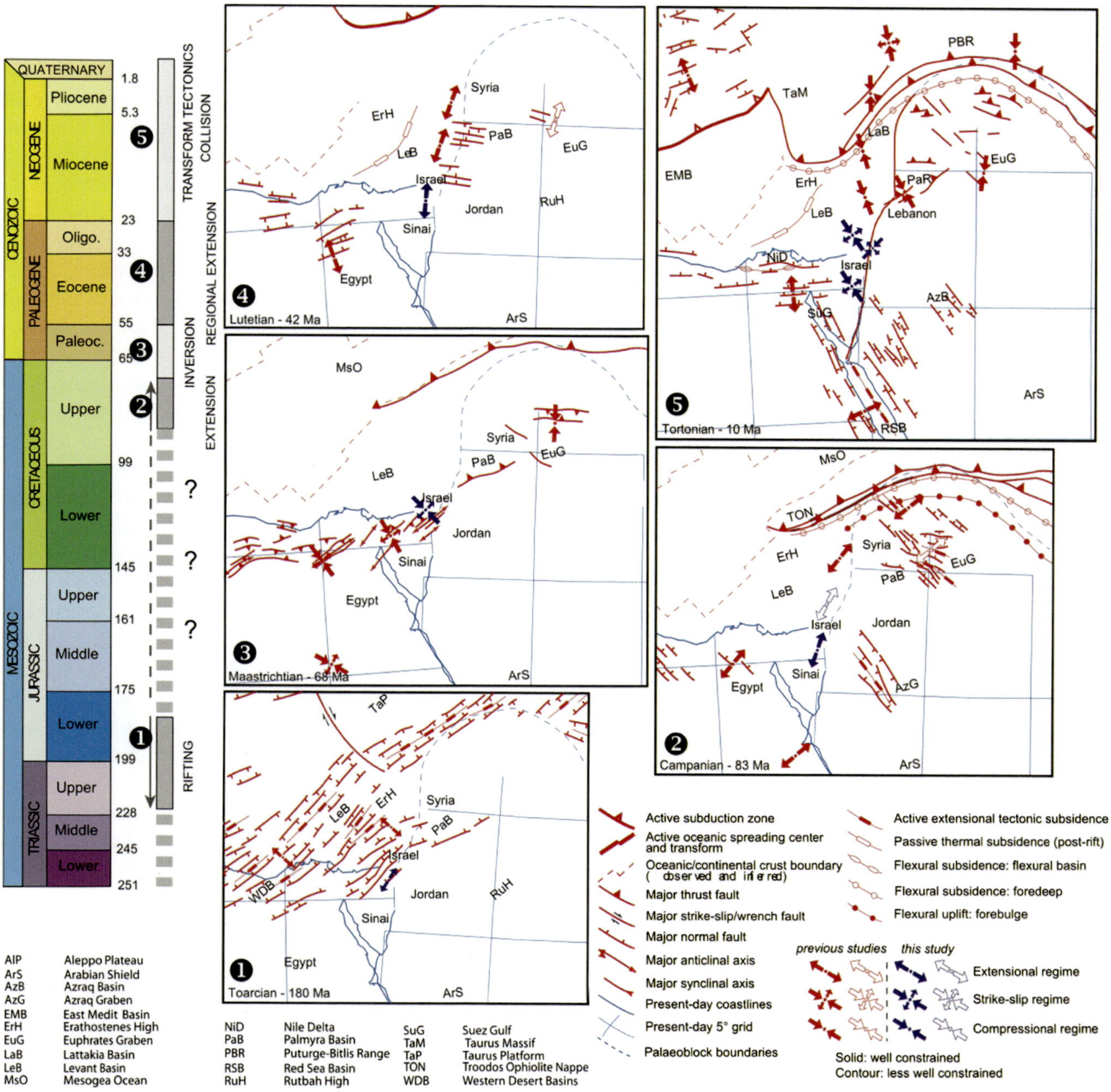

Fig. 1.3 Synthesis of major tectonic events in Israel and surrounding areas (Hardy et al. 2010, based on Barrier and Vrielynck 2008) with associated stress states. (1) Extension of the Lower Mesozoic leading to the development of the Levantine and Palmyrid basins. (2) Extension of the Upper Cretaceous, leading to the rifting of the Euphrates and Azrak grabens. (3) Senonian tectonic reversal in Israel and North Africa. (4) Eocene extension in the whole Levantine domain. (5) Neogene Trans-pressure leading to the inversion of the Palmyrid and Lebanon chains. The arrows show the directions of the principal stress

2008; Montadert et al. 2014). These domains also contain important crustal edifices (e.g. Eratosthenes Seamount and the Levitan and Jonah Heighs). On the eastern edge of the Levantine Basin, up to 1.5 km of Cenozoic sediments are preserved onshore in Lebanon (Nader 2011). According to seismic facies observed in this basin and stratigraphic models, rifting persists up to the Middle Jurassic followed by significant subsidence characterized by Upper Jurassic to Lower Cretaceous deposits from deep environments (Hawie 2014). Jurassic syn-rift deposits are exposed in the Palmyrids in Lebanon and Syria (Brew et al. 2001). From the Cenomanian to the Turonian, shallow depositional environments have been identified in the north contrasting with deep marine environments around Beirut and the south Lebanese coastal region (Hawie et al. 2013). In the Turonian, Afro-Arabian and Eurasian convergence gives rise to several compressive events and induces the formation of a foreland basin in the north in front of the Latakia Ridge and the progressive southward migration of the central deposits that continues during the Miocene.

Onshore, Mesozoic and Cenozoic outcrops show several hiatuses dating from Late Turonian to Late Santonian, Maastrichtian to Paleocene, Lower/Upper Eocene and Late Burdigalian (Nader 2011). The collision of the Afro-Arab and Eurasian plates during the Cenozoic is believed to have caused the propagation of the Levant Fault northwards and induced the rapid orogene of Mount Lebanon (Beydoun 1999). At sea in the northern part of the basin, a network of receding faults marks the entire pre-Messinian sequence (Ghalayini et al. 2014). During the Oligo-Miocene, local sources (Levantine margin) provided sediments in the northern part of the basin through canyon systems incising the margin as well as through turbidite channels, while regional drainage systems such as the Nile and Lattakia rivers transported sediments from more distant sources (Hawie 2014). During the Messinian salinity crisis, large evaporitic deposits accumulate, locally more than 3 km thick. In most of the basins, the filling is a little deformed. The sequences from the Pliocene to the present are affected by a ductile deformation of the underlying Messinian series leading to the formation of diapirs observed in the Levant basin and a portion of the Herodotus basin (Garfunkel 1998). Hence, the structural scheme and understanding of the impact of major geodynamic and magmatic events on the architectural evolution and sedimentary infilling of this province are summarized in Fig. 1.4 (Segev et al. 2018).

In the Eastern Mediterranean, however, kinematic models involving the opening and closing of the Neo-Tethys still depend largely on the determination of the nature, age, and deformation of the crust forming the Cyprus, Herodotus, and Levant basins and many questions are still debated.

A first model assumes that the opening of the Eastern Mediterranean began at the end of the Paleozoic (Stampfli et al. 1991; Stampfli and Borel 2004). A second model proposes an opening during the Cretaceous (Dercourt et al. 1993; Ricou 1994). Then, Segev and Rybakov (2010) and Segev et al. (2018) suggest that rifting was facilitated by a mantle plume in the Lower Cretaceous, followed by accretion of an ocean floor during the period of magnetic calm between CM0n ($\sim$122 Ma) and C34n ($\sim$84 Ma, Finally, the magnetic anomaly has been interpreted in the Herodotus basin suggesting a 340 Ma old oceanic floor (Granot 2016). The direction of opening of these basins is also debated: is the Levant margin a transform margin in direct connection with the Neo-Tethys (Stampfli et al. 1991; Keeley 1994; Stampfli and Borel 2004) or a conjugate normal margin of the micro-continental block of Eratosthenes and Cyprus (Garfunkel 1998; Barrier and Vrielynck 2008; Gardosh et al. 2010).

Schattner and Ben-Avraham (2007, 2012) propose that the Carmel structure south of the Galilee graben separates a basin in the north (Phoenician) from the Levantine basin in the south. The northern basin would be rifted in the Permian of the Eratosthenes block and followed by N-S oceanic accretion, forming a transform margin along the continental fringe of north-Leventine (Galilean and Lebanese); the southern basin is derived from the NW migration of the Eratosthenes block. Montadert et al. (2014) propose a paleo-tectonic scheme common to all the basins in the Eastern Mediterranean, separated by transfer/transform faults (Fig. 1.5). Would the Levantine Basin be rather a back-arc basin formed by the subduction of the Meso-Tethys (the Herodotus Basin) during the Senonian-Maastrichtian (Segev and Rybakov 2010). Was this extension followed by rifting and accretion of oceanic crust (Frizon de Lamotte et al. 2011). The processes active during rifting, such as crustal thinning, and exhumation of lower crust and/or upper mantle are poorly characterized, and the location of the ocean-continent transition in these basins also remains uncertain (Gardosh et al. 2010).

Finally, the role of the nature and crustal structure of these basins in the tectonic episodes that follow their formation is debated: does the Sinai Rift extend N-W in the Levantine and Herodotus basins to the Eratosthenes seamount and the Florence Ridge, individualizing a Sinai microplate between the African and Arabian plates, before being incorporated into the Aegean-Anatolian plate (Mascle et al. 2000). The northern boundary of this domain, in the active subduction/collision constituting the island of Cyprus and the Hecataeus and Latakia Ridges, is still poorly characterized. The most important geological unit in Cyprus is the ophiolite of the Troodos Massif, with a U–Pb zircon age of 91 $\pm$ 1.4 Ma. The core of the complex is composed of serpentinized mantle peridotite, surrounded by plutonic dykes, and then topped with layered lava and sediments (Mackenzie et al. 2006). The southeastern to Cyprus extension of the Troodos ophiolitic complex is debated. In order to answer these questions, the acquisition of new deep seismic data in the East Mediterranean/Levant province is essential.

1.2.2 The Western Mediterranean (Alboran-Algerian-Adria Basins)

The Western Mediterranean formed in an overall convergence context during the slow Africa-Iberia and Africa-Europe convergence (e.g., Seton et al. 2012; Torsvik et al. 2012; Vissers and Meijer 2012), which included subduction initiation and slab fragmentation with an associated intense but well-studied history of crustal deformation (e.g., Lonergan and White 1997; Gueguen et al. 1998; Rosenbaum

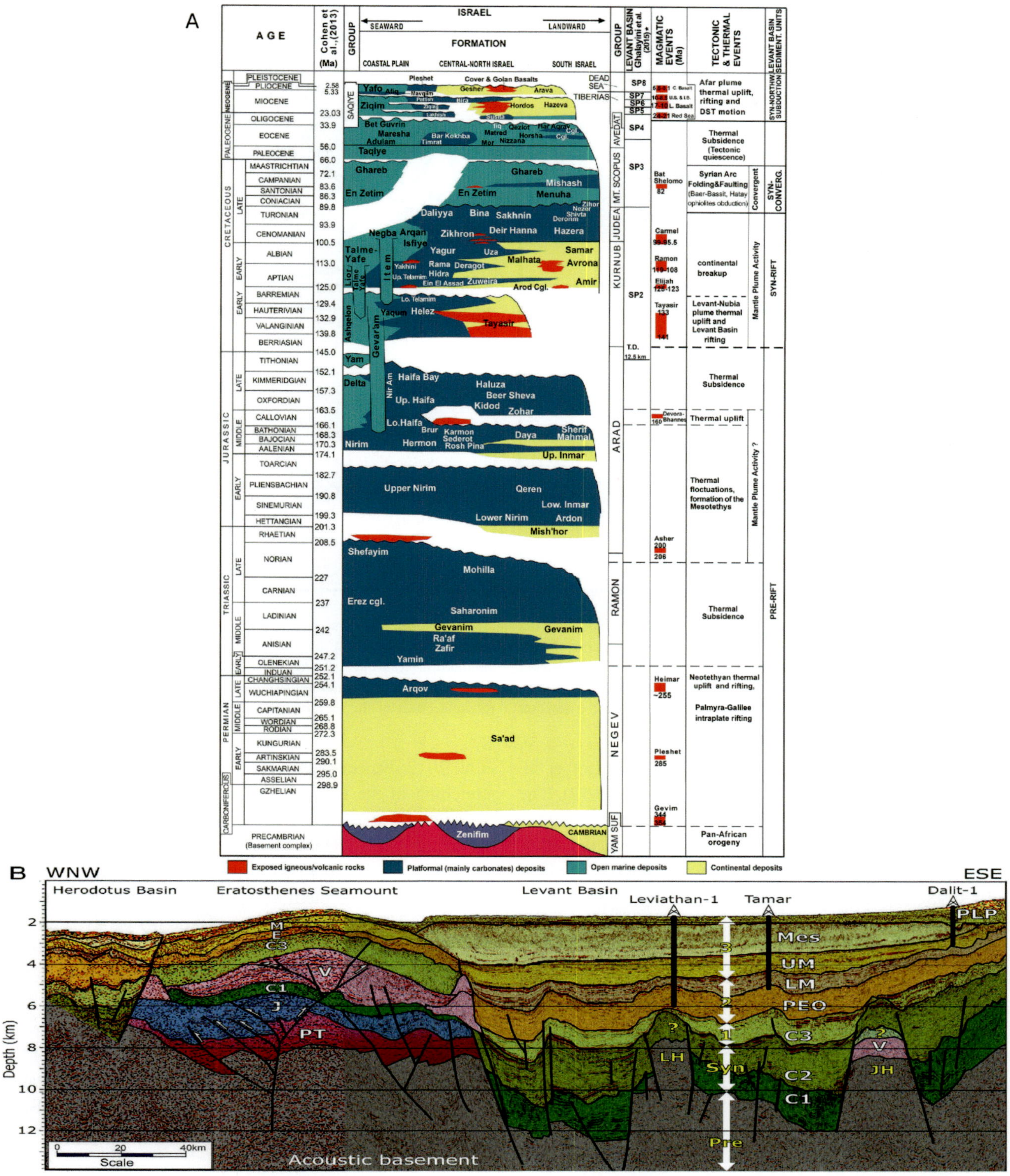

Fig. 1.4 **A** Schematiqc tectono-stratigraphic chart of the Levantine Basin and Israel. SP2 = carbonates and marls, SP3 = hemipelagic/pelagic sediments, SP4, SP5, et SP6 = clastic et hemipelagic deposits, SP7 = evaporites, SP8 = turbidity/hemipelagic inter-bedded. **B** Reference seismic profile and stratigraphic interpretation (localization Fig. 1.1). After Segev et al. (2018)

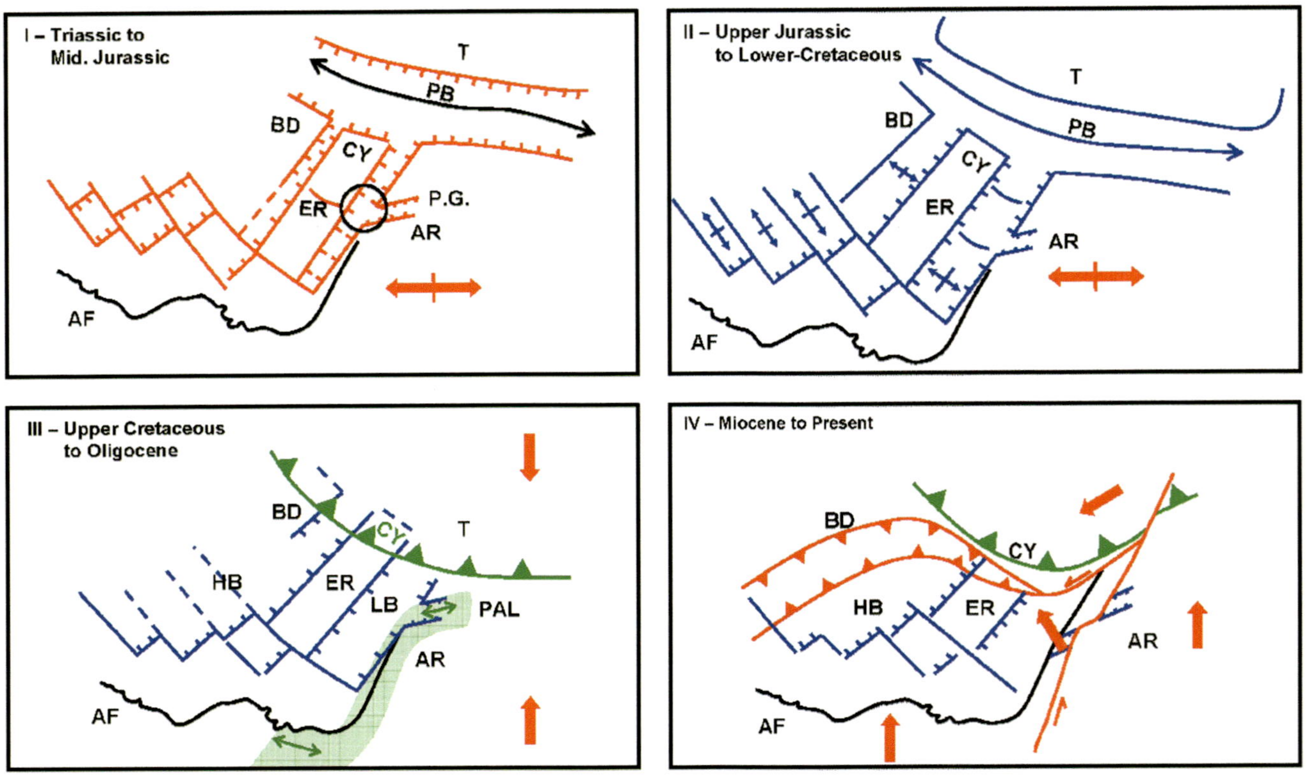

Fig. 1.5 Paleo-tectonic diagram of the Eastern Mediterranean since the Triassic (ER = Eratosthenes continental Block, CY = Cyprus, BD = Bey Daglari, HB = Herodotus Basin, PB = Pamphyliein Basin, T = Taurus, AR = Arabia, AF = Africa: MR = Mediterranean Ride). (I) Rift from the Triassic to the Middle Jurassic. (II) Oceanic extension and accretion from the Late Jurassic to the Cretaceous. (III) Upper Cretaceous: formation of the Cyprus Arc and its ophiolitic belt, subduction continues during the Paleogene. (IV) Miocene to present day, activity of the Mediterranean Ridge, Latakia Ridge, and Transforming Fault of the Levant. After Montadert et al. (2014)

et al. 2002a; Faccenna et al. 2004, 2014; Mauffret et al. 2004; Jolivet et al. 2006, 2009; Carminati et al. 2012). The seismic tomography images show a structurally complex mantle with clear evidence of remnants of subducted lithosphere (Spakman 1991; Spakman et al. 1993; Carminati et al. 1998a, b; Wortel and Spakman 2000; Gutscher et al. 2002; Piromallo and Morelli 2003; Spakman and Wortel 2004; Wortel et al. 2009; Bezada et al. 2013). Nevertheless, during the Oligocene, both extension and compression simultaneously affected the Mediterranean area leading to: (1) a significant continental extension in the Alboran and northern Tyrrhenian basins, with continental crust thinning and possible oceanization in the Ligurian-Provencal, Algerian, and, later (Late Miocene), in the southern Tyrrhenian, and (2) the formation of mountain ranges in the Mediterranean region, such as the Alps, the Maghrebides, the Apennines, and the Bético-Rif Cordillera (Lustrino et al. 2009; Fig. 1.6).

The Algero-Provencal Basin opened during the late Oligocene-early Miocene times in a convergent context behind a Tethyan oceanic lithosphere subducting with different polarities (Jolivet and Faccenna 2000; Gelabert et al. 2002; Speranza et al. 2002; Lustrino et al. 2009; Carminati et al. 2012; Faccenna et al. 2014). In the early Miocene, the stretching of the European plate caused the rifting and the drifting of the eastern margin of the Greater Iberia, now occurring as isolated continental blocks (e.g., Sardinia-Corsica Block) and continental fragments accreted to Southern Spain-Northern Morocco (Alboran Block), Algeria (Lesser and Greater Kabylies) and Southern Italy (Peloritan Mts. And Calabria). The latter accreted blocks were originally grouped under the acronym AlKaPeCa (for Alboran, Kabylies, Peloritan, and Calabria (Bouillin 1986); see Fig. 1.8).

The opening of the Algerian Basin remains controversial regarding the kinematic evolution and the nature of the margins. The disagreement stems first from the interpretation of the opening direction of the oceanic Algerian Basin according to two models (Fig. 1.7):

- The first one assumes a dominant N-S opening leading to a presently inactive slab under the North African margin (e.g., Gueguen et al. 1998; Carminati et al. 1998a, b; Frizon de Lamotte et al. 2000; Wortel and Spakman 2000; Gelabert et al. 2002; Rosenbaum et al. 2002b; Rosenbaum and Lister 2004; Schettino and Turco 2006). The westward rollback in the Gibraltar region is thus

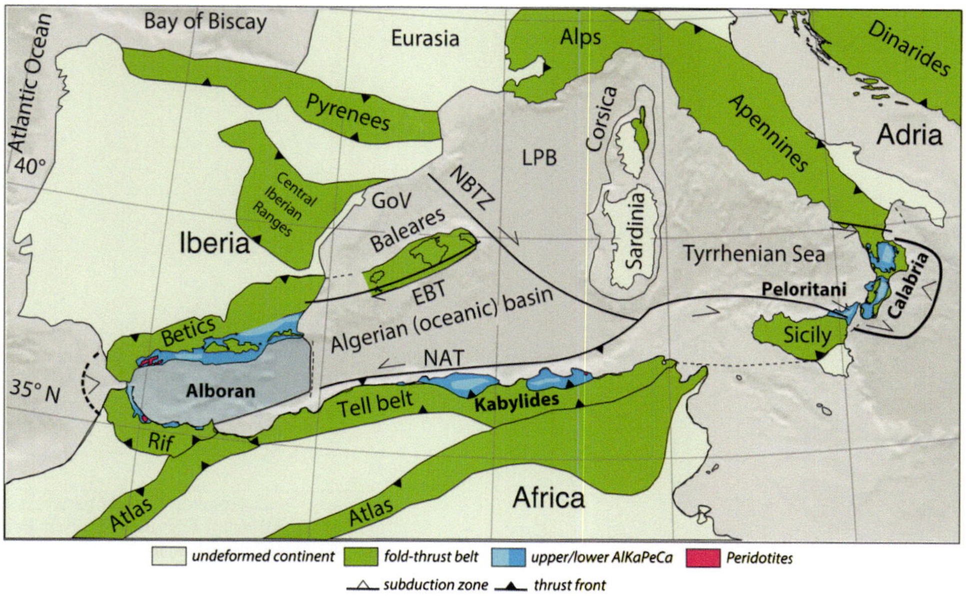

Fig. 1.6 Schematic tectonic map of the western Mediterranean region. EBT: Emile Baudot Transfom; LPB: Liguro-Provencal Basin; GoV: Gulf of Valencia; NAT: North African Transform; NBTZ: North Balearic Transform Zone. From Van Hinsbergen et al. (2014)

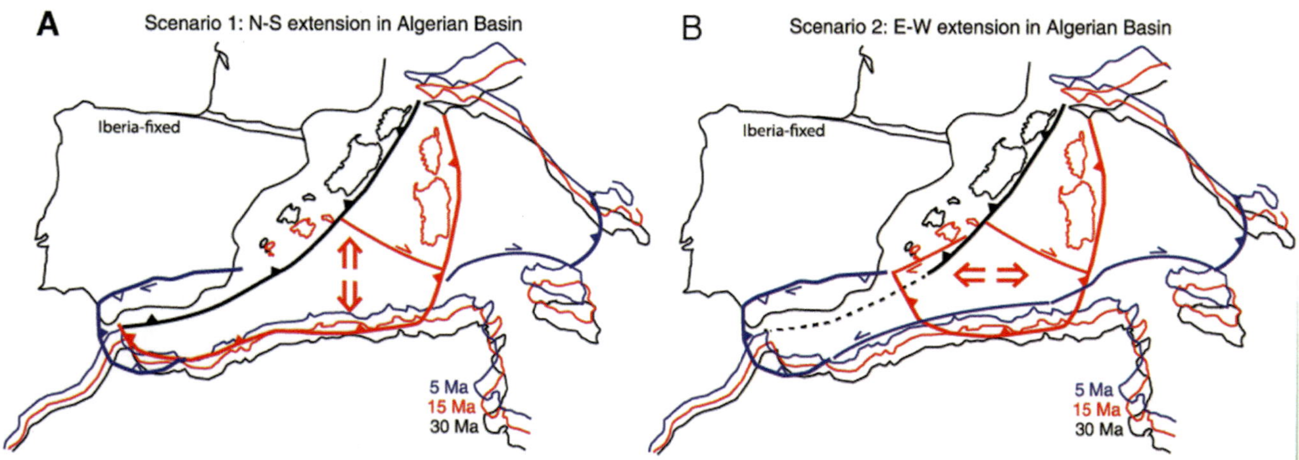

Fig. 1.7 Two types of scenarii for the opening of the Algerian Basin (Van Hinsbergen et al. 2014)

accommodated with no significant displacement (i.e., less than 200 km) of the Alboran block (Faccenna et al. 2004; Jolivet et al. 2009; Rossetti et al. 2013) (Fig. 1.7A). In this hypothesis, the slab started to rollback from Gibraltar to Corsica and the subduction started along an E-W trending collisional zone.

- A second model promotes a dominant E-W opening (Malinverno and Ryan 1986; Royden 1993; Lonergan and White 1997; Rosenbaum et al. 2002a) associated with slab rupture along and removal of slab under the North African margin (Gutscher et al. 2002; Mauffret et al. 2004; Spakman and Wortel 2004; Duggen et al. 2003, 2004, 2005) (Fig. 1.7B). In this hypothesis, the slab is restricted to the Balearic-Sardinia/Corsica segment. This westward migration of the Alboran block would have induced a left-lateral deformation along the Western and Central Algerian margins and right-lateral deformation along the Balearic Promontory (Camerlenghi et al. 2009). This scenario argues for subduction starting along a fossil transform (subduction transform edge propagator (STEP)) fault.

Whatever model chosen, the westernmost Algerian margin can be assumed to represent a purely strike-slip type margin (Domzig et al. 2006), having formed as a STEP-fault system (Govers and Wortel 2005; Medaouri et al. 2014).

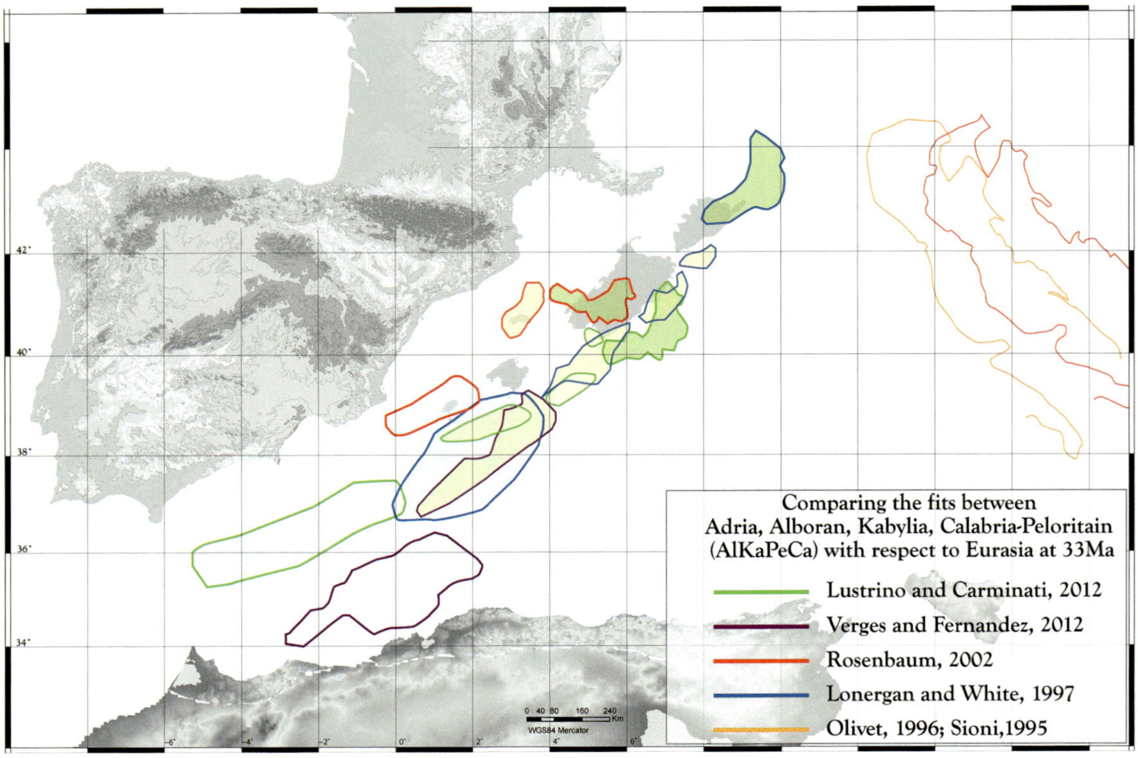

Fig. 1.8 Different positions of the AlKaPeCa blocks according to various authors

Recent evolution (Etheve et al. 2016; Lepretre et al. 2018) favors a two-step opening scenario including a first westward migration and a second one towards the edges of the basin, which is accompanied by detachment or delamination. Last, while the rollback process is the most widely accepted mechanism to explain the coincidence of compression and expansion in this region, other models such as the collapse of the previously thickened continental crust during the Alpine stage (Dewey et al. 1989; Platt and Vissers 1989) or lithospheric delamination (Roure et al. 2012) are also proposed.

This great variety of evolution hypotheses arises from the differences in study scale, chosen approach and type of data, in particular geological and/or geophysical, onshore and/or offshore data. This leads to very different positions of the same element when the Cenozoic evolution of the Western Mediterranean system proposed by different authors is compared. This is illustrated with the position of the AlKaPeCa metamorphic units at 33 Ma (Fig. 1.8) and even the proposal of the absence of the Alboran terrane in its pre-drift position (Carminati et al. 2012). At the same time, the kinematic and amount of displacement between the different segments of the AlKaPeCa terrane is a key data to determine origin and role of the observed limit between each onshore and offshore domain developed during the opening of the Western Mediterranean. Depending of the chosen model, it raises major questions as the crustal nature (oceanic/continental/ "transitional") and the segmentation of the Algerian Basin (the number of the involved segments, the role of Hannibal ridge, the significance of the magnetic anomalies, and so on), the original position and the kinematic motion of AlKaPeCa block and therefore the complete geodynamic evolution of the Western Mediterranean Sea.

Towards the East, the Adria domain is also a place of debate (Fig. 1.9): it has been considered as either a microplate independent from the Africa Plate, or as an advanced promontory from Africa. These two theories give rise to many uncertainties. Were the deep Ionian and Neotethysian Basins connected, or was there a threshold between the two domains? (see Vezzani 2010; Carminati et al. 2012). At the Adria microplate scale, and before the initiation of the Apennine-Maghrebian arc, several uncertainties are observed concerning the position, dimension, and nature of the Mesozoic paleogeographic domains. The connection between the S.A.B. (South Adriatic Basin) and the deep Ionian-Levantine domains is complex. These uncertainties are linked to the Westward development of the Albanic arc, the latter being connected to the Westward extrusion of the Anatolian Plate and the establishment of the Aegean and Balkan areas. This domain is the focus of the *Transect* project and will not be integrated in the following paragraph.

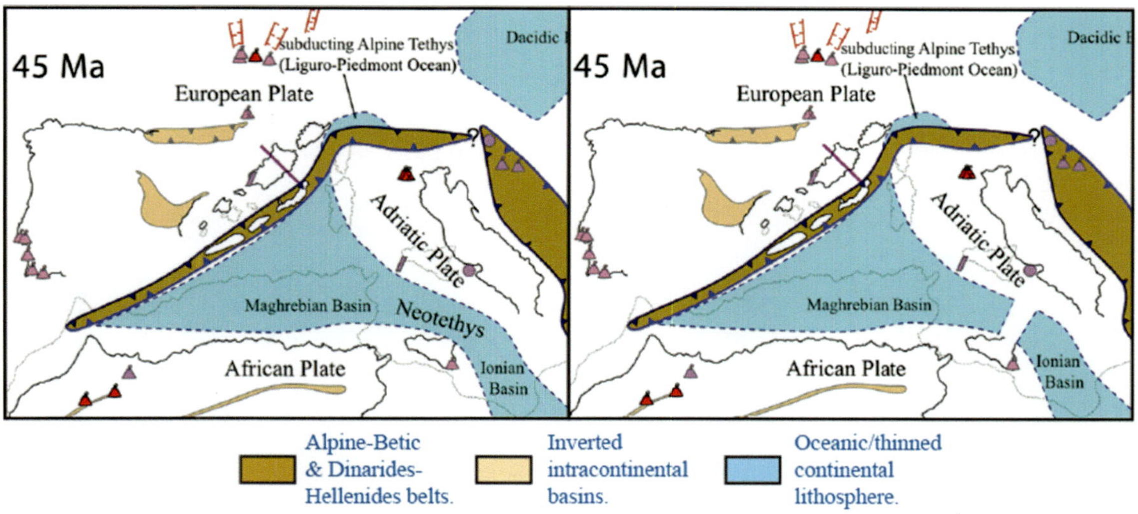

Fig. 1.9 Two types of scenarios concerning the link between the Maghrebian and Ionian domains (Carminati et al. 2012). On the left, the scenario adopted by Carminati et al. (2012). On the right, the second scenario seems to be adopted by Van Hinsbergen et al. (2014)

1.3 Pre-existing Wide-Angle and Reflection Seismic Surveys

1.3.1 Levant Basin

Numerous seismic reflection surveys have been conducted in the Eastern Mediterranean; those acquired in the academic field offer insufficient penetration to characterize the base of the basins and the crust. Shalimar (complemented by Shalibande; Carton et al. 2009) image about 2.5 s twt of the deposits on the continental slope and the deep basin of the Levant. Two geophysical acquisition companies have carried out seismic surveys throughout the Eastern Mediterranean, PGS (www.pgs.com/data-library/europe/mediterranean/) and Spectrum (www.spectrumgeo.com/interactive-map), and made part of these profiles available to research and teaching institutes (e.g. Al-Balushi et al. 2016; Bertoni et al. 2017; Inati et al. 2016, 2018; Papadimitriou et al. 2018; Gao et al. 2019). These profiles image the roof of the basement located between 7 and 18 km deep but rarely the Moho; the crustal thickness is generally inverted from the geoide and free-air gravity anomaly (Fig. 1.10, Al-Balushi 2015; Steinberg et al. 2018).

In our study area, 7 wide-angle seismic surveys have been published (Fig. 1.11). The oldest profile extends from southern Cyprus through the Eratosthenes Seamount to the coast of Israel (Makris et al. 1983). The acquisition consists of 3 OBS, 9 seismic stations in Cyprus and 14 on the Israeli coast; 33 detonations of 800 kg of dynamite were fired along the profile. The authors observe a 35 km thick continental crust in Cyprus, which continues southward thinning to about 25 km below the Eratosthenes Seamount; in the deep Levant basin where the sedimentary cover reaches 12 to 14 km, the crust is thinning to 8 km thick and is characterized by a velocity of 6.5 km/s and interpreted as oceanic in nature. These velocities are confirmed along two wide-angle profiles on the margin of the Levant off Israel by Ben-Avraham et al. (2002), who observe a thinner continental crust and a narrower transition zone along the northern profile—a constriction zone at 20 and 100 km offshore—and characterized by an acoustic velocity of 6.7 km/s in the lower crust and varying from 6.3 km/s in northern Israel to 6.0 km/s in the south in the upper crust. The acquisition consists of 12 + 10 OBS, 90 seismic stations on the Israeli coastline, 3 detonations of 250 kg and a 32 l seismic array fired at 150 m intervals. Ginzburg et al. (1994) extended the northern profile with 90 seismometers, 3 detonations of 250 kg (+31 km of vibroseis reflection seismic), and observed a continental crust thickening to 25 km in Galilea. Two other profiles are acquired off Israel with 15 OBS over 150 km and 19 OBS over 158 km (and 1 seismometer), along which Netzeband et al. (2006) do not observe an ocean-continent transition and conclude that the basement of the Levantine Basin is very thinned continental in nature (β-factor $\sim$ 3).

In a short wide-angle line on the Troodos Ophiolite in Cyprus, Mackenzie et al. (2006) observed velocities of about 7 km/s at depths between 5 and 10 km, characteristic of ophiolites. In March 2010, a 650 km long N-S oriented sea-land profile was completed from the Tuz Gölü Basin in the north to the Eratosthenes Seamount in the south, crossing Cyprus (Feld et al. 2017). A total of 246 land stations were deployed and 2 detonations of 1125 kg each were fired in the Taurids: 50 stations at $\sim$0.9 km intervals over Cyprus and 196 stations at 1.55 km intervals in southern Turkey;

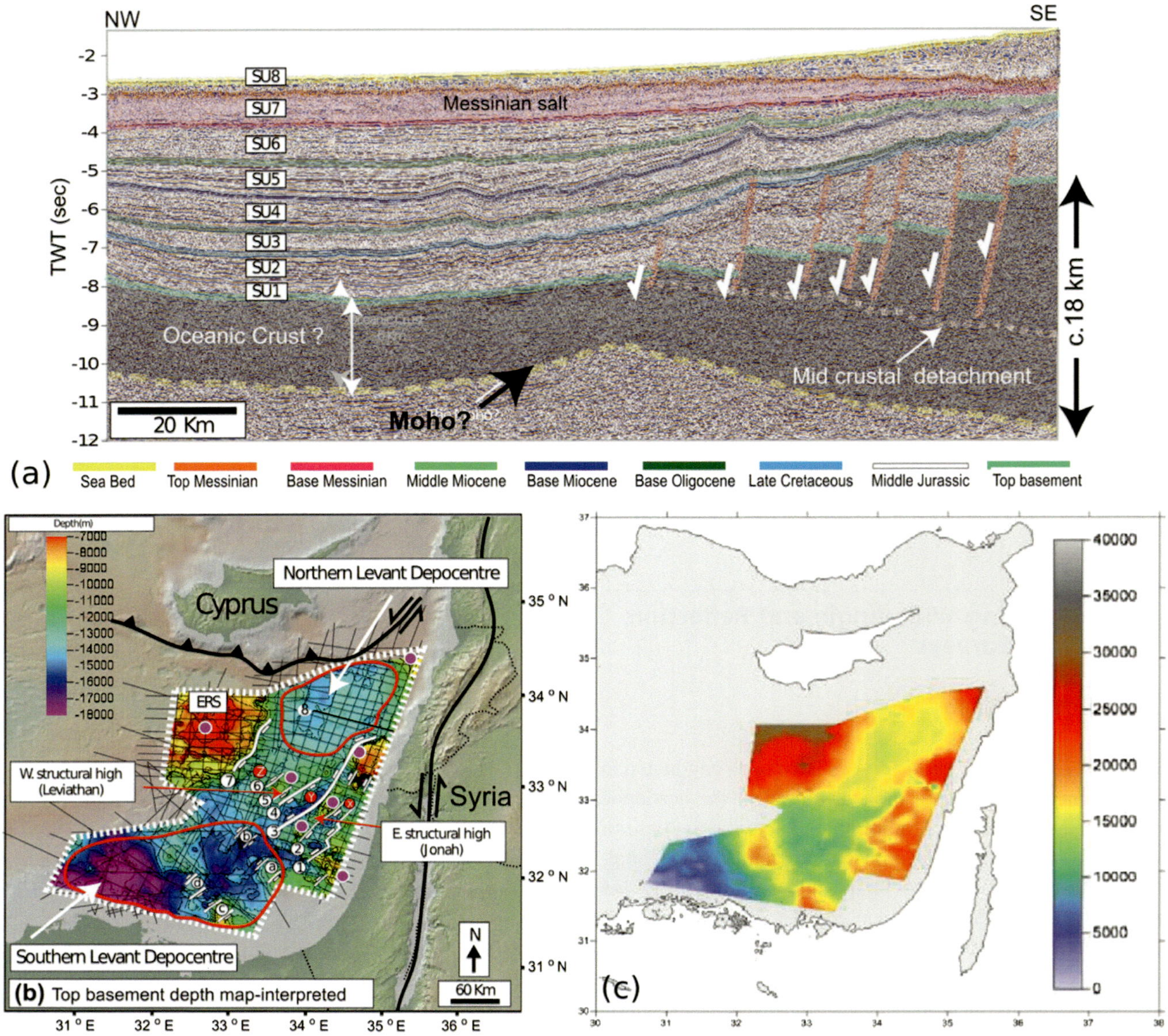

Fig. 1.10 **a** P8 seismic profile and stratigraphic interpretation; **b** Spectrum profile location and basement depth map; **c** Moho depth map from the inversion of the gravity anomaly. Based on Al-Balushi (2015)

South of Cyprus, marine shots with a total volume of 120 l were fired, recorded by 34 OBSs at 5–6 km interval. Feld et al. (2017) observe a Moho at 38–45 km depth under the Anatolian Shelf and an upper and lower continental crust with large lateral variations in thickness and acoustic velocity. Below Cyprus, the subducted slab is imaged, overlain by an anomalously fast layer (6.6–7.6 km/s), representing either an ancient thinned continental margin or an oceanic section comprising the gabbroic layer 3 of the oceanic crust and mantle, on which the 12 km thick ophiolite complex was obducted. Toward the south, the crustal thickness below the accretionary prism reaches 37 km; it varies between 35 and 28 km at the Eratosthenes Seamount, with velocity generally below 6.5 km/s. Three other wide-angle profiles complete this survey at sea: lines 1 (23 OBS over 230 km) and 1a (15 OBS over 140 km) cross the Eratosthenes Seamount, the Cyprus Arc, and the Hecataeus Ridge (Welford et al. 2015a), and line 2 (15 OBS over 75 km), sub-perpendicular, crosses the northern Levantine Basin and the Hecataeus Ridge (Welford et al. 2015b). These authors conclude that Eratosthenes Seamount has a velocity structure that is compatible with a thinned continental crust but is also laterally variable with a thicker middle crust and a thinner lower crust towards its north-eastern limit; the Hecataeus Ridge presents a thick sedimentary cover (5–8 km) over a crust of undetermined origin

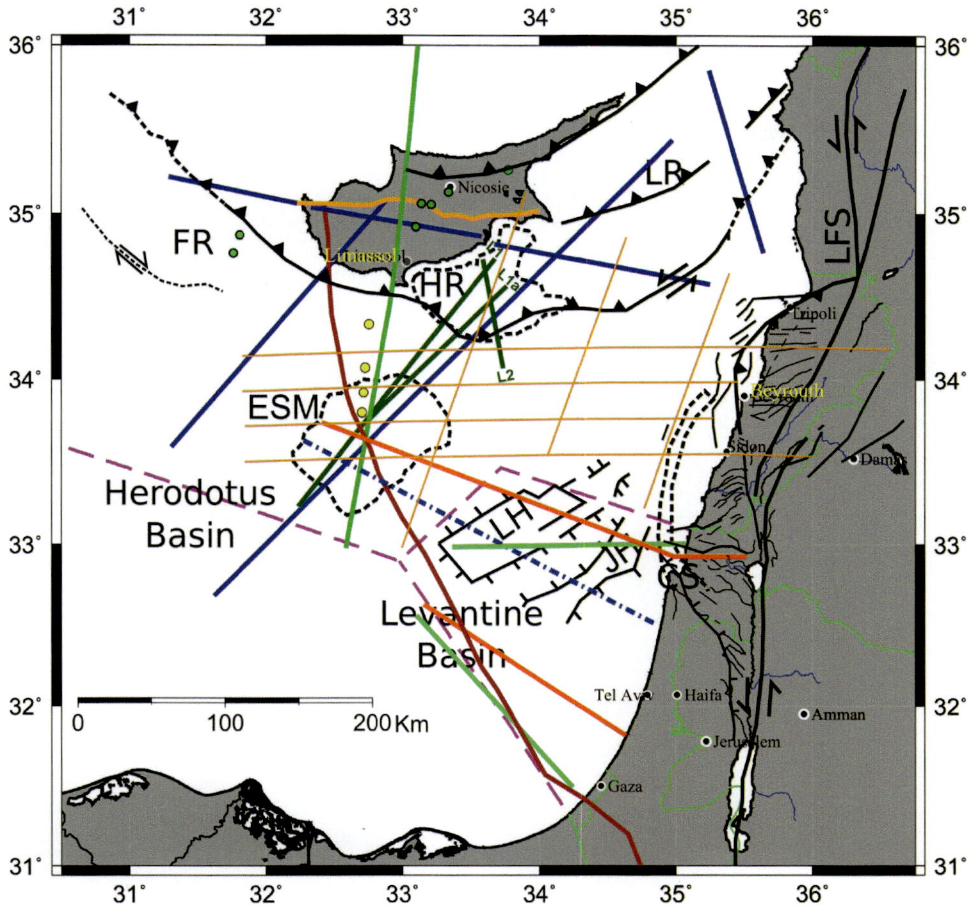

Fig. 1.11 Location map of previously acquired boreholes and Wide-Angle Seismic profiles. WAS: 0 orange = this survey, 1 green = Feld et al. (2017), 2 dark green = Welford et al. (2015a, b), 3 dark blue = Zverev (2010), 4 light orange = Mackenzie et al. (2006), 5 light green = Netzeband et al. (2006), 5 dark orange = Ben-Avraham et al. (2002), 7 brown = Makris et al. (1983). Boreholes are indicated by circles: yellow = ODP (Robertson 1998), green = DSDP, dark green = Geological survey of Cyprus. Baseline seismic profile (Fig. 1.4) in blue dashed lines. ESM = Eratosthenes Seamount, FR = Florence Ridge, HR = Hecataeus Ridge, LR = Latakia Ridge, LH = Levitan High, JH = Jonah High, CS = Carmel Structure, LFS = Levant Fault System. The borders between states are represented by green lines. The limits of Israeli and Egyptian waters are indicated by magenta dotted lines

(probably continental), and a Moho at about 20 km depth. Between these 2 structures, blocks of lower crust characterized by velocity reaching 7.2 km/s are imaged: they could correspond to deformed remains of Tethysian oceanic crust or mafic intrusions, trapped in the nascent continental collision. These profiles do not show evidence of shallow ophiolite complexes such as those observed in Cyprus (Welford et al. 2015b).

In addition, 4 wide-angle seismic profiles were acquired in the Cyprus region by the oceanographic vessels Akademik Boris Petrov and Akademik Nikolai Strakhov (Zverev and Ilinsky 2005; Zverev 2010), but their descriptions, analyses, and interpretations are not available in the literature. Finally, in the Herodotus Basin from 1 ESP (expanding spread profile), de Voogd et al. (1991) observed 10 km of sedimentary cover over the Mesozoic oceanic crust about 10 km thick.

1.3.2 The Western Mediterranean Sea

The West part of the Western Mediterranean Sea (Fig. 1.12) is divided into several segments.

To the north (Valencia, Menorca and Provencal Basins), the limits of these segments as well as their geodynamics are well-defined:

- In the Provencal Basin, wide angle data (Moulin et al. 2015; Afilhado et al. 2015) and 3D-grid of seismic, drillings and numerical stratigraphic modeling (Leroux et al. 2015a, b, 2018), identified 5 domains with different subsidence and crustal nature: seaward tilting subsidence on the platform and slope (11: continental domain and necking) and purely homogeneous vertical subsidence in the deep basin made of exhumed lower continental crust (10) and proto-oceanic crust (9).

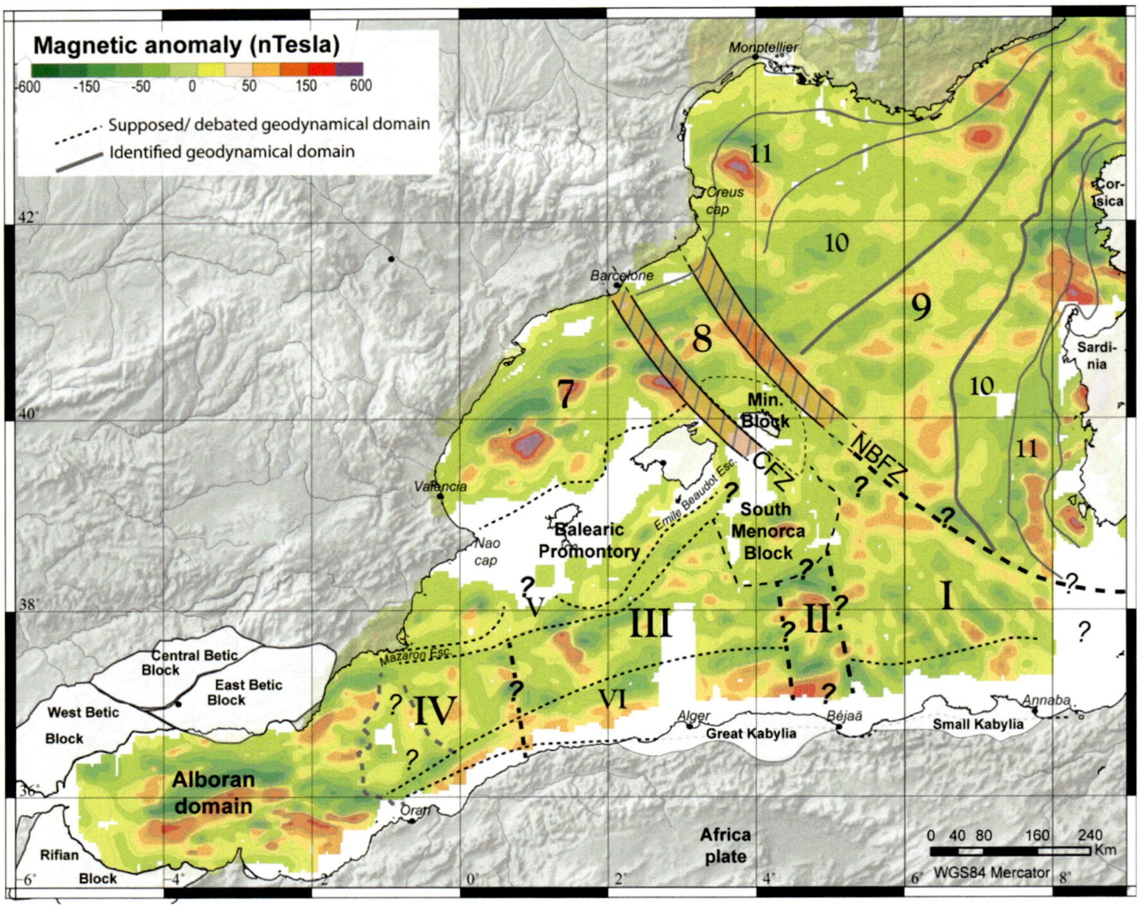

Fig. 1.12 Segmentation of the west part of the Western Mediterranean Sea. Left gravity map. Right: magnetic map. Thick lines and Arabic numbers: well-identified segments. Dashed lines and roman numbers: supposed segments on the base of gravimetry and magnetic data. Small-sized number: variation in the same segment. I: East Algerian Basin (EAS) with fan-shaped magnetic anomalies, mostly NW–SE; II: Hannibal ridge (HR); III: Central Algerian segment (CAS) and IV: Western Algerian Segment (WAS) without clear aligned magnetic anomalies; V and VI: Thinned continental crust/intermediate domain on both sides of the Algerian Basin based on wide angle data (Leprêtre et al. (2013), Aidi et al. (2013), Mihoubi et al. (2014), Bouyahiaoui et al. (2015), Badji (2014), Medaouri et al. (2014), Hamai et al. (2015), Arab et al. (2016a, b) et Aidi et al. (2013); 7: Mesozoic Valencia Basin (VB) and 8: Neogene Menorca Basin (MB) (after Pellen et al. 2016 and Etheve et al. 2016); 9-10-11: Provencal Basin with Necking zone (11), Exhumed continental crust (10) and 9 proto-oceanic crust (after Moulin et al. 2015 and Afilhado et al. 2015).NBFZ: North Balearic Fracture Zone. Note the presence of a supposed continental block south of Menorca microplate (Driussi et al. 2015). Note also the microblocks in Alboran area (after Crespo-Blanc et al. 2016) and the presence of Great Kalybie a and Small Kabylia blocks along the Algerian margins, on both sides of the Hannibal Ridge

- Detailed analysis of data from seismic surveys and boreholes in the Valencia Basin (VB) highlight a differentiated basin, the Menorca Basin (MB), lying between the old Mesozoic Valencia Basin s.s. (VBss) and the young Oligocene Liguro-Provencal Basin (LPB; Etheve et al. 2016; Pellen et al. 2016). The Central and North Balearic Fracture Zones (CFZ and NBFZ) that border the MB represent two morphological, structural, and geodynamic thresholds that created an accommodation in steps between the three domains. Little to no horizontal Neogene movements are found for the Ibiza and Mallorca Islands. In contrast, the counter-clockwise movement of the Sardinia-Corsica blocks induced a counter-clockwise movement of the Menorca block towards the SE along the CFZ and NBFZ, during the exhumation of lower continental crust in the LPB.

Southwards the Balearic Promontory and the NBFZ, according to the gravity and magnetic data, the Algerian Basin can be divided into five segments, from east to west:

- East Algerian Basin (EAS), between the NBZF and the South Menorca Block-Hannibal Ridge. This triangle-shaped segment presents in the south-east NNW-SSE aligned magnetic anomalies, which disappear to the North, near the SMB.

- The South Menorca block (SMB) is proposed to be of continental nature based on analysis of reflection seismic data and gravity modeling (Driussi et al. 2015)
- The Hannibal short segment (HS), in the prolongation of the SMB, with strong magnetic anomaly and a relative negative gravity anomaly with respect to the adjacent segments.
- The Central Algerian Segment (CAS), west to the SMB-HS. This segment may be divided into two sub-segments as the gravity and magnetic pattern change on both side of a limit more-less located between Alger and the western end of Emile Beaudot escarpment.
- The West Algerian Segment (WAS), located south to the Mazarron Escarpment. The Valencia Basin (Fig. 1.13), as the Provencal Basin (see the "*Action-Marges*" website - http://actionsmarges.fr/medocc/), is well covered by a huge set of seismic data. This is also true for both sides of the Algerian Basin, but the central part presents an important lack of seismic data.

Several wide-angle profiles (Fig. 1.14) have been analyzed on both sides of the Algerian basin (Leprêtre et al. 2013; Aidi et al. 2013; Mihoubi et al. 2014; Bouyahiaoui et al. 2015; Badji 2014; Medaouri et al. 2014; Hamai et al. 2015; Arab et al. 2016a, b).

They all show offshore a less than 5 km thick crust in the Algerian Basin, all interpreted as anomalously thin oceanic crust, but with very different seismic velocity structure (Fig. 1.14) where the first crustal layer velocities ranges from 4.3 km/s to 5.5 km/s. Note that no profile is connected to the others (no crossing points, no cross interpretation), and that no profile is doubled to avoid lateral effects and to confirm the proposed model.

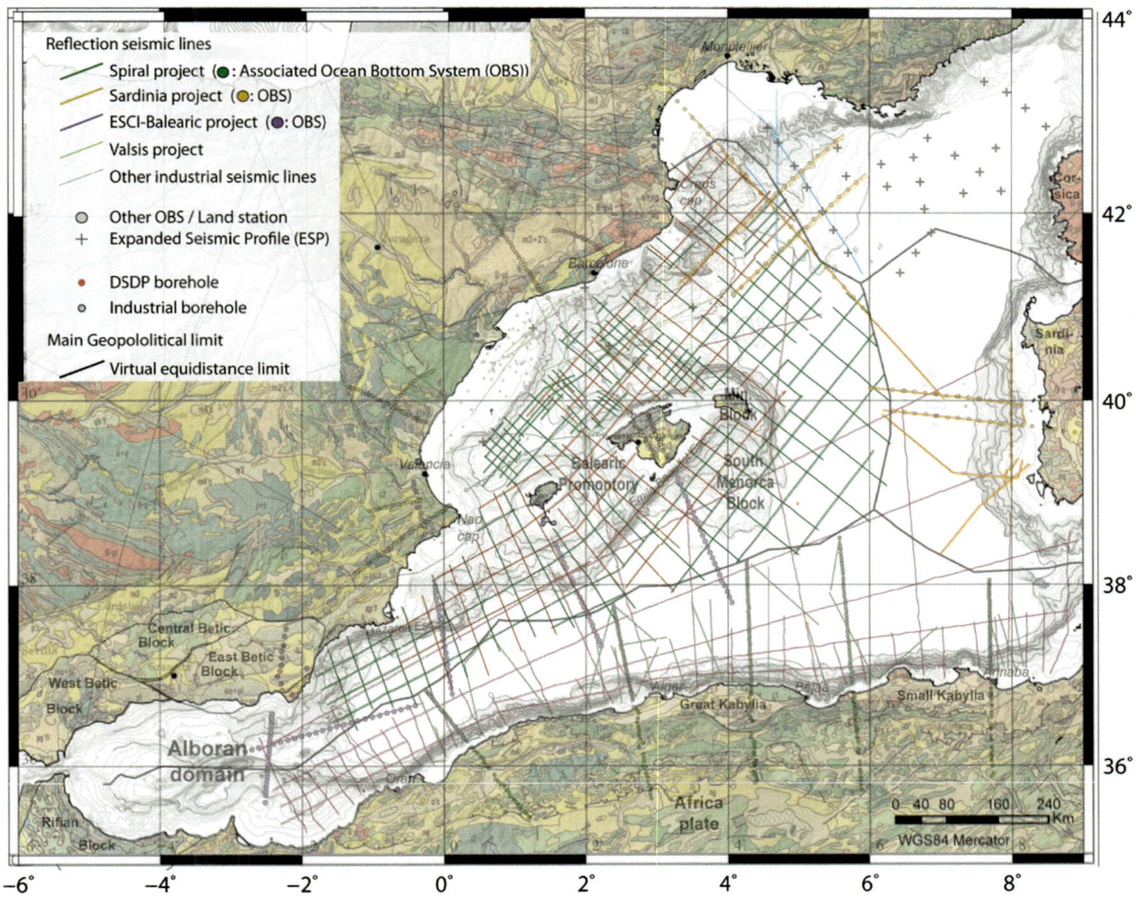

Fig. 1.13 Deep Multi-Channel and wide-angle seismic data in the Algerian basin and the Valencia basin and Sardinia wide-angle experiment in the Provencal Basin (Compilation: Romain Pellen). The wide-angle experiments are represented with colored circles (position of the OBSs). Note the lack of connection between these wide-angle data and the gap of seismic profiles in the middle of the Algerian Basin

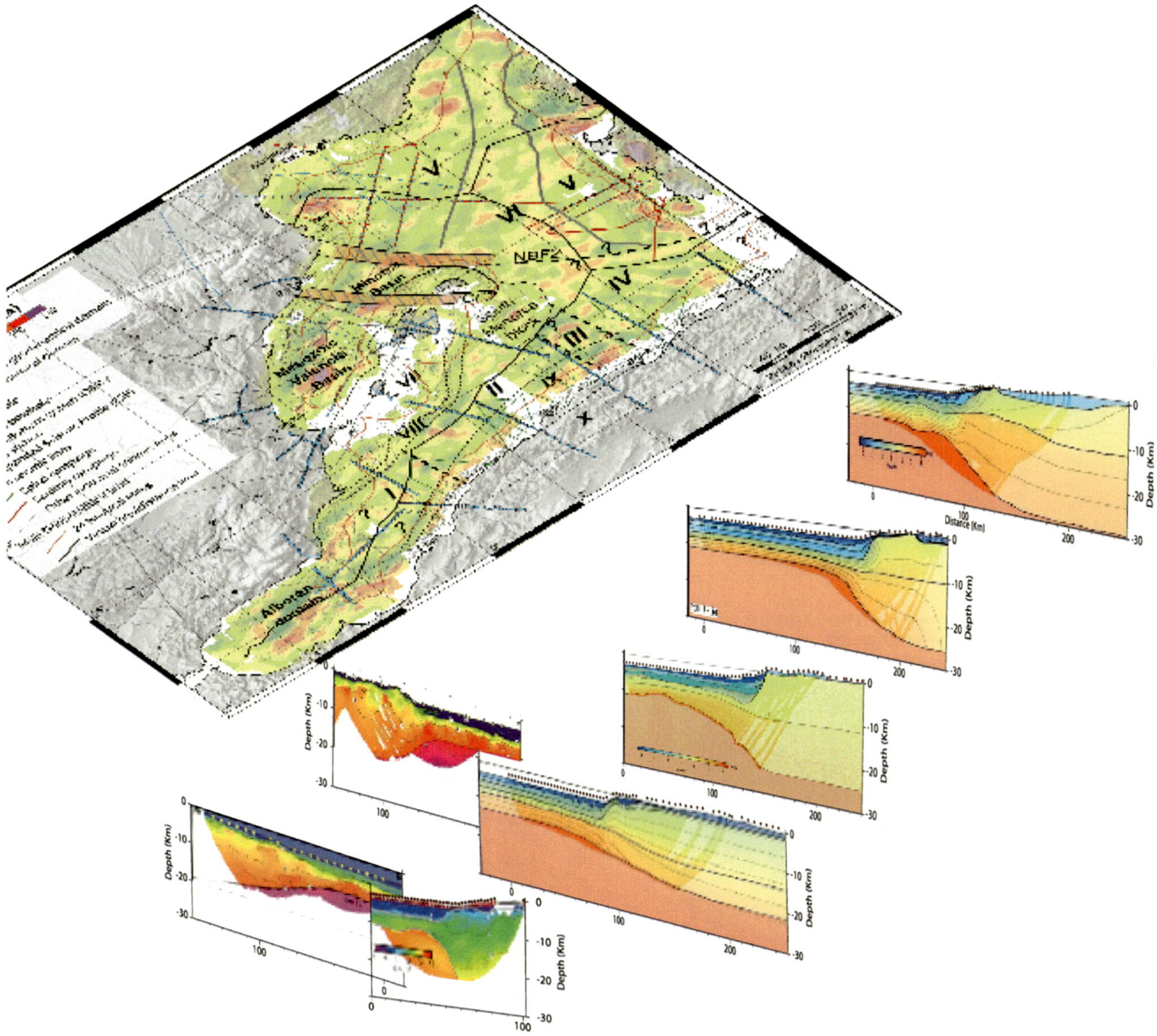

Fig. 1.14 Published wide-angle seismic model in the Algerian Basin: Afilhado et al. (2015); East Algerian Segment: Bouyahiaoui et al. (2015) and Mihoubi et al. (2014); Central Algerian Segment: Aidi et al. (2013), Leprêtre et al. (2013) and Viñas Gaza (2016); West Algerian Segment: Viñas Gaza (2016) and Badji (2014). Note that the processing methodology is not always the same (Tomography Versus Forward modeling)

1.4 Similar Variations on Other Margins: The Proto-Oceanic Crust

The same type of lateral and longitudinal variations were described (Fig. 1.15) in the Equatorial and South Atlantic Brazilian margins (Aslanian and Moulin 2012, Aslanian et al. 2021a, b; Dias et al. 2016; Evain et al. 2015; Klingelhoefer et al. 2015; Loureiro et al. 2018; Moulin et al. 2012; Moulin et al. 2021; Pinheiro et al. 2018) as in the Provencal Basin (Fig. 1.15; Moulin et al. 2015; Afilhado et al. 2015; Leroux et al. 2015a, b).

In the Santos Basin, the two 700 km long parallel wide-angle profiles exhibit in the middle of the basin very different velocity structure, while they are only spaced by 50 km. The southernmost profile as a structure of thin exhumed continental crust while the northern shows an anomalous and probably heterogeneous crust, which can be inferred as proto-oceanic crust on a failed rift.

On the Ceara-Maranhão margin, the 4–5 km thick crust of the intermediate domain is interpreted as exhumed lower continental crust. Seawards and parallel to the coast, an about 50 km-wide band of proto-oceanic crust is described before reaching a more normal but only of 5 km thick

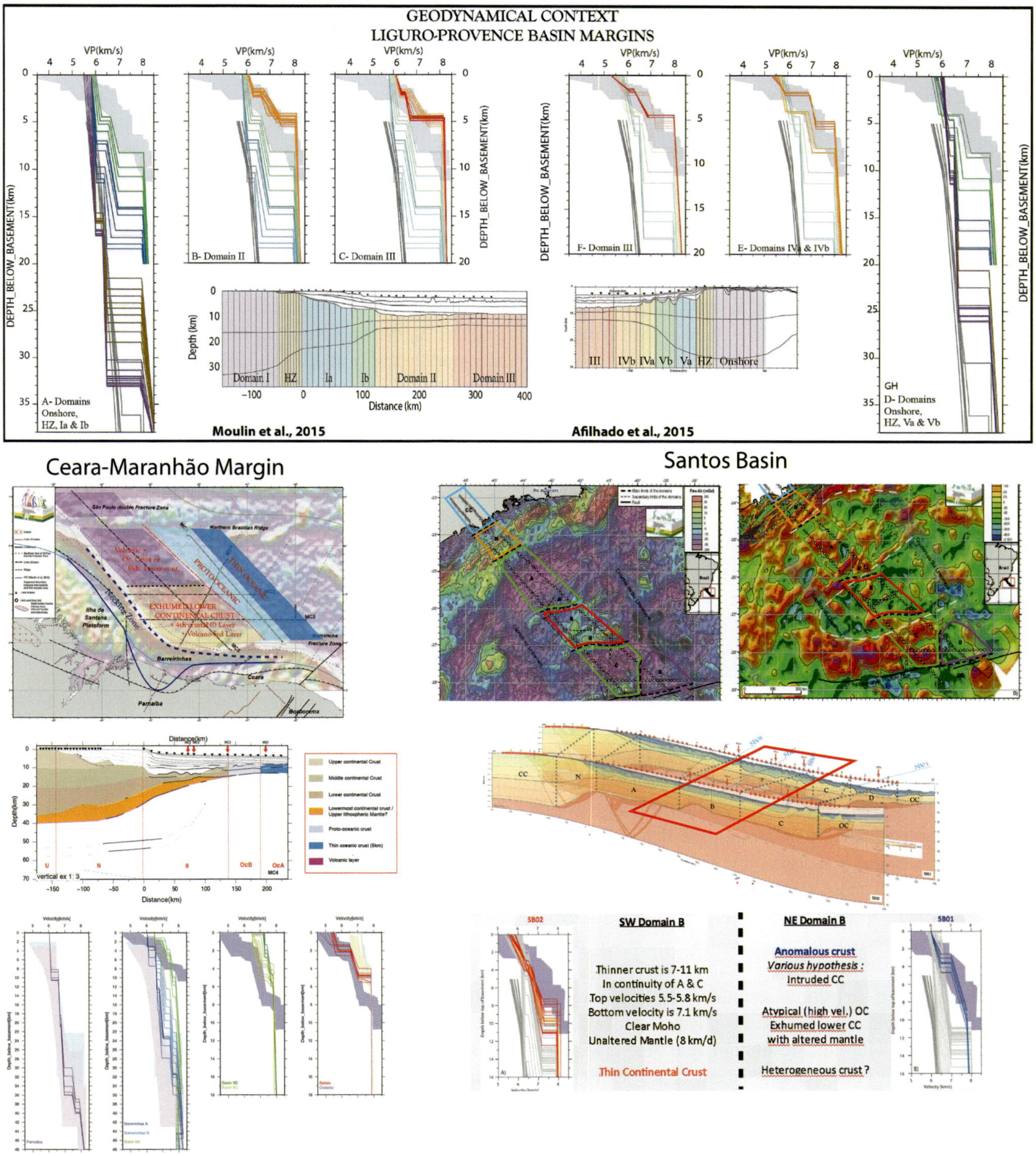

Fig. 1.15 10 km spaced VZ profiles evolution along: Bottom) the two wide-angle profiles in the Provencal basin (Moulin et al. 2015; Afilhado et al. 2015). Top left) one of the E-W profile of the Ceara-Maranhão Margin, Brazil (Aslanian et al. 2021a, b; Moulin et al. 2021). Top right) the Santos Margin (Evain et al. 2015). Note the segmentation in each Basin and the slight but sharp evolution of the VZs at the segments boundary. Note also the lateral variations in the central part of the Santos Basin

oceanic crust. Note that on the three examples (Provencal, Equatorial and Santos Basin), strong and abrupt jumps exist in the evolution of the VZ profiles when reaching the boundaries in-between domains (Fig. 1.15).

1.5 The ARCMAL Survey

The understanding of the formation of passive continental margins in their diversity (transform, pull-apart or divergent), as well as the impact of major kinematic reorganizations, are major issues of current research in Geosciences. These issues cannot be addressed without constraining primordial elements such as:

(1) the paleogeographic context and the kinematics of the opening of the margin;
(2) the evolution of the structure, its geometry, and the definition of the nature of the crust from the continent to the ocean;
(3) the impact of the legacy of pre-, syn- and post-rift tectonic events and volcanism (Fig. 1.4);
(4) the evolution of sedimentation from pre-rift to post-rift: the subsidence and the associated sedimentary records;
(5) the Messinian salinity crisis—erosion surface, the thickness of deposits, and post-depositional deformation.

While the Levantine margin and the Levant Basin south of the Carmel structure, the Eratosthenes Seamount, the Cyprus Arc, and the Hecataeus Ridge have been the subject of several wide-angle seismic surveys, the North Levantine Basin (also called Phoeniciean Basin), the Latakia Ridge as well as the Lebanese margin are essentially devoid of wide-angle surveys (Fig. 1.11). These structures individualize areas where tectonic events (Fig. 1.3) are numerous and diverse, and their structural boundaries are still poorly constrained.

Wide-angle and reflection seismic surveys allow both imaging and characterization of the physical properties of geological units with fine resolution and true geometry. These profiles will allow the refinement of the stratigraphy of the sedimentary series of the Cyprus, Herodotus, and Levant basins. We will be able to quantify the thickness of the deposits, their environment in particular pre-, syn, and post-rift, and the recorded deformations, with sufficient penetration to characterize the base of the basins. It will thus be possible, for example, to clearly distinguish magmatic events from sedimentary deposits (in particular syn-rift SDRs and post-rift sills and dykes), and evaporites from mobile units in the Messinian sequence. The main objectives of this survey concern the deep structure and crustal architecture. Our first objective is therefore geophysical, then geological and kinematic. It will be important to carry out an acquisition dimensioned to the spatial resolution and precision sufficient to characterize the structures present. In order to image the thickest continental domains, notably the Levant margin, the profiles will be extended on land (Fig. 1.11). The acquisition of several profiles on the same object is important: parallel profiles make it possible to evaluate the lateral variability on a segment of the Levant margin having a priori the same origin and tectonic evolution; perpendicular profiles make it possible to locate the fine segmentation of the margin, but also to limit the biases related to the acquisition (length of the profiles, quality of the data) to the processing and the interpretation by a fine analysis of the crossings between profiles. It will then be necessary to apply all the available tools to the joint analysis of the wide-angle and reflection seismic data. The analysis of horizontal speed gradients at crossings (at different scales and depths) should permit to better constrain the direction of the structure's emplacement. Finally, the interpretation of intra-crustal reflectors related to petrological alteration or shear of the rocks in place, frequently interpreted as Moho, will be differentiated by determining the acoustic propagation velocity. In cases where the petrological nature of a structure is poorly constrained, the imaging of shear-converted waves and the determination of the S-wave propagation velocity may be decisive. Our next objective is to re-evaluate the large seismic reflection data-set existing in the region in the light of the observations delivered by the wide-angle survey: geophysical and petroleum companies interested in this province lack information on the basement of those basins, and commonly share their seismic profiles. Finally, our objective is to provide new geometric and petrological constraints to the numerical modeling of the formation and evolution of the Levantine Basin, in collaboration with the IFP-EN "basins" group (Hawie et al. 2013; Hawie 2014; Inati et al. 2016, 2018).

In order to achieve our objectives, we propose to acquire 7 vertical reflection seismic profiles using Ifremer's multi-channel seismic (MCS), and to simultaneously acquire wide-angle seismic data recorded by seabed seismometers (OBS) at sea and mobile stations (LSS) on land (Fig. 1.16).

In total, 3386 mn (5450 km) would be covered, including 1075 nm (1730 km) of MCS acquisition. A total of 106 OBSs will be deployed. On the Levantine margin, the Levantine Basin, the Eratosthenes Seamount, and the Herodotus Basin, we propose to carry out 4 parallel profiles oriented E-W: These profiles comprise 28, 7, 7, and 26 OBS respectively, regularly spaced at 7 nautical miles (nm) at sea. The key partnerships of the ARCMAL survey for the terrestrial component will bring together the Instituto Dom Luiz/Faculdade de Ciências da Universidade de Lisboa (FCUL) and Instituto Superior de Engenharia de Lisboa (ISEL), in Portugal, and the National Centre for Geophysical Research, CNRSL, Bhannes, El Metn, in Lebanon. These partners will install from the

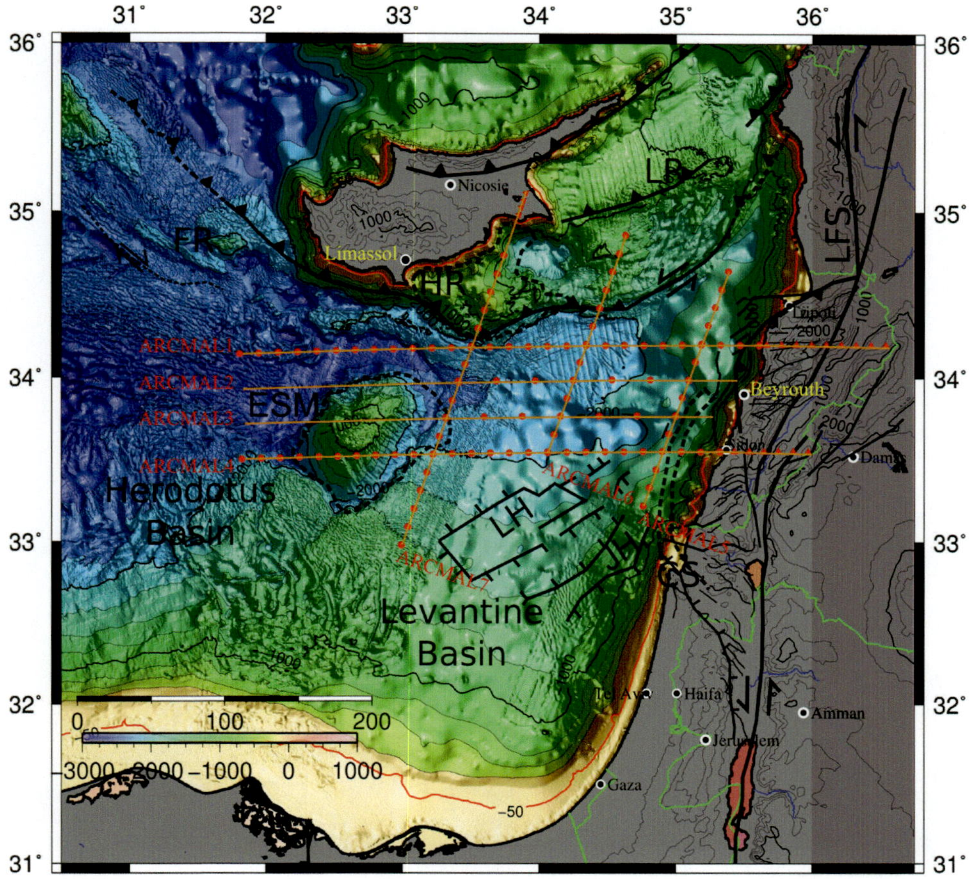

Fig. 1.16 Location map of the ARCMAL (ARchitecture Crustale de la MArge du Levant) survey. The orange lines represent this survey's profiles. The red circles represent the 106 OBS deployed. Red triangles symbolize the 20 land seismometers. The bathymetry is illuminated from the SE and contoured at 250 m intervals. The boundaries between states are represented by green lines

coast to the Syrian border (if possible), at 10 km intervals, short period seismometers: 9 and 6 mobile stations spaced 5 km by 85 km and 30 km would and extended onland the ARCMAL1 and ARCMAL4 profiles, in order to record the arrivals of shots fired at sea. Finally, the James Mechie group of the GeoForschungsZentrum (GFZ) in Potsdam, Germany, proposes to deploy seismometers in Cyprus along the ARCMAL7 profile.

The ARCMAL1 profile, 189 nm long (304 km) and located at 34°12'N, starts SE of the Florence Ridge, crosses the northern extension of the Eratosthenes Seamount, skirts the deformation front of the Hecataeus Ridge, crosses the North-East basin and the continental slope, and extends ashore south of Mount Lebanon. The onshore stations will be deployed in the vicinity of profile SLEB-2D-01-13, 50 km long between the coast at Batroun and the Yammouneh fault at Ainata (Nader and Browning-Stamp 2016). The geological section proposed by Nader and Browning-Stamp (2016; Fig. 1.17) includes all sedimentary units up to Triassic evaporites that overlie the Permian to Carboniferous basement. The lithological description on this profile should make it possible to construct, a-priori, the terrestrial part of the ARCMAL1 velocity model. Indeed, the long travel times from the arrival of the shots at sea to the passage of sedimentary units make the determination of the thickness of the underlying crust particularly dependent on these units. Moreover, in the first 20 km from the coast, these units reach ~5 km thickness: near-shore firing should lead to refracted arrivals in the sedimentary portion that will constrain the acoustic velocity of these layers.

The ARCMAL2 and ARCMAL3 profiles, 177 and 159 nm long (285 and 256 km) and located at 33°59'N and 33°45'N respectively, cross a portion of the Herodotus Basin, the Eratosthenes Seamount, and the Levantine Basin; Since Ifremer's Marine Geoscience Unit pool includes 68 MicrOBS, on these 2 profiles our effort is more focused on the Leventin Basin and only the 84 nm (135 km) to the east are equipped with OBS spaced 7 nm apart. On this eastern portion, the coverage is therefore identical to the rest of the mission.

The ARCMAL4 profile, 175 nm long (282 km) and located at 33°33'N, crosses a portion of the Herodotus Basin, the southern flank of the Eratosthenes Seamount, the Levantine Basin and the continental slope, and extends onshore south of Lake Qaraoun and north of the Hermon Mountains. On this profile, OBS01 is located at the Calypso drill hole where in 2018 significant gas reserves were drilled in the carbonate reefs of the ESM. On the Cyprus Arc and

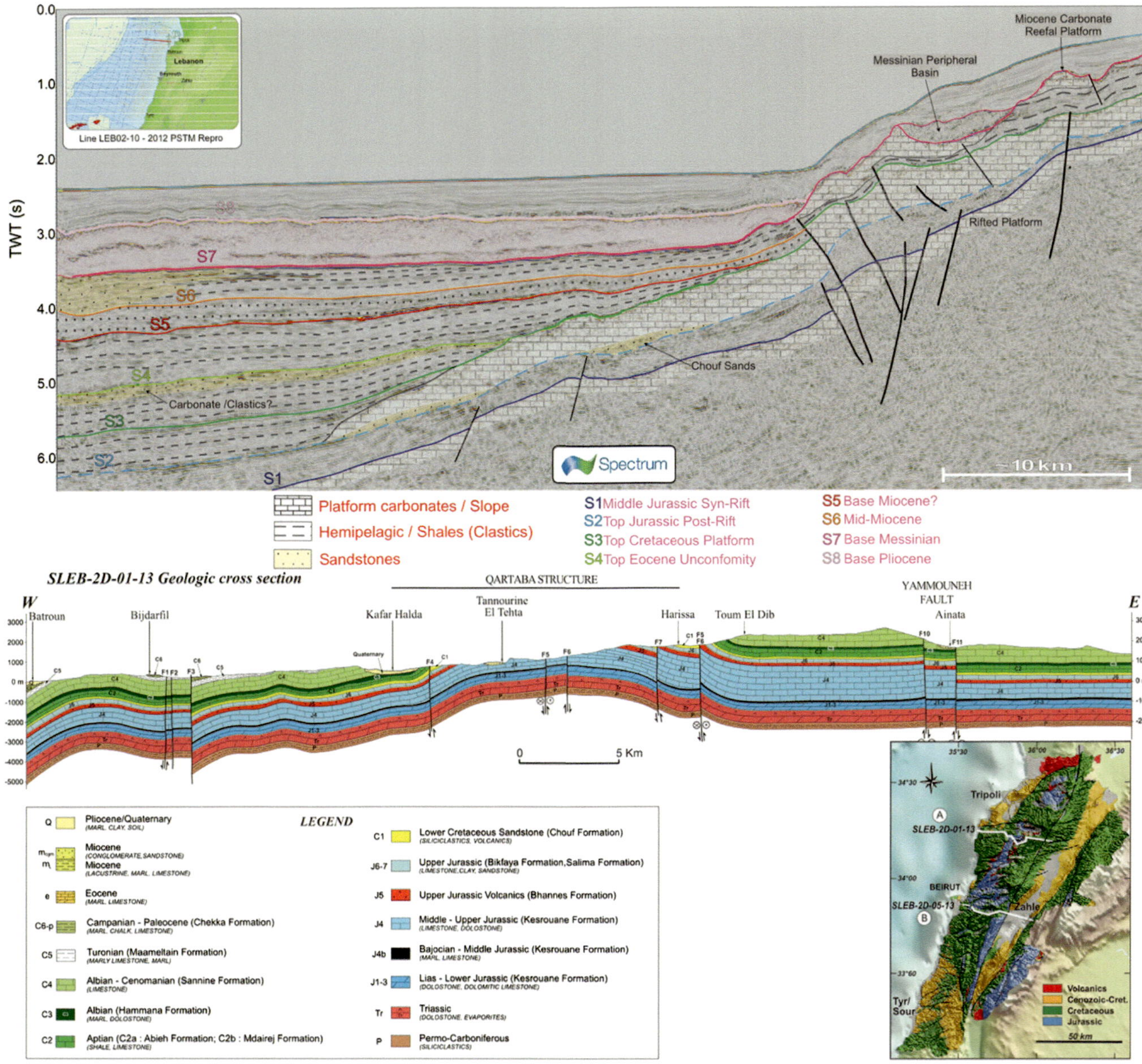

Fig. 1.17 **a** E-W seismic profile offshore northern Lebanon and stratigraphic interpretation (Spectrum Geo Ltd). **b** Geological section along the SLEB-2D-01-13 seismic profile crossing the Qartaba structure, built on the basis of geological maps of Lebanon. The insertion map shows the location of the sections and seismic profiles. According to Nader and Browning-Stamp (2016)

the Levantine Basin, we propose to carry out 3 profiles ARCMAL5, ARCMAL6, and ARCMAL7, at sea only, with a regular spacing of 7 nm between OBS and NNE-SSW oriented sub-parallel profiles: The ARCMAL5 profile includes 14 OBS over 91 nm (170 km), and follows the continental slope from the Syrian border in the north and the Israeli border in the south.

The ARCMAL6 profile includes 13 OBS over 84 nm (155 km), and intersects the Latakia Ride and the Levantine Basin in its central axis. Finally, the ARCMAL7 profile includes 17 OBS over 112 nm (207 km), intersects the Hecataeus Ridge in southern Cyprus and the Levantine Basin near the southern flank of the Eratosthenes Seamount and extends to the limit of the Israeli-Egyptian waters. To the north, it is extended onshore in the Ayia Napa province between the villages de Xylofagou and Frenaros by 4 mobile stations over ~15 km.

This acquisition scheme has proven to be suitable for seismic imaging of a wide range of continental margins. These previously acquired wide-angle seismic surveys permit to build of a synthetic model (Fig. 1.18) of structure and acoustic velocity along an E-W transect at 34°12′, intersecting the Herodotus Basin, the Eratosthenes Seamount, the Levant Basin, and the continental margin of Lebanon.

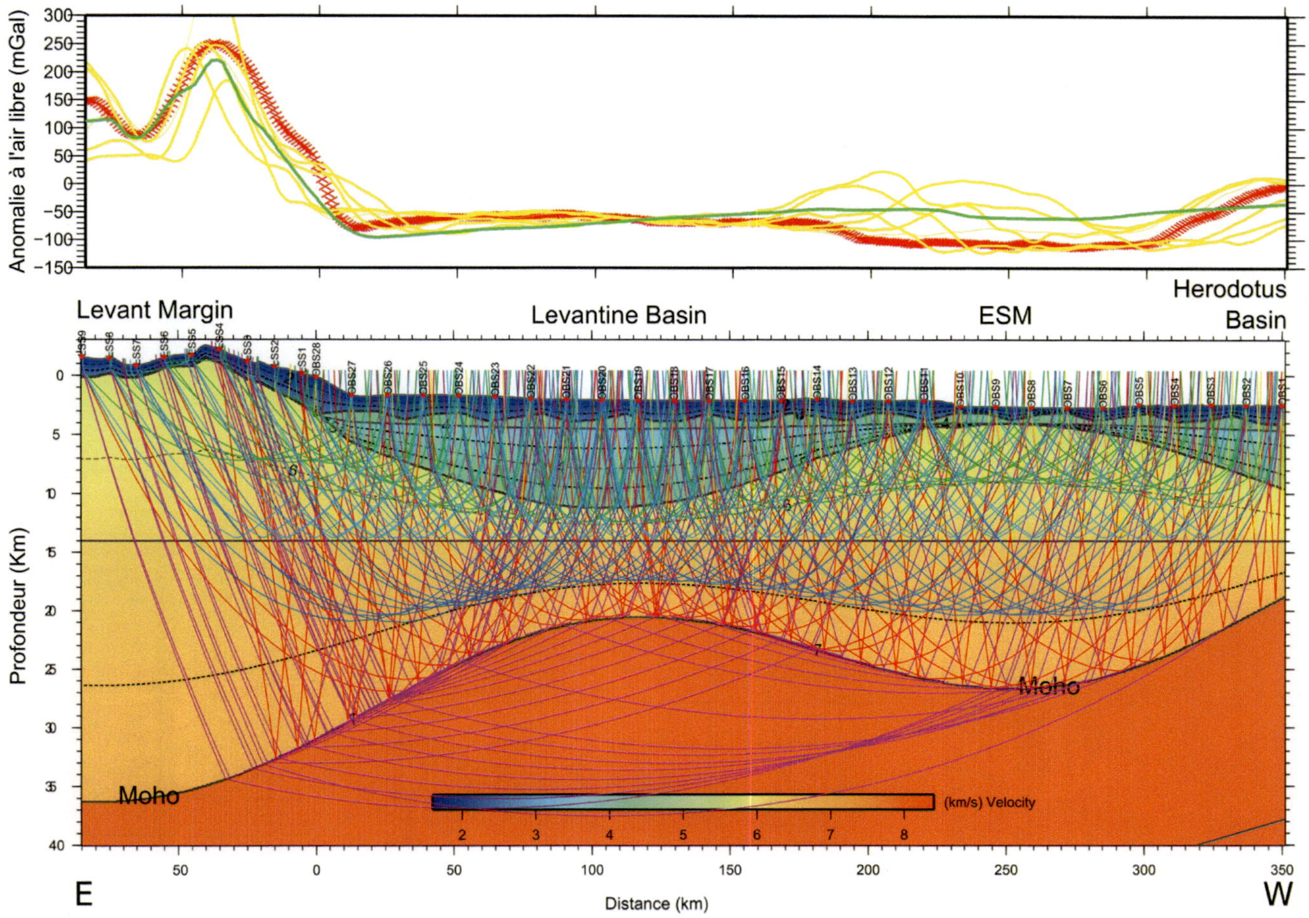

Fig. 1.18 Bottom: Synthetic profile of the structure, acoustic velocity, and ray paths along an E-W transect at 34°12'. The five main units (pre-Messinian sediment, evaporite, post-Messinian sediment, upper crust, lower crust, and lithospheric mantle) are separated by grey boundaries. The velocity is contoured by a dotted line every km/s. The path of the refracted rays in the upper crust (green), reflected at the upper crust / lower crust interface. (cyan), refracted in the lower crust (blue), reflected at the Moho (red), and refracted in the upper mantle (magenta) are shown. Top: Modeling of the free-air gravity anomaly; The satellite-derived anomaly along the profile is represented by red crosses, those at 10, 20, and 30 km on either side of the profile by yellow lines; The calculated anomaly is represented by a green line; The conversion of acoustic velocity to density obeys Brocher (2005)

Modeling of the paths of the rays refracted in the lower crust reflected at Moho, and refracted in the upper mantle in the synthetic profile shows that: (1) the proximity of land mobile stations to the coast will constrain the crustal velocities of the Lebanon margin and the geometry of the necking zone in the first 40 km of the coastline. (2) The regular spacing of 7 nm between OBS results in sufficient coverage to image with adequate resolution the base of the Levantine Basin; vertical seismic imaging (of the reflected events of the MCS and OBS data) will be essential to constrain the crustal geometry of the summit of the Eratosthenes Seamount and the Herodotus Basin, as well as the geometry of the salt domes. (3) The maximum depth of investigation should reach 40 km on profiles ARCMAL1 to ARCMAL4, 25 km on ARCMAL5 and ARCMAL6, and 30 km on ARCMAL7.

In 2021, the french Commission Nationale de la Flotte Hauturière has placed the ARCMAL cruise project in priority 1, with a possible timeframe between 2023 and 2025.

1.6 The BasAlg Survey

Geodynamic evolution and crustal issues are closely linked: the presence of an ocean crust does not have the same meaning as that of a thinned or even stretched continental crust (Aslanian et al. 2009; Aslanian and Moulin 2012). The formation of the Algerian basin remains a subject of debate, and radically opposed hypotheses are proposed, both on the nature of the substratum and on the geodynamic evolution. This is the consequence of a lack of data in the central part of the Algerian Basin that could constrain its segmentation,

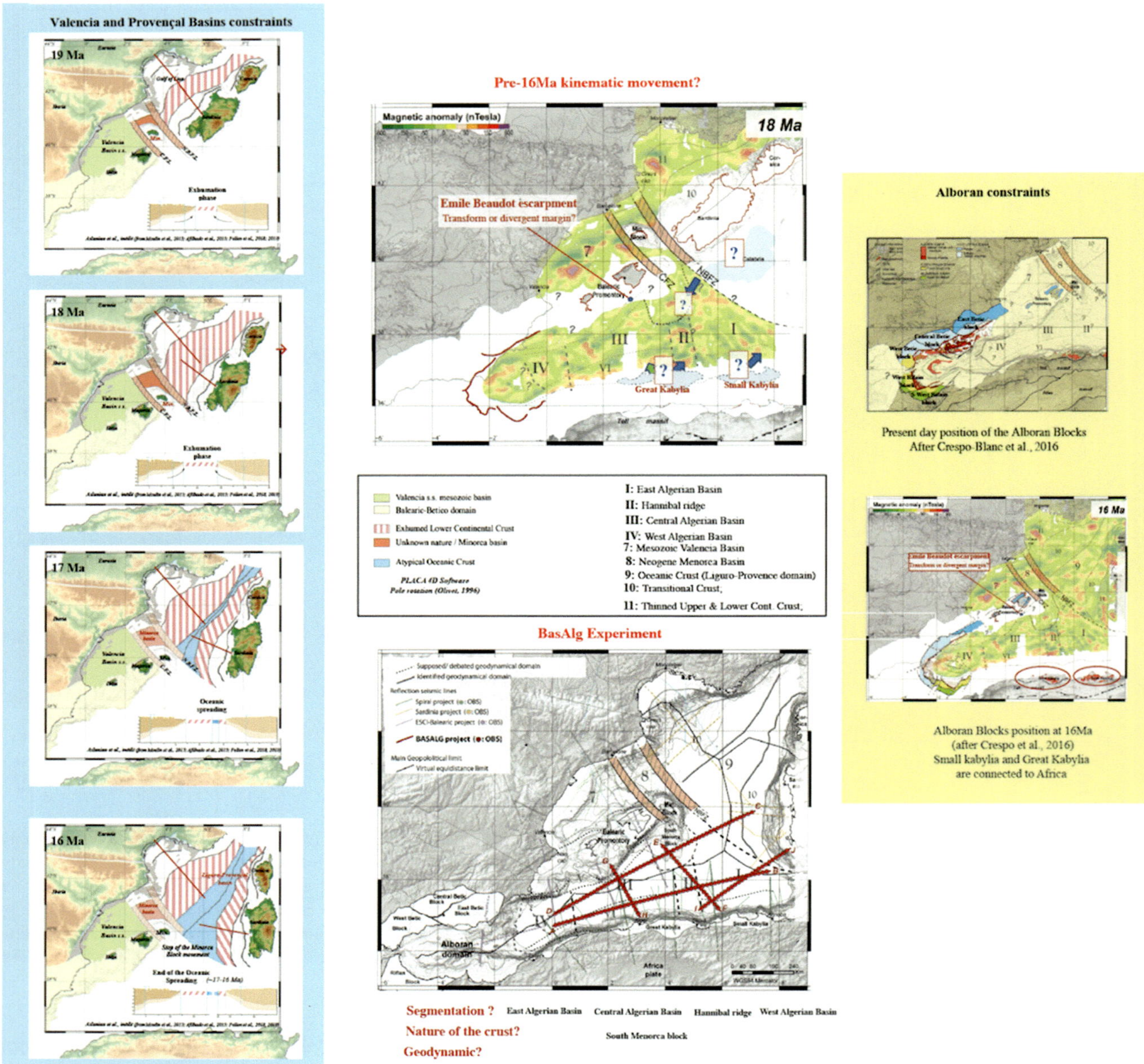

Fig. 1.19 Unsolved geodynamic questions and aims of the BasAlg Project. Left: geodynamic evolution and constraints in the Valence and Provencal Basins (after Moulin et al. 2015; Afilhado et al. 2015; Pellen et al. 2016, 2019; Leroux et al. 2018). Right: Geodynamic evolution of the Alboran blocks (after Crepso-Blanc et al. 2016, Pellen et al., in preparation). Center top: Unsolved geodynamic positions (questions marks) Center bottom: The BasAlg experiment: position of the proposed profiles in the present day setting

which seems to be apparent from the reading of gravity and magnetism (Leroux et al. 2018; Fig. 1.12), as well as its crustal nature. The geodynamic constraints and the unsolved geodynamic questions described above are shown in Fig. 1.19.

We propose to conduct a wide-angle and multi-trace reflection seismic survey along 5 profiles (171 deployments of ocean bottom seismometers). These profiles will allow determining the geometry, the structure, and the acoustic velocity of the different segments of the Algerian basin, their base, and the underlying lithospheric mantle:

- 2 profiles crossing The West Algerian Segment (segment IV),
- 3 profiles crossing Central Algerian Segment (segment III),
- 2 profiles at the southern limit of the South Minorca block,
- 2 profiles crossing the Hannibal short segment (segment II) in prolongation of the SMB with strong magnetic anomaly and a relative negative gravity anomaly,
- 2 profiles crossing the boundary between the Algerian Basin and

- 2 profiles crossing East Algerian Basin (segment I), with the NNW-SSE aligned magnetic anomalies.

These profiles intersect therefore the main structural edifices South Menorca block, the well-aligned magnetic anomalies in the easternmost segment, and cross the different boundaries of the supposed segmentation. They intersect also the existing wide-angle profiles of both sides of the Algerian basin, and even the wide-angle profile offshore the west Sardinian margin. The series of crossing points will ensure a coherence and a continuity between the interpretations (as in Magic or Sanba experiments) and will allow following of the transversal and longitudinal crustal variations (discrete 3D image). Gravity and magnetic data will be used, as usual, to test the seismic models. Magnetic data are crucial where oceanic crust is proposed.

1.7 Expected Results

Characterizing the sequence of processes leading from continental rifting to oceanic spreading during the genesis of passive margins is a key to our fundamental knowledge of the Earth's structure and history, and has important consequences for the environment and geo-resources.

Through the analysis of data from the ARCMAL survey, we believe we can:

- fill in the areas of the northern part of the Eastern Mediterranean with deep seismic surveys acquired in the past;
- specify the geometry, tectonic structure, and petrological nature of these domains; characterize the nature of the basement and upper mantle;
- to define the position and the extent of the necking and transitional zones of proto-oceanic and oceanic accretionary;
- quantify the active processes (and their orientation and segmentation) during rifting and subsequent spreading, such as crustal thinning factor, delamination and/or exhumation of the lower crust and/or upper mantle, and magmatism;
- refine the lithology/stratigraphy of the deep sedimentary series of the Cyprus, Herodotus, and Levant basins;
- examine the role of the nature and crustal structure of these basins in the tectonic episodes following their emplacement;
- contribute to a paleo-tectonic scheme of the Eastern Mediterranean.

The project BasAlg aims are to:

- establish the nature and geometry of the different segments, their limits, and their connection;
- replace this new information in a geodynamic context in order to answer some unsolved questions about the western Mediterranean Sea;
- connect the different information and results thanks to the two long profiles crossing all existent wide-angle profiles in the area and see, if they exist, the different segments of the Algerian Basin;
- constrain the initial position of the Great and Small Kabylies,
- reconcile the different observations (onshore/offshore),
- set up the different geodynamical setting of each segment, and propose a new kinematic evolution of the area.

On both sides of the Strait of Sicily, itself of debatable crustal nature and evolution, in the old Eastern Mediterranean sea and in the younger Western Mediterranean sea, two basins present, despite their different ages and geological contexts, the same uncertainties about their crustal nature and geodynamic evolution, including the position and the movements of their continental micro-blocks. Without additional wide-angle seismic data, assumptions about the nature of the subsurface, especially in the so-called intermediate domain, and about geodynamic evolution are profoundly model-dependent; on the other hand, these new data must be located in precise paleo-geographic maps in order to test and validate the evolution models in a global framework, avoiding "conceptual kinematic models" on small sketches without correctly calculated Eulerian poles. These issues have a considerable impact on our understanding of the full evolution of the Mediterranean Sea. The resolution of these controversies, as was recently done in the Indian Ocean for the breakup of Gondwana (Thompson et al. 2019; Moulin et al. 2020) requires these integrated studies with new dense wide-angle seismic acquisitions.

References

Afilhado A, Moulin M, Aslanian D, Schnürle P, Klingelhoefer F, Nouzé H, Rabineau M, Leroux E, Beslier M-O (2015) Deep crustal structure across a young passive margin from wide-angle and reflection seismic data (The SARDINIA Experiment) – II. Sardinia's margin. Bull Soc Géol Fr 186:331–351

Aidi C, Klingelhoefer F, Yelles A, Beslier MO, Bracène R, Schnurle P, Djallit H, Galves A, Bounif A, Schenini L, Sage F, Charvis F (2013) Deep structures of the Algerian continental margin in the great kabylies- Insights from wide-angle seismic modeling:

AGU2013. T21A-2522. Continental Rifts and Rifted Margins, Posters. CONTROL ID: 1794050
Al-Balushi AN (2015) Tectono-Stratigraphic evolution of the eastern mediterranean and the impact of the Messinian salinity crisis on the petroleum systems. Ph.D. thesis. Imperial College, London, pp 273. http://hdl.handle.net/10044/1/44964
Al-Balushi AN, Neumaier M, Fraser AJ, Jackson CA-L (2016) The impact of the Messinian salinity crisis on the petroleum system of the Eastern Mediterranean: a critical assessment using 2D petroleum system modelling. Pet Geosci 22:357–379. https://doi.org/10.1144/petgeo2016-054
Arab M, Rabineau M, Bracene R, Déverchère J, Belhai D, Roure F, Marok A, Bouyahiaoui B, Granjeon D, Andriessen P, Sage F (2016a) Origin and Tectono-Sedimentary Evolution of the Eastern Algerian Basin (offshore) from Upper Oligocene to Present-Day. Marine Petrol Geol 77:1355–1375
Arab M, Belhai D, Granjeon D, Roure F, Arbeaumont A, Rabineau M, Bracene R, Belhai D, Lassal A, Sulzer C, Deverchère J (2016b). Coupling stratigraphic and petroleum system modelling tools in complex tectonic domains: Case study in the North Algerian Offshore. Arabian J Geosci https://doi.org/10.1007/s12517-015-2296-3
Aslanian D, Moulin M (2010) Comments on « A new scheme for the opening of the south Atlantic Ocean and the dissection of an Aptian salt Basin » from Torsvik et al, 2009. Geophys J Int. https://doi.org/10.1111/j.1365-246X.2010.04727.x
Aslanian D, Moulin M (2012) Paleogeographic consequences of conservational models in the South Atlantic Ocean. In: Mohriak WU, Danforth A, Post PJ, Brown DE, Tari GC, Nemcok M, Sinha ST (eds) Conjugate divergent margins. Geological Society, London, Special Publications, vol 369. https://doi.org/10.1144/SP369.5
Aslanian D, Moulin M (2021) South and equatorial atlantic margins, Chapter. In: Continental rifted margins book, Peron-Pinvidic editor, ISTE-Wiley. https://doi.org/10.51926/ISTE.5061.9
Aslanian D, Galais F, Afilhado A, Schnurle P, Moulin M, Evain M, Dias N, Soares J, Fuck R, Pessoa Neto O, Viana A (2021a) The MAGIC team. Deep structure of the Pará-Maranhão/Barreirinhas passive margin in the Equatorial Atlantic (NE Brazil). J South Am Earth Sci 110
Aslanian D, Moulin M, Evain M, Schnurle P (2021b) Comment on 'The challenge in restoring magma-rich rifted margins: the example of the Mozambique-Antarctica conjugate margins' by Tomasi S, Kusznir N, Manatschal G, Despinois F, Gondwana Research
Aslanian D, Moulin M, Olivet J-L, Unternehr P, Matias L, Bache F, Rabineau M, Nouzé H, Klingelhoefer F, Contrucci I, Labails C (2009) Brasilian and Angolan passive margins: the kinematic constraints. Tectonophysics, Special Issue: Role of the Magmatism 468:98–112
Badji R (2014) Structure profonde de la croûte et potentiel pétrolier des bassins sédimentaires à l'ouest de l'Algérie. Phd Thesis, pp 1–221
Barrier E, Vrielynck B (2008) Paleotectonic maps of the middle east. Comm de la Carte Geol. du Monde, Paris
Bellucci M, Aslanian D, Moulin M, Rabineau M, Leroux E, Pellen R, Poort J, Del Ben A, Gorini C, Camerlenghi A (2021) Salt morphologies and crustal segmentation relationship: new insights from well-known salt-bearing margins. Earth Sci Rev 222. https://doi.org/10.1016/j.earscirev.2021.103818
Ben-Avraham Z, Ginzburg A, Makris J, Eppelbaum L (2002) Crustal structure of the Levant Basin, eastern Mediterranean. Tectonophysics 346(1–2):23–43. https://doi.org/10.1016/S0040-1951(01)00226-8
Bertoni C, Kirkham C, Cartwright J, Hodgson N, Rodriguez K (2017) Seismic indicators of focused fluid flow and cross-evaporitic seepage in the Eastern Mediterranean. Mar Pet Geol 88:472–488. https://doi.org/10.1016/j.marpetgeo.2017.08.022
Beydoun ZR (1999) Evolution and development of the Levant dead sea Rift transform system: a historical-chronological review of a structural controversy. Geol Soc London Spec Publ 164:239–255
Bezada MJ, Humphreys ED, Toomey DR, Harnafi M, Davila JM, Gallart J (2013) Evidence for slab rollback in the westernmost Mediterranean from improved upper mantle imaging. Earth Planeary Sci Lett 368:51–60
Bosworth W, El-Hawat H, Helgeso DE, Burke K (2008) Cyrenaican "shock absorber" and associated inversion strain shadow in the collision zone of northeast Africa. Geology 36(9):695–698. https://doi.org/10.1130/G24909A.1
Bouillin JP (1986) Le bassin maghrébin: une ancienne limite entre l'Europe et l'Afrique à l'ouest des Alpes
Bouyahiaoui B, Sage F, Abtout A, Klingelhoefer F, Yelles-Chaouche K, Schnürle P, Marok A, Déverchère J, Arab M, Galve A, Collot JY (2015) Crustal structure of the eastern Algerian continental margin and adjacent deep basin: implications for late Cenozoic geodynamic evolution of the western Mediterranean: Geophys J Int 201:1912–1938
Brew G, Barazangi M, Al-Maleh K, Sawaf T (2001) Tectonic and geologic evolution of Syria. GeoA rabia 6(4)
Brocher T (2005) Empirical relations between elastic wavespeeds and density in the earth's crust. B Seismol Soc Am 95(6):2081–2092. https://doi.org/10.1785/0120050077
Camerlenghi A, Accetella D, Costa S, Lastras G, Acosta J, Canals M, Wardell N (2009) Morphogenesis of the SW Balearic continental slope and adjacent abyssal plain, Western Mediterranean sea. Int J Earth Sci 735–750
Carminati E, Lustrino M, Doglioni C (2012) Geodynamic evolution of the central and western Mediterranean: tectonics versus igneous petrology constraints. Tectonophysics 579:173–192
Carminati E, Wortel MJR, Meijer PT, Sabadini R (1998b) The two-stage opening of the western-central Mediterranean basins: a forward modeling test to a new evolutionary model. Earth Planet Sci Lett 160:667–679
Carminati E, Wortel, MJR, Spakman W, Sabadini R (1998a) The role of slab detachment processes in the opening of the western-central Mediterranean basins: some geological and geophysical evidence. Earth Planet Sci Lett 160:651–665
Carton H, Singh SC, Tapponnier P, Elias A, Briais A, Sursock, Jomaa R, King GCP, Daëron M, Jacques E, Barrier L (2009) Seismic evidence for Neogene and active shortening offshore of Lebanon Shalimar cruise. J Geophys Res Solid Earth 114: B07407https://doi.org/10.1029/2007JB005391
Cavazza W, Roure F, Spakman W, Stampfli GM, Ziegler PAE (2004) The TRANSMED atlas—The mediterranean region from crust to mantle. Springer, Berlin Heidelberg
Chauvet F, Dumont T, Basile C (2009) Structures and timing of Permian rifting in the central Oman mountains Saih Hatat. Tectonophysics 475:563–574. https://doi.org/10.1016/j.tecto.2009.07.008
Contrucci I, Matias L, Moulin M, Géli L, Klingelhoefer F, Nouzé H, Aslanian D, Olivet J-L, Réhault J-P, Sibuet J-C (2004) Deep structure of the West African continetal margin (Congo, Zaïre, Angola), between 5°S and 8°S, from reflexion/refraction seismic and gravity data. Geophys J Int
Crespo-Blanc A, Comas M, Balanyà JC (2016) Clues for a Tortonian reconstruction of the Gibraltar Arc: Structural pattern, deformation diachronism and block rotations. Tectonophysics 683
Dercourt J, Ricou LE, Vrielynck B (1993) Atlas Tethys of Paleoenvironmental Maps, Gauthier-Villars, Paris Gardosh, M. A., Z. Garfunkel, Y. Druckman, and B. Buchbinder 2010: Tethyan rifting in the Levant region and its role in Early Mesozoic crustal

evolution. In: Homberg C, Bachmann M (eds) Evolution of the levant margin and western Arabia platform since the Mesozoic. Geological Society of Special Publications, vol 341, pp 9–36

De Voogd B, Nicolich R, Olivet JL, Fanucci F, Burrus J, Mauffret A, Pascal G, Argnani A, Auzende JM, Bernabini M, Bois C, Carmingani L, Fabri A, Fi-netti I, Galdeano A, Gorini CY, Labaume P, Lajat D, Patriat P, Pinet B, Ravat J, Ricci Lucchi F, Vernassa S (1991) First deep seismic reflection transect from the gulf of Lions to Sardinia (ECORS-CROP profiles in western Mediterranean). In: American Geophysical Union (Ed), Continental lithosphere: deep seismic reflec-tions. Geodynamics 22:265–274

Dewey J, Helman M, Turco E, Hutton D, Knott S (1989) Kinematics of the western mediterranean: In: Coward MP, Dietrich D (eds) Alpine tectonics. Geological Society of London Special Publication, vol 45, pp 265–283

Dias NA, Afilhado A, Schnürle P, Gallais F, Moulin M, Aslanian D, Soares J, Fuck R, Cupertino JA, Viana A, MAGIC team (2016) Deep segmentation from 2D forward modeling and 3D tomography of the Maranhão-Barreirinhas-Ceará margin, NW Brazil. SBGF, Porto Alegre

Doglioni C, Gueguen E, Sàbat F, Fernandez M (1997) The Western Mediterranean extensional basins and the Alpine orogeny. Terra Nova 9:109–112

Domzig A, Le Roy C, Yelles K, Deverchère J, Bouillin J-P, Bracene R, Mercier de Lepinay B, Le Roy P, Calais E, Kherroubi A, Gaullier V, SAvoye B, Pauc H (2006) Searching for the Africa-EUrasia Miocene boundary offshore Western Algeria (MARADJA'03 cruise). Compte Rendu de Géosci 338:80–91

Driussi O, Briais A, Maillard A (2015) Evidence for transform motion along the south Balearic margin and Implications for the kinematics of opening of the Algerian basin. Bull Soc Géol Fr 186:49–51

Duggen S, Hoernle K, Bogaard PVD, Rüpke I, Morgan J (2003) Deep roots of the messinian salinity crisis. Nature 422:602–606

Duggen S, Hoernle K, Harris C, Van Den Bogaard P (2004) Magmatic evolution of the Alboran region: the role of subduction in forming the western Mediterranean and causing the Messinian salinity crisis. Earth Planet Sci Lett 218:91–108

Duggen S, Hoernle K, Van der Boogaard P, Garbe-Schonberg C-D (2005) Post-collisional transition from subduction to intraplate type magmatism in the westernmost Mediterranean: Evidence for continental-edge delamination of subcontinental lithosphere. J Petrol Geol 46:1155–1201

Etheve N, Frizon De Lamotte D, Mohn G, Martos R, Roca E, Blanpied C (2016) Extensional versus contractional Cenozoic deformation in Ibiza (Balearic Promontory, Spain): integration in the west Mediterranean back-arc setting. Tectonophysics 682:35–55

Evain M, Afilhado A, Rigotti C, Loureiro A, Alves D, Klingelhoefer F, Schnürle P, Feld A, Fuck R, Soares J, Vinicius de Lima M, Corela C, Matias L, Benabdellouahed M, Baltzer A, Rabineau M, Viana A, Moulin M, Aslanian D (2015) Deep structure of the Santos Basin-São Paulo Plateau System, SE Brazil. J Geophy Res - Solid Earth. https://doi.org/10.1002/2014JB011561

Faccenna C, Becker TW, Auer L, Billi A, Boschi L, Brun JP, Capitanio FA, Funiciello F, Horvath F, Jolivet L, Piromallo C, Royden L, Rossetti F, Serpelloni E (2014) Mantle dynamics in the Mediterranean. Rev Geophys 52:283–332

Faccenna C, Bellier O, Martinod J, Piromallo C, Regard V (2006) Slab detachment beneath eastern Anatolia: a possible cause for the formation of the North Anatolian fault. Earth Planet Sci Lett 242(1–2):85–97. https://doi.org/10.1016/j.epsl.2005.11.046

Facenna C, Piromallo C, Crespo-Blanc A, Jolivet J, Rossetti F (2004) Lateral slab deformation and the origin of the western Mediterranean arcs. Tectonics 23:TC1012

Feld C, Mechie J, Hübscher C, Hall J, Nicolaides S (2017) Crustal structure of the Eratosthenes Seamount, Cyprus and S. Turkey from an amphibian wide-angle seismic profile. Tectonophysics 700–701:32–59.https://doi.org/10.1016/j.tecto.2017.02.003

Frizon de Lamotte D, Raulin C, Mouchot N, Wrobel-Daveau J-C, Blanpied C, Ringenbach J-C (2011) The southernmost margin of the Tethys realm during the Mesozoic and Cenozoic: Initial geometry and timing of the inversion processes. Tectonics 30

Frizon de Lamotte D, Saint Bezar B, Bracène R, Mercier E (2000) The two main steps of the Atlas building and geodynamics of the Western Mediterranean. Tectonics 19:740–761

Frizon de Lamotte D, Leturmy P, Missenard Y, Khomsi S, Ruiz G, Saddiqi O, Guillocheau F, Michard A (2009) Mesozoic and Cenozoic vertical movements in the Atlas system (Algeria, Morocco, Tunisia): An overview. Tectonophysics 475:9–28. https://doi.org/10.1016/j.tecto.2008.10.024

Gaina C, Gernigon L, Ball P (2009) Palaeocene– Recent plate boundaries in the NE Atlantic and the for- mation of the Jan Mayen microcontinent. J Geol Soc London 166:601–616. https://doi.org/10.1144/0016-76492008-112

Gao H, Tonga X, Wen Z, Wang Z (2019) The tectonic evolution of the eastern Mediterranean basin and its control on hydrocarbon distribution. J Pet Sci Eng 178:389–407. https://doi.org/10.1016/j.petrol.2019.03.029

Gardosh MA, Garfunkel Z, Druckman Y, Buchbinder B (2010) Tethyan rifting in the Levant region and its role in Early Mesozoic crustal evolution, in evolution of the levant margin and western Arabia Plat- form Since the Mesozoic, edited by C. Homberg and M. Bachmann. Geol Soc Spec Publ 341:9–36

Gardosh M, Druckman Y, Buchbinder B, Calvo R (2008) The Oligo-Miocene deepwater system of the Levant basin. Geol Sur Israel 33:1–73

Garfunkel Z (1981) Internal structure of the Dead Sea leaky transform (rift) in relation to plate kinematics,Tectonophysics 80(1–4):81–108. https://doi.org/10.1016/0040-1951(81)90143-8

Garfunkel Z (1998) Constrains on the origin and history of the Eastern Mediterranean basin. Tectonophysics

Gelabert B, Sabat F, Rodriguez-Perea A (2002) A new proposal for the late Cenozoic geodynamic evolution of the Western Mediterranean. Terra Nova 14:93–100

Ghalayini R, Daniel J, Homberg C, Nader F, Comstock J (2014) Impact of Cenozoic strike-slip tectonics on the evolution of the northern Levant Basin offshore Lebanon. Tectonics 33(11):2121–2142. https://doi.org/10.1002/2014TC003574

Ginzburg A, Ben-Avraham Z, Makris J, Hubral P, Rotstein Y (1994) Crustal structure of northern Israel. Mar Pet Geol 11(4)

Govers R, Wortel MJR (2005) Lithosphere tearing at step faults: response to edges of subduction zones. Earth Planeary Sci Lett 236:505–523

Granot R (2016) Palaeozoic oceanic crust preserved beneath the eastern Mediterranean. Nat Geosci 9(9):701–705. https://doi.org/10.1038/ngeo2784

Gueguen E, Doglioni C, Fernandez M (1998) On the post-25 Ma geodynamic evolution of the western Mediterranean. Tectonophysics 298:259–269

Gutscher M-A, Malod JA, Rehault J-P, Contrucci I, Klingelhoefer F, Men-des-Victor L, Spakman W (2002) Evidence for active subduction beneath Gibraltar. Geology 30:1071–1074

Hamai L, Petit C, Abtout A, Yelles-Chaouche A, Déverchère J (2015) Flexural behaviour of the north Algerian margin and tectonic implications. Geophys J Int 201(3):1426–1436

Handy MR, Schmid SM, Bousquet R, et al (2010) Reconciling plate-tectonic reconstructions of Alpine Tethys with the geological-geophysical record of spreading and subduction in the Alps. Earth-Sci Rev 102(3–4):121–158. Published: OCT

He L, Moulin M, Aslanian D, Evain M, Schnurle P, Leprêtre A, Yong T, Jiabiao L (2021) Seismic evidence for crustal architecture and stratigraphy of the Limpopo Corridor: new insights into the evolution of the sheared margin offshore southern Mozambique. Mar Geol

Hardy C, Homberg C, Eyal Y, Barrier É, Müller C (2010) Tectonic evolution of the southern Levant margin since Mesozoic. Tectonophysics 494(3–4):211–225. https://doi.org/10.1016/j.tecto.2010.09.007

Hawie N (2014) Architecture, geodynamic evolution and sedimentary filling of the levant basin: a 3D quantitative approach based on seismic data. Earth Sciences. Université Pierre et Marie Curie - Paris VI

Hawie N, Deschamps R, Nader FH, Gorini C, Müller C, Desmares D, Hoteit A, Granjeon D, Montadert L, Baudin F (2013) Sedimentological and stratigraphic evolution of northern Lebanon since the Late Cretaceous: implications for the Levant margin and basin. Arab J Geosci. https://doi.org/10.1007/s12517-013-0914-5

Inati L, Lecomte J-C, Zeyen H, Nader FH, Adelinet M, Rahhal ME, Sursock A (2018) Crustal configuration in the northern Levant basin based on seismic interpretation and numerical modeling. Mar Pet Geol 93:182–204.https://doi.org/10.1016/j.marpetgeo.2018.03.011

Inati L, Zeyen H, Nader FH, Adelinet M, Sursock A, Rahhal ME, Roure R (2016) Lithospheric architecture of the Levant basin eastern Mediterranean region): a 2D modeling approach. Tectonophysics 693:143–156. https://doi.org/10.1016/j.tecto.2016.10.030

Jolivet L, Brun JP (2010) Cenozoic geodynamic evolution of the Aegean. Int J Earth Sci 99:109–138. https://doi.org/10.1007/s00531-008-0366-4

Jolivet L, Faccenna C (2000) Mediterranean extension and the Africa-Eurasia collision. Tectonics 19:1095–1106

Jolivet L, Auger R, Robin C, Suc JP, Rouchy JM (2006) Lithospheric-scale geodynamic context of the Messinian salinity crisis. Sediment Geol 188–189:9–33

Jolivet L, Faccenna C, Piromallo C (2009) From mantle to crust: stretching the Mediterranean: Earth Planeary Sci Lett 285:189–209

Keeley ML (1994) Phanerozoic evolution of the basins of northern Egypt and adjacent areas. Geol Rundsch 83(4):728–742. https://doi.org/10.1007/BF00251071

Klingelhoefer F, Evain M, Afilhado A, Rigoti C, Loureiro A, Alves D, Lepretre A, Moulin M, Schnurle P, Benabdellouahed M, Baltzer A, Rabineau M, Feld A, Viana A, Aslanian D (2015) Imaging proto-oceanic crust off the Brazilian continental margin. Geophys J Int 200(1):471–488. https://doi.org/10.1093/gji/ggu387

Leprêtre R, Frizon de Lamotte D, Combier V, Gimeno-Vives O, Mohn G, Eschard R (2018) The Tell-Rif orogenic system(Morocco, Algeria, Tunisia) and the structural heritage of the southern Tethys margin. BSGF - Earth Sci Bull 189:10

Leprêtre A, Klingelhoefer F, Graindorge D, Schnurle P, Beslier MO, Yelles K, Déverchère J, Bracene R (2013) Multiphased tectonic evolution of the central Algerian margin from combined wide-angle and reflection seismic data off Tipaza. Algeria: J Geophys Res: Solid Earth 118:3899–3916

Leprêtre A, Schnurle P, Evain M, Verrier F, Moorcroft D, de Clarens P, Corela C, Afilhado A, Loureiro A, Leroy S, d'Acremont E, Thompson J, Aslanian D, Moulin M, MOZ3/5 Team (2021) Deep structure of the North Natal Valley (Mozambique) using combined wide-angle and reflection seismic data. J Geophys Res: Solid Earth 126. https://doi.org/10.1029/2020JB02117Z

Leroux E, Aslanian D, Rabineau M, Moulin M, Granjeon D, Gorini C, Droz L (2015a) Sedimentary markers in the Provencal Basin (western Mediterranean): a window into deep geodynamic processes. Terra Nova 27(2):122–129. Publisher's official version. https://doi.org/10.1111/ter.12139. Open Access version. https://archimer.ifremer.fr/doc/00253/36466/

Leroux E, Aslanian D, Rabineau M, Moulin M, Granjeon D, Gorini C, Droz L (2015b) Sedimentary markers in the Provencal Basin (western Mediterranean): a window into deep geodynamic processes. Terra Nova 27:122–129

Leroux E, Aslanian D, Rabineau M, Gorini C, Rubino J-L, Blanpied C, Poort J (2018) Atlas of the Stratigraphic markers in the western Mediterranean sea with focus on the Gulf of Lion. Commission de la Carte Géologique du Monde

Lonergan L, White N (1997) Origin of the betic-Rif moutain belt. Tectonics 16:504–522

Loureiro A, Schnurle P, Klingelhofer F, Afilhado A, Pinheiro JM, Evain M, Gallais F, Dias NA, Rabineau M, Baltzer A, Benabdellouahed M, Soares J, Fuck R, Cupertino JA, Viana A, Matias L, Moulin M, Aslanian D (2018) Imaging exhumed lower continental crust in the distal Jequitinhonha basin, Brazil. J South Am Earth Sci 84:351–372. Publisher's official version. https://doi.org/10.1016/j.jsames.2018.01.009. Open Access version. https://archimer.ifremer.fr/doc/00428/53928/

Lustrino M, Morra V, Fedele L., Franciosi, L., 2009. Beninning of the Apennine subduction system in the central western Mediterranean: constraints from Cenozoic "orogenic" magmatic activity of Sardinia, Italy. Tectonics 28

Mackenzie G, Maguire P, Coogan L, Khan M, Eaton M, Petrides G (2006) Geophysical constraints on the crustal architecture of the Troodos ophiolite: results from the IANGASS project. Geophys J Int 167(3):1385–1401. https://doi.org/10.1111/j.1365-246X.2006.03144.x

Makris J, Abraham ZB, Behle A (1983) Seismic refraction profiles between Cyprus and Israel and their interpretation. Geophys J Res Astr Soc. https://doi.org/10.1111/j.1365-246X.1983.tb05000.x

Malinverno A, Ryan WBF (1986) Extension in the Tyrrhenian Sea and shortening in the Apennines as a result of arc migration driven by the sinking of the lithosphere. Tectonics 5:2227–2245

Mascle J, Benkhelil J, Bellaiche G, Zitter T (2000) Marine geologic evidence for a Levantine-Sinai plate, a new piece of the Mediterranean puzzle. Geology 28(9):779–782. https://doi.org/10.1130/0091-7613

Mauffret A, Frizon de Lamotte D, Lallemant S, Gorini C, Maillard A (2004) E–W opening of the Algerian Basin (Western Mediterranean). Terra Nova 16:257–264

McClusky S, Reilinger R et al. (2010) Kinematics of the southern Red Sea–Afar Triple Junction and impli- cations for plate dynamics. Geophys Res Letter 37:L05301.https://doi.org/10.1029/2009GL041127

Medaouri M, Deverchère J, Graindorge D, Bracene R, Badji R, Ouabadi A, Yelles-Chaouche K, Bendiab F (2014) The transition from Alboran to Algerian basins (Western Mediterranean Sea): Chronostratigraphy, deep crustal structure and tectonic evolution at the rear of a narrow slab rollback system. J Geodyn 77:118–205

Mihoubi A, Schnurle P, Benaissa Z, Badsi M, Bracene R, Djelit H, Geli L, Sage F, Agoudjil A, Klingelhoefer F (2014) Seismic imaging of the eastern Algerian margin off Jijel: integrating wide-angle seismic modelling and multichannel seismic pre-stack depth migration. Geophys J Int 198:1486–1503

Montadert L, Nicolaides S, Semb PH, Lie Ø (2014) Petroleum systems offshore Cyprus. In: Marlow L, Kendall C, Yose L (eds) Petroleum systems of the Tethyan region: AAPG Memoir 106, pp 301–334

Moulin M, Aslanian D, Olivet J-L, Contrucci I, Matias L, Géli L, Klingelhoefer F, Nouzé H, Réhault J-P, Unternehr P (2005) Geological constraints on the evolution of the Angolan margin based on reflection and refraction seismic data (ZaïAngo project), Geophys J Int 162:793–810. https://doi.org/10.1111/j.1365-246X.2005.02668.x

Moulin M, Aslanian D, Unternher P (2010) A new starting point for the South and Equatorial Atlantic Ocean. Earth Sci Rev 98(1):1–37. https://doi.org/10.1016/j.earscirev.2009.08.001

Moulin M, Aslanian D, Evain M, Leprêtre A, Schnurle P, Verrier F, Thompson J, De Clarens P (2020) Dias, and the PAMELA-MOZ35 team, Gondwana breakup and passive margin genesis: messages from the natal valley. Terra Nova. https://doi.org/10.1111/ter.12448

Moulin M, Aslanian D, Rabineau M, Patriat M, Matias L (2012) Kinematic keys of the santos - namibe basins. In: Mohriak WU, Danforth A, Post PJ, Brown DE, Tari GC, Nemcok M, Sinha ST (eds) Conjugate divergent margins. Geological Society, London, Special Publications, vol 369. https://doi.org/10.1144/SP369.3

Moulin M, Gallais F, Afilhado A, Schnurle P, Dias N, Soares J, Fuck R, Viana A, Aslanian D, MAGIC Team (2021) Image of a first oceanic crust in the Equatorial Atlantic Ocean: insights from the MAGIC wide-angle experiment. J South Am Earth Sci 111. https://doi.org/10.1016/j.jsames.2021.103493

Moulin M, Klingelhoefer F, Afilhado A, Aslanian D, Schnürle P, Nouzé H, Rabineau M, Beslier MO, Feld A (2015) Deep crustal structure across a young 1 passive margin from wide-angle and reflection seismic data The SARDINIA Experiment)—I. Gulf of Lion's margin. Bull Soc Geol Fr 186:4–5. https://doi.org/10.2113/gssgfbull.186.4-5.309

Mueller CO, Jokat W, Schreckenberger B (2016) The crustal struc- ture of Beira High, central Mozambique—Combined investigation of wide-angle seismic and potential field data. Tectonophysics 683:233–254. https://doi.org/10.1016/j.tecto.2016.06.028

Nader FH (2011) The petroleum prospectivity of Lebanon: an overview. J Pet Geol. https://doi.org/10.1111/j.1747-5457.2011.00498.x

Nader FH, Browning-Stamp P (2016) Geological interpretation of 2d seismic reflection profils onshore lebanon: implications for petroleum exploration. J Petrol Geol 39(4):333–356. https://doi.org/10.1111/jpg.12656

Netzeband GL, Gohl K, Hübscher CP, Ben-Avraham Z, Dehghani GA, Gajewski D, Liersch P (2006) The Levantine Basin—crustal structure and origin. Tectonophysics 418(3–4):167–188. https://doi.org/10.1016/j.tecto.2006.01.001

Olivet JL (1996) La cinématique de la plaque Ibérique. Bulletin des Centres de Recherches Exploration-Production Elf-Aquitaine 20:131–195

Olivet JL, Bonnin J, Beucart P, Auzende JM (1982) Plate kinematics and palaeogeography—a review. Bull Soc Géol Fr 24:875–892

Papadimitriou N, Gorini C, Nader FH, Deschamps R, Symeou V, Lecomte JC (2018) Tectono-stratigraphic evolution of the western margin of the Levant Basin offshore Cyprus. Mar Pet Geol 91:683–705. https://doi.org/10.1016/j.marpetgeo.2018.02.006

Pellen R, Aslanian D, Rabineau M, Leroux E, Gorini C, Silenziario C, Blanpied C, Rubino J-L (2016) The Minorca Basin: a buffer zone between the Valencia and Liguro-Provencal Basins (NW Mediterranean Sea). Terra Nova 28: 245–256

Pellen R, Aslanian D, Rabineau M, Suc JP, Gorini C, Leroux E, Blanpied C, Silenziario C, Popescu SM, Rubino JL (2019) The Messinian Ebro river incision. Global Planet Change 181:102988 (15p.). https://doi.org/10.1016/j.gloplacha.2019.102988

Pillevuit A, Marcoux J, Stampfli G, Baud A (1997) The Oman exotics: a key to the understanding of the Neotethyan geodynamic evolution. Geodin Acta 10:209–238

Pinheiro JM, Schnurle P, Evain M, Afilhado A, Gallais F, Klingelhoefer F, Loureiro A et al (2018) Lithospheric structuration onshore-offshore of the Sergipe-Alagoas passive margin, NE Brazil, based on wide-angle seismic data. J South Am Earth Sci 88:649–672. Publisher's official version. https://doi.org/10.1016/j.jsames.2018.09.015. Open Access version. https://archimer.ifremer.fr/doc/00461/57263/

Piromallo C, Morelli A (2003) P wave tomography of the mantle under the Alpine-Mediterranean area. J Geophys Res 108:2065

Platt JP, Vissers RLM (1989) Extensional collapse of thickened continental lithosphere: a working hypothesis for the Alboran Sea and Gibraltar Arc. Geology 17: 540–543

Ricou LE (1971) Le croissant ophiolitique péri-arabe: Une ceinture de nappes mises en place au Crétacé supérieur. Rev Geogr Phys Geol Dyn 13:327–350

Ricou LE (1994) Tethys reconstructed: plates, continental fragments and their boundaries since 260 Ma from Central America to South-eastern Asia. Geodinamica Acta 7:169–218

Robertson A (1998) Mesozoic-Tertiary tectonic evolution of the easternmost Mediterranean area: integration of marine and land evidence. In: Proceedings of the ocean drilling program vol 160

Rosenbaum G Lister GS (2004) Neogene and quaternary rollback evolution of the Tyrrhenian Sea, the Apennines and the Sicilian Maghrebides. Tectonics 23

Rosenbaum G, Lister GS, Duboz C (2002a) Reconstruction of the tectonic evolution of the western Mediterranean since the Oligocene. J Virtual Explorer 8:107–130

Rosenbaum G, Lister GS, Duboz C (2002b) Relative motions of Africa, Iberia and Europe during Alpine orogeny. Tectonophysics 359:117–129

Rossetti F, Dini A, Lucci F, Bouybaouenne M, Faccenna C (2013) Early Miocene strike-slip tectonics and granite emplacement in the Alboran Domain (Rif Chain, Morocco) and its impact on the geodynamic evolution of Western Mediterranean. Tectonophysics 608: 774–791

Roure F, Casero P, Addoum B (2012) Alpine inversion of the North African margin and delamination of its continental lithosphere. Tectonics 31:TC3006

Royden LH (1993) Evolution of retreating subduction boundaries formed during continental collision. Tectonics 12:629–638

Sahabi M, Aslanian D, Olivet J-L (2004) Un nouveau point de départ pour l'histoire de l'Atlantique central: Comptes Rendus. Geosciences 336:1041–1052

Schnürle P, Leprêtre A, Evain M, Verrier F, de Clarens P, Thompson J, Dias N, Afilhado A, Loureiro A, Leroy S, d'Acremont E, Aslanian D, Moulin M (In press) Crustal structure of the South Mozambique Margin to South Mozambique Ridge from combined wide-angle and reflection seismic data. Earth Space Sci

Schattner U, Ben-Avraham Z (2012) Tectonic development of the Levant continental margin, eastern Mediterranean, Researchgate. https://www.researchgate.net/publication/281763542.

Schattner U, Ben-Avraham Z (2007) Transform margin of the northern Levant, eastern Mediterranean: from formation to reactivation. Tectonics 26:5. https://doi.org/10.1029/2007TC002112

Schettino A, Turco E (2006) Plate kinematics of the Western Mediterranean region during the Oligocene and early Miocene. Geophys J Int 166:1398–1423

Schettino A, Turco E (2011) Tectonic history of the western Tethys since the Late Triasic. Geol Soc Am 123:89–105

Segev A, Sass E, Schattner U (2018) Age and structure of the Levant basin, Eastern Mediterranean. Earth Sci Rev 182:233–250. https://doi.org/10.1016/j.earscirev.2018.05.011

Segev A, Rybakov M (2010) Effects of Cretaceous plume and convergence, and early tertiary tectonomagmatic quiescence on the central and southern Levant continental margin. J Geol Soc London 167(4):731–749. https://doi.org/10.1144/0016-76492009-118

Seton M, Müller RD, Zahirovic S, Gaina C, Torsvik T, Shephard G, Talsma A, Gurnis M, Turner M, Maus S, Chandler M (2012) Global

continental and ocean basin reconstructions since 200 Ma. Earth Sci Rev 113: 212–270
Sichler B (1980) La biellette Danakil, un modèle pour l'eévolution ge ′odynamique des Afars. Bulletin de la Société Géologique de France 22(6):925–932
Spakman W (1991) Delay-time tomography of the upper mantle below Europe, The Mediterranean and Asia Minor. Geophys J Int 107:309–332
Spakman W, Wortel MJR (2004) Tomographic view on western Mediterranean geodybamics, in the TRANSMED Atlas, The Mediterranean Region from Crust to Mantle, edited by W. Cavazza et al.: Springer-Verlag, Berlin-Heidelberg, Germany, pp 31–52
Spakman W, Van der Lee S, Van der Hilst R (1993) Travel-time tomography of the European-Mediterranean mantle down to 1400 km. Phys Earth Planet Int 79:3–74
Speranza F, Villa IM, Sagnotti L, Florindo F, Cosentino D, Cipollari P, Matei M (2002) Age of the Corsica-Sardinia rotation and Liguro-Provencal Basin spreading: new paleomagnetic and Ar/Ar evidence. Tectonophysics 347: 231–251
Stampfli GM, Borel GD (2004) The TRANSMED transects in space and time: constraints on the paleotectonic evolution of the Mediterranean domain. In: Cavazza W et al (eds) The TRANSMED Atlas: the mediterranean region from crust to mantle, vol 1. Springer, Berlin, pp 53–80
Stampfli G, Marcoux J, Baud A (1991) Tethyan margins in space and time. Palaeogeogr Palaeoclimatol Palaeoecol 87:373–409. https://doi.org/10.1016/0031-0182(91:90142-E
Steinberg J, Roberts AM, Kusznir NJ, Schafer K, Karcz Z (2018) Crustal structure and post-rift evolution of the Levant Basin. Marin Petrol Geol. https://doi.org/10.1016/j.marpetgeo.2018.05.006
Talwani M, Eldholm O (1977) Evolution of the Nor-wegian – Greenland Sea. Geol Soc America Bull 88:969–999. https://doi.org/10.1130/0016-7606
Thompson J, Moulin M, Aslanian D, Guillocheau F, de Clarens P (2019) New starting point for the Indian Ocean: Second phase of breakup for the Gondwana. Earth Sci Rev 191:26–56. https://doi.org/10.1016/j.earscirev.2019.01.018
Torsvik TH, Van der Voo R, Preeden U, Mac Niocaill C, Steinberger B, Doubrovine PV, van Hinsbergen DJJ, Domeier M, Gaina C, Tohver E, Meert JG, McCausland PJA, Cocks LRM (2012) Phanerozoic polar wander, palaeogeography and dynamics. Earth Sci Rev 114:325–368
Van Hinsbergen DJJ, Vissers RLM, Spakman W (2014) Origin and consequences of western Mediterranean subduction, rollback, and slab segmentation. Tectonics 33:393–419
Viñas Gaza M (2016) Tectonic structure and formation kinematics of the estern Mediterranean basins: Phd Thesis, pp 1–226
Vissers RLM, Meijer PT (2012) Iberia plate kinematics and Alpine collision in the Pyrennes.Earth Sci Rev 114:61–83
Vezzani L, Festa A, Ghisetti FC (2010) Geology and tectonic evolution of the Central-Southern Apennines, Italy. Geol Soc America Special Papers 469:1–58
Watremez L, Leroy S, d'Acremont E, Roche V, Evain M, Leprêtre A, Verrier F, Aslanian D, Dias N, Afilhado A, Schnurle P, Castilla R, Despinois F, Moulin M (2021) The Limpopo Magma-Rich transform margin, south Mozambique: 1. Insights from deep-structure seismic imaging. Tectonics 40(12). https://doi.org/10.1029/2021TC006915
Welford J, Hall J, Hubscher C, Reiche S, Louden K (2015a) Crustal seismic velocity structure from Eratosthenes Seamount to Hecataeus Rise across the Cyprus Arc, eastern Mediterranean. Geophys J Int 200(2):933–951. https://doi.org/10.1093/gji/ggu447
Welford J, Hall J, Rahimi A, Reiche S, Hübscher C, Louden K (2015b) Crustal structure from the Hecataeus Rise to the Levantine Basin, eastern Mediterranean, from seismic refraction and gravity modelling. Geophys J Int 203(3):2055–2069. https://doi.org/10.1093/gji/ggv422
Wortel MJR, Spakman W (2000) Subduction and Slab detachment in the Mediterranean-Carpathian region. Science 290:1910–1917
Wortel MJR, Govers R, Spakman W (2009) Continental collision and the STEP-wise evolution of convergent plate boundaries: From structure to dynamics. In: Subduction Zone Geodynamics, (ed) Lallemand S, Funiciello F (pp 47–59), Springer-Verlag, Berlin Heidelberg
Zverev S (2010) Blocks and faults of earth crust on the levant basin: DSS results of cruise of RAS r/v 'Akademik Boris Petrov' and 'Akademik Nikolai Strakhov'. Institute of Physics of Earth RAS.-M: Svetoch Plus
Zverev SM, Ilinsky DA (2005) The deep structure of Eratosthenes Seamount from seismic refraction data. In: Hall JK, Krasheninikov VA, Hirsch F, Benjamini C, Flexer A (eds) Geological framework of the levant, vol 2. Historical Publications-Hall. The Levantine Basin and Israel, Jerusalem, pp 73–112

2 The Westernmost Tethyan Margins in the Rif Belt (Morocco), A Review

André Michard, Ahmed Chalouan, Aboubaker Farah, and Omar Saddiqi

Abstract

The Rif belt is the westernmost segment of the Maghrebides and the southern branch of the Gibraltar Arc connecting North Africa to Iberia. The Rif belt formed coevally with the Betic Cordilleras (northern branch of the Arc) during the Cenozoic due to the Africa-Eurasia convergence associated with the subduction of the westernmost Tethys lithosphere of the Ligurian-Maghrebian basin. In this work, we describe the remnants of the margins of the latter basin as exposed in the Rif belt. The External Zones of the belt expose remnants of the Jurassic southern Ocean-Continent Transition (OCT) of the Maghrebian Tethys and a Triassic volcanic-rich segment of the NW African passive margin. These consist, respectively, of serpentinites and gabbros slivers included in the accretionary prism derived from the inversion of the African passive margin. The northern margin of the Maghrebian Ocean is classically represented by the Dorsale Calcaire and Predorsalian Triassic-Paleogene units at the external border of the Internal Zones (Alboran Domain). The latter mainly consists of two complexes of basement nappes, from top to bottom, the Ghomarides (Malaguides in Spain) and the Sebtides (Alpujarrides in Spain). The Dorsale sedimentary units are transitional between the Ghomarides-Malaguides coeval sequences and the Maghrebian Flyschs deposits. They likely detached from the Sebtides-Alpujarrides thinned crust domain. Marbles of probable Triassic age overlie the granulites (kinzigites) envelope of the Beni Bousera peridotites included in the Lower Sebtides units. Thus, the mantle of the Sebtides-Alpujarrides domain would have been exhumed close to the surface as early as the Triassic during the incipient formation of a Jurassic magma-poor margin bordering the Maghrebian Tethys to the north.

Keywords

Tethys • North African transform • Magma-poor passive • Peridotites • Oceanic core complexes • Exhumation • Orogenic wedge

C'est surtout l'étude des Montagnes qui peut accélérer les progrès de la Théorie de la Terre.
Horace-Bénédict de Saussure (1779), p. II.

Les montagnes maghrébines détiennent la clef de l'histoire géotectonique de la Méditerranée occidentale.
Rudolf Trümpy (1983), p. 197.

2.1 Introduction

The western Mediterranean Sea began to form 30 million years ago by a breathtaking magic trick, replacing another sea, the Alpine-Ligurian-Maghrebian Tethys, sometimes labeled shortly Alpine Tethys. The Alpine Tethys developed at the western tip of the Tethys ocean, which widely extended eastward between Eurasia and Gondwana. When the Alpine Tethys progressively vanished to give birth to the Western Mediterranean, the West Mediterranean Alpine Belts grew coevally. Modern geology has fully verified the pioneering views of Argand (1916, 1924) on the Alpine belts ("déferlement des nappes par le rapprochement des môles anciens de l'Eurasie et de l'Indo-Afrique [qui comprime la Téthys comme] entre les deux mâchoires d'un étau"; see Escher and Masson 1984). However, the nature of the "inland sea" named Tethys by Suess (1885) remained poorly

A. Michard (✉)
Em. Pr, Université Paris-Sud (Orsay), 10, Rue Des Jeûneurs, 75002 Paris, France
e-mail: andremichard@orange.fr

A. Chalouan
Earth Science Department, Faculty of Sciences, Mohamed V University, BP 1014, Rabat-Agdal, Morocco

A. Farah · O. Saddiqi
Geosciences Laboratory, Faculty of Sciences Aïn Chock, Hassan II University, BP 5366 Maârif, Casablanca, Morocco

S. Khomsi and F. Roure (eds.), *Geology of North Africa and the Mediterranean: Sedimentary Basins and Georesources*, Regional Geology Reviews, https://doi.org/10.1007/978-3-031-18747-6_2

understood until the advent of plate tectonics, despite the works of Wegener (1915) and Choubert (1935; see Kornprobst et al. 2018) dealing with the opening of the Atlantic Ocean, and regardless of the suggestion by Argand (1924) himself that the Tethys could have formed through extensional tectonics ("La traction continue-t-elle…, le sial finit de s'étirer et le sima apparaît au fond de l'alvéole. Sur les diamètres où cela arrive, la condition géosynclinale fait place à la condition océanique; si le fait se généralise, il n'y a plus qu'un océan."). Lemoine et al. (1987) have clearly connected the opening of the Alpine Tethys with that of the Central Atlantic through a Maghrebian transform domain (Fig. 2.1 A), an idea that is still currently accepted (Fig. 2.1B).

However, recognizing the extent of the Alpine Tethys and describing its evolution based on its scarce, deformed and more or less metamorphic remains in the surrounding Alpine belts is not an easy task. Different reconstructions have been proposed in the early 2000s, based on an increasing wealth of data (sedimentology and paleontological dating of the marginal and oceanic sedimentary formations; petrology and isotopic dating of the ophiolites; structural and paleomagnetic analyses of block rotations; seismic tomography, as well as global plate tectonics). Limited to the last decade, not less than a dozen works (sometimes contradictory) have been dedicated to the Alpine Tethys and/or its transition to the bordering continents in the western Mediterranean realm (Handy et al. 2010; Schettino and Turco 2011; Puga et al. 2011; Frizon de Lamotte et al. 2011; Carminati et al. 2012; Vergès and Fernàndez 2012; Guerrera and Martín-Martín 2014; van Hinsbergen et al. 2014; Jolivet et al. 2016; Leprêtre et al. 2018; Fernàndez 2019; Dilek and Furnes 2019). The interest of the geologic community in the Gibraltar Arc was reinforced by the presence of vast massifs of subcontinental peridotites, the Ronda and Beni Bousera massifs, in Spain and Morocco, respectively. In the last two years, i.e., between the early form of this chapter and its present, edited form, not less than ten major papers were devoted to the West Algerian Basin, Alboran Sea and bordering belts (Civiero et al. 2020; Marrone et al. 2021; Jabaloy Sánchez et al. 2021; Kumar et al. 2021; Bessière et al. 2022; Haidar et al. 2022; Schito et al. 2022; Gomez de la Peña et al. 2022; Poulaki and Stockli 2022; Porkoláb et al. 2022).

The present work focuses on the remains of the Alpine Tethys extended margins and Ocean-Continent Transition (OCT) domains, and not on the sedimentary infilling of the Tethyan oceanic basin, i.e., the Jurassic-Cenozoic pelagic-turbiditic deposits of the Maghrebian Flyschs (e.g., Guerrera et al. 2020; Abbassi et al. 2022 and references therein). We will concentrate on the remains of the southern (African) and northern (European/Iberian) margins, that are preserved in the Rif Belt, with only limited mentions of the Betic Cordilleras.

The Rif is the westernmost segment of the E-trending Maghrebide belt (Durand-Delga 1980) or Tell-Rif belt (Wildi 1983). It shares with the Kabylias and Calabria segments of this belt the privilege of showing transported and deformed exposures of the two conjugated passive margins of the Maghrebian Tethys. This comes from the origin of these segments: they all include elements of the so-called AlKaPeCa (or Alkapeca) terranes ("Alboran-Kabylias-Peloritani-Calabria"; Bouillin et al. 1986), which belonged to the southeastern margin of Iberia until the Liassic (Fig. 2.1B). Alkapeca rifting was coeval with that of the Central Atlantic (Stampfli and Hochard 2009; Labails et al. 2010) and of the Atlas basin or "aborted rift" (Fernàndez 2019). Afterwards, the Alkapeca domain would have been separated from Iberia by an oceanic corridor, labeled "Betic ocean" (Puga et al. 2004, 2011; Chalouan and Michard 2004) or "West Ligurian ocean" (Handy et al. 2010; Leprêtre et al. 2018) until the Late Eocene (Fig. 2.2). The Internal Rif units and their equivalent in the Betic Cordilleras that form the present Alboran Domain derive from the southwestern tip of Alkapeca. Alkapeca was first described as the "Alboran microplate" (Andrieux et al. 1971), and is also labeled "Mesomediterranean terrane" (Guerrera et al. 1993, 2020),

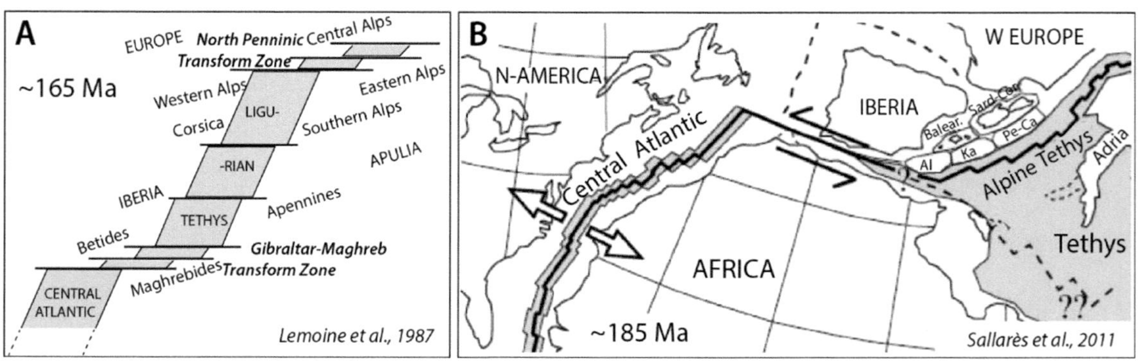

Fig. 2.1 The Central Atlantic-Alpine/Ligurian Tethys connection after Lemoine et al. 1987 (**A**) and after Sallarès et al. 2011 (**B**). In both models, a sinistral transform fault extends between Iberia and Northern Africa

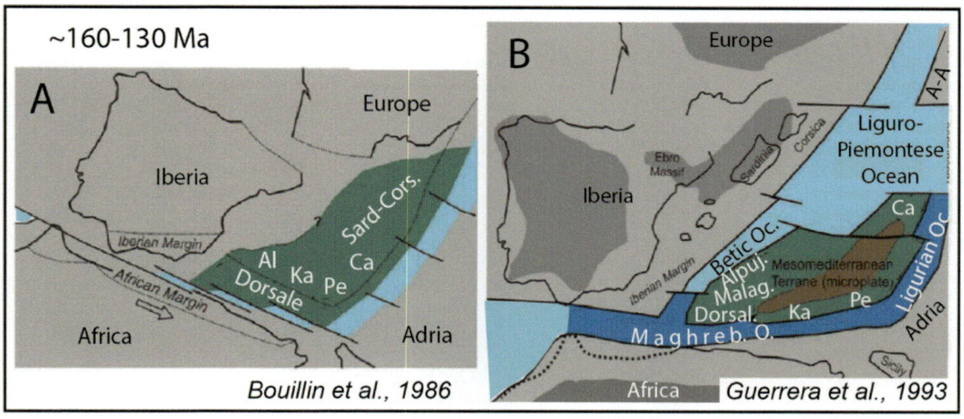

Fig. 2.2 Setting of the Alboran-Kabylias-Peloritani-Calabria (Al-Ka-Pe-Ca) domains during the Late Jurassic-Early Cretaceous, after Bouillin et al. 1986 (**A**) and after Guerrera et al. 1993 (**B**). Redrawn after Guerrera et al. (2020). Notice that the position of the Alpujarrides north of the Malaguides is debatable (see Sect. 2.5.3)

but it could have been rather a bunch of continental islands progressively drifted away from Iberia.

During the Late Cretaceous-Cenozoic, the SE-NW trending Africa-Eurasia convergence (Rosenbaum et al. 2002) was accommodated by the NW-ward subduction of the Ligurian-Maghrebian Tethys lithosphere and the correlative back-arc extension opened progressively the western Mediterranean basins between Iberia and the Corsica-Sardinia/Alkapeca domains (Lonergan and White 1997; Leprêtre et al. 2018, and references therein). During the Early Miocene ($\sim$20–17 Ma) and due to the subduction roll-back process, the Alkapeca elements collided against North Africa, Sicily, and Adria, and the pelagic/turbiditic infilling of the Maghrebian Tethys formed the Maghrebian Flyschs nappes at the front of the allochthons. The closure of the Betic "Ocean", or better said, "oceanic corridor" or "branch" gave birth to the Nevado-Filabrides Complex, overlain by the Alboran Domain Complex (Puga et al. 2017; Jabaloy Sánchez et al. 2019a, b, 2021; Poulaki and Stockli 2022; Porkoláb et al. 2022). The Nevado-Filabrides are not represented in the Moroccan branch of the Gibraltar Arc, and then the remains of the Betic oceanic corridor are not considered in the present review.

2.2 Geological Setting

2.2.1 General

The Rif Mountains display curved structural zoning from the Alhoceima-Nador area to the Tangier-Ceuta area (Fig. 2.3). Between the Nador and Tangier salients of the Meseta-Atlas foreland, the Rif Belt forms a Miocene accretionary prism thrust toward the SSW and WSW according to the transect (Chalouan et al. 2008; Leprêtre et al. 2018). The backstop of the prism (Internal Zones) is made up of the Alboran Domain metamorphic complex, which extends in the Betic Cordilleras and forms the basement of the Miocene-Pliocene Alboran Sea in between (Booth-Rea et al. 2007). In front of the backstop, the crust of the former African passive margin is likely shortened by a system of duplexes (Fig. 2.3D). After Petit et al. (2015), the depth of the Moho decreases from >30 km at the southern border of the belt to $\sim$25 km beneath the backstop.

2.2.2 External Zones

They are classically divided into three zones, from top to bottom (Fig. 2.3A, B, D), the Intrarif, Mesorif, and Prerif (Suter 1965, 1980a, b). *The Intrarif* includes, starting from the base, (i) the Ketama Unit, which mainly consists of thick Upper Jurassic clastic alternations ("ferrych"; Wildi 1983) and Lower Cretaceous pelites and siliciclastic turbidites, and (ii) the Tangier and Aknoul units, which consist of the Late Cretaceous-Cenozoic, dominantly marly series detached from the Ketama Unit. The Tangier series show a broad continuity with the Ketama Unit, whereas the Aknoul series form a gravity-driven nappe thrust onto the eastern Mesorif and Prerif zones (Frizon de Lamotte 1985; Jabaloy-Sánchez et al. 2015; Gimeno-Vives et al. 2020b). To the west of the belt, the Habt and Ouezzane nappes would be detached from the Tangier-Ketama system (Suter 1980a, b; Martin-Martin et al. 2022).

The Mesorif includes allochthonous units (nappes) that display Triassic to Lower Miocene formations, and para-utochthonous windows whose series ends with Middle-Upper Miocene turbidites and olistostromes (e.g., Tamda and Izzarene windows). The Nekor Breccia cropping along the Nekor sinistral fault (Leblanc 1980; Frizon de Lamotte 1985; Gimeno-Vives et al. 2020a, b) can be interpreted as a Mesorif window intruded by a Triassic diapir. In the Central Rif, the Senhadja Nappe was the first nappe ever described in the Rif Mountains (Russo and Russo 1929; Lacoste and Marçais 1938). Lower-Middle Jurassic carbonates and Upper Jurassic turbidites ("ferrysch") units

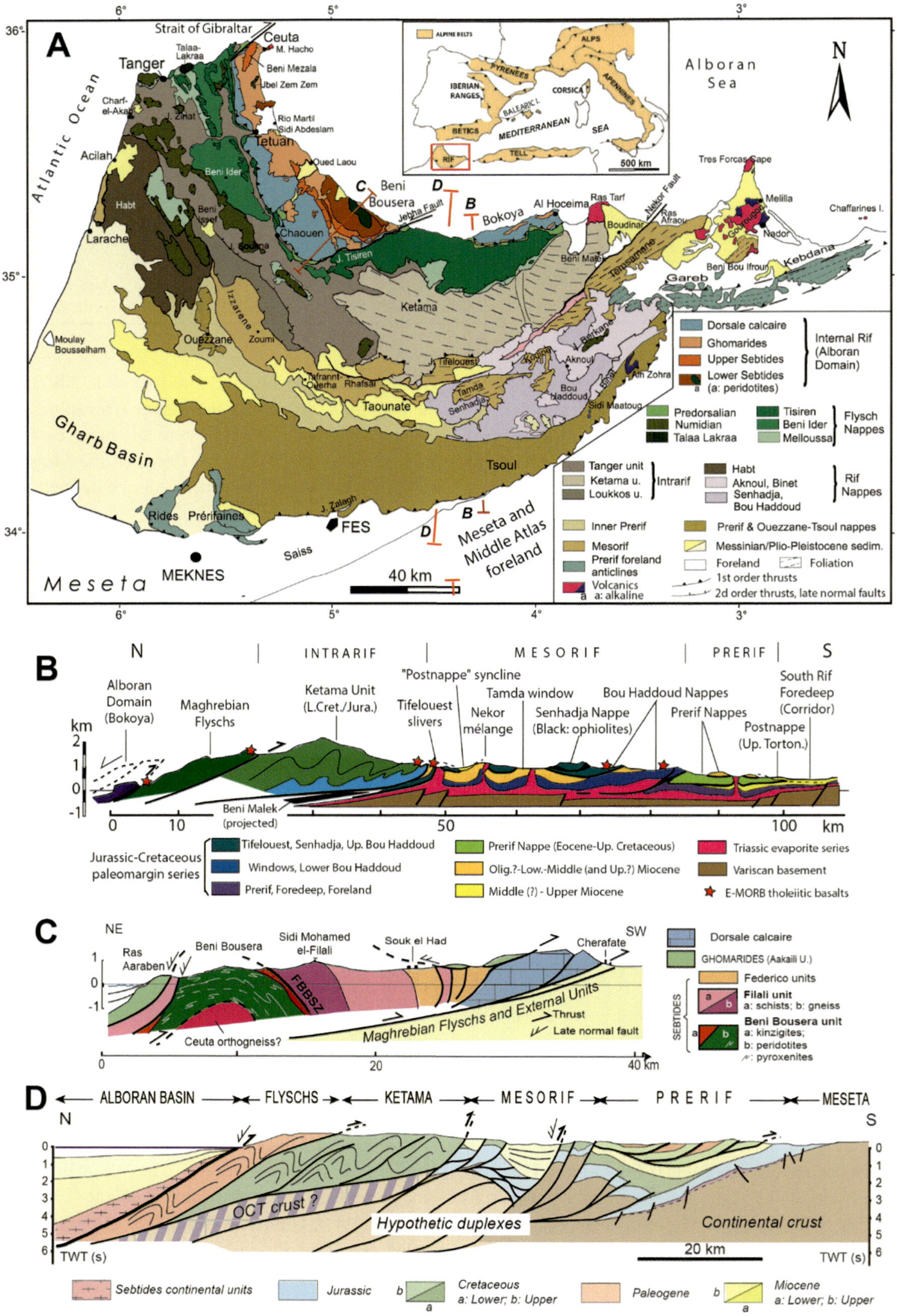

Fig. 2.3 **A** Structural map of the Rif Alpine belt, after Chalouan et al. (2008), modified. **B** Geological profile of the Central Rif, after Michard et al. (2014). **C** Cross-section of the Internal Zones, after Chalouan et al. (2008), modified. FBBSZ: Filali-Beni Bousera Shear Zone. **D** Cross-section of the Central Rif with interpretation of the African crust structure, after Leprêtre et al. (2018), modified

mainly form this nappe. The Senhadja nappe is thrust either onto the Mesorif parautochthonous windows or onto the Mesorif Bou Haddoud nappe (Leblanc 1979, 1983). In the Eastern Rif, two extensional allochthons on top of the Temsamane massif and beneath the Aknoul nappe are the metamorphic equivalent of the Senhadja nappe sensu stricto (Jabaloy-Sánchez et al. 2015; Gimeno-Vives et al. 2020b). The Bou Haddoud nappe is typified by thick "ferrysch" deposits topped by more or less brecciated Upper Jurassic-Berriasian carbonates with frequent E-MORB basalts flows (Ben Yaïch et al. 1989; Benzaggagh et al. 2014), followed upward by thick Cretaceous marls and sandstones overlain in turn by Lower Miocene conglomerates and marls. Outcrops of evaporitic Triassic formations associated with hydrothermally altered dolerites of the Central Atlantic Magmatic Province (CAMP) are scattered in the Mesorif tectonic contacts as well as in diapiric intrusions piercing the Mesorif windows.

The Prerif consists dominantly of Jurassic-Middle Miocene units detached on the Triassic evaporites and thrust toward the foreland. The Jurassic carbonates and "ferrysch" sequences crop out as extrusive slivers ("sofs") in the Internal Prerif, whereas the detached Cretaceous-Eocene and Miocene marly sequences moved further to the south. The front of the Prerif Nappe is interstratified within the Upper Miocene sandy marls of the Rif foredeep (Gharb Basin). The front of the External Rif displays evidence of Quaternary shortening (Bargach et al. 2004; Agharroud 2022).

Metamorphism is lacking in the External Zones, except in the deepest units of the Central and Eastern Rif, i.e., the Ketama Unit in the Intrarif, and the Tifelouest and Temsamane units in the Mesorif. The Ketama Unit has been folded in low-grade greenschist-facies conditions during the Cenozoic (Andrieux 1971; Gimeno-Vives et al. 2020a, b). The occurrence of a Cretaceous recrystallization event is debated yet (Vázquez et al. 2013; Michard et al. 2015). The age of the syntectonic, low-grade greenschist-facies metamorphism in the Tifelouest window is established at ~28–23 Ma (Favre 1992). A spaced cleavage developed afterward in the Tifelouest and the more external Tamda window during the Serravallian-early Tortonian. In the footwall of the Temsamane detachment (Booth-Rea et al. 2012) east of the Nekor fault, the Temsamane units exhibit metamorphic recrystallizations whose grade increases upward due to post-metamorphic stacking. In the uppermost units, chloritoid and Si-rich engine characterize medium pressure, low-temperature conditions suggesting an incipient subduction at 30–33 Ma (Negro et al. 2007, 2008; Jabaloy-Sánchez et al. 2015). The collisional/transpressive shortening of the Temsamane units triggered the uplift of the tectonic prism and the detachment of the Ketama Unit and Aknoul nappe at 15–10 Ma (Negro et al. 2008) or 10–6 Ma (Jabaloy-Sánchez et al. 2015).

The main thrust contacts of the External Zones are sealed by the transgression of Upper Tortonian conglomerates and sandy marls (molasses) now preserved within large "post-nappes synclines", better labeled wedge-top synclines (Leprêtre et al. 2018; Gimeno-Vives et al. 2020b). Their marine series was folded during the Messinian-Pliocene, together with the underlying nappes, in a thick-skinned tectonic style (Capella et al. 2017).

2.2.3 Maghrebian Flyschs

The Maghrebian Flysch nappes correspond to the detached, Upper Jurassic-Burdigalian infilling of the Tethyan basin (Bouillin et al. 1986; Chalouan et al. 2008; Frizon de Lamotte et al. 2011; Guerrera and Martín-Martín 2014). In the Rif Belt, only a few remains of tholeiitic basalts (Durand-Delga et al. 2000) occur beneath the pelagic and mostly turbiditic deposits of the Flyschs, south of the Bokkoya massif (Fig. 2.3B). However, ophiolites occur in Algeria beneath the Lesser Kabylia massif (Texenna or Rekkada-Metletine serpentinites, gabbros, pillow basalts and Jurassic radiolarites; Bouillin et al. 1977; Boukaoud et al. 2021), which makes a transition toward the well-developed Ligurian ophiolites of the southern Apennines (Vitale et al. 2019).

Two groups of nappes are recognized in the Rif belt (Fig. 2.3A), and named as in the Algerian Tell (Bouillin et al. 1970), from top to bottom, (i) *the Mauretanian nappes* whose Upper Jurassic-Albian sequence (Tisiren nappe; Durand-Delga et al. 1999) overlie the Upper Cretaceous-Burdigalian diverticulated sequence (Beni Ider nappe; Chalouan et al. 2006; De Capoa et al. 2007), and (ii) *the Massylian nappes* whose Cretaceous sequence (Chouamat-Mellousa nappe) occurs as sheared slivers beneath the Late Eocene-Burdigalian detached sequence (Numidian "nappe") (Thomas et al. 2010; El Talibi et al. 2014; Abbassi et al. 2022). All these Flysch units root beneath the Alboran Domain and the strongly sheared Predorsalian Zone (Olivier 1990). They form large outliers over the External Zones and scarce inliers backthrust onto the Internal Zones (e.g., J. Zemzem, north of Tetouan).

During the Barremian-Albian, the major detrital input toward the Chouamat and Tisiren basinal areas was most likely sourced in the NW African uplifted platform (Vila 1980, his Fig. 180; El Talibi et al. 2014; Azdimousa et al. 2019a). The Early Cretaceous flyschs clearly correlate with the coeval deposits of the External Zones (Mesorif and Intrarif), suggesting the northward development of a huge, Africa-sourced detrital prism at that time. However, from the Campanian onward, the sources of the Maghrebian flyschs are twofold. The Beni Ider sequence (calciturbidites, breccias, sandy-micaceous flysch) contains elements originating from the Alboran Domain, which was progressively uplifted and deformed between ~80–20 Ma (El Kadiri et al. 2006; De Capoa et al. 2007). In contrast, the Oligo-Miocene fluxo-turbidites and quartz almond-bearing sandstones of the Numidian Flysch can be definitively ascribed to African craton sources (Thomas et al. 2010; Abbassi et al. 2022).

2.2.4 Internal Zones: The Alboran Domain

The Alboran Domain of the Rif Belt includes the Sebtides high-grade units at the bottom, the Ghomarides low-grade nappes on top, and the Dorsale Calcaire non-metamorphic slivers in front of the complex (Fig. 2.3A, C).

The Sebtides (named after Ceuta city, Sebta in Arabic) are subdivided into four main units following Kornprobst (1974), i.e., from bottom to top, (i) the Ceuta orthogneiss unit (Monte Hacho Lower Unit; Homonnay et al. 2018); (ii) the Beni Bousera Unit, which includes the Beni Bousera peridotites and the overlying migmatitic granulites and kinzigites (Reuber et al. 1982; Saddiqi et al. 1988; Bouybaouene et al. 1998; Haissen et al. 2004; Rossetti et al. 2020; Farah et al. 2021); (iii) the Filali gneiss and mica-schist unit (Michard et al. 1983; El Maz and Guiraud 2001), and (iv) the Federico Permian–Triassic metasedimentary units (Bouybaouene et al. 1995; Michard et al. 1997; El Bakili et al. 2021; Marrone et al. 2021). All these rock units have their counterparts in the Alpujarrides of the Betic Cordilleras, with the Ronda peridotites sandwiched between crustal units.

The Ghomarides (labeled after the name of a large Berber tribe of the Chaouen-Jebha area) consist of four stacked nappes each with folded Paleozoic metasediments from the Variscan upper crust (Chalouan and Michard 1990; Sanz de Galdeano et al. 2006; Esteban et al. 2017) unconformably overlain by thick Middle Triassic red beds, Jurassic limestones, restricted Cretaceous-Middle Eocene limestones and Eocene-Early Miocene clastics (Perrone et al. 2006; El Kadiri et al. 2006; Serrano et al. 2007; Hlila et al. 2008; Olivier and Paquette 2019). Next to Malaga in the Betics, the Malaguides and the underlying Alpujarrides are crosscut by a swarm of dolerite dykes dated at 33 Ma (Jabaloy Sánchez et al. 2019b).

The Dorsale Calcaire also consists of stacked units whose rock material compares with that of the Mesozoic terms of the Ghomarides, except that the Middle Triassic series exhibit shallow marine carbonate facies instead of red beds and that deep marine facies with radiolarites and synsedimentary normal faults occur in the Jurassic series. The Dorsalian and adjacent Predorsalian units are referred to as the passive margin of the Alboran Domain north of the Maghrebian Flyschs basin (Olivier 1990; El Hatimi et al. 1991; Chalouan et al. 2008; El Kadiri et al. 2006, 2009).

Most of the Alboran Domain suffered metamorphic recrystallization during its Alpine deformation. Only the Dorsale Calcaire units and the uppermost Ghomarides nappes escaped Alpine metamorphism. Greenschist-facies recrystallization is observed in the lower part of the lowest Ghomaride nappe where the temperature exceeded 450 °C (Negro et al. 2006; Schito et al. 2022). In this nappe, the Alpine recrystallization occurred at ~25–30 Ma (Chalouan and Michard 1990), and overprinted the Variscan greenschist-facies metamorphism.

The lower Federico units display blueschist/eclogite facies conditions of peak metamorphism before retrogression under a high geothermal gradient (Bouybaouene et al. 1995; Michard et al. 1997; Janots et al. 2006). In the southern Federico units, Marrone et al. (2021) recognized two metamorphic events at 37–34 Ma and 21–25 Ma, respectively ($^{40}Ar/^{39}Ar$ geochronology), whereas El Bakili et al. (2021) observe the effects of a Triassic event superimposed by the Miocene one. The Betic equivalents of these units also record HP-LT recrystallization (Goffé et al. 1989; Booth-Rea et al. 2005), thus suggesting they all have been involved in a subduction zone whose location is debated (see Sect. 2.5). Ruiz Cruz and Sanz de Galdeano (2012) discussed the reality of the HP-LT event in the Federico units, possibly because of sampling bias.

In the granulitic envelope of the Beni Bousera and Ronda peridotites, a Variscan high-pressure migmatitic event occurred at ~300–290 Ma (Montel et al. 2000; Sánchez-Rodriguez and Gebauer 2000; Rossetti et al. 2010, 2020). Alpine high-temperature (HT), medium to low-pressure (MP/LP) metamorphism began as early as ~30 Ma (Massonne 2014; Homonnay et al. 2018) or even before 33 Ma (Jabaloy Sánchez et al. 2019b) and lagged in HT-MP to LP conditions until ~20 Ma (Michard et al. 2006; Rossetti et al. 2010, 2013; Gueydan et al. 2015; Homonnay et al. 2018; El Bakili et al. 2020). A swarm of granite veins and dykes emplaced in the Beni Bousera peridotites and kinzigites and in the Filali gneisses at 22–20 Ma (Rossetti et al. 2010), coeval with greenschist-facies normal shear zones (El Bakili et al. 2020). The HT-LP Alpine event is also registered in the ultramafics (Sm–Nd and Lu–Hf Cpx-Grt ages at ~20–25 Ma; see Michard et al. 2006, and references therein), although their evolution began certainly much earlier in the deep mantle (graphite pseudomorphs after diamond, e.g., El Atrassi et al. 2011).

2.3 The Southern Margin of the Maghrebian Tethys

The southern margin of the Maghrebian Tethys, i.e., the OCT crust adjacent to the North African continental margin, has been inverted during the Alpine orogeny together with the continental margin itself. Then, recognizing the remains of the pre-orogenic architecture in the present-day External

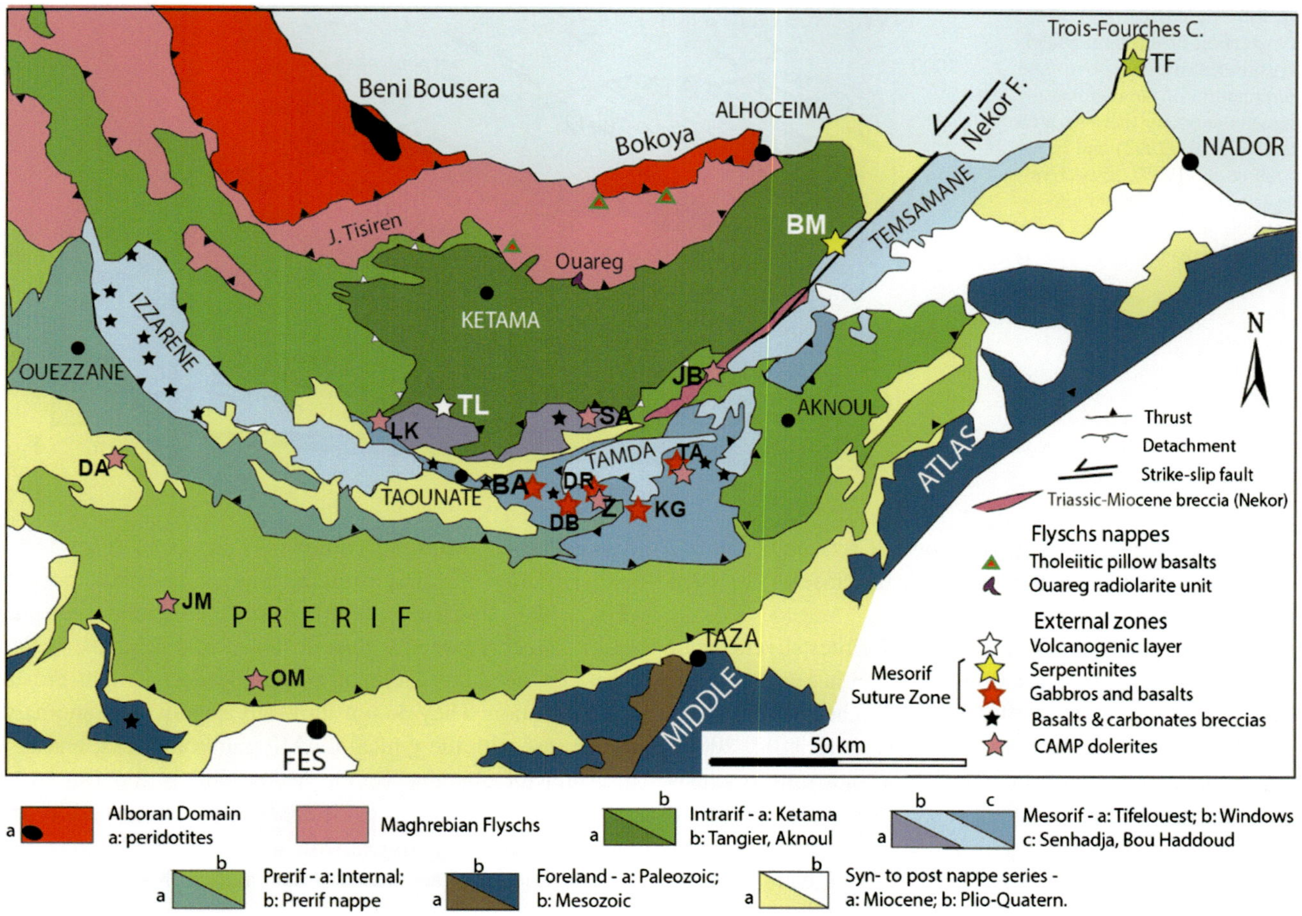

Fig. 2.4 Geological map of the central and eastern parts of the Rif Belt, with emphasis on the location of the mafic/ultramafic rocks of the Flysch Nappes and the External Zones, after Michard et al. (2018) and Haissen et al. (2021). BA: Bou Adel; BM: Beni Malek; DA: Douar Alami; DB: Dar bou Aza (Ain Akreb); DR: Douar Roukba; JB: Jbel Baio; JM: Jorf el Melha; KG: Kef-el-Ghar; LK: Laklaaia (Klaia); OM: Oued Melha; R: Roukba (Douar Roukba); SA: Sof Aouzzai; TA: Taineste; TF: Trois-Fourches (Tres-Forcas); TL: Taounat Lechkar; Z: Zitouna

nappe stack (Fig. 2.4) is currently debated (Michard et al. 2020; Gimeno-Vives et al. 2020a). In the following sections, we summarize the data and suggest a tentative solution.

2.3.1 Ocean-Continent Transition in the Eastern Rif

2.3.1.1 Beni Malek Serpentinites and Metabasites

A witness of the Maghrebian Tethys OCT has been recognized for long in the Beni Malek serpentinized lherzolite massif of the Eastern Rif (Michard et al. 1992, 2007; El Azzab et al. 1997). The serpentinite massif and the neighboring outcrops of greenschist-facies basalts (Skifat village, 6 km SSW of Beni Malek) are overlain by a thin oceanic-type sedimentary cover (limestones with serpentinite and spinel clasts) supposed to be Late Jurassic in age. These units are overlain by the Lower Cretaceous series of the Ketama Unit (Fig. 2.5). In a recent work, Gimeno-Vives et al. (2019) propose that the Ketama Lower Cretaceous clastics were deposited unconformably onto the exhumed mantle serpentinites in the Beni Malek area and onto the Triassic-Jurassic continental margin series a few kilometers further to the SW. However, the alleged unconformity is no longer observable in the Beni Malek area, where calcareous breccias underline the contact between the serpentinites and the Ketama Unit.

2.3.1.2 Correlative Outcrops

The southwestern Ketama Unit (Taounat Lechkar, Fig. 2.4) displays a significant volcanogenic layer interbedded within the lowest Barremian beds (Suter 1964). Within this 3 m-thick layer, diabase and gabbro pebbles or boulders with E-MORB affinities are associated with pelagic limestone pebbles in a pyroxene and olivine-rich matrix (Zaghloul et al. 2003). This occurrence is reminiscent of the Skifat metabasites next to the Beni Malek serpentinites. This

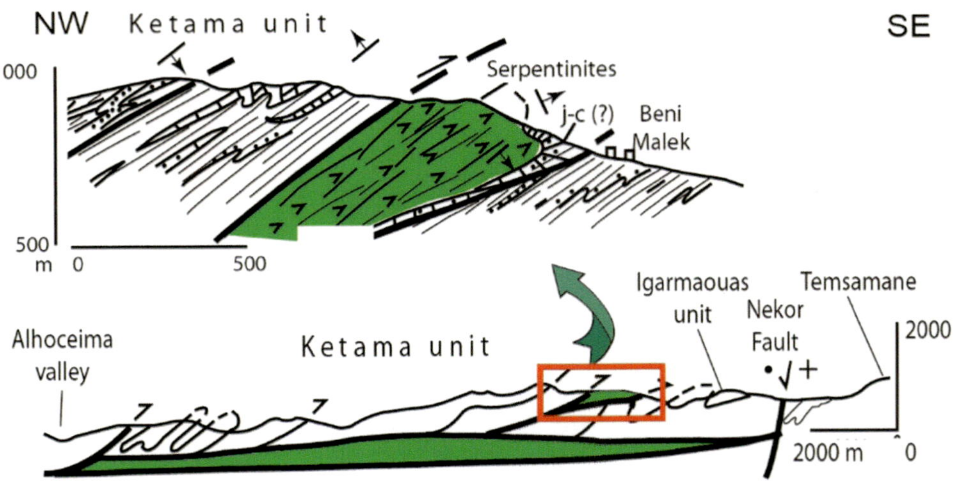

Fig. 2.5 Geological cross-section of the Beni Malek serpentinites massif (above) and interpretation of the regional magnetic anomaly (below), after Michard et al. (1992) and El Azzab et al. (1997), respectively. The outcropping serpentinites would be detached from a larger subsurface sliver

suggests that the Upper Jurassic-Lower Cretaceous Ketama beds were deposited in the OCT domain close to the Beni Malek exhumed mantle ridge.

It is worth noting that serpentinite/chloritites slivers occur at Cape of Trois-Fourches (Fig. 2.4; García-Dueñas et al. 1995; Azdimousa et al. 2019b) above the eastern equivalent of the uppermost Temsamane Unit (Chalouan et al. 2008). Further to the east, serpentinite and chloritoid-bearing schists crop out in the Oran Mountains of western Algeria (Michard et al. 2007; Gimeno-Vives et al. 2019). These remains of the Jurassic OCT partly concealed beneath the Ketama Intrarif unit and obducted onto the Mesorif Temsamane units were labeled the "Mesorif Suture Zone" (MSZ) by Michard et al. (2007).

2.3.2 Tracking the Southern Tethys Margin in the Central Mesorif

2.3.2.1 The MSZ Gabbros and Their Envelopes

The MSZ concept defined above acquired a larger dimension when oceanic-type gabbro massifs were recognized in the Mesorif zone of the Central Rif (Benzaggagh et al. 2014; Michard et al. 2014). The gabbro massifs of Bou Adel and Kef-el-Ghar described by Benzaggagh et al. (2014) are the largest and most typical of this SW-ward extended MSZ (Fig. 2.4). They form kilometer-sized, lensoid klippes whose visible thickness hardly reaches ~200 m, located at the bottom of the Senhadja nappe. The base of the Bou Adel gabbro is not exposed but corresponds to the central Mesorif Triassic-Cretaceous series that form the Taounate E-trending anticline. The Kef-el-Ghar massif overlies directly the Serravallian-Lower Tortonian mélange of the Tamda window (Vidal 1983; Leblanc 1983).

Two petrographic characters of the Bou Adel gabbros (Benzaggagh et al. 2014) strongly suggest their oceanic origin: (i) they are mostly layered gabbros, troctolitic to ferrogabbroic, recording cumulative processes (Fig. 2.6A), and (ii) they exhibit trondhjemite veins and pockets (Fig. 2.6B). Consistently, the Bou Adel gabbros display multi-elements patterns close to those of E-MORB, with a weak Eu negative anomaly and evidence of slight crustal contamination. They do not show any negative Nb anomaly, in contrast with the typical CAMP gabbroic rocks, and in a discriminant diagram Th/Yb versus Nb/Yb, the Bou Adel gabbro samples fall within the area of the oceanic basalts and mid-ocean ridge ophiolites (Michard et al. 2020a).

The Bou Adel and Kef-el-Ghar massifs have been dated from the lower Liassic as follows. Michard et al. (2018) reported on Laser Ablation Inductively Coupled Plasma Mass Spectrometry (LA-ICP-MS) U–Pb analysis of zircons grains from a trondhjemite vein crosscutting the Bou Adel gabbro. The less cracked, less contaminated grains provided a concordant age of 190 ± 2 Ma. Gimeno-Vives et al. (2019) published four U–Pb zircon dates, selecting the oldest ages obtained from each sample: 196 ± 4 Ma for a gabbro sample from Bou Adel; 192 ± 4 Ma for a gabbro sample from Kef-el-Ghar, and 195 ± 4 Ma for a dolerite sample from Kef-el-Ghar. The authors assume that these ages fit with the peak age of the CAMP. However, as noted by Michard et al. (2020a), all the reported ages are significantly younger than the 201 Ma peak of the CAMP (Davies et al. 2017; Marzoli et al. 2018, 2019), with a mean age value at 193.2 ± 4.5 Ma (Fig. 2.7). In contrast, they match the 195–190 Ma age of the Central Atlantic opening (Labails et al. 2010; Sibuet et al. 2012). Likewise, Stampfli and Hochard (2009) argue that the Central Atlantic Ocean connected to the Ligurian Tethys between 200 and 180 Ma.

Each gabbro massif displays its own volcanic-sedimentary envelope. At Bou Adel, this envelope includes metabasalts and basaltic metabreccias (1 to 20 m-thick), fine- to coarse-grained marbles (1–10 m) with metabasalt clasts and a few meters of layered marble-volcanoclastic/terrigeneous alternations (Fig. 2.8A–E). At Kef-el-Ghar, the gabbro is

Fig. 2.6 Typical lithologies of the Bou Adel gabbro. **A** Layered cumulative troctolite; dip is to the NNW. **B** Close view of the trondhjemite dyke dated at 190 Ma (photo A. Mokhtari in Michard et al. 2018)

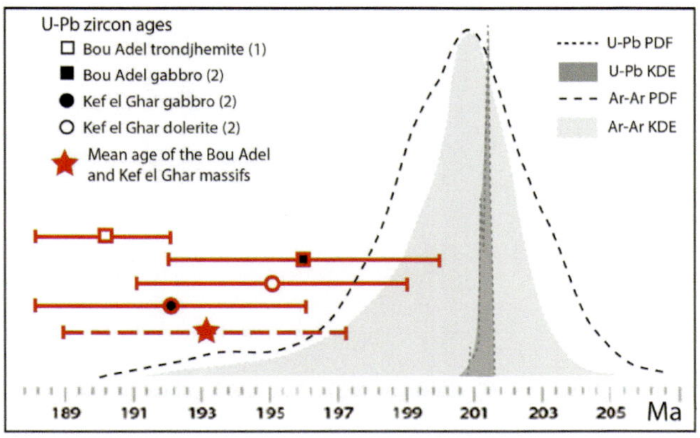

Fig. 2.7 Plot of the ages of the Mesorif samples (red symbols) compared to the CAMP ages (grey signatures). The Bou Adel and Kef-el-Ghar ages are from Michard et al. 2018 (1) and Gimeno-Vives et al. 2019 (2). The CAMP ages are compiled by Marzoli et al. (2018) and shown by the probability density functions (PDF) and Kernel density estimates (KDE) of the $^{40}Ar/^{39}Ar$ and U–Pb age

overlain by microgabbroic meta breccias (~ 10 m), followed upward by radiolaritic metacherts and recurrent basalts (~ 25 m) and by pelite-calcschist alternations (Fig. 2.8E, F). A first tentative of paleontological dating of the metacherts failed: radiolarian fossils were observed, but remained undeterminable (written comm. Pr. Chiari, in Michard et al. 2018).

Hence, the Bou Adel and Kef-el-Ghar examples of MSZ gabbros exhibit two different kinds of overlying envelope. The Kef-el-Ghar envelope evokes the volcanic-sedimentary cover of the Alpine-Ligurian ophiolites where the radiolarites are mostly Bathonian to Oxfordian in age (Lagabrielle and Lemoine 1997; Lagabrielle et al. 2015; Balestro et al. 2019). In contrast, the Bou Adel envelope evokes a marginal accumulation of basalts, volcanoclastics and terrigeneous deposits subsequently hydrothermally altered and sheared. As the geochemistry of some Bou Adel metabasalts shows CAMP affinities (Haissen et al. 2021), this envelope could represent relics of a Triassic Volcanic Passive Margin (VPM; Geoffroy et al. 2022) within which the gabbro would have emplaced. The lack of crustal rocks in the Bou Adel envelope does not support an emplacement within the continental crust itself (Gimeno-Vives et al. 2020a). The VPM margin illustrated at Bou Adel would have subsequently evolved toward the hyperextended, magma-poor transform margin type during the Middle and Late Jurassic, coevally with the magma-poor margin more to the east, as sampled at Kef-el-Ghar. Exhumation of the Liassic gabbros (Bou Adel, Kef-el-Ghar) would have occurred, possibly as early as the Middle Jurassic in the frame of core-complex structures as observed in the Alpine ophiolites during the Late Jurassic (Lagabrielle et al. 2015; Balestro et al. 2019).

2.3.2.2 The Mesorif Basalt-Limestone Breccias

A rosary of basalt lava flows, mostly brecciated and associated with limestone breccias (Fig. 2.9) have been observed in the Mesorif and along the Mesorif-Internal Prerif boundary (Fig. 2.4). They are well-dated by their intercalation in the Kimmeridgian-Berriasian sequence, but some volcanoclastics also occur locally in the Upper Oxfordian (Benzaggagh and Habibi 2006; Benzaggagh 2011; Benzaggagh et al. 2014). This "Mesorif Basalt-Limestone Breccias" (MBLB) sequence occurs on top of the kilometer-thick, turbitic Callovian–Oxfordian "ferrysch" that seals the Lower-Middle Jurassic tilted-block and half-graben structures of the North African paleomargin (Suter 1980a; Favre and Stampfli 1992). The shallow water calcareous elements have been likely resedimented in mass-flows triggered by the volcanic events along a relatively steep slope. The chaotic

Fig. 2.8 Volcano-sedimentary envelopes above the Bou Adel (**A–C**) and Kef-el-Ghar (**D–F**) gabbros. **A** Thin, sheared sequence gabbro, brecciated metabasalt, marble. **B** Calcareous metabreccia with metabasalt clasts, ~2 m above the contact seen in (A). **C** Marble-volcanoclastic alternations a few meters above the gabbro. **D** Overview of the Kef-el-Ghar envelope. 1: gabbro; 2: microgabbroic meta breccia; 3: radiolaritic metacherts with recurrent basaltic layers; 4: calcschists. **E** Close view of the microgabbroic meta breccia (2). **F** Close view of the radiolaritic metacherts; notice the early boudinage of the lowest red and black beds

breccias also contain pebbles of ferrysch-like sandstones whose abundance and size increase upward. Olistoliths of ferrysch beds occur locally in the Berriasian breccias (Benzaggagh et al. 2014).

It is tempting to ascribe the MBLB event to an abrupt increase of the extensional/transtensional evolution of the North African margin (Ben Yaïch et al. 1989). Its age (~160–145 Ma) compares with that of the opening of the Alpine-Ligurian Tethys (Lagabrielle and Lemoine 1997; Bill et al. 2001), where the syn-extensional magmatism develops basically during the Kimmeridgian (Balestro et al. 2019). Accordingly, Michard et al. (2018) suggested that the MBLB event and the associated faults have been linked to the activity of the North African Transform (NAT) zone connecting the Alpine-Ligurian Tethys to the Central Atlantic (Fig. 2.1). The MBLB zone evokes an intra-margin structure recording the extension of the North African margin and marking the landward border of the OCT and distal margin domains.

2.3.2.3 Latest Stages of the Passive Margin Evolution

As reported above (Sect. 2.2.2), the MSZ gabbros occur beneath the Senhadja nappe, which mainly consists of Lower Jurassic carbonates, Middle Jurassic marly limestones and Upper Jurassic turbidites (ferrych). The gabbros and their metamorphic volcanic-sedimentary envelope represent tectonic slivers within the sole thrust of the Senhadja nappe in the frame of the Cenozoic Rif belt. These slivers are separated from the overlying carbonates by a brittle thrust marked by vacuolar carbonate breccias (cargneules; Benzaggagh et al. 2014). Taking into account the setting of the

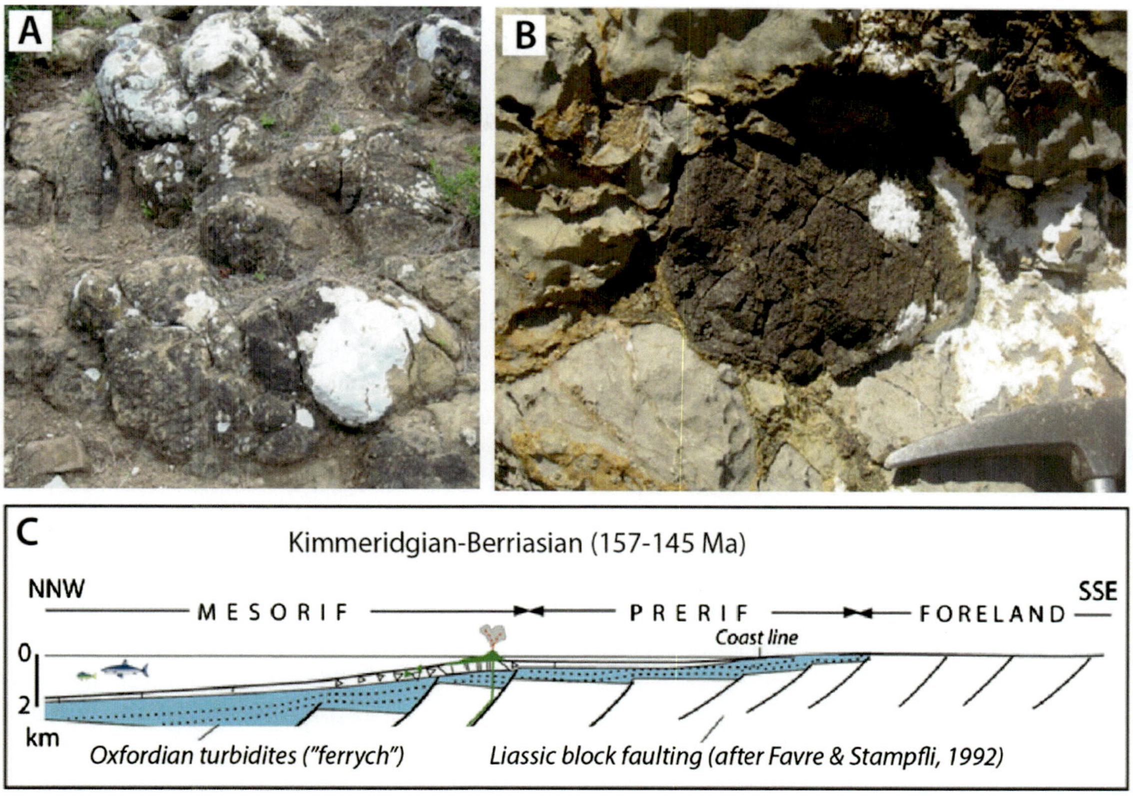

Fig. 2.9 The Mesorif Basalts and Breccias Zone. **A** Tithonian pillow basalts from the area 10 km ESE of Ouezzane, after Benzaggagh (2011). **B** Block of pillow basalts included in a Tithonian carbonate breccia from the Mesorif nappe ESE of Taineste, after Benzaggagh et al. (2014). **C** Interpretation of the Mesorif association of basalts and carbonate breccias, after Ben Yaïch et al. (1989) and Michard et al. (2014)

Senhadja Jurassic series, which form kilometer-scale tilted blocks above their basal thrust contacts, Gimeno-Vives et al. (2019, 2020b) proposed that these blocks were detached from a proximal part of the margin and slid onto Triassic evaporites toward the distal margin. The tectonic contact at the base of the blocks ("extensional allochthons" or "rafts", well known in the Pyrenees, cf. Jammes et al. 2009; Asti et al. 2019; Lagabrielle et al., 2019a, b) would be responsible for the exhumation of the gabbros up to the surface during the Middle Jurassic rifting.

However, an alternative scenario can be proposed (here favored) involving, (i) exhumation of the gabbro bodies by core-complex mechanism in the distal margin (see above, Sect. 2.3.2.1), and (ii) sliding of the rafts during the Late Jurassic-Early Cretaceous (Fig. 2.10). In the Pyrenean examples quoted above, gravity sliding of the rafts was

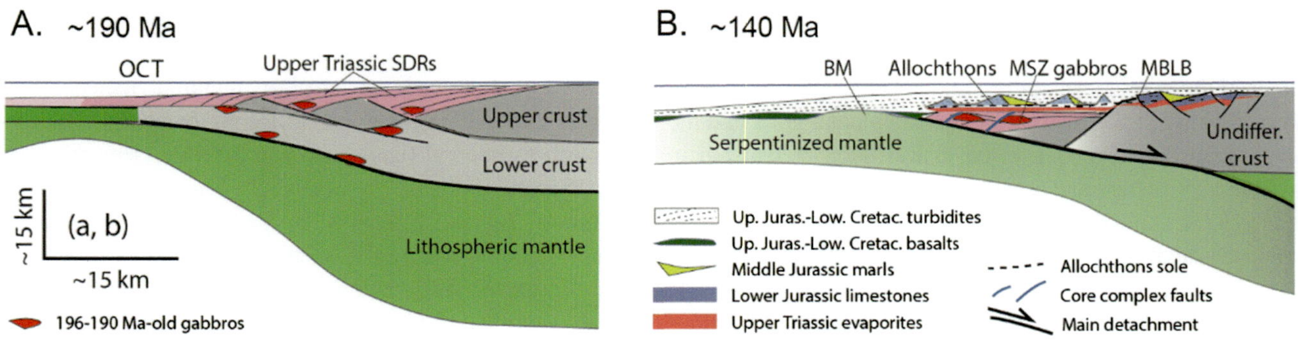

Fig. 2.10 A new proposal for the early evolution of the NW African passive margin as sampled in the sole thrust of the Senhadja nappe of the Central Rif (Bou Adel example). **A** Emplacement of gabbro bodies in a VPM margin (SDRs: seaward dipping reflectors). **B** The Jurassic-Early Cretaceous extension allowed the exhumation of the Liassic gabbros up to the seafloor in core-complex structures, followed by gravity-driven emplacement of the Senhadja rafts triggered by the post-rift detrital infilling

coeval with the influx of the thick Albo-Cenomanian detrital sedimentation and the formation of coarse conglomerates along the faulted boundary of the proximal margin. A similar process could have operated on the slope of the African paleomargin, with the sliding of the rafts triggered by the increasing faulting along the MBLB trend and the input of the Late Jurassic-Early Cretaceous detrital prism (ferrych and sandy turbidites of the Intrarif and Massylian Flyschs). In both the Pyrenees and Rif cases, raft gliding was favored by the occurrence of Triassic evaporites below the carbonate sequence of the paleomargin. In the external Rif, Upper Triassic sequences including evaporitic red beds and altered basalts from the CAMP are widespread (Figs. 2.3B and 2.4).

It is worth noting that gravity sliding or gliding processes are also illustrated in non-inverted passive margin settings where they are generally combined with gravity spreading (Peel 2014). Examples include the Niger and Nile deltas, the west coast of Africa, and the Gulf of Mexico (Damuth 1994; Fort et al. 2004; Morley et al. 2011; King and Morley 2017). These systems have either weak shale or salt detachments levels. Shortening values from the restorations presented by Rowan et al. (2004) are 6 km for the Mississippi Fan Fold Belt and 27–30 km for the Kwanza Basin. The temporal relationship between major depositional episodes and phases of accelerated gravity-driven movement can be expected due to increasing load and pore fluid pressures throughout the sediment body. In northern Tunisia, such a gliding gravitary process is described by Jaillard et al. (2017) during the Albian.

2.4 The Northern Margin of the Maghrebian Tethys

2.4.1 General

The paleogeographic location of the Maghrebian Flysch basin to the south of the Kabylian and Internal Rif massifs was progressively established in the sixties and seventies (Durand-Delga 1964, 1969, 1980; Bouillin et al. 1973; Raoult 1975). In the meanwhile, the plate tectonics concepts renewed the geological studies, and a former passive margin domain between the Internal Zones continental units and the Maghrebian Flyschs oceanic basin was soon recognized in the Dorsale Calcaire and associated Predorsalian units (Wildi et al. 1977; Durand-Delga and Fontboté 1980; Nold et al. 1981; Durand-Delga and Olivier 1988; Olivier 1990; El Hatimi et al. 1991). However, the relationship of the Dorsale unrooted units with the juxtaposed Ghomarides and Sebtides basement units remained controversial (Chalouan and Michard 2004): some authors argued that the Dorsale nappes could have been detached from the Sebtides, while others underscore the continuous stratigraphic variations between the Triassic-Liassic series of the Ghomarides and that of the Dorsale. We turn back to this issue in the next section (Sect. 2.6.1).

Prior to the Alpine orogeny, the Dorsale Calcaire marginal domain and the North African margin were located on each side of the Maghrebian Tethys (Figs. 2.2A, B and 2.11A). However, their present-day aspect is deeply contrasting from the Paleocene onward, the first belonged to the overriding plate, and the second to the subducting plate of the Ligurian-Maghrebian subduction system (Fig. 2.11B; Leprêtre et al. 2018). As a result, the Dorsale Calcaire units overlie the Flyschs through a major thrust contact within which the more distal Predorsalian units are pinched. The basement of the former passive margin is not directly exposed there, and thus diverging hypotheses coexist on whether it was connected to the Ghomarides or the Sebtides (see Sect. 2.6.1).

2.4.2 Stratigraphic Record of the Rift-Drift Stages

The scenic Dorsale Calcaire is made of a number of minor units stacked one above each other, and altogether set down above the Predorsalian units, the Flyschs nappes, and the External units through a major sole thrust (Figs. 2.3 and 2.12).

Restoring the original paleogeography of the passive margin requires simultaneous stratigraphic and structural studies. Within the numerous minor units that can be mapped in the Dorsale, two types of stratigraphic sequence have been distinguished for long (Wildi et al. 1977; Durand-Delga and Olivier 1988). The *Internal Dorsale* units are typified by Late Triassic greyish dolostones (100–200 m), followed upward by Hettangian-Sinemurian white limestones and dolostones ($\sim$20 m) and Pliensbachian cherty limestones (10–150 m). Contrastingly, the *External Dorsale* units are typified by thick Carnian-Norian stromatolitic dolostones and limestones (up to 600 m thick), Rhaetian-Hettangian limestones/dolostones alternations and alveolar limestones ($\sim$200 m), followed upward by hundred meters thick Liassic cherty limestones and siliceous breccias. This corresponds to internal (landward) and external (seaward) passive margin domains, respectively. The paleogeographic continuity of the Internal Dorsale with the Ghomaride domain is indicated by the local occurrence of Middle Triassic Verrucano facies at the very base of the Dorsale and by some outcrops of Upper Triassic dolostones and Liassic limestones preserved on top of the Ghomaride Verrucano series (El Kadiri et al. 1992). The main stratigraphic

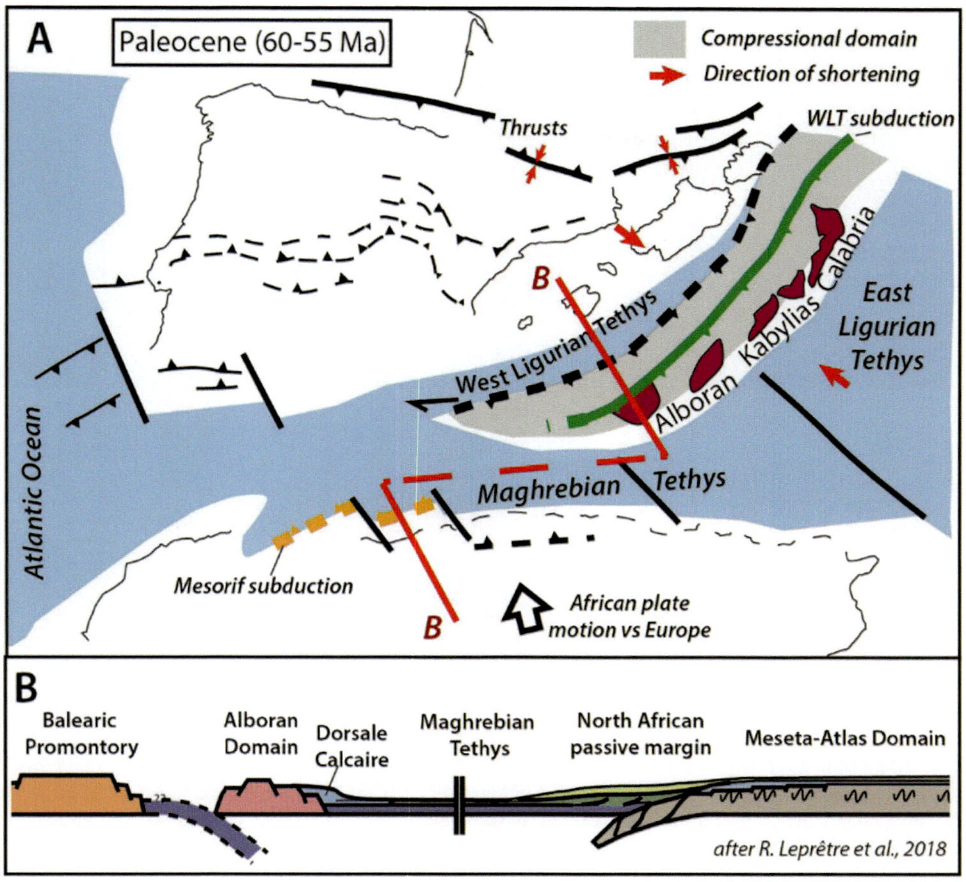

Fig. 2.11 Geodynamic setting of the westernmost Tethys area during the Paleocene, after Leprêtre et al. (2018), modified. **A** Generalized map of the land masses and intervening oceanic branches. The West Ligurian Tethys (also named Betic Ocean) is already closing in Alpine Corsica. **B** Sketch crustal-scale profile showing the conjugated margins of the Maghrebian Tethys (see A for location of the northern and southern segments of the profile)

difference between the Ghomarides and Internal Dorsale domains is the greater Eocene –Oligocene erosion of the former with respect to the latter.

In the Internal Dorsale units, the relatively thick Triassic-Liassic formations are unconformably followed by thin (~10–15 m), condensed pelagic formations that correspond to the Middle Jurassic-Lower Cretaceous span of time ("calcaires à filaments", ammonitico rosso, radiolarites, Calpionellids limestones). The condensed series that mark the demise of the proximal carbonate platform were particularly studied by El Kadiri et al. (1989, 1992), El Kadiri (2000–2002) and El Kadiri et al. (2009). These are characterized not only by their pelagic facies, but also by the frequent gaps associated with fracturing and dissolution of the calcareous substrate ("paleokarst" filled with neptunian dykes), and with Fe-Mn-rich hardgrounds. The main gaps are observed, first, during the Domerian-Toarcian (~185–175 Ma), and second, between the Berriasian and the Campanian–Maastrichtian (~140–80 Ma).

In the External Dorsale and Predorsalian units, the gaps of the Internal Dorsale Jurassic series no longer exist or they are much shorter. The corresponding intervals are mostly occupied by condensed facies such as ammonitico rosso, cherty limestones, and radiolarites. Carbonate breccias with slumps and ammonitico rosso develop as early as the Hettangian and Sinemurian, respectively (~200–190 Ma). Radiolarites develop from the Aalenian onward in the External Dorsale and the Predorsalian units, whereas they occur not before the Upper Kimmeridgian in the Internal Dorsale. This diachronism evidence the progressive extension and deepening of the margin during the Jurassic. Synsedimentary normal faulting is recorded at the front of the External Dorsale of the Jbel Cherafat by coarse radiolaritic breccias (Fig. 2.13). The underlying Predorsalian units exhibit similar, although thinner sequences.

2.4.3 Maghrebian Tethys Conjugate Margins

The chronology of extension and deepening of the Dorsalian margin compares perfectly with that of the North African margin, as described by Favre (1992) and Chalouan et al. (2008). Rifting begins during the Late Triassic-Early Liassic, then resumes during the Middle-Late Liassic. The breakup unconformity corresponds to the Aalenian-Bajocian in both margins. The northern margin differs from its southern counterpart by the lack of the clastic post-rift "ferrysch" prism sourced in the Atlas-Meseta or Saharan domains.

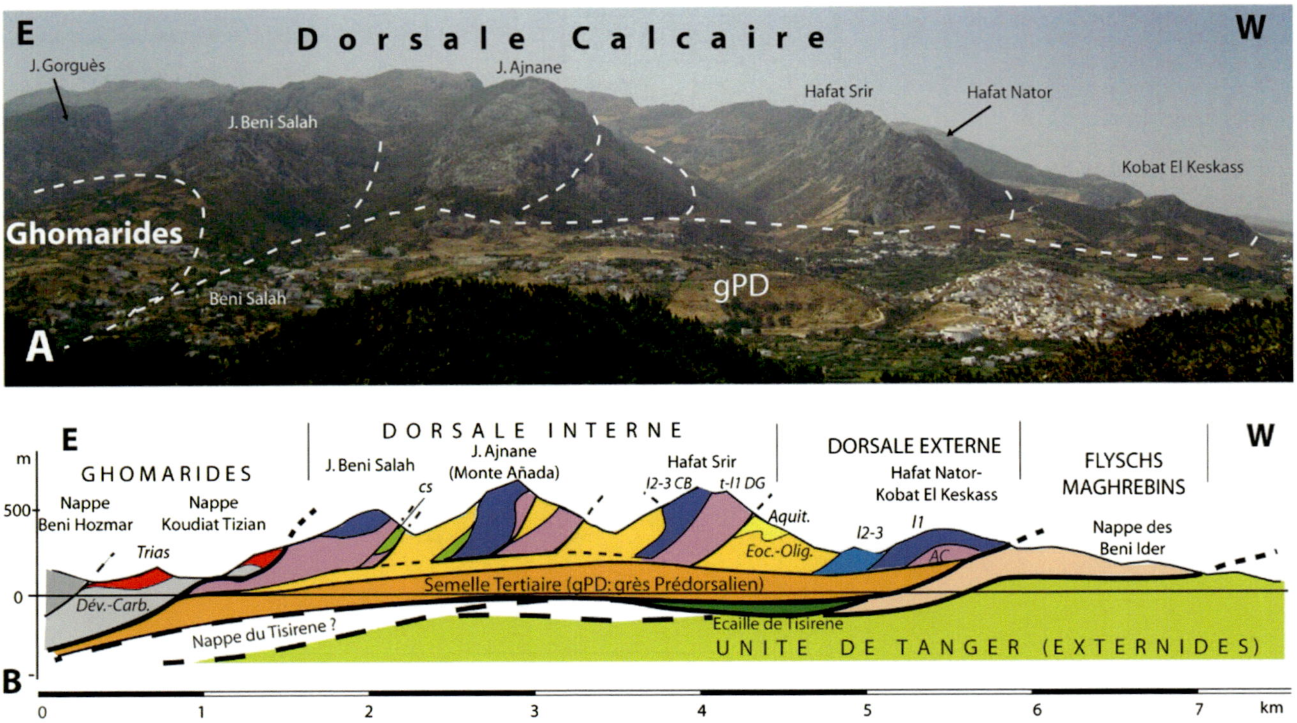

Fig. 2.12 Structural setting of the *Dorsale Calcaire* in the Tetouan transect (see Fig. 2.3A for location), after Chalouan et al. (2011). **A** Panoramic view from Tetouan city center. **B** Geological interpretation after the geological map, scale 1:50,000, by Kornprobst and Durand-Delga (1985).- Internal Dorsale: t-l1 DG, *Dolomie grise*, Upper Triassic-Hettangian; l2-3 CB: *Calcaires blancs*, Hettangian-Sinemurian (Pliensbachian p.p. ?); cs: Upper Cretaceous.- External Dorsale: AC: *Alternances calcaréo-dolomitiques*, Rhaetian; l1: *Calcaires massifs*, Hettangian; l2-3: *Calcaires à silex*, Hettangian-Pliensbachian

In the southern margin domain, the emplacement of the Central Mesorif gabbros is coeval with the Early Liassic rifting, and their exhumation would correspond to the Middle Jurassic phase of extension, as discussed above (Sects. 2.3.2, 2.3.4). The latter is responsible for the earliest exhumation of the serpentinized mantle rocks of the Western Alpine-Ligurian ophiolites (Bill et al. 2001; Balestro et al. 2019). In the Rif transect, the only record of serpentinized peridotite is the Beni Malek massif (Sect. 2.3.1). However, south of the Bokoya massif, E-MORB basalts associated with Middle-Upper Jurassic radiolarites occur as tectonic slices in the Lower Cretaceous Mauretanian Flysch nappe, and form an olistolith in the Upper Cretaceous Massylian nappe (Fig. 2.4; Durand-Delga et al. 2000). Although restricted, these basalts and the Beni Malek OCT slivers support the idea that the Maghrebian Flyschs accumulated upon an oceanic crust. Only the northern part of the Mauretanian Tisiren Flysch could have had a substrate comparable to the Predorsalian units. This is suggested by the occurrence of the Ouareg unit, pinched between the Tisiren and Chouamat Flysch nappes (Olivier et al. 1996; Chalouan et al., 2011).

2.5 The Beni Bousera Incipient Paleomargin

In this section, we consider the puzzling occurrence of marbles linked to the Beni Bousera Unit of the Lower Sebtides, and the new interpretation (Michard et al 2020b; Farah et al. 2021) of these marbles as recording the incipient formation of an oceanic margin during the Triassic-Early Jurassic (?). The location of this "Beni Bousera paleomargin" (or, enlarging to the Betics, "Alpujarrides-Sebtides paleomargin") with respect to the two branches of the westernmost Tethys, i.e., the Maghrebian and Betic branches (Fig. 2.11), is discussed in the next section (Sect. 2.6).

2.5.1 The Beni Bousera Marbles

As reported above (Sect. 2.2.4), the Alboran Domain of the Rif exposes beneath the poorly metamorphic Ghomarides-Malaguides nappes another complex of nappes (Fig. 2.3A), i.e., the Sebtides-Alpujarrides nappes strongly affected by polyphase Alpine recrystallizations, except the

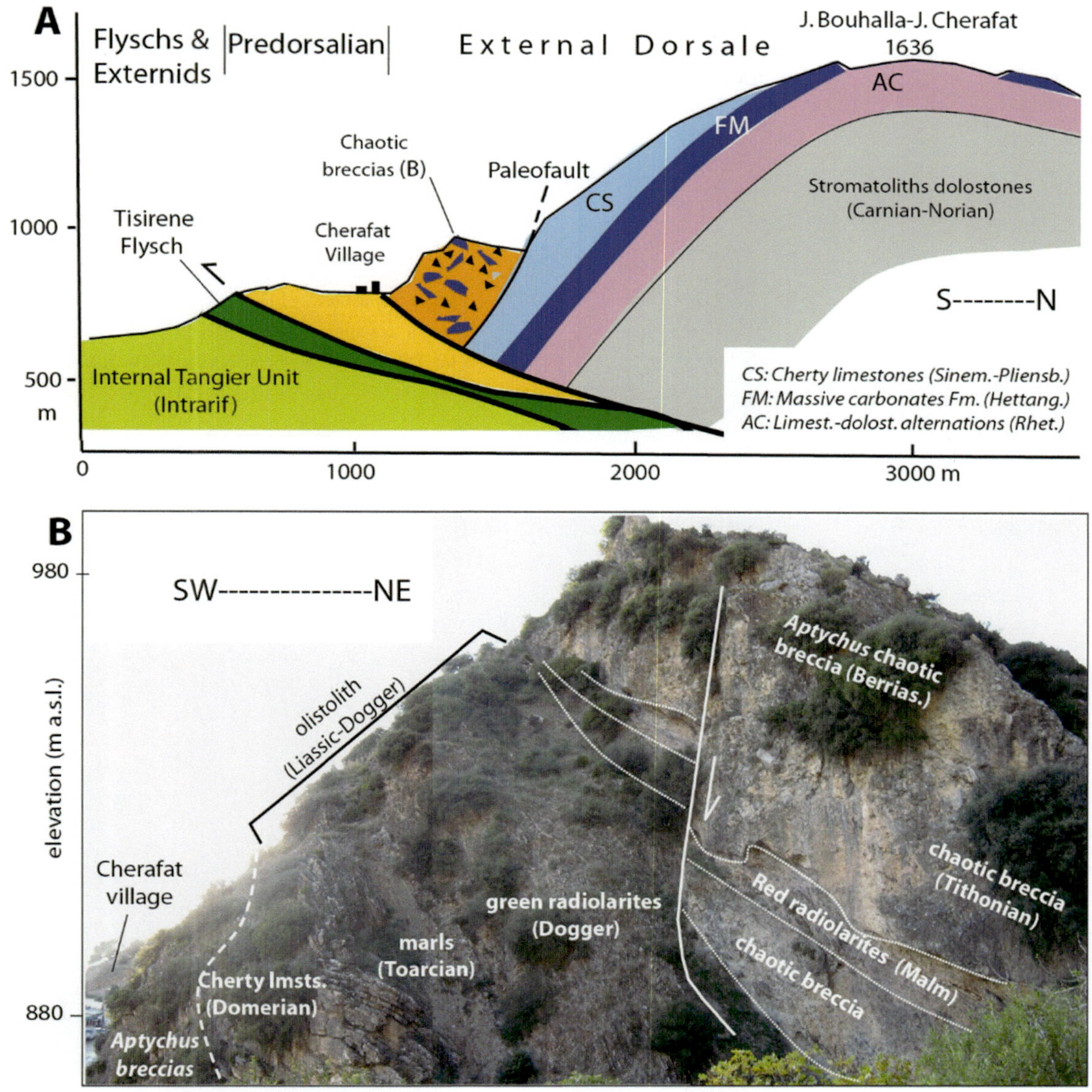

Fig. 2.13 The Cherafat chaotic breccias, an example of Tithonian-Berriasian syn-extensional sedimentation in the External Dorsale domain, after Chalouan et al. (2011)

uppermost units (upper Federico units). A significant particularity of the Sebtides is the presence of subcontinental peridotites homologous of the Ronda peridotites in the Alpujarrides and similarly sandwiched between crustal units. They constitute the core of the Beni Bousera antiform (Fig. 2.3A, C). The serpentinites slices exposed above the Monte Hacho gneisses of Ceuta (Fig. 2.3A; Homonnay et al. 2018) represent a marker between the Beni Bousera and Ronda massifs.

The ultramafics of the Beni Bousera massif (spinel and garnet peridotites, pyroxenites harzburgites and dunites, serpentinites) are draped by metapelitic granulites described as kinzigites (Kornprobst 1974). They are locally associated with basic granulites (Bouybaouene et al. 1998) and marbles (Kornprobst 1974; Saddiqi 1988). In his pioneering work, Kornprobst (1974) briefly described the mineralogy of the marbles (occurrence of diopside, scapolite, wollastonite, K-feldspar), but he did not map them, and concluded that these marbles were carbonate intercalations in the sedimentary protolith of the kinzigites. This point of view was adopted subsequently by all of the geologists who studied the Beni Bousera massif, included by us. However, it becomes clear now that the marbles are not intercalated in the kinzigites, but pinched within the contact between these rocks and the overlying gneisses of the Filali Unit (Fig. 2.14). As the kinzigites belong to the Beni Bousera Unit (Kornprobst 1974), this marble-bearing contact was labeled "Filali-Beni Bousera Shear Zone" (FBBSZ; Farah et al. 2021). In the Oued Amter valley (Figs. 2.16 and 2.17A, C), kinematic indicators linked to the FBBSZ suggest a

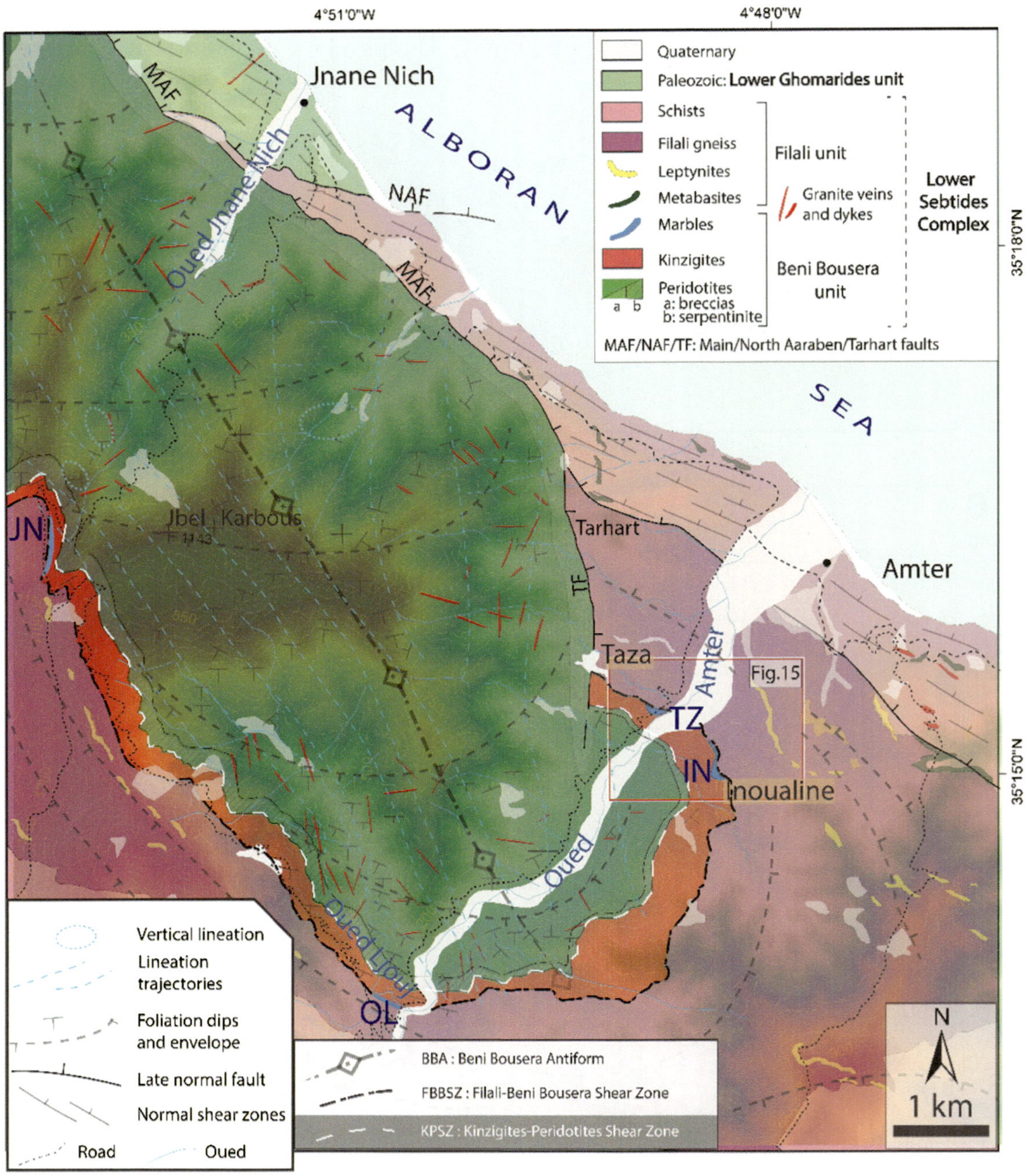

Fig. 2.14 Geologic map of the southeastern part of the Beni Bousera massif, after the Geological map of Morocco, scale 1:50,000, sheets Bou Ahmed and Bab Berred (mapping by J. Kornprobst; see Kornprobst 1974), with additions from Reuber et al. (1982), Elbaghdadi et al. (1996), Afiri et al. (2011), El Bakili et al. (2020). The marble outcrops of Taza (TZ), Inoualine (IN), Oued Ljouj (OL) and Jnane Nich (JN) underline the Filali-Beni Bousera Shear Zone (FBBSZ)

top-to-the WNW displacement in high-temperature mylonitic conditions.

Not only do the Beni Bousera marbles mark the FBBZS, but they reveal the occurrence of a stratigraphic sequence involving distinct sedimentary facies, now dismembered within the shear zone (Fig. 2.15).

The basal beds of the sequence display clastic marbles with gneissic pebbles; they overlie unconformably the kinzigite envelope of the peridotites (Fig. 2.16A). Upward in the sequence, meta-dolomitic limestones and meta-dolostones occur in the form of silicate marbles with forsterite, Mg–Al-spinel, phlogopite and geikielite ($MgTiO_3$). The sequence includes marbles with thin, boudinaged and folded silicate layers (Fig. 2.16B, C) mainly consisting of diopside, epidote, scapolite and wollastonite. The protoliths of these marbles can be interpreted as impure

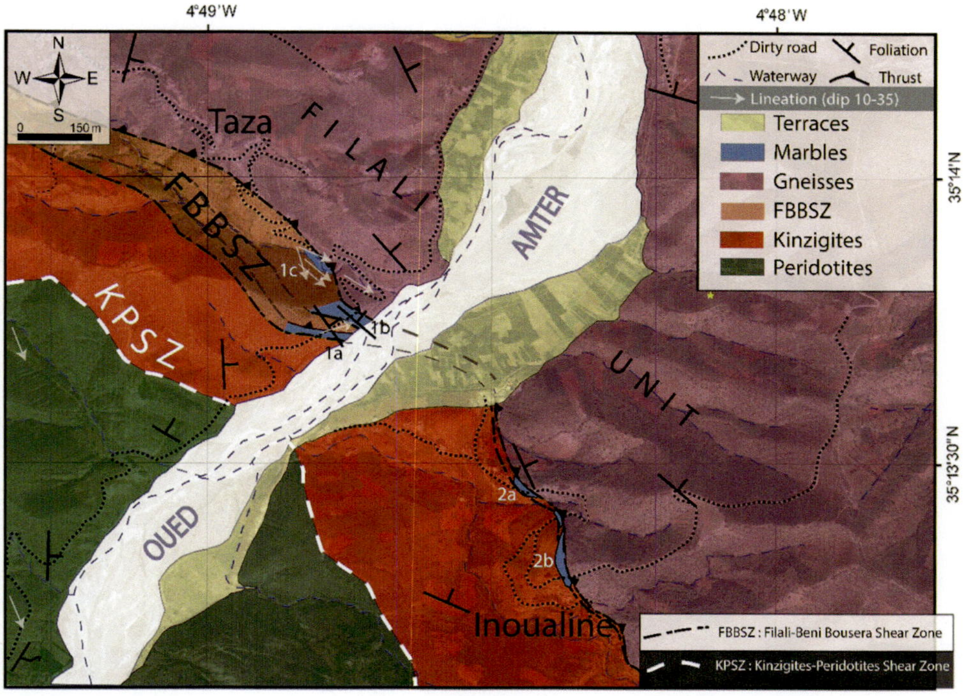

Fig. 2.15 Map of the marble outcrops, ~7 km SW of Amter-village (see Fig. 2.14 for location), after Farah et al. (2021), simplified. 1a-1c: Taza outcrops. 2a-2b: Inoualine outcrops. The Filali-Beni Bousera Shear Zone (FBBSZ) has not been mapped west of 1a-1c

limestones with sandy-arkosic layers. Another facies of thinly bedded alternations of meta-dolomitic limestones and biotite-vermiculite (altered phlogopite) schists (Fig. 2.16D) occurs toward the top of the dismembered sequence.

There is no hope of discovering any fossil in these marbles that recrystallized at 700–750 °C, ≤4.5 kbar (Farah et al. 2021). However, the lithofacies of the protoliths, and particularly the importance of dolostones and dolomitic limestones associated with pure or sandy limestones, are evocative of the Alpine-type Middle to Upper Triassic series of the Alpujarrides-Sebtides crustal units (Fallot et al. 1954; Sanz de Galdeano et al. 1999; Chalouan et al. 2008; Booth-Rea et al. 2005; Martin-Rojas et al. 2009, 2012). Moreover, SHRIMP U-Th-Pb dating of the zircon grains from the marbles allowed Farah et al. (2021) to confirm the likeliness of a Triassic age: first, the scatter of dates from their cores from ~270 Ma to ~3000 Ma points to a detrital origin for these grains, and second, the youngest age cluster of the cores of the grains at ~270 Ma calls for a late or post- Permian age of the original carbonate deposits.

2.5.2 Geodynamic Significance of the Beni Bousera Marbles

What could mean the occurrence of these likely Triassic carbonates onto the granulitic envelope of the Beni Bousera peridotites? To answer this question, it is first necessary to withdraw the dramatic effects of Alpine orogeny. The marbles are affected by a pervasive mylonitic deformation associated with sheath folds and asymmetric folds pointing to a NW transport direction. This deformation and the coeval HT-LP metamorphism are similar to those observed in the Filali Unit. Therefore, we can assume that the northwestward thrusting of the Filali Unit (probably overloaded by some other crustal units) onto the Beni Bousera Unit was at the origin of the recrystallization of the Triassic (?) carbonate series. As for the occurrence of these Triassic (?) carbonates on top of the Beni Bousera kinzigites before the Alpine orogeny, it might be explained in two ways, either by the transgression of the Triassic sea or by the tectonic emplacement of extensional blocks or rafts involving Triassic sequences (Fig. 2.17).

The "transgression hypothesis" envisions the transgression of the Triassic sea onto the exhumed deep crustal rocks in relation to the early rifting of the eastern Iberian plate (Fig. 2.17A). In the more complex "raft hypothesis," the Triassic shallow water sequence was deposited onto the mid-crustal units of the proximal margin before being carried as continental allochthons (rafts) onto the kinzigites of the distal margin by low-angle normal faulting during the maturation of the magma-poor margin (Fig. 2.17B). This hypothesis is inspired by the examples of the Adriatic paleomargin (e.g., Manatschal 2004; Epin et al. 2017) and North Pyrenean trough (Jammes et al. 2009; Lagabrielle et al. 2019a, b). The hypothetic rafts could have been emplaced at any time after the Triassic, during the increasing extension of the margin, i.e., probably during the Late Jurassic-Early Cretaceous as those of the Adriatic margin.

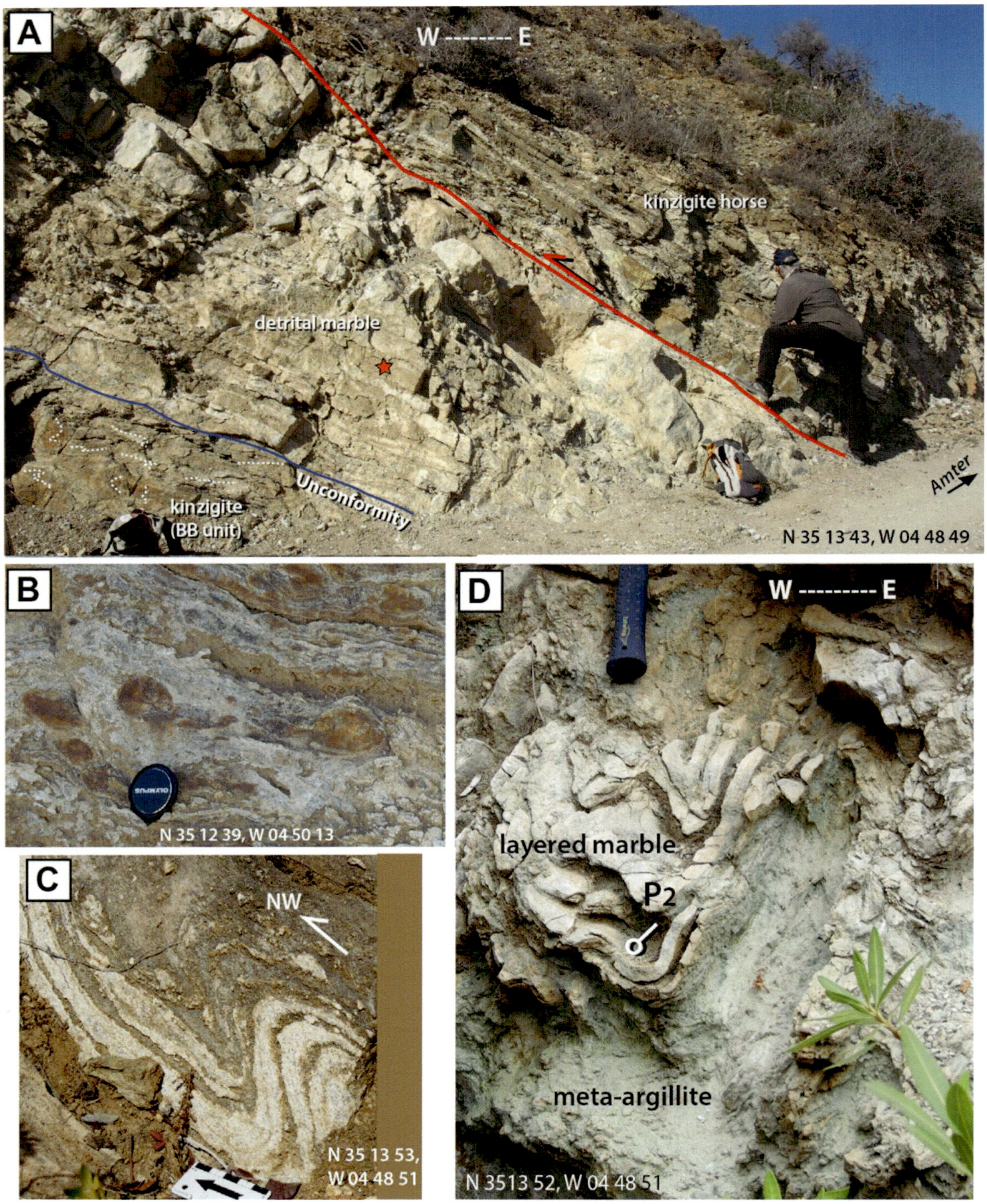

Fig. 2.16 Typical aspects of the Beni Bousera marbles, after Farah et al. (2021), modified. **A** Basal clastic marbles overlying unconformably the main kinzigite sequence and overlain by a WNW-verging kinzigite horse. Location: 1a, Fig. 2.15. **B** Close view of the banded siliceous marbles of the upper Jnane Nich outcrops (JN, Fig. 2.14). The former sandy layers are boudinaged within the calc-mylonite. **C** Boudinaged and folded siliceous layers in the mylonitic zone on top of outcrops 1c, Fig. 2.14. **D** Thinly bedded marble - meta-argillite alternations affected by superimposed folds; the P2 folds here exposed deform a foliation associated with isoclinal P1 folds observed at a short distance in the same outcrop 1c, Fig. 2.14

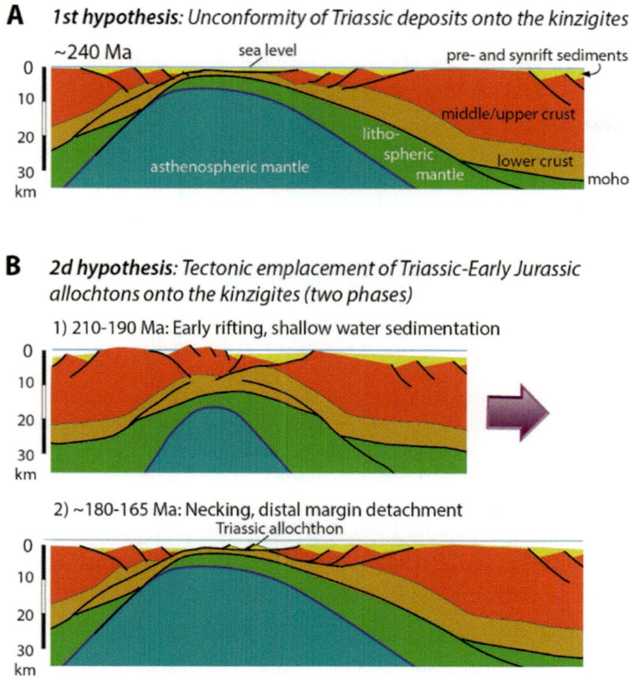

Fig. 2.17 Alternative hypotheses accounting for the occurrence of the Triassic (?) marbles in between the Beni Bousera and Filali units of the Alboran Domain (after Farah et al. 2021). **A** The marbles were deposited on the kinzigite envelope of the peridotites. **B** The marbles have been carried there as rafts by extensional faulting

The choice between the transgression *versus* raft hypotheses is not easy, as the field data are scarce. In particular, the unconformity (either stratigraphic or tectonic) of the marbles onto the kinzigites is only visible for a few meters in length. Anyway, these marbles document an early uplift of the mantle beneath the Alboran domain, consistent with the development of the paleomargin at the northern border of the Maghrebian Tethys. Comparisons with the marble slices associated with the Ronda peridotites are not developed here as their interpretation is presently controversial (Sánchez-Gómez et al. 1995; Sanz de Galdeano and Ruiz Cruz 2016).

2.6 Discussion

2.6.1 Paleogeography of the Alboran Domain

There is a consensus to consider the Alboran Domain as the southwesternmost part of the Alkapeca continental block (Fig. 2.2), but the relative position of the Malaguides-Ghomarides and Alpujarrides-Sebtides inside the Alboran Domain has been under debate for years (Wildi 1983; Chalouan and Michard 2004; Guerrera et al. 2020). In the Betics, it has been suggested that the Malaguides are simply the stratigraphic upward continuation of the Alpujarrides (Tubía et al. 1993). However, this hypothesis is precluded as each complex owns its Triassic cover sequence, transitional from one to the other (Sanz de Galdeanoa et al. 2001). The Malaguides-Ghomarides display thick Middle Triassic conglomeratic red beds ("Verrucano"-type facies; Baudelot et al. 1984; Perrone et al. 2006) and restricted Middle-Upper Triassic dolostones (El Kadiri et al. 1992) whereas the Alpujarrides-Sebtides show a more developed Middle-Upper Triassic carbonate series ("Alpine-type" facies; Fallot et al. 1954; Martin-Rojas et al. 2009, 2012). Hence, the Malaguides-Ghomarides and Alpujarrides-Sebtides domains were clearly located side by side during the Triassic, and likely formed a complex Alboran domain in between the two branches of the Alpine Tethys (Fig. 2.11). By the end of the Variscan orogeny (Bashkirian), the Malaguides and Alpujarrides terranes would have been both located along the southeastern boundary of Iberia from the NE to the SW, respectively, with the Nevado-Filabrides extending between Iberia and the Alpujarrides (Jabaloy-Sánchez et al. 2021). However, Poulaki and Stockli (2022) suggest that the Eastern and Western Alpujarrides were located differently during the Jurassic, i.e., to the north and southwest of the Nevado-Filabrides, respectively, and separated by major transform faults.

In the Rif belt, many authors argued that the post-Triassic series of the Ghomarides (i.e., Jurassic shallow water, then pelagic limestones; Cretaceous-Middle Eocene limestones and Eocene-Early Miocene clastics; El Kadiri et al. 2006; Serrano et al. 2007; Hlila et al. 2008) are transitional to those of the Dorsale units, and then they assumed that the Dorsale units are detached from the external border of the Ghomarides (e.g., Didon et al. 1973; Sanz de Galdeanoa et al. 2001; Durand-Delga 2006; El Kadiri et al. 2009; Guerrera et al. 2020). Therefore, the Sebtides would have been more internal inside Alkapeca than the Ghomarides, and the Alpujarrides-Sebtides domain could have been located to the N or NE of the Ghomarides, possibly at the western border of the Betic branch of the Tethys (Fig. 2.2B).

On the contrary, Trümpy (1973), Wildi et al. (1977), and Nold et al. (1981) observed that the stratigraphic series of the Sebtides ends with Middle Triassic carbonates whereas that of the Dorsale begins with Upper Triassic dolostones and limestones, suggesting that the Sebtides would represent the former basement of the detached Dorsale units. In the Betics, Bourgois (1980) assumed that the Dorsale-type, but metamorphic Nieves Unit is linked to the Alpujarrides. Bessière (2019) interpreted the basal contact of the Dorsale-type Nieves sequence over the Ronda peridotites as a major extensional contact responsible for the high-temperature, low-pressure metamorphism of the Nieves lowest formations. If correct, the Alpujarrides-Sebtides would have been located beneath the Dorsale sedimentary formations, at the northern border of the Maghrebian Tethys (Fig. 2.18). This

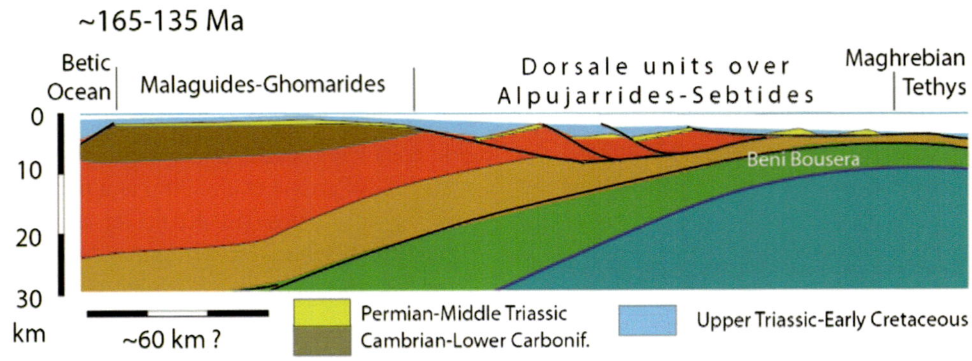

Fig. 2.18 Tentative restoration of the hyperextended northern margin of the Maghrebian Tethys on the transect of the Alboran Domain during the Late Jurassic-Early Cretaceous. (see location on Fig. 2.11, northern part of cross-section B). Crust and mantle signatures as Fig. 2.17. The Triassic (?) bodies on top of the Beni Bousera granulites are shown according to the raft hypothesis (Fig. 2.17B)

is consistent with the moderate thickness of the crustal rocks that form the Alpujarrides-Sebtides basement, i.e., a few thousand meters of schists, gneiss and granulites affected by high-grade Variscan metamorphism (e.g., Chalouan et al. 2008; Jabaloy-Sánchez et al. 2019a). Such a lithological pile covered by Permian and Triassic series corresponds to a thinned crust formed during the rifting of Pangea (Martin-Rojas et al. 2009). Porkoláb et al. (2022) also consider the NW-ward subduction of the Alpujarrides marginal complex beneath the more continental Malaguides complex. In such a framework, the occurrence of Triassic unconformable deposits or rafts upon the Beni Bousera granulites, as proposed above (Fig. 2.17), is quite likely.

2.6.2 Up-Down-Up Movements of the Subcontinental Peridotites

The new interpretation of the Beni Bousera marbles (Sect. 2.5.2 and Farah et al. 2021) impacts the old issue of the timing of the exhumation of the Gibraltar Arc peridotites, repeatedly debated for half a century. From Loomis (1972) to Platt et al. (2003), a Cenozoic exhumation of the ("hot") mantle rocks has been the favored concept for long. The idea that exhumation began as early as the Jurassic was first advocated based on West Alpine comparisons (Michard et al. 1991, 1997), then based on U–Pb zircon dates from the Ronda garnet pyroxenites (Sánchez-Rodríguez and Gebauer, 2000). Accordingly, Michard et al. (2002) and Chalouan and Michard (2004) proposed that the Beni Bousera lithospheric mantle has been exhumed to crustal level during the Tethyan rifting. The role of the Permian-Mesozoic extension was progressively accepted, but regarded as subordinate with respect to the Oligocene–Miocene tectonic events (Garrido et al. 2011; Afiri et al. 2011; Tubia et al. 2012; Précigout et al. 2013; Álvarez-Valero et al. 2014; Hidas et al. 2015; Gueydan et al. 2015; Gervilla et al. 2019) or even remained uncited (Williams and Platt 2017). In contrast, Rossetti et al. (2010, 2013, 2020) have given increasing interest to the Variscan petrological-tectonic events that affected the Sebtides granulites and gneisses; they show that high-pressure partial melting of these rocks records the Late Carboniferous-Permian (300–290 Ma) collapse of the thermally weakened Hercynian orogen. The exhumation of the peridotites documented by the Beni Bousera marbles (Sect. 2.5.2) would have been achieved after the early Permian collapse of the Hercynian orogen, i.e., during the late Permian–Triassic Pangea breakup, due to increasing extension and opening of the Ligurian-Maghrebian Ocean. Consistently, El Bakili et al. (2021) dated a Triassic metamorphic event in the Lower Federico metagreywackes on top of the Beni Bousera unit and ascribed this event to the early exhumation of the Sebtides-Alpujarrides mantle. It is worth noting that the crucial importance of the late Permian–Triassic rifting in the Western Tethys-Atlantic borders has been evidenced by compiling existing seismic profiles and geological constraints along the North Atlantic margins and inside Iberia (Angrand et al. 2020).

Obviously, an Early Miocene exhumation of the Gibraltar Arc mantle rocks also occurred, well-documented by the petrological P–T-t data for the ultramafics and overlying crustal rocks (e.g., Montel et al. 2000; Platt et al. 2003; Massonne 2014; Rossetti et al. 2010, 2013; Homonnay et al. 2018; El Bakili et al. 2020, 2021; Marrone et al. 2021). However, this recent exhumation occurred after the Late Eocene–Oligocene subduction of the Maghrebian lithosphere and part of the Alpujarrides-Sebtides and Nevado-Filabrides units, which determined the formation of an orogenic wedge. In the hypothesis here favored, this event of crustal thickening was linked to the thrusting of the Malaguides-Ghomarides onto the Alpujarrides-Sebtides (see also Porkoláb et al. 2022). This compares with the Austroalpine setting with the Higher Austroalpine nappes (hinterland) thrust onto the Middle and Lower Austroalpine nappes derived from the hyperextended paleomargin (e.g., Mohn et al. 2011). Thus, from the Permian–Triassic to the Neogene, the subcontinental peridotites of the Gibraltar Arc were successively exhumed, buried, and again exhumed. Such an evolution is also well-illustrated in the Geisspfad peridotites of the Central Alps, included similarly in a former segment of the European margin (Pastorelli et al. 1995; Pelletier et al. 2008).

The stacked nappes of the Alboran Domain have been dramatically thinned both ductilely and through low-angle extensional faulting (García-Dueñas et al. 1992; Lonergan 1993; Lonergan and Platt 1995; Azañon and Crespo-Blanc 2000; Booth-Rea et al. 2004, 2005; Gueydan et al. 2015). Part of the overburden of the HP-LT units would have been removed in an active accretionary wedge setting by syn-orogenic extension (Avigad et al. 1997) prior to the Burdigalian, whose deposits locally contain detrital carpholite, and glaucophane (Lonergan and Mange-Rajetsky 1994). Another part of the exhumation could be linked to extrusion in the subduction channel, as proposed by Porkoláb et al. (2022) for the Nevado-Filabrides. The final exhumation (re-exhumation) of the mantle rocks occurred during the Early Miocene due to high-temperature back-arc extension of the orogenic wedge (e.g., Garrido et al. 2011; Gueydan et al. 2015, 2019; Williams and Platt 2017). This Oligo-Miocene evolution was accompanied by important rotations of the peridotite bodies and juxtaposed units, anticlockwise in the Beni Bousera and clockwise in the Ronda massif, respectively (e.g., Feinberg et al. 1996; Berndt et al. 2015; Crespo-Blanc et al. 2016). Further discussion of the Alpine events that affected the Gibraltar Arc would be beyond the scope of this paper (see Jabaloy-Sanchez et al. 2019b; Rossetti et al. 2020; El Bakili et al. 2020).

2.7 Conclusion

The Gibraltar Arc formed at the expense of the area of a triple junction between the Central Atlantic and the two branches of the westernmost Alpine Tethys, i.e., the Ligurian-Maghrebian and the Betic or West Ligurian branches. Continental blocks detached from the southeastern margin of Iberia formed the Alboran Domain in between the two branches. The closure of this Central Atlantic-Alpine Tethys bypass resulted in the building of the Betic Cordilleras and Rif belt of Spain and Morocco, respectively. The Rif geology described above offers the opportunity to recognize the remnants of two passive margins seemingly conjugated on both sides of the Ligurian-Maghrebian Tethys. These remnants are extremely disrupted and their interpretation certainly needs additional studies. However, the available data presented above lead to the following conclusions.

(1) Southern margin

Tectonic slivers sampled at the southern margin of the Maghrebian Tethys occur in the External Rif accretionary prism. They exposes from NE to SW serpentinites from the Late Jurassic OCT and gabbro bodies intruded during the Early Jurassic in the distal African margin or the OCT. Altogether, these slivers define a transported Mesorif Suture Zone. The westernmost gabbros would have intruded the SDRs of the Late Triassic Volcanic Passive Margin whereas the gabbros located more to the NE could have been emplaced in the OCT. They would have been exhumed by core-complex mechanism during increasing extension of the passive margin, which evolved toward the magma-poor transform type during the Middle-Late Jurassic. By the Late Jurassic-Early Cretaceous, the exhumed gabbros and the metamorphic relics of their envelope were overlain by carbonate rafts sliding from the uplifted, proximal margin above Upper Triassic evaporitic red beds. The faulted boundary between the uplifted and downwarped parts of the margin is marked by a Basalts-Limestone Breccias lineament.

(2) Northern margin

The sedimentary sequences derived from the northern passive margin of the Maghrebian Tethys have been recognized for long in the Dorsale Calcaire and Predorsalian units located at the external boundary of the Internal Zones (Alboran Domain). The available structural and stratigraphic data suggest that these units have been detached from the thinned Alpujarrides-Sebtides crustal domain to the south of the Ghomarides-Malaguides continental hinterland. In the Lower Sebtides, a probably Triassic sequence of marbles on top of the granulitic envelope of the Beni Bousera peridotites could derive either from the transgression of the Triassic sea onto the exhumed granulites or from the tectonic transport of Triassic rafts over low-angle normal faults during the Jurassic evolution of the magma-poor margin. The HT-MP/LP metamorphism of these marbles can be referred to as their Cenozoic burial beneath other crustal units of the tectonic prism. Thus, the Beni Bousera peridotites and their Ronda equivalent were almost exhumed up to the surface in the frame of the early dislocation of Pangea.

Acknowledgements We warmly thank Mohamed Ouazzani-Touhami for his providential help at Tetouan. We acknowledge the crucial help of Christian Chopin, coauthor of our "twin-paper" (Farah et al. 2021), for the petrological study of the Beni Bousera marbles. We thank Michel Corsini, coauthor of the same paper, for his reading and pertinent criticism of the first draft of this chapter. Thanks are due to François Roure and Samy Khomsi for their editorial remarks and support. We are greatly indebted to Guillermo Booth-Rea for his accurate, constructive, and helpful review of the submitted manuscript, and to the anonymous Reviewer who encouraged us to develop a general discussion section.

References

Abbassi A, Cipollari P, Fellin M.G., Zaghloul M.N., Guillong M, El Mourabet M, Cosentino D (2022) The Numidian sand event in the Burdigalian foreland basin system of the Rif, Morocco, in a source-to-sink perspective. Geol Soc Am Bull.https://doi.org/10.1130/B36136.1

Afiri A, Gueydan F, Pitra P, Essaifi A, Précigout J (2011) Oligo-Miocene exhumation of the Beni-Bousera peridotite through a lithosphere-scale extensional shear zone. Geodin Acta 24:49–60. https://doi.org/10.3166/ga.24.49-60

Agharroud K (2022) Tectonique active du front sud de la chaîne rifaine (tronçon Fès-Meknès, Nord du Maroc): Approche morphochronologique. PhD thesis, Abdelmalak Essaadi University, Tetuan, 233 pp. Agharroud%2022%20Thèse%20Rif%20southern%20front.pdf

Álvarez-Valero AM, Jagoutz O, Stanley J, Manthei C, El Maz A, Moukadiri A, Piasecki A (2014) Crustal attenuation as a tracer for the emplacement of the Beni Bousera ultramafic massif (Betico-Rifean belt). Geol Soc Am Bull 126 :1614–1624

Andrieux J, Fontboté JM, Mattauer M (1971) Sur un modèle explicatif de l'Arc de Gibraltar. Earth Planet Sci Lett 12:191–198

Andrieux J (1971) La structure du Rif central : étude des relations entre la tectonique de compression et les nappes de glissement dans un tronçon de la chaîne alpine. Notes et Mémoires du Service Géologique, 235

Angrand P, Mouthereau F, Masini E, Asti R (2020) A reconstruction of Iberia accounting for W-Tethys/N-Atlantic kinematics since the late Permian-Triassic. Solid Earth Discuss 1–24https://doi.org/10.5194/se-2020-24

Argand E (1916) Sur l'arc des Alpes Occidentales. Eclogae Geol Helv 14:145–191

Argand E (1924) La tectonique de l'Asie. Congr Géol Int 13e s. (Belgique 1922) 1/5:171–372

Asti R, Lagabrielle Y, Fourcade S, Corre B, Monié P (2019) How do continents deform during mantle exhumation? Insights from the northern Iberia inverted paleo-passive margin, western Pyrenees (France). Tectonics 38:1666–1693. https://doi.org/10.1029/2018TC005428

Avigad D, Garfunkel Z, Jolivet L, Azañon JM (1997) Back arc extension and denudation of Mediterranean eclogites. Tectonics 16:924–941. https://doi.org/10.1029/97TC02003

Azañon JM, Crespo-Blanc A (2000) Exhumation during a continental collision inferred from the tectonometamorphic evolution of the Alpujarride complex in the cetral betics (Alboran Domain, SE Spain). Tectonics 19:549–565. https://doi.org/10.1029/2000TC900005

Azdimousa A, Jabaloy-Sánchez A, Talavera C, Asebriy L, González-Lodeiro F, Evans NJ (2019a) Detrital zircon U-Pb ages in the Rif Belt (northern Morocco): paleogeographic implications. Gondwana Res 70:133–150. https://doi.org/10.1016/j.gr.2018.12.008

Azdimousa A, Jabaloy-Sánchez A, Münch P, Martínez-Martínez JM, Booth-Rea G, Vázquez-Vílchez M, Asebriy L, Bourgois J, González-Lodeiro F (2019b) Structure and exhumation of the Cap des Trois Fourches basement rocks (eastern Rif, Morocco). J Afr Earth Sc 150:657–672. https://doi.org/10.1016/j.jafrearsci.2018.09.018

Balestro G, Festa, A, Dilek Y (2019) Structural architecture of the Western Alpine ophiolites, and the Jurassic seafloor spreading tectonics of the Alpine Tethys. J Geol Soc London. https://doi.org/10.1144/jgs2018-099

Bargach K, Ruano P, Chabli A, Galindo-ZaldÍvar J, Chalouan A, Jabaloy A, Akil M, Ahmamou M, Sanz de Galdeano C, Benmakhlouf M (2004) Recent tectonic deformations and stresses in the frontal part of the Rif Cordillera and the Saïss Basin (Fes and Rabat regions, Morocco). Pure Appl Geophys 161(3):521–540. https://doi.org/10.1007/s00024-003-2461-6

Baudelot S, Bouhdadi S, Durand-Delga M (1984) Datation palynologique du Trias moyen au sein des grès rouges "permo-triasiques" des environs de Tétouan (Rif septentrional, Maroc). C R Acad Sci Paris 299:1061–1068

Ben Yaïch A, Hervouët Y, Duée G, El Hatimi N (1989) Age jurassique et réinterprétation des roches basiques au Nord de Ouezzane (Rif externe, Maroc): significations géodynamique. C R Acad Sci Paris 309, sér. II:1197–1202

Benzaggagh M (2011) Activités volcaniques sous-marines à la limite Jurassue-Crétacé dans le Rif externe (Maroc). Age et relation avec la sédimentation et la paléogéographie du sillon rifain externe. C R Geosci 343:302–311. https://doi.org/10.1016/j.crte.2010.12.009

Benzaggagh M, Habibi M (2006) Les séries carbonatées du Jurassique supérieur et les niveaux de passage au Crétacé inférieur dans la partie occidentale du Rif externe (Prérif interne et Mésorif, Maroc). Stratigraphie, paléogéographie et évolution tectonosédimentaire. Newsl Stratigr 42:115–141. https://doi.org/10.1127/0078-0421/2006/0042-0115

Benzaggagh M, Mokhtari A, Rossi P, Michard A, El Maz A, Chalouan A, Saddiqi O, Rjimati E (2014) Oceanic units in the core of the External Rif (Morocco): intramargin hiatus or South-Tethyan remnants? J Geodyn 77:4–21. https://doi.org/10.1016/j.jog.2013.10.003

Berndt T, Ruíz-Martínez VC, Chalouan A (2015) New constraints on the evolution of the Gibraltar Arc from palaeomagnetic data of the Ceuta and Beni Bousera peridotites (Rif, northern Africa). J Geodyn 84:19–39. https://doi.org/10.1016/j.jog.2014.09.014

Bessière E (2019) Évolution géodynamique des Zones Internes des Cordillères Bétiques (Andalousie, Espagne): apports d'une étude pluridisciplinaire du Complexe Alpujárride. Sci. Terre. Univ. Orléans. https://tel.archives-ouvertes.fr/tel-02392008

Bessière E, Scaillet S, Augier R, Jolivet L, Azanon JM, Booth-Rea G et al (2022) 40Ar/39Ar age constraints on *HP*/*LT* metamorphism in extensively overprinted units: the example of the Alpujarride subduction complex (Betic Cordillera, Spain). Tectonics 41: e2021TC006889. https://doi.org/10.1029/2021TC006889

Bill M, O'Dogherty L, Guex J, Baumgartner PO, Masson H (2001) Radiolarite ages in Alpine-Mediterranean ophiolites: constraints on the oceanic spreading and the Tethys-Atlantic connection. Bull Geol Soc Am 113:129–143. https://doi.org/10.1130/0016-7606(2001)113%3c0129:RAIAMO%3e2.0.CO;2

Booth-Rea G, Ranero CR, Grevemeyer I, Martínez-Martínez JM (2007) Crustal types and tertiary tectonic evolution of the Alborán sea, western Mediterranean. Geochem Geophys Geosyst 8:1–25. https://doi.org/10.1029/2007GC001639

Booth-Rea G, Azañon JM, Martinez-Martinez JM, Vidal O, V G-D (2005) Contrasting structural and P-T evolution of tectonic units in the southeastern Betics: key for understanding the exhumation of the Alboran Domain HP/LT crustal rocks (western Mediterranean). Tectonics 24.https://doi.org/10.1029/2004TC001640

Booth-Rea G, Azañon JM, García-Dueñas V (2004) Extensional tectonics in the northeastern Betics (SE Spain): case study of extension in a multilayered upper crust with contrasting rheologies. J Struct Geol 26:2039–2058. https://doi.org/10.1016/j.jsg.2004.04.005

Booth-Rea G, Jabaloy-Sánchez A, Azdimousa A, Asebriy L, Vázquez-Vílchez M, Martínez-Martínez JM (2012) Upper-crustal extension during oblique collision: the Temsamane extensional detachment (eastern Rif, Morocco). Terra Nova 24:505–512. https://doi.org/10.1111/j.1365-3121.2012.01089.x

Bouillin JP, Durand-Delga M, Gélard JP, Leikine M, Raoult JF, Raymond D, Tefiani M, Vila JM (1970) Définition d'un flysch massylien et d'un flysch maurétanien au sein des flyschs allochtones de l'Algérie. C R Acad Sci Paris 270:2249–2262

Bouillin JP, Durand-Delga M, Gélard JP, Leikine M, Raoult JF, Raymond D, Tefiani M, Vila JM (1973) Les olistostromes d'âge miocène inférieur lies aux flyschs allochtones kabyles de l'orogène alpin d'Algérie. Bull Soc Géol Fr 7(15):340–344. https://doi.org/10.2113/gssgfbull.S7-XV.3-4.340

Bouillin JP, Durand-Delga M, Olivier P (1986) Betic-Rifian and Tyrrhenian arcs: distinctive features, genesis and development stages. In: Wezel JF (ed.) The origin of arcs. Elsevier Publishing, Amsterdam, pp 281–304. https://doi.org/10.1016/B978-0-444-42688-8.50017-5

Bouillin J-P, Kornprobst J, Raoult J-F (1977) Données préliminaires sur le complexe volcano-sédimentaire de Rekkada Metletine (ex-Texenna) en Petite Kabylie (Algérie). Bull Soc Géol Fr 7 (19):805–813. https://doi.org/10.2113/gssgfbull.S7-XIX.4.805

Boukaoud EL, Godard G, Chabou MC, Bouftouha Y, Douk S (2021) Petrology and geochemistry of the Texenna ophiolites, northeastern Algeria: implications for the Maghrebian flysch suture zone. Lithos 386–387.https://doi.org/10.1016/j.lithos.2021.106019

Bourgois J (1980) De l'origine ultra-bétique des Malaguides (zones internes bétiques, Espagne). Geol Romana 19:151–170. http://151.100.51.154/Volumi/VOL%2019/GR_19_151_170_Bourgois.pdf

Bouybaouene ML, Goffé B, Michard A (1995) High pressure, low-temperature metamorphism in the Sebtides nappes, northern Rif Morocco. Geogaceta 17:117–119

Bouybaouene ML, Goffé B, Michard A (1998) High-pressure granulites on top of the Beni Bousera peridotites, Rif belt, Morocco: A record of an ancient thickened crust in the Alboran domain. Bull Soc Geol Fr 169:153–162

Capella W, Matenco L, Dmitrieva E, Roest WMJ, Hessels S, Hssain M, Chakor-Alami A, Sierro FJ, Krijgsman W (2017) Thick-skinned tectonics closing the Rifian Corridor. Tectonophysics 710–711:249–265. https://doi.org/10.1016/j.tecto.2016.09.028

Carminati E, Lustrino M, Doglioni C (2012) Geodynamic evolution of the central and western Mediterranean: Tectonics versus igneous petrology constraints. Tectonophysics 579:173–192. https://doi.org/10.1016/j.tecto.2012.01.026

Chalouan A, Michard A, El Kadiri Kh, Negro F, Frizon de Lamotte D, Soto JI, Saddiqi O (2008) The Rif belt. In: Michard A, Saddiqi O, Chalouan A, Frizon de Lamotte D (eds) Continental evolution: the geology of morocco; structure, stratigraphy and tectonics of the Africa–Atlantic–Mediterranean triple junction. Lecture Notes Earth Science, vol 116, pp 203–302

Chalouan A, Michard A, El Kadiri Kh, Saddiqi O (2011) Rif central et nord-occidental. In: Michard et al (eds) Nouveaux Guides géologiques et miniers du Maroc, vol 5. Notes Mém. Serv. Géol. Maroc 560, pp 11–90

Chalouan A, Michard A (2004) The Alpine Rif belt (Morocco): a case of mountain building in a subduction-subduction-transform fault triple junction. Pure Appl Geophys 161:489–519. https://doi.org/10.1007/s00024-003-2460-7

Chalouan A, El Mrihi A, El Kadiri K, Bahmad A, Salhi F, Hlila R (2006) Mauretanian flysch nappe in the northwestern Rif Cordillera (Morocco): deformation chronology and evidence for a complex nappe emplacement. Geol Soc London Spec Publ 262:161–175. https://doi.org/10.1144/GSL.SP.2006.262.01.10

Chalouan A, Michard A (1990) The Ghomarides nappes, Rif coastal ranges, Morocco: a Variscan chip in the Alpine belt. Tectonics 9:1565–1583. https://doi.org/10.1029/TC009i006p01565

Choubert B (1935) Recherche sur la genèse des chaînes paléozoïques et antécambriennes. Rev Géogr Phys Géol Dyn 8:5–50

Civiero C, Custódio S, Duarte JC, Mendes VB, Faccenna C (2020) Dynamics of the Gibraltar arc system: a complex interaction between plate convergence, slab pull, and mantle flow. J Geophys Res: Solid Earth 125:e2019JB018873

Crespo-Blanc A, Comas M, Balanyá JC (2016) Clues for a Tortonian reconstruction of the Gibraltar arc: structural pattern, deformation diachronism and block rotations. Tectonophysics 683:308–324

Damuth JE (1994) Neogene gravity tectonics and depositional processes on the deep Niger Delta continental margin. Mar Pet Geol 11.https://doi.org/10.1016/0264-8172(94)90053-1

Davies JHFL, Marzoli A, Bertrand H, Youbi N, Ernesto M, Schaltegger U (2017) End-Triassic mass extinction started by intrusive CAMP activity. Nat Commun 8:1–8. https://doi.org/10.1038/ncomms15596

De Capoa P, Di Staso A, Perrone V, Zaghloul MN (2007) The age of the foredeep sedimentation in the Betic-Rifian Mauretanian units: a major constraint for the reconstruction of the tectonic evolution of the Gibraltar Arc. Comptes Rendus - Geosci 339:161–170. https://doi.org/10.1016/j.crte.2007.01.003

De Saussure HB (1779) Voyages dans les Alpes, précédés d'un essai sur l'histoire naturelle des environs de Genève. Neuchâtel, Samuel Fauche, t. 1

Didon J, Durand Delga M, Kornprobst J (1973) Homologies géologiques entre les deux rives du détroit de Gibraltar. Bull Soc Géol Fr (7) 15:77–105

Dilek Y, Furnes H (2019) Tethyan ophiolites and Tethyan seaways. J Geol Soc London, 15 pp. https://doi.org/10.1144/jgs2019-129

Durand-Delga M (1964) Essai sur la structure des domaines émergés autour de la Méditerranée occidentale. Geol Rundsch 53:534–535

Durand-Delga M (1980) La Méditerranée occidentale, étapes de sa genèse et problèmes structuraux liés à celle-ci. Mém Soc Géol Fr 10:203–224

Durand-Delga M (2006) Geological adventures and misadventures of the Gibraltar Arc. Zeitschrift der Deutschen Gesellschaft für Geowissenschaften 687–716.https://doi.org/10.1127/1860-1804/2006/0157-0687

Durand-Delga M, Fontboté JM (1980) Le cadre structural de la Méditerranée occidentale. In : 26th International Geological Congress, Paris, Collaboration. C5, Géologie des haînes alpines issues de la Téthys, pp 67–85

Durand-Delga M (1969) Mise au point sur la structure du Nord-Est de la Bérbérie. Publ Serv Géol Algérie 39:89–131

Durand-Delga M, Gardin S, Olivier P (1999) Datation des flyschs éocrétacés maurétaniens des Maghrébides: la formation du jbel Tisirene (Rif, Maroc). Comptes Rendus De L'académie Des Sciences - Serie IIa: Sciences De La Terre Et Des Planètes 328:701–709. https://doi.org/10.1016/S1251-8050(99)80180-3

Durand-Delga M, Olivier P (1988) Evolution of the Alboran block margin from Early Mesozoic to early Miocene time. In: Jacobshagen VH (ed) The Atlas system of Morocco. Lecture Notes Earth Science, vol 15, pp 465–480

Durand-Delga M, Rossi P, Olivier P, Puglisi D (2000) Situation structurale et nature ophiolitique de roches basiques jurassiques associees aux flyschs maghrebins du Rif (Maroc) et de Sicile (Italie). Comptes Rendus De L'academie De Sciences - Serie IIa:

Sciences De La Terre Et Des Planetes 331:29–38. https://doi.org/10.1016/S1251-8050(00)01378-1

El Atrassi F, Brunet F, Bouybaouene M, Chopin C, Chazot G (2011) Melting textures and microdiamonds preserved in graphite pseudomorphs from the Beni Bousera peridotite massif, Morocco. Eur J Miner 23:157–168. https://doi.org/10.1127/0935-1221/2011/0023-2094

El Azzab D, Galdeano A, Feinberg H, Michard A (1997) Subsurface extent of an ultrabasic thrust sheet: aeromagnetic data analysis and 3D modelling of the Beni-Malek peridotites (eastern Rif-Morocco). Bull Soc Geol Fr 168:15–23

El Bakili A, Corsini M, Chalouan A, Münch P, Romagny A, Lardeaux JM, Azdimousa A (2020) Neogene polyphase deformation related to the Alboran Basin evolution : new insights for the Beni Bousera massif (Internal Rif , Morocco). BSGF 0–18https://doi.org/10.1051/bsgf/2020008

El Bakili A, Corsini M, Lardeaux JM, Gallet S, Münch Ph, Chalouan A (2021) Amphibolite facies metamorphic event within the Upper Sebtides tectonic units (Internal Rif, Morocco): a record of a hyperextended margin at the border of the western Tethys. C R Geosci 353:193–208.https://doi.org/10.5802/crgeos.59

El Kadiri K (2000–2002) Jurassic ferruginous hardgrounds of the Dorsale Calcaire and the Jbel Moussa Group (Internal Rif, Morocco): stratigraphic context and paleogeographic consequences of mineralization processes. Geol Romana 36:33–69

El Kadiri Kh, Hlila R, Sanz de Galdeano C, López-Garrido AC, Chalouan A, Serrano F, Bahmad A, Guerra-Merchán A, Liemlahi H (2006) Regional correlations across the Internides-Externides front (northwestern Rif Belt, Morocco) during the Late Cretaceous - Early Burdigalian times: Palaeogeographical and palaeotectonic implications. In: Moratti G, Chalouan A (eds), Tectonics of the western Mediterranean and North Africa. Geological Society Special Publication, vol 262, pp 193–215. https://doi.org/10.1144/GSL.SP.2006.262.01.12

El Kadiri Kh, Horstemeyer M.F, El Kadiri H, Pessagno EAJr (2009) Jurassic radiolarite pulses from the Dorsale Calcaire (internal Rif, Morocco): a clue for correlating and interpreting the Tethyan radiolarites. Stratigraphy 6:277–312

El Kadiri Kh, Linares A, Oloriz F (1989) La Dorsale calcaire interne entre les Accidents de l'Oued Martil et de l'Oued Laou (Rif septentrional, Maroc): évolutions stratigraphique et géodynamique au cours du Jurassique-Crétacé. Commun Serv Geol Portugal 75:39–65

El Kadiri Kh, Linares A, Oloriz F (1992) La Dorsale Calcaire rifaine (Maroc septentrional): évolution stratigraphique et géodynamique durant le Jurassique–Crétacé. Notes Mém Serv Géol Maroc 336:217–265

El Maz A, Guiraud F (2001) Example of low-variance parageneses in the metapelite of Filali (inner Rif, Morocco): Description, interpretation and geodynamic implication. Bull Soc Géol Fr 172:469–485

El Talibi H, Zaghloul MN, Perri F, Aboumaria K, Rossi A, El Moussaoui S (2014) Sedimentary evolution of the siliciclastic Aptian-Albian Massylian flysch of the Chouamat Nappe (central Rif, Morocco). J Afr Earth Sci 100:554–568. https://doi.org/10.1016/j.jafrearsci.2014.08.004

Elbaghdadi M, Tabit A, Kornprobst J, Duthou JL (1996) Les injections acides dans le massif des Beni Bousera et son envelope métamorphique : conséquence sur l'évolution du métamorphisme et la cinématique. Notes Mém Serv Géol Maroc 387:33–44

Epin ME, Manatschal G, Amann M (2017) Defining diagnostic criteria to describe the role of rift inheritance in collisional orogens: the case of the Err-Platta nappes (Switzerland). Swiss J Geosci 110:419–438. https://doi.org/10.1007/s00015-017-0271-6

Escher A, Masson H (1984) Le Cervin: un dessin géologique inédit d'Emile Argand (1929) et son interprétation actuelle. Tra. Com Fr Hist Géol 2ème sér. (t. 2) 95–127. hal-00950445

Esteban JJ, Cuevas J, Tubía JM, Gutiérrez-Alonso G, Larionov A, Sergeev S, Hofmann M (2017) U-Pb detrital zircon ages from the Paleozoic Marbella Conglomerate of the Malaguide Complex (Betic Cordilleras, Spain). Impl Paleotethyan Evol Lithos 290–291:34–47. https://doi.org/10.1016/j.lithos.2017.07.022

Fallot P, Solé Sabaris L, Lemoine M (1954) Observations sur le Trias Bétique et ses Algues Calcaires. Mem Comm Inst Geol Prov Barcelona 11:23–60

Farah A, Michard A, Saddiqi O, Chalouan A, Chopin C, Montero P, Corsini M, Bea F (2021) The Beni Bousera marbles, record of a Triassic-Early Jurassic hyperextended margin in the Alpujarrides-Sebtides units (Rif belt, Morocco). BSGF - Earth Sci Bull 192.https://doi.org/10.1051/bsgf/2021015

Favre P (1992) Géologie des massifs calcaires situés au front sud de l'unité de Kétama (Rif, Maroc). Publ Départ Géol Paléont Univ Genève 11:138 pp

Favre P, Stampfli GM (1992) From rifting to passive margin: the examples of the Red Sea, central atlantic and alpine tethys. In: Ziegler PA (ed) Geodynamics of Rifting, III: Thematic Discussions. Tectonophysics 215:69–97. https://doi.org/10.1016/0040-1951(92)90075-H

Feinberg H, Saddiqi O, Michard A (1996) New constraints about the bending of the Gibraltar Arc from palaeomagnetism of the Ronda peridotites (Betic Cordilleras, Spain). In: Morris A, Tarling DH (eds) Palaeomagnetism and tectonics of the mediterranean region. Geological Society London Special Publication, vol 105, pp 43–52

Fernàndez O (2019) The Jurassic evolution of the Africa-Iberia conjugate margin and its implications on the evolution of the Atlantic-Tethys triple junction. Tectonophysics 750:379–393. https://doi.org/10.1016/j.tecto.2018.12.006

Fort X, Brun JP, Chauvel F (2004) Contraction induced by block rotation above salt (Angolan margin). Mar Pet Geol 21:1281–1294

Frizon de Lamotte D (1985) La structure du Rif oriental (Maroc), rôle de la tectonique longitudinale et importance des fluides. Thèse Doct. Ès-Sci. University Paris VI, 85–03, 436 p

Frizon de Lamotte D, Raulin C, Mouchot N, Wrobel-Daveau JC, Blanpied C, Ringenbach JC (2011) The southernmost margin of the Tethys realm during the Mesozoic and Cenozoic: initial geometry and timing of the inversion processes. Tectonics 30:TC3002. https://doi.org/10.1029/2010TC002691

García-Dueñas V, Balanyá JC, Martínez-Martínez JM (1992) Miocene extensional detachments in the outcropping basement of the northern Alboran Basin (Betics) and their tectonic implications. Geo-Mar Lett 12:88–95

García-Dueñas V, Balanyá JC, Sánchez-Gómez M (1995) El despegue extensional de Lahsene y los jirones de serpentinitas del anticlinal de Taryat (Melilla, Rif). Geogaceta 17:138–139. http://hdl.handle.net/10272/13837

Garrido CJ, Gueydan F, Booth-Rea G, Precigout J, Hidas K, Padrón-Navarta JA, Marchesi C (2011) Garnet lherzolite and garnet-spinel mylonite in the Ronda peridotite: Vestiges of Oligocene backarc mantle lithospheric extension in the western Mediterranean. Geology 39:927–930.https://doi.org/10.1130/G31760.1

Geoffroy L, Chauvet F, Rigenbach J-C (2022) Middle-lower continental crust exhumed at the distal edges of volcanic passive margins. Commun Earth Environ. https://doi.org/10.1038/s43247-022-00420-x

Gervilla F, González-Jiménez JM, Hidas K, Marchesi C, Piña R. 2019. Geology and metallogeny of the upper mantle rocks from the Serranía de Ronda. Soc. Esp. Miner., Granada University, 122 pp

Gimeno-Vives O, Mohn G, Bosse V, Haissen F, Zaghloul MN, Atouabat A, Frizon de Lamotte D (2019) The Mesozoic margin of the Maghrebian Tethys in the Rif Belt (Morocco): evidence for polyphase rifting and related magmatic activity. Tectonics 38:2894–2918. https://doi.org/10.1029/2019TC005508

Gimeno-Vives O, Mohn G, Bosse V, Haissen F, Zaghloul MN, Atouabat A, Frizon de Lamotte D (2020a) Reply to comments by Michard et al. on "The Mesozoic Margin of the Maghrebian Tethys in the Rif Belt (Morocco): evidence for polyphase rifting and related magmatic activity." Tectonics 0–2. https://doi.org/10.1029/2020tc006165

Gimeno-Vives O, Frizon de Lamotte D, Leprêtre R, Haissen F, Atouabat A, Mohn G (2020b) The structure of the central-eastern external Rif (Morocco); Poly-phased deformation and role of the under-thrusting of the North-West African paleo-margin. Earth Sci Rev. in press https://doi.org/10.1016/j.earscirev.2020.103198

Goffé B, Michard A, García-Dueñas V, González-Lodeiro F, Monié P, Campos J, Galindo-Zaldívar J, Jabaloy A, Martínez-Martínez JM, Simancas F (1989) First evidence of high-pressure, low-temperature metamorphism in the Alpujarride nappes, Betic Cordillera (SE Spain). Eur J Miner 1:139–142. https://doi.org/10.1127/ejm/01/1/0139

Gomez de la Peña L, Ranero CR, Gràcia E, Booth-Rea G, Azañón JM, Umberta Tinivella U, Yelles-Chaouche A (2022) Evidence for a developing plate boundary in the western Mediterranean. Nat Commun. https://doi.org/10.1038/s41467-022-31895-z

Guerrera F, Martin-Algarra A, Perrone V (1993) Late Oligocene-Miocene syn-/-late-orogenic successions in Western and Central Mediterranean Chains from the Betic Cordillera to the Southern Apennines. Terra Nov 5:525–544. https://doi.org/10.1111/j.1365-3121.1993.tb00302.x

Guerrera F, Martín-Martín M (2014) Geodynamic events reconstructed in the Betic, Maghrebian, and Apennine chains (central-western Tethys). Bull La Soc Geol Fr 185:329–341. https://doi.org/10.2113/gssgfbull.185.5.329

Guerrera F, Martín-Martín M, Tramontana M (2020) Evolutionary geological models of the central-western peri-Mediterranean chains: a review. Int Geol Rev 00:1–22. https://doi.org/10.1080/00206814.2019.1706056

Gueydan F, Mazzotti ., Tiberi C, Cavin C, Villaseñor A (2019) Western Mediterranean sub-continental mantle emplacement by continental margin obduction. Tectonics 38:2142–2157

Gueydan F, Pitra P, Afiri A, Poujol M, Essaifi A, Paquette JL (2015) Oligo-Miocene thinning of the Beni Bousera peridotites and their Variscan crustal host rocks, internal Rif, Morocco. Tectonics 34:1244–1268. https://doi.org/10.1002/2014TC003769

Haidar S, Déverchere J, Graindorge D, Arab M, Medaouri M, Klingelhoefer F (2022) Back-arc dynamics controlled by slab rollback and tearing: A reappraisal of seafloor spreading and kinematic evolution of the Eastern Algero-Balearic basin (western Mediterranean) in the Middle-Late Miocene. Tectonics 41: e2021TC006877. https://doi.org/10.1029/2021TC006877

Haissen F, Garcia-Casco A, Torres-Roldan R, Aghzer A (2004) Decompression reactions and P-T conditions in high-pressure granulites from Casares-Los Reales units of the Betic-Rif belt (S Spain and N Morocco). J Afr Earth Sci 39:375–383. https://doi.org/10.1016/j.jafrearsci.2004.07.030

Haissen F, Zaghloul MN, Dilek Y, Gimeno-Vives O, Mohn G., Cambeses A,Frizon de Lamotte D, Bosse V (2021) Geochemistry and petrogenesis of Lower Jurassic mafic rock suites in the External Rif Belt, and chemical geodynamics of the Central Atlantic Magmatic Province (CAMP) in Northwest Morocco. J Geol 129https://doi.org/10.1086/716499

Handy MR, M. Schmid S, Bousquet R, Kissling E, Bernoulli D (2010) Reconciling plate-tectonic reconstructions of Alpine Tethys with the geological-geophysical record of spreading and subduction in the Alps. Earth-Science Rev 102:121–158.https://doi.org/10.1016/j.earscirev.2010.06.002

Hatimi N, Duée G, Hervouet Y (1991) La dorsale calcaire du Haouz: ancienne marge continentale passive téthysienne (Rif, Maroc). Bulletin De La Société Géologique De France 162:79–90. https://doi.org/10.2113/gssgfbull.162.1.79

Hidas K, Varas-Reus MI, Garrido CJ, Marchesi C, Acosta-Vigil A, Padrón-Navarta JA, Targuisti K, Konc Z (2015) Hyperextension of continental to oceanic-like lithosphere: the record of late gabbros in the shallow subcontinental lithospheric mantle of the westernmost Mediterranean. Tectonophysics 650:65–79. https://doi.org/10.1016/j.tecto.2015.03.011

Hlila R, Chalouan A, El Kadiri Kh, Sanz de Galdeano C, Martín-Pérez JA, Serrano F, López-Garrido AC, Maaté A, Guerra-Merchán A (2008) New stratigraphic data of the Oligo-Miocene transgressive over of the Ghomarides units (Northern Internal Rif, Morocco): Implications on tectonosedimentary evolution. Rev Soc Geol Esp 21:59–71

Homonnay E, Corsini M, Lardeaux J, Romagny A, Münch P, Bosch D, Cenki-tok B, Ouazzani-touhami M (2018) Tectonophysics Miocene crustal extension following thrust tectonic in the Lower Sebtides units (internal Rif, Ceuta Peninsula, Spain): implication for the geodynamic evolution of the Alboran domain. Tectonophysics 722:507–535. https://doi.org/10.1016/j.tecto.2017.11.028

Jabaloy-Sánchez A, Padrón-Navarta JA, Gómez-Pugnaire M.T, López Sánchez-Vizcaíno V, Garrido CJ (2019b) Alpine orogeny: deformation and structure in the southern iberian margin (Betics s.l.). In Quesada C, Oliveira JT (eds) The geology of Iberia: a geodynamic approach. Springer Nature ed., pp 453–486, https://doi.org/10.1007/978-3-030-11295-0_10

Jabaloy-Sánchez A, Talavera C, Rodríguez-Peces MJ, Vázquez-Vílchez M, Evans NJ (2021) U-Pb geochronology of detrital and igneous zircon grains from the Águilas Arc in the Internal Betics (SE Spain): implications for Carboniferous-Permian paleogeography of Pangea. Gondwana Res 90:135–158. https://doi.org/10.1016/j.gr.2020.10.013

Jabaloy-Sánchez A, Azdimousa A, Booth-Rea G, Asebriy L, Vazquez-Vilchez M, Martinez-Martinez JM, Gabites J (2015) The structure of the Temsamane fold- and-thrust stack (eastern Rif, Morocco): evolution of a transpressional orogenic wedge. Tectonophysics 663:150–176. https://doi.org/10.1016/j.tecto.2015.02.003

Jabaloy-Sánchez A, Martín-Algarra A, Padrón-Navarta JA, Martín--Martín M, Gómez-Pugnaire MT, López Sánchez-Vizcaíno V, Garrido CJ, (2019a) Lithological successions of the internal zones and flysch trough units of the betic chain. In: Quesada C, Oliveira JT (eds) The geology of Iberia: a geodynamic approach. Springer Nature ed., pp 377–432. https://doi.org/10.1007/978-3-030-11295-0_8

Jaillard E, Bouillin JP, Ouali J, Dumont T, Latil JL, Chihaoui A (2017) Albian salt-tectonics in Central Tunisia: evidences for an Atlantic-type passive margin. J Afr Earth Sci 135:220–234. https://doi.org/10.1016/j.jafrearsci.2017.09.009

Jammes S, Manatschal G, Lavier L, Masini E (2009) Tectonosedimentary evolution related to extreme crustal thinning ahead of a propagating ocean: example of the western Pyrenees. Tectonics 28: TC4012. https://doi.org/10.1029/2008TC002406

Janots E, Negro F, Brunet F, Goffé B, Engi M, Bouybaouène ML (2006) Evolution of the REE mineralogy in HP-LT metapelites of the Sebtide complex, Rif, Morocco: monazite stability and geochronology. Lithos 87:214–234. https://doi.org/10.1016/j.lithos.2005.06.008

Jolivet L, Faccenna C, Agard P, De LDF, Menant A, Sternai P, Guillocheau F (2016) Neo-Tethys geodynamics and mantle convection: from extension to compression in Africa and a conceptual

model for obduction. Can J Earth Sci 53:1190–1204. https://doi.org/10.1139/cjes-2015-0118
King RC, Morley CK (2017) Wedge geometry and detachment strength in deepwater fold-thrust belts. Earth-Sci Rev 165:268–279. https://doi.org/10.1016/j.earscirev.2016.12.012
Kornprobst J (1974) Contribution à l'étude pétrographique et structurale de la Zone interne du Rif (Maroc septentrional). Notes Mém Serv Géol Maroc 251:1–256
Kornprobst J, Àbalos B, Barbey P, et al (11 authors), 2018. Boris Choubert: Unrecognized visionary geologist, pioneer of the global tectonics. Bull Soc Geol Fr 189:7https://doi.org/10.1051/bsgf/2018006
Kornprobst J, Durand-Delga M (1985) Carte géologique du Maroc au 1/50.000, feuille Tétouan-Ras Mazari. Notes Mém Serv Géol Maroc, N° 292
Kumar A, Fernàndez M, Vergés J, Torne M, Jiménez-Munt I (2021) Opposite symmetry in the lithospheric structure of the Alboran and Algerian basins and their margins (Western Mediterranean): geodynamic implications. JGR Solid Earth.https://doi.org/10.1029/2020JB021388
Labails C, Olivet JL, Aslanian D, Roest WR (2010) An alternative early opening scenario for the central Atlantic Ocean. Earth Planet Sci Lett 297:355–368. https://doi.org/10.1016/j.epsl.2010.06.024
Lacoste J, Marçais J (1938) Sur la nature et l'importance des chevauchements entre le Haut-Ouergha et le Haut Leben (Nord Marocain). C R Somm Soc Géol Fr 7:112–113
Lagabrielle Y, Asti R, Fourcade S, Corre B, Poujol M, Uzel J, et al. (2019a) Mantle exhumation at magma-poor passive continental margins. Part I. 3D architecture and metasomatic evolution of a fossil exhumed mantle domain (Urdach lherzolite, northwestern Pyrenees, France). BSGF – Earth Sci Bull 190:8. https://doi.org/10.1051/bsgf/2019007
Lagabrielle Y, Asti R, Fourcade S, Corre B, Poujol M, Uzel J, et al. (2019b). Mantle exhumation at magma-poor passive continental margins. Part II. Tectonic and metasomatic evolution of large-displacement detachment faults preserved in a fossil distal margin domain (Saraillé lherzolites, northwestern Pyrenees, France). https://doi.org/10.1051/bsgf/2019013
Lagabrielle Y, Lemoine M (1997) Alpine, Corsican and Apennine ophiolites: the slow spreading ridge model. C R Acad Sci, Paris Sér II 325:909–920
Lagabrielle Y, Vitale Brovarone A, Ildefonse B (2015) Fossil oceanic core complexes recognized in the blueschist metaophiolites of Western Alps and Corsica. Earth Sci Rev 141:1–26. https://doi.org/10.1016/j.earscirev.2014.11.004
Leblanc D (1979) Etude géologique du Rif externe oriental au nord de Taza (Maroc) Notes Mém Serv Géol Maroc 281:159 pp
Leblanc D (1983) Carte géologique du Maroc au 1/50.000, feuille de Taïneste. Notes Mém Serv Géol Maroc, 305
Leblanc D (1980) L'accident du Nekor et la structure du Rif oriental (Maroc). Rev Géol Dyn Géogr Phys 22:267–277
Lemoine M, Tricart P, Boillot G (1987) Ultramafic and gabbroic ocean floor of the Ligurian Tethys (Alps, Corsica, Apennines): in search of a genetic model. Geology 15:622–625. https://doi.org/10.1130/0091-7613(1987)15%3c622:UAGOFO%3e2.0.CO;2
Leprêtre R, Frizon de Lamotte D, Combier V, Gimeno-Vives O, Mohn G, Eschard R (2018) The Tell-Rif orogenic system (Morocco, Algeria, Tunisia) and the structural heritage of the southern Tethys margin. Earth Sci Bull 189:10. https://doi.org/10.1051/bsgf/2018009
Lonergan L, Platt J (1995) The Malaguide-Alpujarride boundary: a major extensional contact in the internal zone of the eastern Betic Cordillera, SE Spain. J Struct Geol 17:1665–1671
Lonergan L, White N (1997) Origin of the Betic-Rif Mountain belt. Tectonics 16:504–522. https://doi.org/10.1029/96TC03937
Lonergan L, Mange-Rajetsky MA (1994) Evidence for internal zone unroofing from foreland basin sediments, Betic Cordillera SE Spain. J Geol Soc London 151:515–529. https://doi.org/10.1144/gsjgs.151.3.0515
Lonergan L (1993) Timing and kinematics of deformation in the malaguide complex, internal zone of the betic Cordillera, Southeast Spain. Tectonics 12:460–476. https://doi.org/10.1029/92TC02507
Loomis TH (1972) Diapiric emplacement of the Ronda high-temperature ultramafic intrusion, southern Spain. Geol Soc Am Bull 83:2475–2496. https://doi.org/10.1130/0016-7606(1972)83[2475:DEOTRH]2.0.CO;2
Manatschal G (2004) New models for evolution of magma-poor rifted margins based on a review of data and concepts from West Iberia and the Alps. Int J Earth Sci 93:432–466. https://doi.org/10.1007/s00531-004-0394-7
Marrone S, Monié P, Rossetti F, Lucci F, Theye T, Bouybaouene LM, Zaghloul MN (2021) The pressure–temperature–time–deformation history of the Beni Mzala unit (Upper Sebtides, Rif belt, Morocco): Refining the Alpine tectono-metamorphic evolution of the Alboran domain of the western Mediterranean. J Metam Geol 00:1–25. https://doi.org/10.1111/jmg.12587
Martín-Martín M, Guerrera F, Maaté A, Hlila R, Serrano F et al (2022) The Cenozoic evolution of the Intrarif (Rif, Morocco). Geosphere 17:1–35. https://doi.org/10.1130/GES02199.1
Martin-Rojas I, Somma R, Delgado F, Estévez A, Iannace A, Perrone V, Zamparelli V (2009) Triassic continental rifting of Pangaea: direct evidence from the Alpujarride carbonates, Betic Cordillera, SE Spain. J Geol Soc London 166:447–458. https://doi.org/10.1144/0016-76492008-091
Martin-Rojas I, Somma R, Delgado F, Estévez A, Iannace A, Zamparelli V (2012) The Triassic platform of the Gador-Turon unit (Alpujarride complex, Betic Cordillera, southeast Spain): Climate versus tectonic factors controlling platform architecture. Facies 58:297–323. https://doi.org/10.1007/s10347-011-0275-z
Marzoli A, Bertrand H, Youbi N et al (2019) (18 authors) The central atlantic magmatic province (CAMP) in Morocco. J Petrol 60:945–996. https://doi.org/10.1093/petrology/egz021
Marzoli A, Callegaro S, Dal Corso J, Davies JHFL, Chiaradia M, Youbi N, Bertrand H, Reisberg L, Merle R, Jourdan F (2018) The Central Atlantic magmatic Province (CAMP): a review, in The late triassic world, edited by L. Tanner. Topics Geobiol 46:91–125. https://doi.org/10.1007/978-3-319-68009-5_4
Massonne HJ (2014) Wealth of P-T-t information in medium-high grade metapelites: example from the Jubrique Unit of the Betic Cordillera, S Spain. Lithos 208:137–157. https://doi.org/10.1016/j.lithos.2014.08.027
Michard A, Chalouan A, Feinberg H, Goffé B, Montigny R (2002) How does the Alpine belt end between Spain and Morocco? Bulletin De La Société Géologique De France 173:3–15. https://doi.org/10.2113/173.1.3
Michard A, Chalouan A, Montigny R, Ouazzani-Touhami M (1983) Les nappes cristallophylliennes du Rif (Sebtides, Maroc), témoins d'un édifice alpin de type pennique incluant le manteau supérieur. C R Acad Sci Paris 296, sér. II :1337–1340
Michard A, Feinberg H, El-Azzab D, Bouybaouene M, Saddiqi O (1992) A serpentinite ridge in a collisional paleomargin setting: the Beni Malek massif, external Rif, Morocco. Earth Planet Sci Lett 113:435–442. https://doi.org/10.1016/0012-821X(92)90144-K
Michard A, Frizon de Lamotte D, Negro F (2007) Serpentinite slivers and metamorphism in the external Maghrebides: arguments for an intracontinental suture in the African paleomargin (Morocco, Algeria). Rev La Soc Geológica España 20:173–185
Michard A, Goffé B, Bouybaouene ML, Saddiqi O (1997) Late Hercynian-Mesozoic thinning in the Alboran domain: metamorphic

data from the northern Rif, Morocco. Terra Nov 9:171–174. https://doi.org/10.1046/j.1365-3121.1997.d01-24.x

Michard A, Goffé B, Chalouan A, Saddiqi O (1991) Les corrélations entre les chaînes bético-rifaines et les Alpes et leurs conséquences. Bull Soc Géol Fr 162:1151–1160

Michard A, Mokhtari A, Chalouan A, Saddiqi O, Rossi P, Rjimati EC (2014) New ophiolite slivers in the external Rif belt, and tentative restoration of a dual Tethyan suture in the western Maghrebides. Bull Soc Geol Fr 185:313–328. https://doi.org/10.2113/gssgfbull.185.5.313

Michard A, Mokhtari A, Lach P, Rossi P, Chalouan A, Saddiqi O, Rjimati EC (2018) Liassic age of an oceanic gabbro of the external Rif (Morocco): implications for the Jurassic continent–ocean boundary of Northwest Africa. Comptes Rendus - Geosci 350:299–309. https://doi.org/10.1016/j.crte.2018.05.006

Michard A, Negro F, Saddiqi O, Bouybaouene ML, Chalouan A, Montigny R, Goffé B (2006) Pressure-temperature-time constraints on the Maghrebide mountain building: evidence from the Rif-Betic transect (Morocco, Spain), Algerian correlations, and geodynamic implications. Comptes Rendus - Geosci 338:92–114. https://doi.org/10.1016/j.crte.2005.11.011

Michard A, Saddiqi, O, Chalouan A, Chabou MC, Lach P, Rossi P et al (2020a) Comment on «The Mesozoic margin of the Maghrebian Tethys in the Rif Belt (Morocco): evidence for polyphase rifting and related magmatic activity" by Gimeno-Vives et al. Tectonics 38. https://doi.org/10.1029/2019TC006004

Michard A, Saddiqi O, Chalouan A, Farah A (2020b) Early Exhumation of the Beni Bousera Peridotite-Granulite Unit (Internal Rif, Morocco) inferred from its metasedimentary cap, a new view on some marbles and regional implications. In: et al. Advances in Geophysics, Tectonics and Petroleum Geosciences. CAJG 2019. Advances in Science, Technology & Innovation. Springer, Cham. https://doi.org/10.1007/978-3-030-73026-0_125

Michard A, Chalouan A, Negro F, Saddiqi O (2015) Comment on "Evidence of extensional metamorphism associated to Cretaceous rifting of the North-Maghrebian passive margin: the Tanger-Ketama Unit (External Rif, Northern Morocco)" by Vázquez et al., Geologica Acta, 11 (2013), 277–293. Geol Acta 13:69–76. https://doi.org/10.1344/GeologicaActa2015.13.1.5

Mohn G, Manatschal G, Masini E, Müntener O (2011) Rift-related inheritance in orogens: a case study from the Austroalpine nappes in Central Alps (SE-Switzerland and N-Italy). Int J Earth Sci 100:937–961. https://doi.org/10.1007/s00531-010-0630-2

Montel JM, Kornprobst J, Vielzeuf D (2000) Preservation of old U-Th-Pb ages in shielded monazite: example from the Beni Bousera Hercynian kinzigites (Morocco). J Metamorph Geol 18:335–342

Morley CK, King R, Hillis R, Tingay M, Backe G (2011) Deepwater fold and thrust belt classification, tectonics, structure and hydrocarbon prospectivity: a review. Earth-Sci Rev 104:41–91. https://doi.org/10.1016/j.earscirev.2010.09.010

Negro F, Agard P, Goffé B, Saddiqi O (2007) Tectonic and metamorphic evolution of the Temsamane units, external Rif (northern Morocco): Implications for the evolution of the Rif and the Betic-Rif arc. J Geol Soc London 164:829–842. https://doi.org/10.1144/0016-76492006-112

Negro F, Beyssac O, Goffé B, Saddiqi O, Bouybaouène ML (2006) Thermal structure of the Alboran Domain in the Rif (northern Morocco) and the Western Betics (southern Spain). Constraints from Raman spectroscopy of carbonaceous material. J Metamorph Geol 24:309–327. https://doi.org/10.1111/j.1525-1314.2006.00639.x

Negro F, de Sigoyer J, Goffé B, Saddiqi O, Villa IM (2008) Tectonic evolution of the Betic-Rif arc: new constraints from 40Ar/39Ar dating on white micas in the Temsamane units (External Rif, northern Morocco). Lithos 106:93–109. https://doi.org/10.1016/j.lithos.2008.06.011

Nold M, Uttinger J, Wildi W (1981) Géologie de la Dorsale calcaire entre Tétouan et Assifane (Rif interne, Maroc). Notes Mém Serv Géol Maroc 300:1–233

Olivier P, Durand-Delga M, Manivit H, Feinberg H, Peybernès B (1996) Le substratum jurassique des flyschs maurétaniens de l'ouest des Maghrébides: l'unité de Ouareg (région de Targuist, Rif, Maroc). Bull Soc Géol France 167:609–616

Olivier Ph (1990) Les unités de Beni Derkoul (Rif, Maroc). Place et signification dans l'évolution alpine de la marge nord de la Téthys maghrébine. Bull. Soc Géol Fr (8) 6:145–154

Olivier Ph, Paquette JL (2019) Early Permian age of granite pebbles from an Eocene or Oligocene conglomerate of the Internal Rif belt (Alboran domain, Morocco): hypothesis on their origin. BSGF-Earth Sci Bull 189:13.https://doi.org/10.1051/bsgf/2018012

Pastorelli S, Martinotti G, Piccardo GB, Rampone E, Scambelluri M. (1995) The Geisspfad Complex and its relationships with the Monte Leone Nappe (Lower Pennine,Western Alps). In: Polino R, Sacchi R (eds) Atti del Convegno Rapporti Alpi-Appennino 1994. Rend Accad Naz Sci Scritti e Documenti 14:349–358

Peel FJ (2014) The engines of gravity-driven movement on passive margins: quantifying the relative contribution of spreading versus gravity sliding mechanisms. Tectonophysics 633:126–142. https://doi.org/10.1016/j.tecto.2014.06.023

Pelletier L, Müntener O, Kalt A, Vennemann T, Belgya T (2008) Emplacement of ultramafic rocks into the continental crust monitored by light and other trace elements: an example from the Geisspfad body (Swiss-Italian Alps). Chem Geol 255:143–159. https://doi.org/10.1016/j.chemgeo.2008.06.024

Perrone V, Martin-Algarra A, Critelli et al (2006) 'Verrucano' and 'Pseudoverrucano' in the Central-Western Mediterranean Alpine Chains: palaeogeographical evolution and geodynamic significance. In: Moratti G, Chalouan A (eds) Tectonics of the Western Mediterranean and North Africa. Geological Socety London, Special Publication, vol 262, pp 1–43. https://doi.org/10.1144/GSL.SP.2006.262.01.01

Petit C, Le Pourhiet L, Scalabrino B, Corsini M, Bonnin M, Romagny A (2015) Crustal structure and gravity anomalies beneath the Rif, northern Morocco: implications for the current tectonics of the Alboran region. Geophys J Int 202(1):640–652. https://doi.org/10.1093/gji/ggv169

Porkoláb K, Matenco L, Hupkes J, Willingshofer E, Wijbrans J, van Schrojenstein LH, van Hinsbergen DJJ (2022) Tectonic evolution of the Nevado-Filábride complex (Sierra de Los Filábres, Southeastern Spain): insights from new structural and geochronological data. Tectonics. https://doi.org/10.1029/2021TC006922

Poulaki E, Stockli D (2022) The paleotectonic evolution of the western Mediterranean: provenance insights from the internal Betics, southern Spain. Front Earth Sci. https://doi.org/10.3389/feart.2022.929502

Précigout J, Gueydan F, Garrido CJ, Cogné N, Booth-Rea G (2013) Deformation and exhumation of the Ronda peridotite (Spain). Tectonics 32:1011–1025. https://doi.org/10.1002/tect.20062

Platt JP, Argles TW, Carter A, Kelley SP, Whitehouse MJ, Lonergan L (2003) Exhumation of the Ronda peridotite and its crustal envelope: constraints from thermal modelling of a P-T–time array. J Geol Soc 160:655–676. https://doi.org/10.1144/0016-764902-108

Puga E, De Federico AD, Martínez JIM et al (2004) The Canteras-Galifa neogene conglomerates: evidence for an ophiolitic association submerged in the mediterranean sea at the eastern end of the Betic chain. Ofioliti 29:213–230. https://doi.org/10.4454/ofioliti.v29i2.214

Puga E, Díaz de Federico A, Fanning M, Nieto JM, Rodríguez Martínez-Conde JA, Díaz Puga MA, Lozano JA, Bianchini G, Natali C and Beccaluva L (2017) The betic ophiolites and the mesozoic evolution of the western tethys. Geosciences 7.https://doi.org/10.3390/geosciences7020031

Puga E, Fanning M, Díaz de Federico A, Nieto JM, Beccaluva L, Bianchini G, Díaz Puga MA (2011) Petrology, geochemistry and U-Pb geochronology of the Betic Ophiolites: inferences for Pangaea break-up and birth of the westernmost Tethys ocean. Lithos 124:255–272. https://doi.org/10.1016/j.lithos.2011.01.002

Raoult JF (1975) Évolution paléogéographique et structurale de la chaîne alpine entre le golfe de Skikda et Constantine (Algérie orientale). Bull Soc Géol Fr 7(17):394–409. https://doi.org/10.2113/gssgfbull.S7-XVII.3.394

Reuber I, Michard A, Chalouan A, Juteau T, Jermoumi B (1982) Structure and emplacement of the Alpine-type peridotites from Beni Bousera, Rif, Morocco: a polyphase tectonic interpretation. Tectonophysics 82:231–251. https://doi.org/10.1016/0040-1951(82)90047-6

Rosenbaum G, Lister GS, Duboz C (2002) Relative motion of Africa, Iberia and Europe during Alpine orogeny. Tectonophysics 359:117–129. https://doi.org/10.1016/S0040-1951(02)00442-0

Rossetti F, Dini A, Lucci F, Bouybaouene ML, Faccenna C (2013) Early Miocene strike-slip tectonics and granite emplacement in the Alboran Domain (Rif chain, Morocco): significance for the geodynamic evolution of western Mediterranean. Tectonophysics 608:774–791

Rossetti F, Lucci F, Theye T, Bouybaouene ML, Gerdes A, Opitz J, Dini A, Lipp C (2020) Hercynian anatexis in the envelope of the Beni Bousera peridotites (Alboran Domain, Morocco): implications for the tectono-metamorphic evolution of the deep crustal roots of the Mediterranean region. Gondwana Res 83:157–182. https://doi.org/10.1016/j.gr.2020.01.020

Rossetti F, Theye T, Lucci F, Bouybaouene ML, Dini A, Gerdes A, Phillips D, Cozzupoli D (2010) Timing and modes of granite magmatism in the core of the Alboran Domain, Rif chain, northern Morocco: implications for the Alpine evolution of the western Mediterranean. Tectonics 29.https://doi.org/10.1029/2009TC002487

Rowan MG, Peel FJ, Vendeville BC (2004) Gravity-driven fold belts on passive margins. In: McClay KR (ed) Thrust tectonics and hydrocarbon systems. AAPG Memoir, vol 82, pp 157–182

Ruiz Cruz MD, Sanz de Galdeano CS (2012) Amphibole-derived evidence of medium P/T metamorphic ratio in Alpujárride and Federico “HP” units (Western Betic-Northern Rif, Spain and Morocco): possible interpretations. Int J Earth Sci 101:221–238. https://doi.org/10.1007/s00531-011-0672-0

Russo P, Russo L (1929) Recherches géologiques sur le NE du Rif. Mém Soc Sci Nat Maroc 20 :157

Saddiqi O (1988) Tectonique de la remontée du manteau: les péridotites des Beni Bousera et leur enveloppe métamorphique, Rif interne, Maroc. PhD thesis, University Louis-Pasteur Strasbourg, 180 pp

Saddiqi O, Reuber I, Michard A (1988) Sur la tectonique de dénudation du manteau infracontinental dans les Beni Bousera, Rif septentrional, Maroc. C R Acad Sci Paris 307 sér. II:657–662

Sallarès V, Gailler A, Gutscher MA, Graindorge D, Bartolomé R, Gràcia E, Díaz J, Dañobeitia JJ, Zitellini N (2011) Seismic evidence for the presence of Jurassic oceanic crust in the central Gulf of Cadiz (SW Iberian margin). Earth Planet Sci Lett 311:112–123. https://doi.org/10.1016/j.epsl.2011.09.003

Sánchez-Gómez M, García-Dueñas V, Muñoz M (1995) Relations structurales entre les péridotites de Sierra Bermeja et les unités alpujarrides sous-jacentes (Benahavís, Ronda, Espagne). C R Acad Sci. Paris 321 (II), 885–892

Sánchez-Rodríguez L, Gebauer D (2000) Mesozoic formation of pyroxenites and gabbros in the Ronda area (southern Spain), followed by Early Miocene subduction metamorphism and emplacement into the middle crust: U-Pb sensitive high-resolution ion microprobe dating of zircon. Tectonophysics 316:19–44. https://doi.org/10.1016/S0040-1951(99)00256-5

Sanz de Galdeanoa C, Andreo B, García-Tortosa FJ, López-Garrido AC (2001) The Triassic palaeogeographic transition between the Alpujarride and Malaguide complexes. Betic±Rif Internal Zone (S Spain, N Morocco). Palaeogeogr Palaeoclim Palaeoecol 167:157±173

Sanz de Galdeano C, El Kadiri K, Simancas JF, Hlila R, López-Garrido AC, El Mrihi A, Chalouan A (2006) Paleogeographical reconstruction of the Malaguide–Ghomaride complex (Internal Betic-Rifian Zone) based on Carboniferous granitoid pebble provenance. Geol Carpathica 57:327–336. http://hdl.handle.net/10261/31513

Sanz de Galdeano C, López-Garrido AC, Andreo B (1999) The stratigraphic and tectonic relationships of the Alpujarride and Malaguide complexes in the western Cordillera (Casares, prov. of Malaga, South Spain). C R Acad Sci Paris 328:113–119. https://doi.org/10.1016/S1251-8050(99)80006-8

Sanz de Galdeano C, Ruiz-Cruz MD (2016) Formaciones del Paleozoico superior al Triásico depositadas discordantes sobre las peridotitas de Ronda: evidencia de su emplazamiento cortical durante el Herciniano. Estud Geol 72(1):e043. https://doi.org/10.3989/egeol.42046.368

Schettino A, Turco E (2011) Tectonic history of the western Tethys since the late Triassic. Bull Geol Soc Am 123:89–105. https://doi.org/10.1130/B30064.1

Schito A, Atouabat A, Muirhead DK, Calcagni R, Galimberti R, Romano C, Spina A, Corrado S (2022) An insight on the polyphase thermal history of the internal Rif (Northern Morocco) through Raman micro-spectroscopy investigation. Ital J Geosci 141(1):104–119. https://doi.org/10.3301/IJG.2022.01

Serrano F, Guerra-Merchán A, El Kadiri Kh, Sanz de Galdeano C, López-Garrido AC, Martín-Martín M, Hlila R (2007) Tectono-sedimentary setting of the Oligocene-early Miocene deposits on the Betic-Rifian Internal Zone (Spain and Morocco). Geobios 288:1–15

Sibuet JC, Rouzo S, Srivastava S (2012) Plate tectonic reconstructions and paleogeographic maps of the central and north Atlantic oceans. Can J Earth Sci 49:1395–1415

Stampfli GM, Hochard C (2009). Plate tectonics of the Alpine realm. In: From: Murphy JB, Keppie JD, Hynes AJ (eds) Ancient orogens and modern analogues. Geological Society London, Special. Publication, vol 327, pp 89–111. https://doi.org/10.1144/SP327.6

Suess E (1885) Das Antlitz der Erde. Trad. fr. par Emmanuel de Margerie (1918): La Face de la Terre. Armand Colin, Paris, 1897

Suter G (1964) Carte géologique du Rif, feuille de Taounate-Aïn Aïcha au 1/50.000. Notes Mémoire du Service géologique du Maroc, 166

Suter G (1965) La région du Moyen Ouerrha (Rif, Maroc): étude préliminaire sur la tectonique et la stratigraphie. Notes Mém Serv Géol Maroc 24:7–17

Suter G (1980a) Carte géologique du Rif au 1/500.000. Notes et Mémoires du Service géologique du Maroc, 245a

Suter G (1980b) Carte structurale du Rif au 1/500.000. Notes et Mémoires du Service géologique du Maroc,

Thomas MFH, Bodin S, Redfern J, Irving DHB (2010) A constrained African craton source for the Cenozoic Numidian Flysch: Implications for the palaeogeography of the western Mediterranean basin. Earth-Science Rev 101:1–23. https://doi.org/10.1016/j.earscirev.2010.03.003

Trümpy R (1973) Situation au Trias, in Colloque Action Thématique Programmée INAG, 21 fév.- 04 mars 1973 Tanger-Ronda. Bull Soc Géol Fr (7) 15 :160–190

Trümpy R (1983) Le Rif et le Tell: leur place entre les océans et les continents. Rev Géol Dyn Géogr Phys 24:197–199

Tubía JM, Cuevas J, Esteban JJ (2012) Localization of deformation and kinematic shift during the hot emplacement of the Ronda peridotites (Betic Cordilleras, southern Spain). J Struct Geol 50:148–160. https://doi.org/10.1016/j.jsg.2012.06.010

Tubía JM, Navarro-Vila F, Cuevas J (1993) The Malaguide-Los Reales Nappe: an example of crustal thinning related to the emplacement of the Ronda peridotites (Betic Cordillera). Phys Earth Planet Int 78:343–354

Van Hinsbergen DJJ, Vissers RLM, Spakman W (2014) Origin and consequences of western Mediterranean subduction, rollback, and slab segmentation. Tectonics 33:393–419. https://doi.org/10.1002/2013TC003349

Vázquez M, Asebriy L, Azdimousa A, Jabaloy A, Booth-Rea G, Barbero L, Mellini M, González-Lodeiro F (2013) Evidence of extensional metamorphism associated to Cretaceous rifting of the North-Maghrebian passive margin: the Tanger- Ketama Unit (External Rif, northern Morocco). Geol Acta 11:277–293

Vergés J, Fernàndez M (2012) Tethys-Atlantic interaction along the Iberia-Africa plate boundary: the Betic-Rif orogenic system. Tectonophysics 579:144–172. https://doi.org/10.1016/j.tecto.2012.08.032

Vidal JC (1983) Carte géologique du Rif, feuille de Dhar Souk au 1/50.000. Notes Mémoire du Service géologique du Maroc, 298

Vila JM (1980) La chaîne alpine d'Algérie orientale et des confins algéro-tunisiens. Unpubl. PhD Thesis Univ. Paris-Sorbonne, 1–665 (available at Société Géologique de France)

Vitale S, Ciarcia S, Fedele L, Tramparulo FDA (2019) The Ligurian oceanic successions in southern Italy: the key to decrypting the first orogenic stages of the southern Apennines-Calabria chain system. Tectonophysics 750:243–261. https://doi.org/10.1016/j.tecto.2018.11.010

Wegener A (1915) Die Entstehung der Kontinente und Ozeane. Trad. fr.: L'origine des continents et des océans, Blanchard éd., Paris, 1924

Wildi W (1983) La chaîne tello-rifaine (Maroc, Algérie, Tunisie), structure et évolution du Trias au Miocéne. Rev Geol Dyn Geogr Phys 24:201–298

Wildi W, Nold M, Uttinger J (1977) La Dorsale calcaire entre Tetouan et Assifane (Rif interne, Maroc). Eclogae Geol Helv 70:371–415

Williams JR, Platt JP (2017) Superposed and refolded metamorphic isograds and superposed directions of shear during late orogenic extension in the Alborán Domain, southern Spain. Tectonics 36:756–786. https://doi.org/10.1002/2016TC004358

Zaghloul MN, Amri I, El Moutchou B, Mazzoli S, Ouazani-Touhami A, Perrone,V (2003). First Evidence of Upper Jurassic-Lower Cretaceous oceanic-derived rocks in the Ketama Unit (External Rif Zones, Morocco). International Congress 3Ma, Casablanca, pp 70–71, abstract volume

Part II

Second Thematic: Stratigraphy and Sedimentary Record

The Tellian Units, the Sellaoua Window and the High Medjerda Foreland in the Souk Ahras Area, NE Algeria

3

Chabbi Abdallah, Chermiti Asma, Brusset Stéphane, Chouabbi Abdelmadjid, and Benyoussef Mohamed

Abstract

In northeastern Algeria and south of Annaba city, the Souk Ahras region forms the southern Tellian external domain. It exhibits a complex structure characterized by the stacking of Tellian thrust sheets. These units are thrust above the Sellaoua and High Medjerda foreland, and are overthrust by the Numidian thrust sheet. The overall structure is complicated by the presence of Triassic evaporitic bodies, which occur in various positions in the stratigraphic pile. This area has been deformed by several tectonic phases in the Cenozoic. In this work, we have updated the stratigraphy and integrated it with geometric, structural and tectonic surface data to characterize the structural style of the Tellian thrust sheets and their relationship with the surrounding units. The Paleocene to Eocene of the Tellian thrust sheets encompasses two main facies: deep marine globiberinid facies to the north (Ouled Driss Sector), and shallow marine nummulitid facies to the south (Boukebch and Dekma areas), both have been unconformably covered by the Miocene. The Tellian thrust sheets of the Souk Ahras region exhibit a duplex geometry in the northern sector and an imbricate fan of thrusts to the south. These units are thrust over the Sellaoua and High Medjerda parautochthonous foreland with a basal decollement located in the Paleocene series. The tectonics wedge involves also the Sellaoua and High Medjerda foreland strata, leading to a classic fold-and-thrust belt structure.

3.1 Introduction

The Tellian thrust sheets are part of the Magrebides thrust belt that extends through the western Mediterranean region from the Betic Cordilleras in Spain, along the North African margin (Morocco, Algeria and Tunisia), extending eastwards to Sicily and the southern Apennines (Fig. 3.1; Durand-Delga 1969, 1980, Wezel 1970, Vila 1980, Hoyez 1989, Wildi 1983, Bouillin 1986, Chouabi 1997, Guerrera et al. 1993, Frizon de Lamotte et al. 2000, 2006 and 2009, Bracène 2001, Khomsi et al. 2009, 2016, 2019 and 2021, Roure et al. 2012, Leprêtre et al. 2018 and references therein).

In order to better understand the stratigraphy, the architecture of the Tellian thrust sheets and its relationship with their neighboring units (Sellaoua and High Medjerda parautochthous foreland units); a detailed interdisciplinary study has been undertaken, integrating biostratigraphy, petrography and structural geology. This study focuses on the Tellian formation of the southern part of the Maghrebides belt in northeastern Algeria between Djebel M'Cid and Djebel Graout in the Souk Ahras region (Fig. 3.3). This study aims to: (1) review the stratigraphy of the main of Tellian outcrops within the Ouled Driss and Boukebch–Dekma sectors using planktonic and benthic foraminifera, (2) document the specific petrographic characteristics, their depositional environments and the current architecture of this segment of the Maghrebides fold-and-thrust belt.

C. Abdallah (✉)
Mohamed Cherif Messaadia University, Souk Ahras, Algeria
e-mail: chabbiabdellah@yahoo.fr

C. Abdallah · C. Abdelmadjid
Geodynamics and Natural Resources Laboratory, UBMA, Annaba, Algeria

C. Asma · B. Mohamed
Water Researches and Technologies Center, Ecopark of Borj Cedria, Soliman, Tunisia

B. Stéphane
Geosciences Environment Toulouse (GET), Paul Sabatier University - Toulouse III, Toulouse, France

S. Khomsi and F. Roure (eds.), *Geology of North Africa and the Mediterranean: Sedimentary Basins and Georesources*, Regional Geology Reviews, https://doi.org/10.1007/978-3-031-18747-6_3

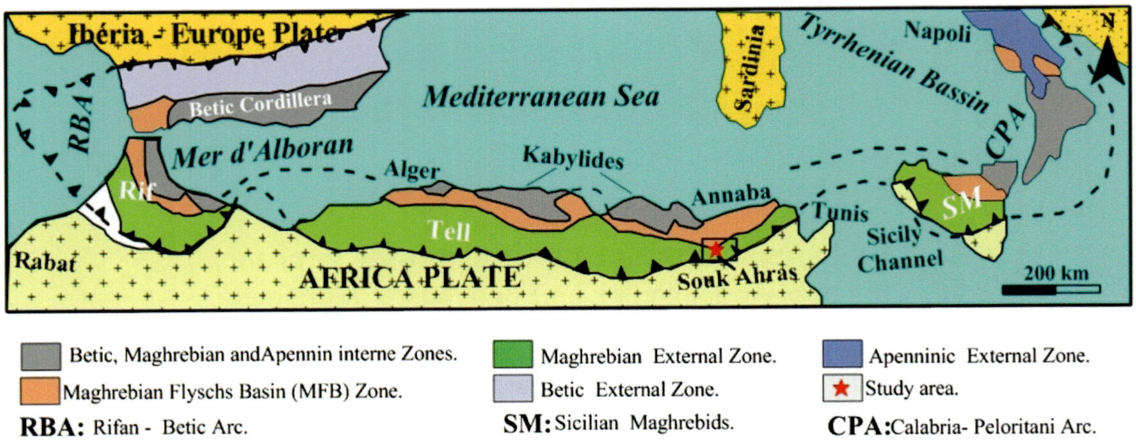

Fig. 3.1 Framework of the different paleogeographic domains of the Maghrebides thrust belt showing the studied area

3.2 Geological Setting

According to numerous studies (Durand-Delga 1969, 1980; Auzende 1978; Vila 1980; Wildi 1983; Bouillin 1986; Frizon de Lamotte et al. 2000; Bracène 2001; Bracène et al. 2002; Roure et al. 2012; Bouyahiaoui 2014; Bouyahiaoui et al. 2015; Leprêtre et al. 2018; Khomsi et al. 2019 and 2021) the Maghrebides fold-and-thrust belt, can be divided into three major domains, i.e., an internal domain, the Flyschs domain, and the external domain (Figs. 3.1, 3.2a and b).

(1) The internal domain is considered of European origin and corresponds to the former northern margin of an Alpine Tethys (Frizon de Lamote et al. 2009). In Algeria, the internal domain is the lesser Kabylia and the greater Kabylia (so-called Kabylides, Raymond 1976), which comprise the Kabylides and the "Dorsale calcaire" units (Fig. 3.2a).

(a) The Kabylides massifs are mainly made up of crystalline Hercynian basement, with erosional remnants of Triassic continental red beds Jurassic to Eocene, marine to shallow marine series (Djelit 1987, Roure et al. 2012). It comprises greenschist facies phyllites of Paleozoic age resting on amphibolitic facies rocks constituting the Kabylian basement. Most authors have considered this basement as rigid blocks weakly deformed during Alpine orogeny (Gélard 1979; Benaouali-Mebarek et al. 2006). The Kabylides massifs are covered by Chattian to Middle Burdigalian molassic deposits, so-called "Oligo-Miocene Kabyle", tectonically overlain by various Flysch units. Finally, the molasses and allochthonous thrusts sheets are overlain by marine sediments of the Langhian-Early Serravallian age (Benaouali-Mebarek et al. 2006). The Kabylides massifs are thrust on top of the "Dorsale calcaire" in the south (Fig. 3.2b).

(b) The "Dorsale calcaire" is considered as the former northern passive margin of a Maghrebian Tethys (Bouillin 1986; Benaouali–Mebarek et al. 2006; Roure et al. 2012; Frizon de Lamotte et al. 2009). It is a thick Lower Jurassic carbonate platform (Khelil et al. 2021) resting on detrital sediments of Upper Paleozoic and Triassic ages and supporting a thin cover of Jurassic–Cretaceous pelagic and Cenozoic siliciclastic sediments (Benaouali et al. 2006).

(2) The Flysch domain is sandwiched between the northern Kabylides and the southern external domain (Fig. 3.2a). This domain is commonly defined as the former cover of a Maghrebian Tethys separating Europe and Iberia from the African plate from Jurassic to Miocene (Bouillin et al., 1986, Durand-Delga 1969, 1986, and Leprêtre et al. 2018). The flyschs are made up of deep water turbiditic deposits containing radiolarians and are divided into three units (Fig. 3.2a): the Mauritanian flysch were deposited in the northern part of the deep basin and are made up of immature turbiditic sandstones derived from the European margin. Its stratigraphic ages span from the Jurassic to Miocene (Raoult 1974). The Massylian flyschs were deposited at the base of the slope of the African margin which fed it with well-sorted sandstone, being coeval in age with the Mauritanian flyschs (Bracène 2001; Vila 1980). Between these two domains, mixed flysch units show both Mauritanian and Massylian characters. The Numidian flysch unit is considered an allochton unit detached above the ductile "Argiles Sous-Numidienne". These allochthonous flysch units have been interpreted as thrust over the Kabylian internal zones to the north and over the Tellian

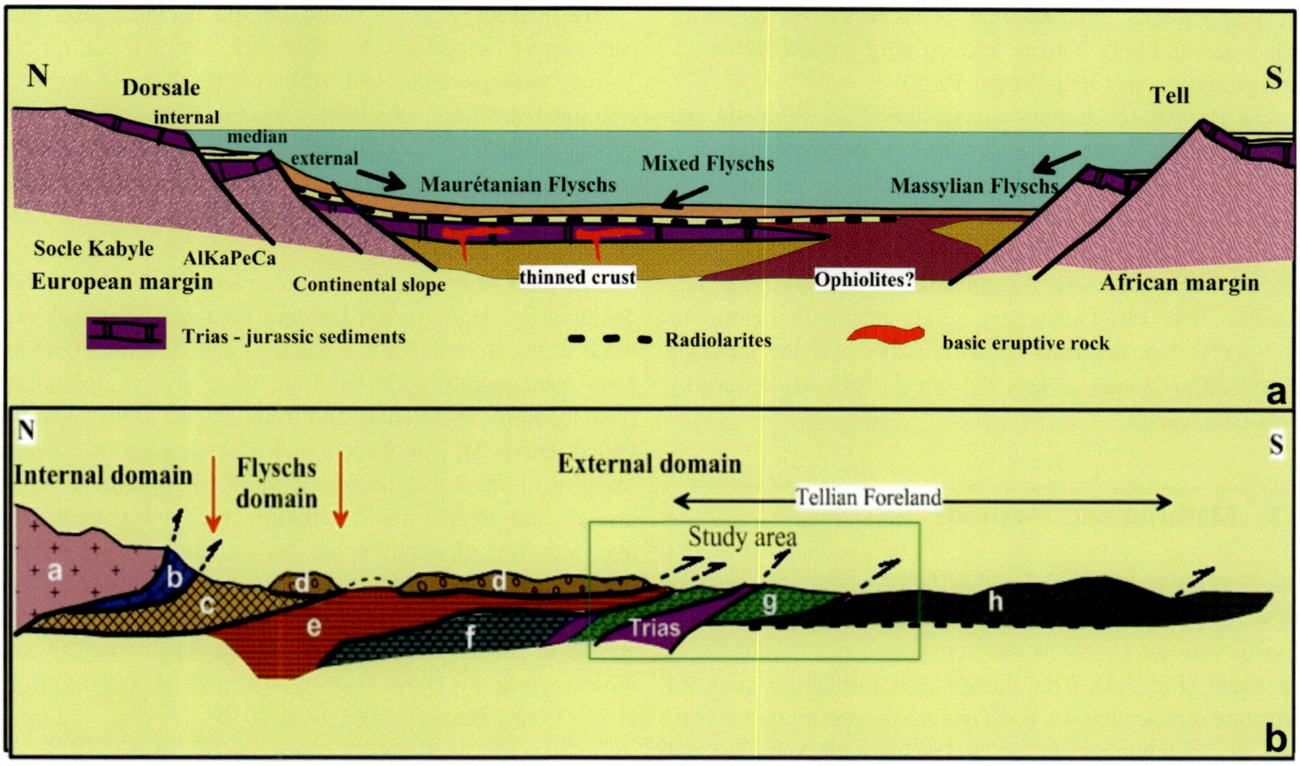

Fig. 3.2 **a** Paleogeographic reconstruction of the main domains of the Maghrebides belt during the Lower Cretaceous (Bouillin 1986), **b** synthetic cross section showing the structure of the Maghrebides thrust belt on its Algerian eastern part (after Peybernès et al. 2002). **Legend**: a. Kabylides massif, b."Dorsale calcaire", c. Mauritanian and Massylian flysch, d. Numidian flysch, e. Tellian thrust sheets, f. Constantinois thrust sheets, g. Sellaoua foreland, h. Pre-Atlasic foreland domain and (High Medjerda and Mellegues)

external area to the south (Fig. 3.2b, Vila et al., 1995). These units extend eastwards to the Tunisian Kroumirie (Rouvier 1977; Talbi et al. 2008; Riahi et al. 2010), Sicily and the southern Apennines (Guerrera et al. 1992; Guerrera et al. 1993). The flysch deposits are mostly made up of a thick series of alternating sandstone and sandy limestone (>1200 m thick), Aquitanian in age, resting on a green mudstone of Oligocene age (120 m thick), topped by a clay and sandstone series of Aquitanian to Burdigalian age (Lahondère et al. 1979; Feinberg et al. 1981; Riahi et al. 2010).

(3) The external domain: this domain is well exposed in Algeria. It has been studied by Leikine (1971), Vila (1980), Obert (1981), Chouabbi (1987) and Chabbi (2017). It represents the former African paleo-margin (Leprêtre et al. 2018; Khomsi et al. 2019 and 2021) (Fig. 3.2a), and is composed of a fan of thrust sheets staked on top of the foreland domain (Sellaoua and High Medjerda domains) in the study area (Chabbi et al., 2019). In eastern Algeria, the external domain comprises the Ultra-Tellian thrust sheets of Medjez Sfa (marly and limestone Barremian to Ypresian series) (Chouabbi 1987; Lahondère 1987) and the Tellian thrusts sheets of the Souk Ahras region. They have been thrust towards the south on top of the Sellaoua foreland domain (Voüte 1967) and of the adjacent Pre-Atlasic foreland domain (High Medjerda–Mellegue). The Tellian thrusts sheets of the Souk Ahras region are the focus of the present study (Fig. 3.2b).

Following Marmi and Guiraud (2006) and Chadi (1991), the Constantinois foreland, between Nador and Debagh cities, includes many allochthonous units among which the principal, in a lower position, corresponds to a carbonate unit represented by marine platform carbonates, extending from Jurassic to Turonian. These series are unconformably overlain by marl-carbonate marine formations of Late Senonian to Middle Miocene age, in which unconformity surfaces and/or sedimentary lacuna were observed. The Constantinois foreland is mainly characterized by thrust sheets verging southward owing to Cenozoic contractional tectonic setting.

The Sellaoua foreland mainly involves a succession of dolomitic-limestone of the Jurassic age, marls of Cretaceous to Paleocene ages and detrital sequences of Miocene age which rest unconformably on the older series. The NE–SW trending Chebka Sellaoua unit is characterized by thrust imbricates and very narrow folds, giving a particular structural pattern to this belt (Voüte 1967).

The Pre-Atlasic domain gathers the High Medjerda and the Mellegues domains. It involves a marine succession represented by marl, limestone, and sandstone formations of lower Cretaceous age, marl and limestone of Upper Cretaceous to Lutetian age. The Miocene cover, represented by marine detrital deposits, unconformably overlies older formations. The Plio-Quaternary series made up lagoon or continental deposits, rests unconformably over the previous series. This domain is less deformed, with south-verging folds (Fig. 3.2b).

3.3 Material and Methods

The study area is situated in northeastern Algeria and represents the southern part of the external domain of the Maghrebides. It extends between Djebel M'Cid in the north and Taoura city in the south (Fig. 3.3). The studied area displays a complex structural architecture. Its north and north-western part constitutes an allochthonous domain, and includes the Numidian and the Tellian thrust sheets and the Sellaoua parautochthonous foreland. Its south and south-eastern part is a para/ or -autochthonous domain, including the High Medjerda foreland, dominated by Cretaceous and Paleocene to Lutetian carbonate deposits unconformably covered by Miocene to Quaternary siliciclastic deposits.

The Triassic lagoon and continental formations made up of gypsum, clays, conglomerates …etc., crop out at the front of the main thrust sheets and between the tectonic units, mainly along the tectonic contacts (Fig. 3.3).

The present work is based on the detailed geological cartography, supported by fieldwork, carried out on the Tellian outcrops of the Souk Ahras region. We also carefully examined the basic documents available, in particular the geological maps and Master/PhD theses, which cover or touch a part of the study area (i.e., David 1956, Vila 1978, Rudis 1985, Kriviakine et al. 1989, Chabbi et al. 2016 and 2019, Chabbi 2017).

The aims of fieldwork were to collect the maximum data on lithology, stratigraphy, bedding attitude and overall tectonic architecture of Tellian outcrops and surrounding units. Data were mainly collected from three sectors, including, from the north to the south and from the east to the west, the Ouled Driss, Dj. Boukebch and Dekma sectors. Several samples of marl and limestone were taken from different sectors and treated for biostratigraphic and petrographic analyses. The biostratigraphic determinations were carried out by M. Benyoussef and A. Chermiti in the Water Research and Technology Laboratory in Tunisia, with reference to the results published on the Mediterranean and other regions by Blow (1969), Aubert et al. (1976), Bolli et al. (1985), Bellier et al. (1995, 2010).

3.4 Stratigraphic Results

3.4.1 Ouled Driss Sector

The results of the various studies carried out in the sector of Ouled Driss located in the north of the region are projected and synthesized in the Ouled Driss cross-section (Fig. 3.3).

This section was measured in the western part of Ouled Driss sector, between Oued Maaden in the south and Dj.

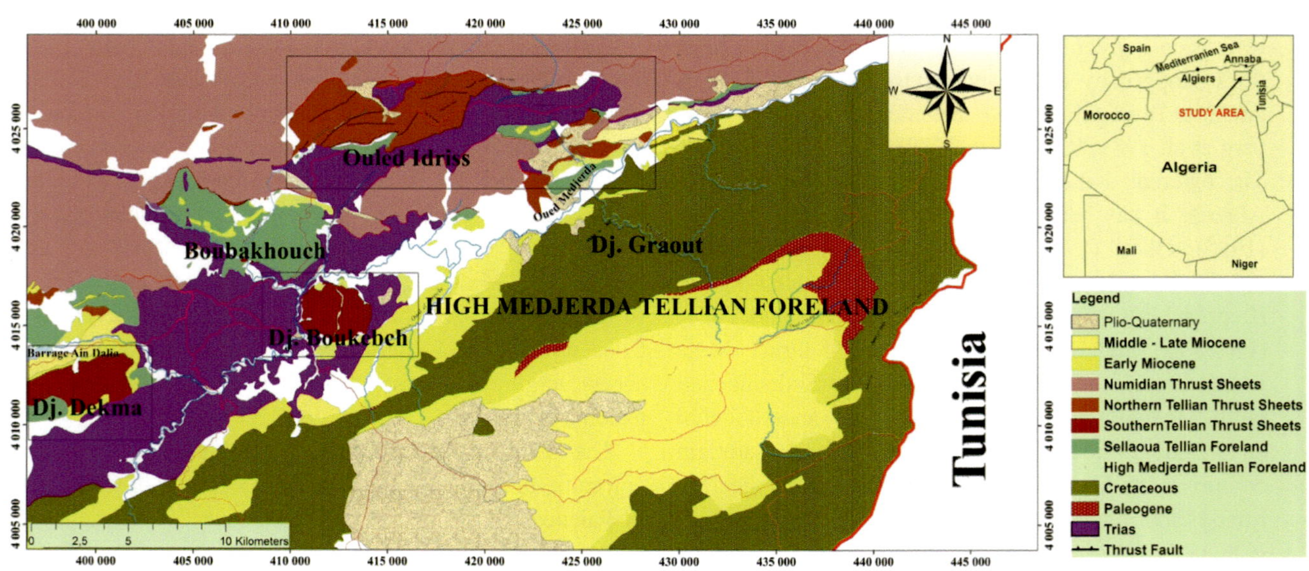

Fig. 3.3 Simplified structural map showing the mains structural units of the study area

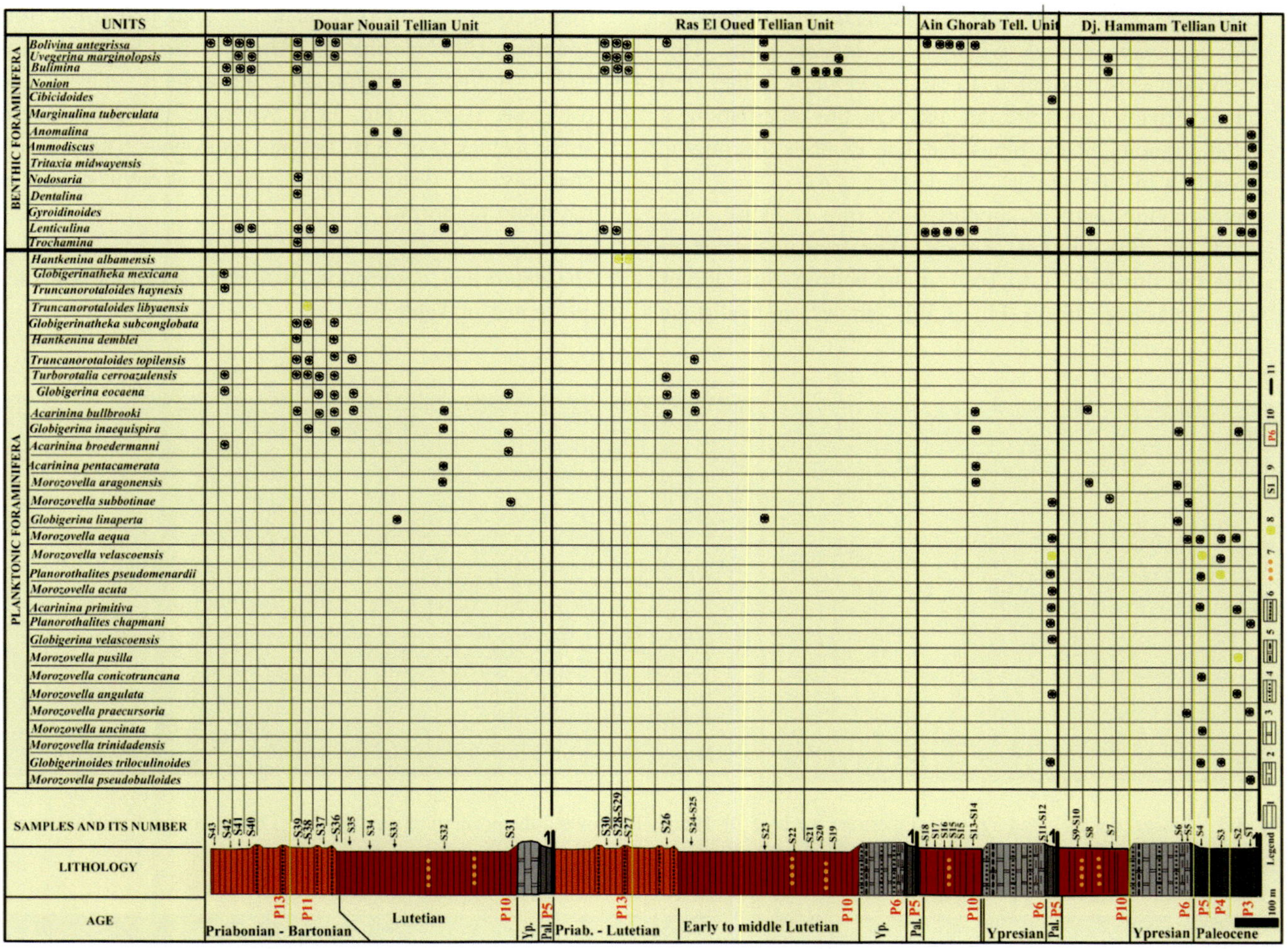

Fig. 3.4 Figure showing the stratigraphy of the different units of the Tellian thrust sheets cuts in Ouled Driss section, showing the main species of benthic and planktonic foraminifers used in biostratigraphic analysis. **Legend**: (1) marl, (2) marly-limestone, (3) limestone, (4) phosphatic limestone, (5) phosphatic level, (6) glauconitic sandstone level, (7) yellow balls, (8) biozone species, (9) samples number, (10) Blow biozone, (1) tectonic contact

M'Cid in the north. The section exhibits a repetition of marls and limestone bars resting tectonically on top of the Ouled Driss Sellaoua foreland unit and Triassic formations and supporting the Numidian thrust sheets of Dj. M'Cid (Fig. 3.3). From bottom to top we can distinguish (Fig. 3.4):

Dj. Hammam Tellian unit: this unit is composed of three members; blackish marls at the base, limestone bar in the middle and blackish marl rich in yellow balls at the top.

- The first member consists of 160 m thick black marl series, which are rich in organic matter and contain rare marl-limestone beds (Fig. 3.5a). Biostratigraphic analysis (Fig. 3.4) indicates that samples 10/2 and 14/6 are rich in a diverse planktonic foraminiferal assemblage including; *Morozovella pseudobulloides, M. praecursoria, M. angulata, M. pusilla, Acarinina primitiva, Planorotalites chapmani*. These species correspond to Globigerina sellii zone (P3 zone) of early to Middle Paleocene age.

  A few tens meters higher, samples 8/2 and 11/2 include *Globigerinoides triloculinoides, Planorotalites pseudomenardii, Morozovella velascoensis, M. aequa*. These species correspond to *Planorotalites pseudomenardii* zone (P4 zone) of Late Paleocene age. Sample 1/22 contains in addition foraminifers of the previous levels: *Morozovella uncinata, M. velascoensis, M. conicotruncana, Planorotalites pseudomenardii*. This level is attributed to *M. velascoensis* zone (P5) characterizing the Late Paleocene (Thanetian) age.

  We note that except for the base of the Middle Paleocene which is rich in benthic foraminifera *(Gyroidinoides, Lenticulina, Dentalina, Nodosaria, Tritaxia midwayensis, Ammodiscus, Anomalinoides, Gavelinella danica)*, the other levels are poor in benthic foraminifera.
- The middle part is made up of a thick limestone bar (140 m), which begins with an alternation of mar-limestone and marl beds. Samples 12/2 and 13/2

contain *M. aequa, Globigerina linaperta, M. subbotinae, M. aragonensis and Globigerina inaequispira*. This assemblage is attributed to *M. subbotinae* zone (P6) characterizing the Ypresian age. On this alternation, a Globigerinous black limestone bar exhibits decimetric beds rich in silex and phosphate (Fig. 3.4).

Petrographic analysis supported by thin sections made from six (06) samples of rock taken from the different levels of this bar (T1 to T6 from the bottom to the top of the limestone bar) (Fig. 3.5b), shows a wakstone (biomicrite) rich in organic matter and planktonic foraminifera (*Globigerina* and *Morozovella*) and rare benthics (*Bolivina, bullimuna*), phosphatic clasts and ostracods debris (Fig. 3.6). We note the presence of shark teethes (Fig. 3.6, T6) and coprolites. These facies characterize a deep marine depositional environment of the distal platform.

The detail and description of the photographic samples are presented in Table 3.1.

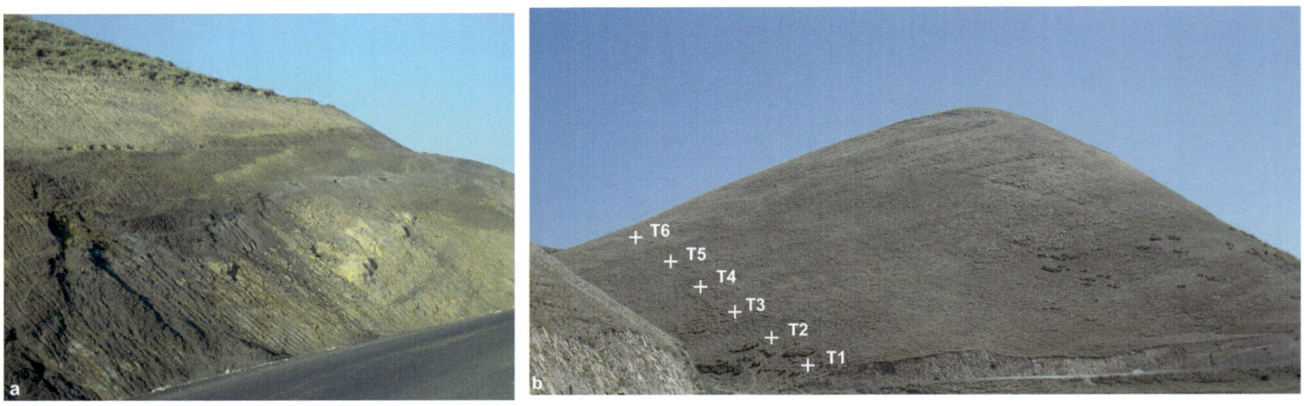

Fig. 3.5 Paleocene marl **a** and Ypresian limestone bar **b** of Dj. Hammam showing the position of the samples (T1 to T6) used for thin sections

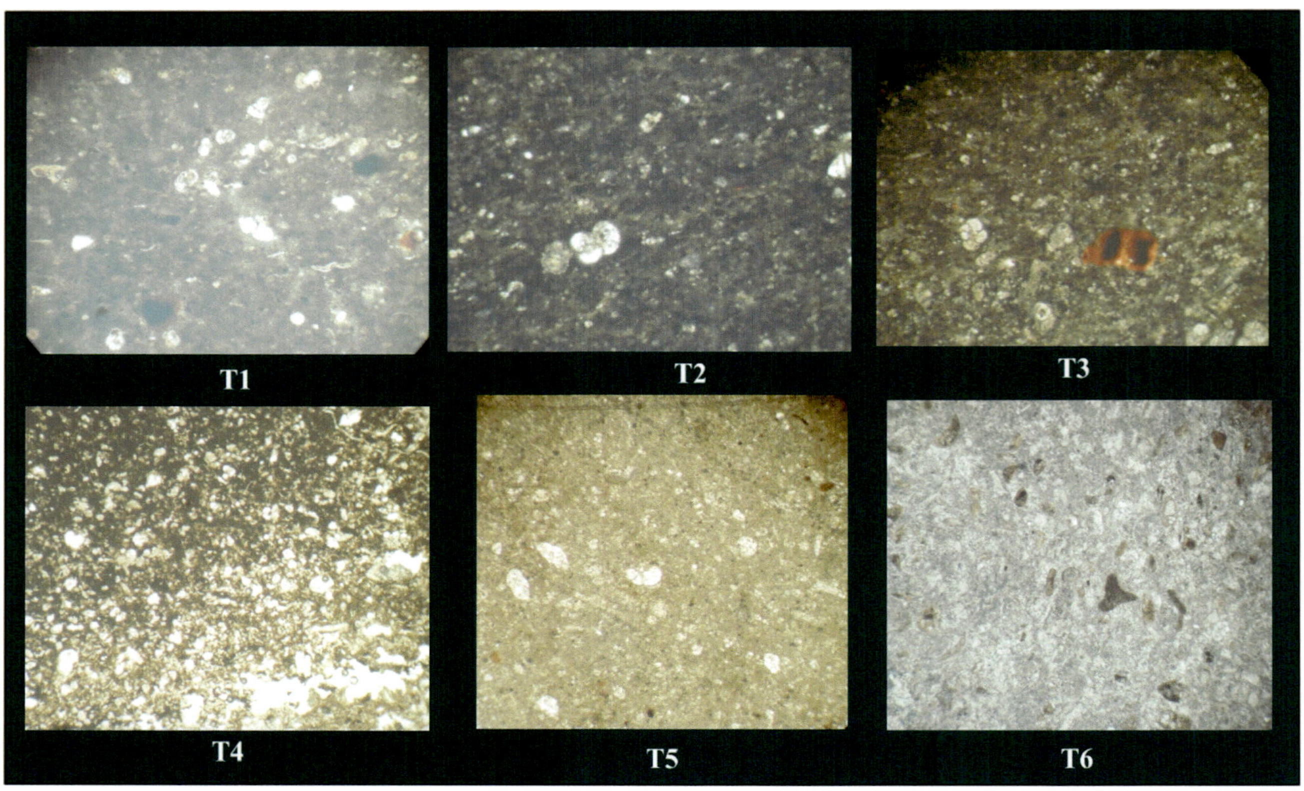

Fig. 3.6 Pictures of thin sections made in the limestone bar of Dj. Hammam

Table 3.1 Description and interpretation of thin sections made in the limestone bar of Dj. Hammam (Ouled Driss)

Thin section number	Description	Interpretation
T1	Wackestone: biomicrite including pyrite pigment, rich in *Morozovelles, Globigerines* and organic matter	Black shale facies of distal platform
T2, T3	Wackestone: biomicrite containing *Morozovella, Globigerina* species and radiolarians, rich in organic matter	Facies of distal platform
T4	Wackestone: biomicrite rich in organic matter, planktonic foraminifera and rare *Bolivina*	Facies of distal platform
T5	Wackestone: Phosphatic biomicrite containing species of *Globigerina*, Bolivina, Bulimina and radiolarians	Facies of distal platform
T6	Wackestone: biomicrite very rich in phosphates, phosphatic clasts, ostracods debris. Only *Globigerines* species are present with some shark teeth and coprolites	Deep marine facies "bathyal domain"

- The upper part of the Dj. Hammam Tellian unit is characterized by a series (240 m thick) of black marl rich in yellow balls and containing a decimetric layer of sandstone at the base. Biostratigraphic analysis (Fig. 3.4) shows that samples 8/20 and 9/20 contain plancktonic foraminiferal assemblages (*Morozovella subbotinae, M. aragonensis, and Acarinina bullbrooki*) associated with benthic foraminifera (*Lenticulina, bulimina and Uvegerina*), outlining a Late Lutetian age.

Aïn Ghorab Tellian unit: above a tectonic contact, this unit starts with about ten (10 m) meters of black marl where biostratigraphic analyses (Fig. 3.4) indicate that samples 1/14 and 1/20 are rich in planktonic foraminiferal assemblages including *Globigerinoides triloculinoides, Globigerina velascoensis, Planorothalites chapmani, P. pseudomenardii, Acarinina primitiva, Morozovella angulata, M. subbotinae, M. aequa, M. velascoensis* and *M. acuta*. These species correspond to *M. velascoensis* zone (P5 zone) characterizing the Late Paleocene (Thanetian) age.

On these Paleocene (Thanetian) marls comes the Ypresian limestone bar (140 m thick) similar to the Dj. Hammam bar described previously. At the top of this bar, the phosphatic (Fig. 3.7a) and silex levels are well exposed (Fig. 3.7b).

Above this bar, the black marl rich in yellow balls spreads out over approximately 120 m in thickness. The base of this level contains a decimetric sandstone bed. Samples taken from these marls (2/20 and 3/20) include *Morozovella aragonensis, Acarinina pentacamerata, A. bullbrooki and Globigerina inaequispira* associated with benthic foraminifera (*Lenticulina and Bolivina antegrissa*), outlining a Lutetian age (Fig. 3.4). Samples 3–7/20 and 11/4 (Fig. 3.4) contain only benthic foraminifera (*Lenticulina and Bolivina antegrissa*).

Ras El Oued Tellian unit (Fig. 3.4): this unit starts with Late Paleocene (Thanetian) marls and Ypresian limestones similar to the previous unit, overlain by thick series of marl rich in yellow balls, but thicker than in the previous units and containing a glauconitic sandstone levels in the upper part (580 m thick). Biostratigraphic analysis shows that

Fig. 3.7 Phosphatic and silex levels of the Ypresian limestone of the Aïn Ghorab Tellian unit

samples 12/3, 15/3, 13/4 and 14/4 contain only benthic foraminiferal assemblages such *Bulimina* and *Bolivina antegrissa*. Sample F2 includes *Globigerina linaperta*, associated with benthics (*Annomalina, Nonion, Bolivina antegrissa* and *Uvegerina marginolopsis*). Few meters beneath the first glauconitic level, samples 10/6 and 11/6 contain *Acarinina bullbrooki*, *Globigerina eocaena* and *Truncanorotaloides topilensis*. These assemblages provide a Late Lutetian age.

In the upper part of marl series which exhibits a glauconitic level, sample 9/6 displays a planktonic foraminiferal association comprising *Acarinina bullbrooki, Globigerina eocaena* and *Turborotalia cerroazulensis*. The assemblage also identified in 1/5 to 4/6 samples is characterized by the presence of *Hantkenina alabamensis* outlining a *Bartonian* age. These marls are rich in benthic foraminiferal assemblage such as *Lenticulina, Bulimina, Bolivina antegrissa* and *Uvegerina marginolopsis*.

Douar Nouail Tellian unit (Fig. 3.4): this section begins with Thanetian marls which are overlain by the Ypresian limestone bar, which is only 30 m thick here. At the base of the black marl rich in yellow balls (360 m thick), sample 1/10 provided *Morozovella subbotinae, Acarinina broedermanni and Globigerina inaequispira,* thus dating the early Lutetian. Benthic foraminifera such *Lenticulina, Bulimina, Bolivina antegrissa* and *Uvegerina marginolopsis* are also present. Samples 3/20, F4, F5 and 10/6 contain planktonic foraminiferal assemblages dating the Lutetian. The planktonic foraminiferal assemblages defined from samples 7/10, 9/6, 12/10 and 7/6 collected from the base of the glauconitic level are characterized by the presence of *Acarinina bullbrooki, Globigerina eocaena, Turborotalia cerroazulensis, Truncanorotaloides topilensis, T. libyaensis, Hantkenina demblei,* and *Globigerinatheka suconglobata* dating the Bartonian age. In the uppermost part, sample 13/10 contains *Truncanorotaloides hayanensis* and *Glaubigerinatheka Mexicana*, indicating a Bartonian-Priabonian age. We note that the Bartonian to Priabonian series are 220 m thick and rich in benthic foraminiferal assemblages.

3.4.2 Boukebch–Dekma Sector

This sector is situated in the south of Ouled Driss region (Fig. 3.3). It includes the outcrops of Dj. Boukebch in the east and Dj. Dekma in the west. The outcrops of this sector are been well studied by Blayac (1902 and 1912), David (1956). In the present study, we covered some gaps in relation to the petrography and structure. The Tellian outcrops of this sector are rich in nummulites, bivales, gastropods and lumachelles with oysters. The stratigraphic results obtained from our field observations, and previous results published by the authors cited above are synthesized as follows (Fig. 3.8).

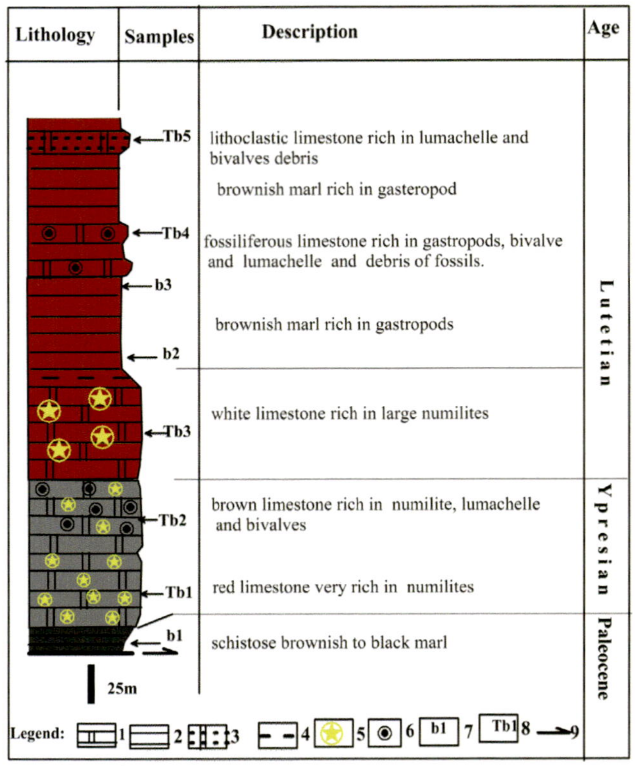

Fig. 3.8 Stratigraphic column of the Tellian thrust sheet of Boukebch–Dekma sector. **Legend**: 1. Limestone, 2. Marl and clay, 3. Lithoclastic limestone, 4. Phosphatic layer, 5. Nummulites, 6. Bivalves, 7. Marl samples 1, 8. Rock sample

The base the Dekma Tellian unit exhibits a deformed, brownish to black marl mostly covered by recent formations. Their thickness estimated by David (1956) is 100 m. At Dj. Boukebch, these marls do not exceed 10 m in thickness. They are in tectonic contact with the Triassic material.

Biostratigraphic analysis of the sample (b1) taken from this marl provided a rich benthic foraminiferal assemblage (*Tritaxia midwayensis, Ammodiscus glabrata, Trochammina abrupta, Trochamina budashvaella)* indicating a Paleocene age. On top of these marls, a thick nummulitic limestone bar develops and is subdivided as follows from bottom to top (Fig. 3.8): 70 m thick of red massif limestone very rich in nummulites (Fig. 3.9), overlain by (20 m thick) of brown limestone rich in nummulites, lumachelles and bivalves. These layers are dated as Ypresian in age by David (1956).

The limestone bar grades upwards into white limestone (70 m thick) rich in large nummulites (Fig. 3.10e) such as (*Nummilite irregularis, N. subirregularis, N. globulus, N. atacicus, N. subatacicus, N. gizehensis),* dating the Late Lutetian age (David 1956).

The nummulitic limestone bar is covered by brownish sandy and marl series (180 m thick) containing some

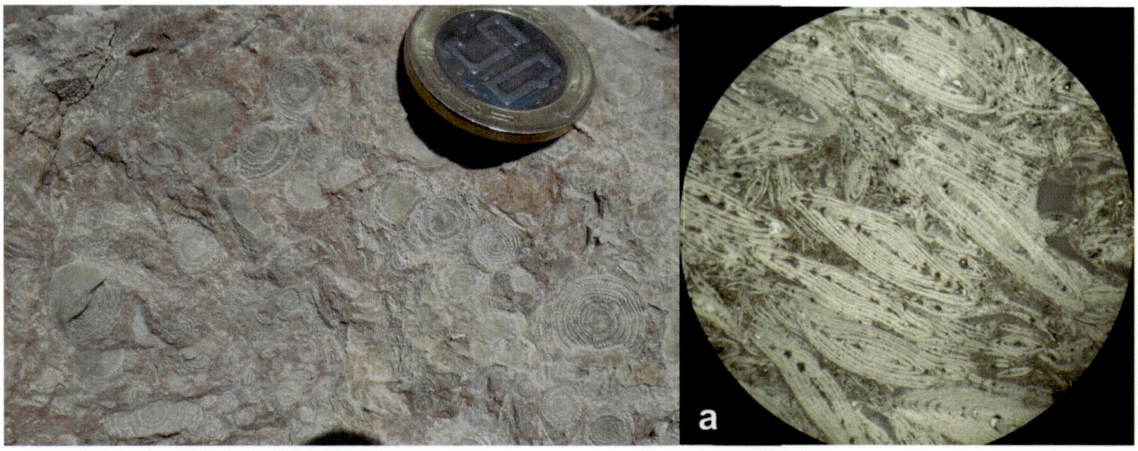

Fig. 3.9 Picture of red limestones very rich in Nummulites of Ypresian bar of Dj. Boukebch (Tb1), **a** microscopic observation from thin section

fossiliferous limestone beds (3–5 m thick) in the middle and lithoclastic limestone beds at the top (Fig. 3.10c, d, g). Marl levels are rich in gastropods (*Phasianella sp. fig.).* Biostratigraphic analysis of samples (b2 and b3) taken from these marls provided *Lenticulines,* rare *Morozovelles, Globigerines* and Ostracods (*Loculocyteretta*) dating the Lutetian age. The fossiliferous limestone beds in the middle of marl series are rich in gastropods (*Turritella carinifera, Phasianella sp.*), lumachelle and clams (Fig. 3.10a, b).

The Dekma series are mostly similar to the Boukebch series but are richer in Oysters fossils, the Paleocene marl being also thicker than in the Boukebch section.

The Tellian thrust sheets of Dj. Boukebch–Dj. Dekma sector is unconformably covered by late Burdigalian-Langhian siliciclastic formations and late Miocene series, similar to the Neogene series exposed in underlying thrust sheets farther to the south. Notice that Dekma Miocene series are better conserved than the Boukebch Miocene series, the latter being mostly eroded and preserved in only small outcrops resting directly on top of the Lutetian series.

3.4.3 Interpretation and Correlation of Stratigraphic Results

Stratigraphic results show that the Tellian thrust sheets of the Souk Ahras region are Paleocene to Priabonien in age. Two facies types are documented: globigerinous facies in the north and nummulitic facies in the south.

The globigerinous facies is recognized at the Ouled Driss sector in the northeast of the study area and extends westwards to Djebel Alia. This facies is characterized by (1) black Paleocene marl, (2) black Ypresian limestone rich in silex and phosphate layers, (3) Lutetian marl rich in yellow balls and (4) Bartonian—Priabonian marl rich in glauconite.

Most part of the Paleocene and Ypresian series of this facies displays a diverse planktonic foraminiferal assemblage indicating a deep marine depositional environment. The occurrence of benthic foraminifera associated with a planktonic foraminifera assemblage at the base of Paleocene marls, indicates that the water level was shallower during the early Paleocene and began to deepen during the lower Paleocene, reaching its maximum depth dating the Ypresian stage when only the Globigerina were present (Fig. 3.5, Table 3.1). The Lutetian marls containing yellow balls indicate an unstable depositional environment, like a slope. The Lutetian to Priabonian series is rich in benthic foraminiferal assemblages with rare planktonic foraminifera, thus indicating a shallow marine depositional environment.

Nummulitic facies is mapped south of the Medjerda River at Djebel Boukebch and Djebel Dekma.

This facies indicates a shallow marine depositional environment. This facies is represented by (1) Paleocene marl, (2) Ypresian—early Lutetian nummulitic limestone and (3) marl and limestone rich in nummulites, gastropod, lummachelle and bivalves. We note that the formations of these facies are thinner than the globigerinous facies series (Fig. 3.11).

These two facies are correlated with those situated in the west of the study area between Guelma and Sedrata city and in the Setif region, as described by Chouabbi (1987) and Vila (1980). According to previous descriptions made by Ben Ismail-Latrech (2000) and Mesrouhi (2006), the Paleocene and Eocene Tellian series of Tunisia at the Anssarine Plateau, Dj. Meftah and Matter look different (both from their thicknesses and structural position) from our Algerian sections, despite the occurrence of the same two facies, i.e., the deep water globigerina and shallower water nummilitic facies.

Fig. 3.10 Picture of Lutetian facies of Dj. Boukebch–Dj. Dekma sector; (Macroscopic). **a** *Gastropods phasianella sp.* **b** *Gastropods turritella carinifera*, **c** fossiliferous limestone, **d** lithoclastic limestone, (Microscopic). **e** brown limestone rich in nummulites, bivalves and lumachelle taken from Tb2, **f** taken from fossiliferous limestone Tb4, **g** lithoclastic limestone Tb5

3.5 Surface Structural Data

Due to a lack of subsurface data (geophysics and wells), we have based our work only on surface data. They were obtained from geological maps (Souk Ahras, Oued Mougras and M'daourouche maps at 1/50000 scale) published by the national geological map agency of Algeria (ASGA), both published and unpublished theses and also from several field surveys carried out in (2012–2016 and 2019–2020) in this area.

Here the orogen contains a stack of the northeast-southwest-trending thrust sheets imbricating the Cenozoic carbonate strata in a fold-and-thrust domain called the Sellaoua and High Medjerda foreland. The architecture and tectonic style of the Tellian thrust sheets are summarized in the following two sectors: Ouled Driss sector in the north and Boukebch—Dekma sector in the south.

3.5.1 Ouled Driss Sector

The Ouled Driss sector exhibits a stack of Tellian thrust sheets resting tectonically on top of the Sellaoua foreland unit and Triassic salt, and below the Numidian thrust sheet

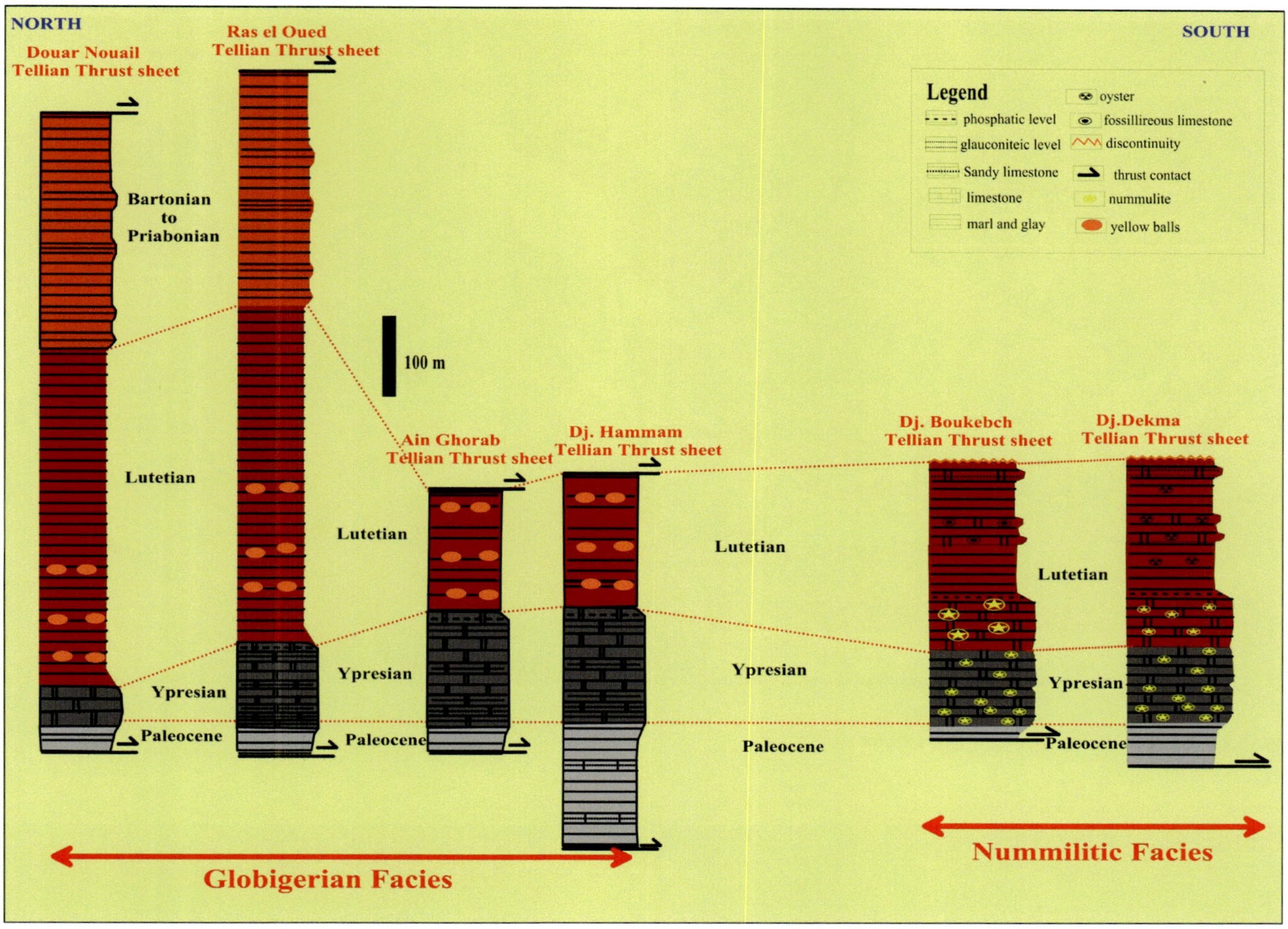

Fig. 3.11 Stratigraphic correlations between the different Tellian thrust sheets

(Figs. 3.12 and 3.13). This sector is dominated by duplex structures with a roof thrust located in the Oligocene mudstone at the base of the Numidian thrust sheet and a sole thrust in the Paleocene black shale. Ouled Driss Tellian thrust sheets have a diamond shape, being 15 km wide, with a maximum elevation reaching 1200 m.

The western part of the sector shows a hinterland dipping duplex structure, made up of four NE-SW trending thrust sheets. From the base to the top, they comprise (Figs. 3.12, 3.14):

- **Dj. El Hammam Tellian thrust sheet** exhibits Paleocene-Lutetian marls and limestone strata (550 m thick), dipping 20°–40° towards the north above Ouled Driss Sellaoua foreland, the later involving upper Maastrichtian strata dipping 20°–60° towards the south.
- Above this unit, the **Aïn Ghorab Tellian thrust sheet** also exhibits Thanetian-Lutetian marls and limestone strata with 310 m thickness, dipping 45°–70° towards the north.
- **Ras el Oued Tellian thrust sheet** has an arcuate shape and shows similar series as in the Aïn Ghrab unit. The Lutetian marls are thicker here than in the previous units and contain Bartonian -Priabonian series. The thickness of this thrust unit reaches up to 730 m. The limestone bar is dipping 70°–80° towards the northwest.
- **Hdeb—M'Cid Tellian thrust sheet** is the higher Tellian unit, directly below the Numidian thrust sheet/ It shows similar series as in the Ras el Oued thrust sheet. In this unit, however, the Ypresian limestone bar is thinner, being only 30 m thick. The total thickness of this thrust sheet reaches 590 m.

The eastern part of the Ouled Driss sector exhibits a foreland dipping duplex formed by two to four thrust sheets involving Paleocene to Lutetian carbonate strata (Fig. 3.14) dipping 40°–60° towards the south. These Tellian units are tectonically overlain by the Numidian thrust sheet in the north. The base of this duplex is dominated by the Triassic salt.

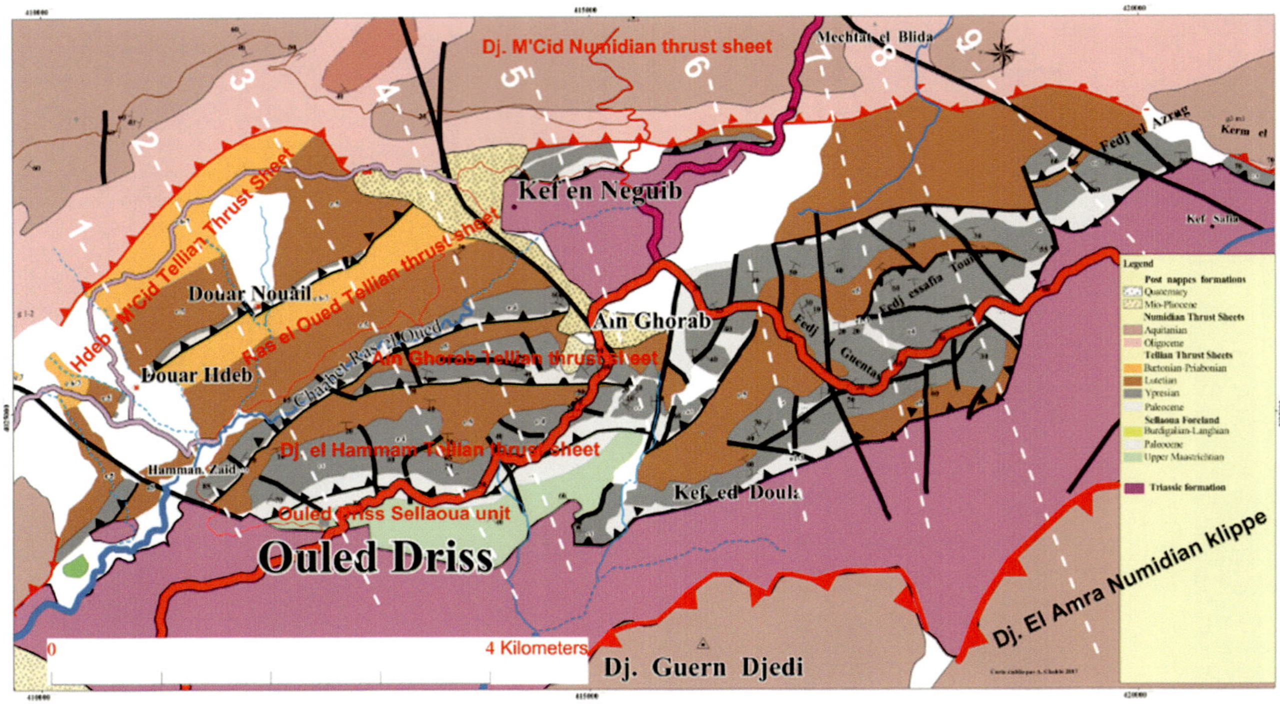

Fig. 3.12 Structural map of Ouled Driss sector

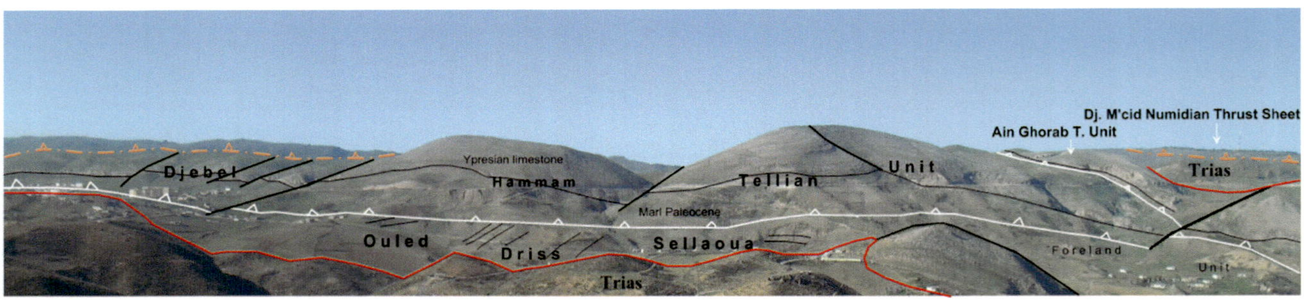

Fig. 3.13 Picture of the southern Ouled Driss Tellian thrust front, looking northward

3.5.2 Boukebch–Dekma Sector

The Boukebch–Dekma Nummulitic thrust sheets are situated to the south of the Globigerinous Tellian thrust sheets, south of the Medjerda River. Two thrust sheets are well exposed, i.e. the Boukebch Nummulitic Tellian thrust sheet in the east and the Dekma Nummulitic thrust sheet in the west.

3.5.2.1 Boukebch Nummulitic Tellian Thrust Sheet

The Boukebch thrust sheet is exposed in the south of Souk Ahras city, and has a circular shape of about five kilometers large with a maximum elevation reaching 850 m (Fig. 3.15). Most part of this thrust sheet is surrounded by Triassic salt. To the south, the Boukebch thrust sheet is thrust over the Dj. Graout foreland of the High Medjerda, resting tectonically on top of the Miocene (Serravallian) siliciclastic series covering the northern flank of the Graout anticline. To the north, this unit is in turn under thrust below the Bouallegue Sellaoua foreland unit. The Boukebch Tellian thrust sheet involves Paleocene-Lutetian nummulitic carbonate strata dipping towards the north (Fig. 3.15).

3.5.2.2 Dekma Tellian Thrust Sheet

In the southwest of the study area, the SW-NE trending Dekma Tellian thrust sheet extends over a surface that is 4 km wide and 8 km long with a maximum elevation reaching 1050 m (Fig. 3.16a). The structural map (Fig. 3.16a) and the two geologic cross sections (Fig. 3.16b, c) show that the Dekma Tellian thrust sheet is thrust over early to middle

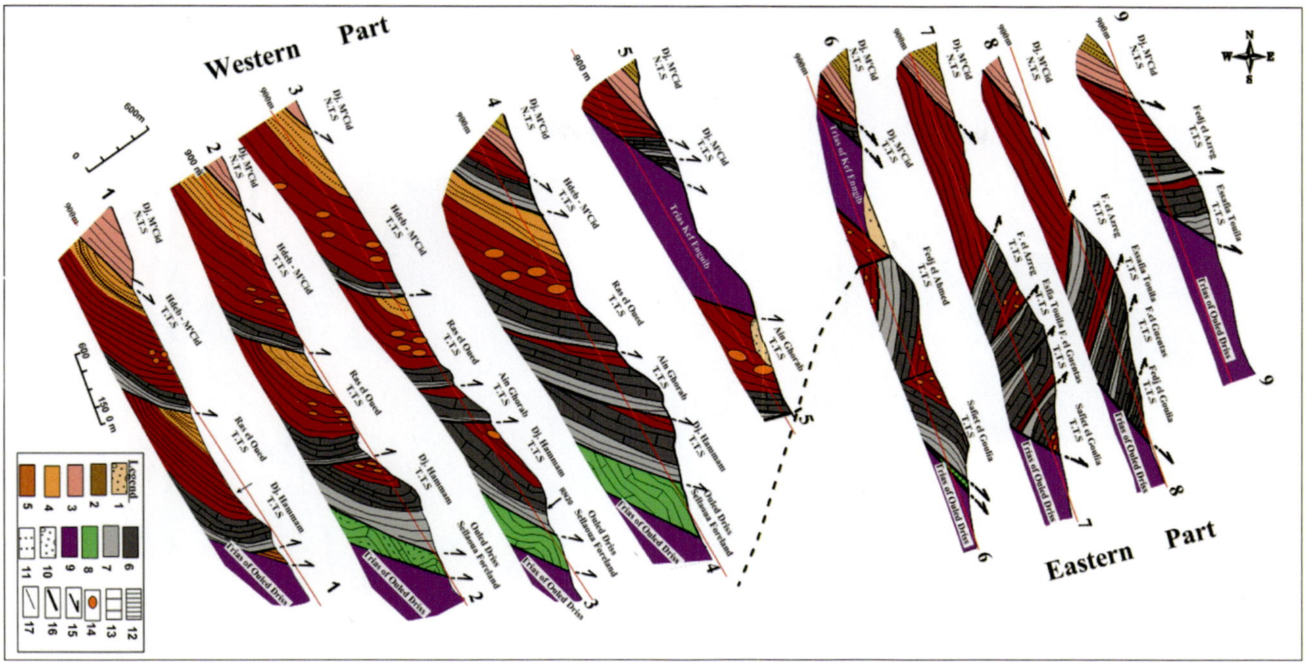

Fig. 3.14 Structural cross sections from Ouled Driss sector. **Legend**: 1. Pliocene, 2. Oligo-Miocene, 3. Oligocene, 4. Bartonian-Priabonian, 5. Lutetian, 6. Ypresian, 7. Paleocene, 8. Maastrichtian, 9. Triassic, 10. Conglomerates, 11. Sandstone, 12. Mudstone, 13. Limestone, 14. Yellow balls, 15. Thrust contact, 16. Tectonic contact, 17. Stratigraphic limit.

Miocene series covering the Sellaoua Foreland and Triassic salt of Argoub El Djemel in the east. In the south, the nummulitic Tellian series forms with the Cretaceous series the anticline of Dj. Serrou, which involves Campanian to Maastrichtian and Paleocene carbonate strata unconformably covered by early-middle Miocene detrital series. The Dekma thrust sheet unit involves Paleocene to Lutetian carbonate strata rich in nummulites and oysters. These series are also unconformably covered by Miocene strata dipping northwestwards. The Dekma Tellian thrust sheet and the Serrou anticline are thrust towards the south on top of the Miocene along a decollement level located within the Campanian-Santonian series. In the north, in contrast, these units are under trusted beneath the Mechta Ech Cheurfa Sellaoua foreland unit (Campanian-Santonian). The ductile Triassic series, made up of clays, breccias and gypsum, is always present at the front of this thrust sheet. This thrust unit shows a main decollement level within the Paleocene and Santonian marls.

3.6 Discussion

Stratigraphic results and structural diagrams (maps and geological sections) lead us to discuss the following points:

Stratigraphic analyses allow us to distinguish two very distinctive facies within the Paleocene- Priabonian carbonate series: (1) a northern facies rich in globigerina, of deep marine origin and (2) a southern facies with nummulites and oysters, of shallow marine origin. Currently, the Ouled Driss Tellian thrust sheets (deep marine facies), which were deposited further north and in a deeper marine domain, are at the same elevation (1200 m) or topographically higher than the southern Dj. Boukebch and Dekma Tellian thrust sheet (1050 m) which display nummulitic facies. This accounts for a thrust wedge involving ramps and flats (Boyer and Elliott 1982), with distinct decollement levels, i.e., in the Paleocene marls for the so-called Tellian thrust sheets and in Santonian series for the so-called parautochthonous units. These thrusts result from Alpine compressional phases inducing the tectonic inversion of the former North African passive margin of the Tethys during long-lasting episodes of convergence between Eurasian and African plates (Bracène et al. 2002, Khomsi et al. 2006, Frizon De Lamote et al. 2009, Roure et al. 2012 and Leprêtre et al. 2018).

Two thrust systems are recognized in the Souk Ahras region: the upper thrust system involves the Tellian units and the lower system involves the Sellaoua and High Medjerda parautochthonous units. The upper thrust system made up of the Tellian thrust sheets rides over the Sellaoua parautochthon domain, with the coeval formation of duplex structures in the northern part of the study area (Ouled Driss sector and large thrust sheets in the south (Boukebch–Dekma sector). The lower thrust system involves the parautochthonous foreland strata, and lead to a fold-and-thrust structure (Serrou anticline).

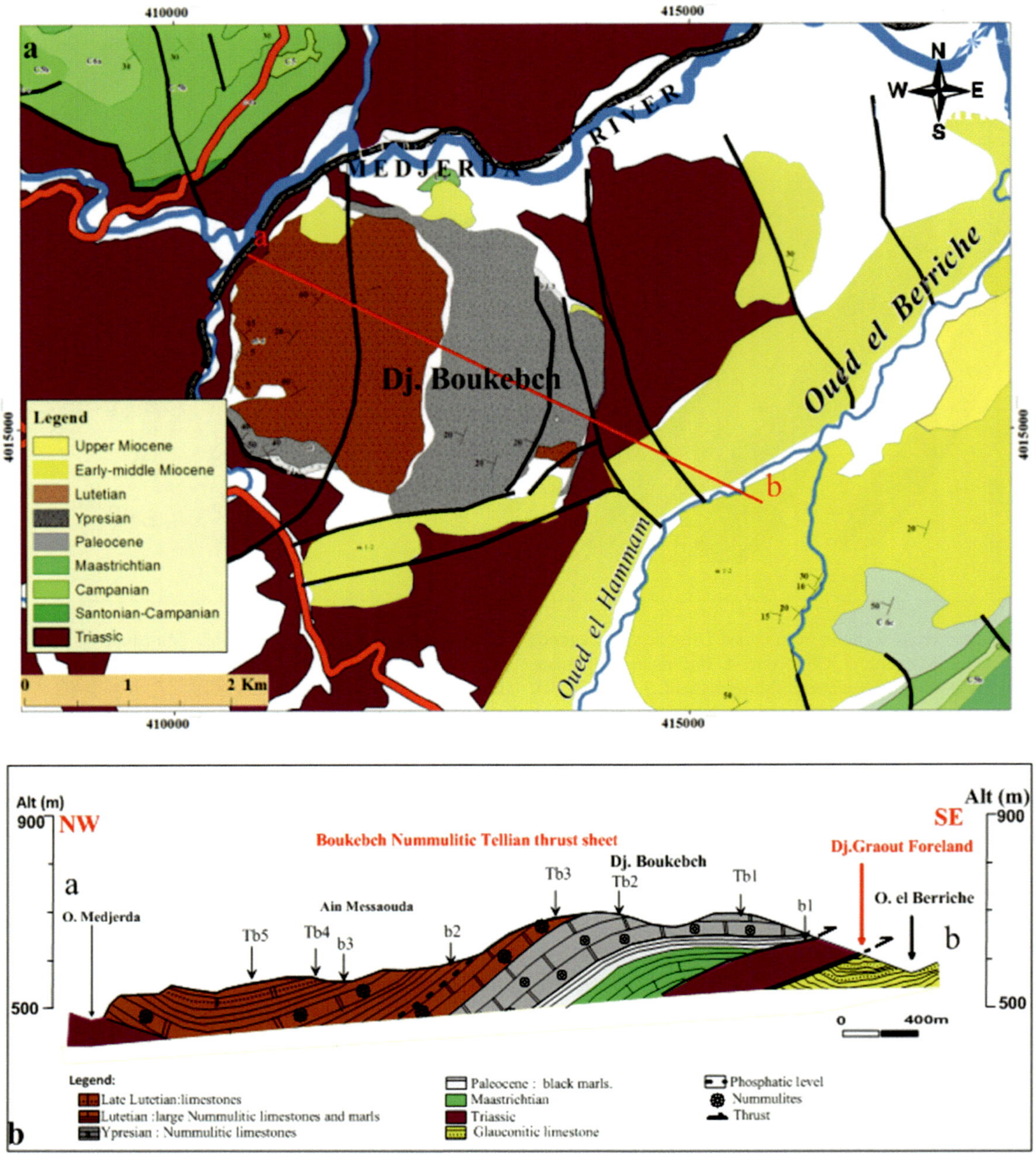

Fig. 3.15 **a** Geologic map of the Boukebch Nummulitic Tellian thrust sheet indicating the location of the cross section. **b** NW–SE geologic cross section (a-b) across the Boukebch thrust sheet

The northern Tellian thrust sheets of Ouled Driss are made up of Paleocene to Priabonian imbricated series and are in tectonic contact with the Numidian thrust sheet of Dj. M'Cid while the southern Tellian thrust sheets of Dj. Boukebch and Dekma are unconformably overlain by the Miocene series. This stratigraphic architecture indicates that: (1) the Tellian domain began to accreted in the tectonic wedge from the late Priabonian onward. This event corresponds to the Atlassic phase which is recognized in the entire Maghrebides domain (Guiraud 1975, Vila 1980, Bracène 2000, Khomsi et al. 2006, Frizon de La Motte et al. 2009). (2) During the Miocene the return of the sea is recorded in certain areas of the Tell domain and its foreland (already folded). The Miocene deposits on the Tellian formations are older (Burdigalian) than in the parautochthonous foreland (late Burdigalian–early Langhian). Therefore the Miocene sea returned first to the Tellian domain and then on the foreland but the withdrawal of the Miocene sea occurred earlier in the northern Tellian area. The sea remained in the southern Tellian area during the thrust emplacement and probably until the Tortonian when the Sea ultimately retired permanently.

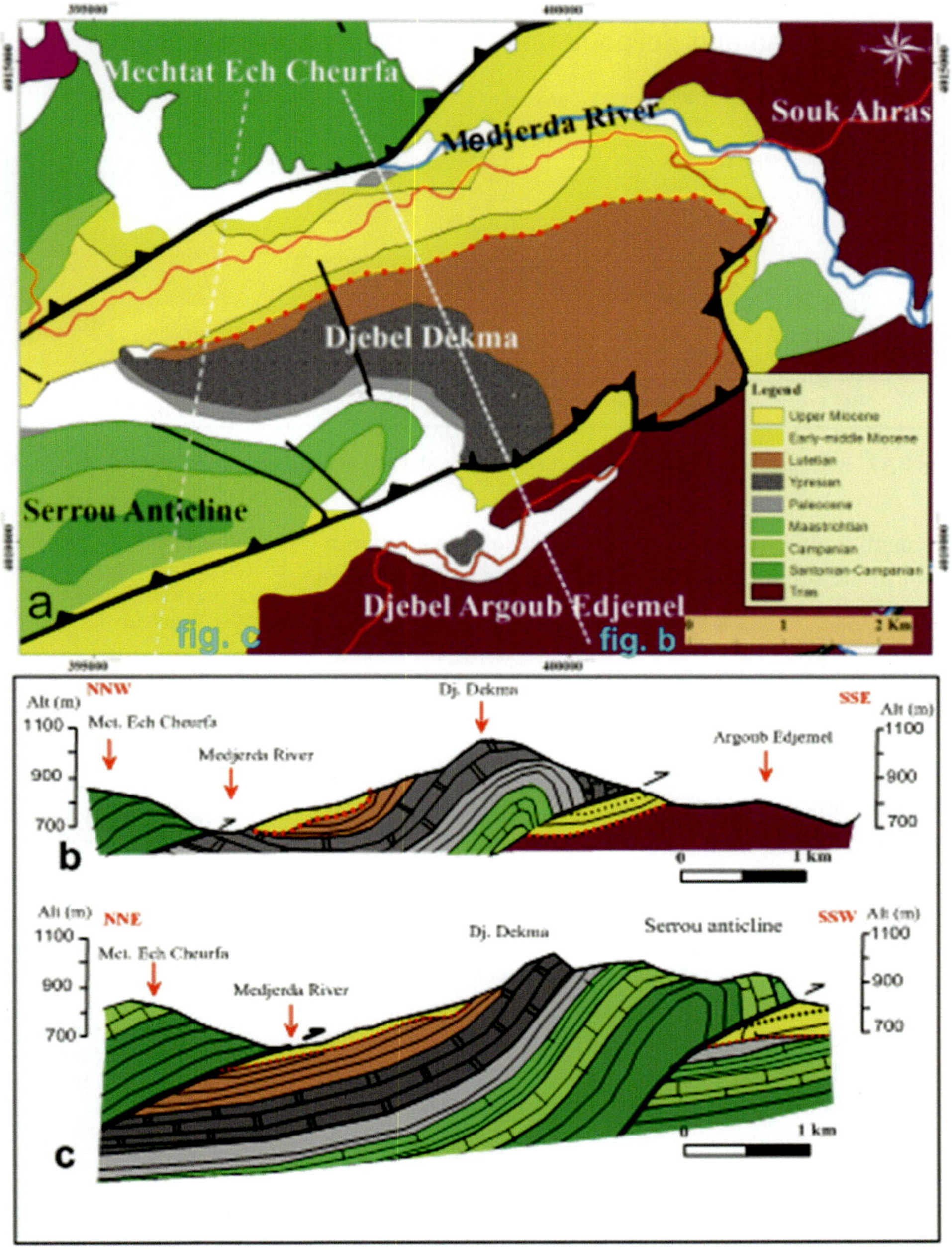

Fig. 3.16 Structural map of Dj. Dekma Tellian thrust sheet (**a**) and schematic geological cross sections (**b** and **c**) cross sections located on the map

The Miocene series stacked between the Tellian thrust sheets and the foreland are Burdigalian—Langhian (−20.4 MA to −13.8 MA) in age, while the Miocene strata below the Sellaoua parautochthonous foreland units are Upper Miocene in age (Tortonian −11.6 MA to −7.2 MA). These allow us to put the chronology of the thrusts setting. The Tellian thrust sheets were first detached from their Cretaceous series along the Paleocene marls, being the thrust over the Sellaoua parautochthonous foreland during the Langhian, then the entire bloc (Tellian thrust sheets and Sellaoua parautochthonous foreland units) were thrust on top of the High Medjerda foreland using Santonian and Campanian marls as the main decollement level during the Tortonian.

3.7 Conclusion

Surface data together with the biostratigraphic analysis and construction of geological cross sections have been integrated to study stratigraphy and structural architecture of Souk Ahras Tellian thrust sheets and their relationship with the Sellaoua and High Medjerda parautochthonous foreland units. The main conclusions are as follows.

The Tellian thrust sheets of Souk Ahras area are made up of Paleocene to Priabonian carbonate series divided into two facies, i.e., a globigerinian facies of deep marine depositional environment in the north (Ouled Driss sector) and a

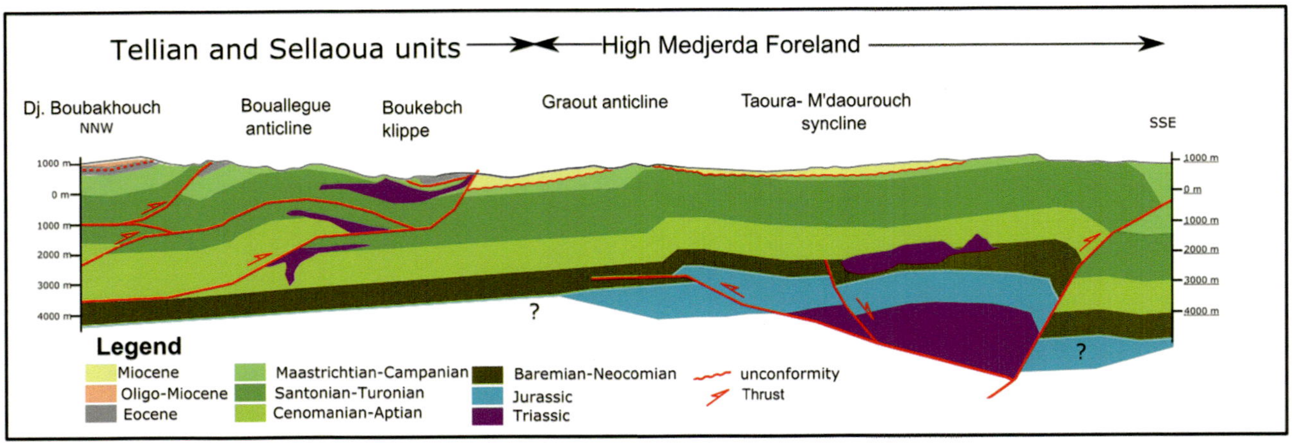

Fig. 3.17 Regional interpretative cross section showing the two contractional systems recognized in the Souk Ahras region

nummulitic facies of shallow marine depositional environment in the south (Boukebch–Dekma sector). The southern facies is covered unconformably by the Miocene series indicating Atlassic tectonic events. They are in tectonic contact with both the underlying Sellaoua and high Medjerda parautochthonous foreland units, and overlying Numidian thrust sheets.

The northern part (Ouled Driss sector) is characterized by a duplex structure involving four thrust sheets with a roof thrust in the Numidian mudstone, and a sole thrust in the Paleocene marls. In the southern part (Boukebch Dekma sector), the Nummulitic Tellian thrust sheets are thrust directly on top of the Sellaoua and High Medjerda parautochthonous foreland units along a Paleocene decollement.

Two contractional systems are recognized in the Souk Ahras region (Fig. 3.17). The Northern thrust system displays an imbricate fan of thrusts involving local and infraformational décollement driven by the specific mechanical stratigraphy. Large Triassic bodies cannot be geometrically linked to this thrust system and should rather be the record of previous salt tectonics (diapirism, canopies…). The southern inversion structures are probably basement-involved pop-ups. The leading fault corresponds to an inverted north-facing normal fault, which controlled the sedimentation during Jurassic and Lower Cretaceous times as the southern Tethys margins developed. Salt tectonic is also associated with such extensional settings. In the Northern thrust systems, break-forward thrust propagation can be considered as a rule, even if synchronous or out-of-sequence thrusting cannot be excluded. As in many orogens, it is probable that inversion occurred early in the evolution as soon as the geodynamic setting switched from extensional to contractional.

Acknowledgements Dr. Khomsi is warmly acknowledged for his editorial effort. The paper benefited from the reviews of Dr. Roure. Petex provided their MOVE suite software to the University Paul Sabatier, Toulouse 3.

References

Aubert J, Berggren W (1976) Paleocene benthic foraminiferal biostratigraphy and paleoecology of Tunisia. Bull C R Pau-SNPA10(2):379–469

Auzende JM (1978) Histoire Tertiaire de la Méditerranée occidentale. PhD thesis, University of Paris VII, 232 pp

Bellier JP, Dépêche F, Mathieu R (1995) Introduction à la Micropaléontologie: educational document of micro-paleontology, University of Pierre et Marie Curie, Paris, N° 1, 50 pp.

Bellier J-P, Mathieu R, Granier B (2010) Court traité de foraminiférologie (L'essentiel sur les foraminifères actuels et fossiles) [Short Treatise on Foraminiferology (Essential on modern and fossil Foraminifera)]. Carnets de Géologie - Notebooks on Geology, Brest, Livre 2010/02 (CG2010_B02), 104 pp.

Benaouali-Mebarek N, Frizon de Lamotte D, Roca E, Bracène R, Faure J-L, Sassi W, Roure F (2006) Post-Cretaceous kinematics of the Atlas and Tell systems in central Algeria: Early foreland folding and subduction-related deformation. C R Geosci 338:115–125

Ben Ismail-Lattrache K (2000) Précision sur le passage Lutétien-Bartonien dans les dépôts éocènes moyens en Tunisie centrale et nord-orientale : Revue de micropaléontologie 43:1–2, 3–16

Blayac J (1902) Sur la présence de l'Eocène moyen dans la région de Souk Ahras (Province de Constantine) : Bulletin du service géologique de France II(4):42–43

Blayac J (1912) Esquisse géologique du bassin de le Seybouse et de quelques régions voisines . PhD, Paris, B.S.C.G. Algérie, 6, 490 pp.

Blow WH (1969) Late middle Eocene to recent planktonic foramineral biostratigraphie. In: Proceedings of the first international conference, of Plankt. Microfossils, Genève, 1967, pp199–421

Bolli HM, Saunders JB, Perch-Nielsen K (1985) Planktonic foraminifera, calcareous nannofossils and calpionellids: 1, 2nd edn. Cambridge University Press, New York, p 1032

Bouillin JP (1986) Le basin maghrébin: une ancienne limite entre l'Europe et l'Afrique à l'ouest des alpes : Bulletin de la Société Géologique de France II(8):4, 547–558

Bouyahiaoui B (2014) Structure profonde et réactivation de la marge Est-algérienne et du bassin adjacent (secteur d'Annaba), contraintes par sismique réflexion multi-trace et grand-angle terre-mer. PhD thesis, University of Nice-Sophia Antipolis, France, 373 pp.

Bouyahiaoui B et al (2015) Crustal structure of the eastern Algerian continental margin and adjacent deep basin: implications for late Cenozoic geodynamic evolution of the western Mediterranean. Geophys J Int 201:1912–1938

Boyer SE, Elliott D (1982) Thrust systems. Bull Am Assoc Pet Geol 66 (9):1196–1230

Bracène R (2001) Géodynamique du Nord de l'Algérie: impact sur l'exploration pétrolière. PhD thesis, University of Cergy-Pontoise, France, 101 pp.

Bracène R, Frizon de Lamotte D (2002) The origin of intraplate deformation in the Atlas system of western and central Algeria: from Jurassic rifting to Cenozoic-Quaternary inversion. Tectonophysics 357(1–4):207–226

Chabbi A, Chouabbi A, Chermiti A, Ben Youssef M, Kouadria T, Ghanmi M (2016) La mise en évidence d'une nappe de charriage en structure imbriquée: Cas de la nappe tellienne d'Ouled Driss. Souk-Ahras, Algérie: Courrier Du Savoir 21:149–156

Chabbi A (2017) Les nappes telliennes de la région nord de Souk Ahras (Nord-Est algerien) : étude géologique et structural. PhD thesis, University of Badji Mokhtar Annaba, Algeria, 172 pp.

Chadi M (1991) Géologie structurale des monts d'Aïn M'Lila (Algérie orientale). PhD thesis, University of Nancy I, 191 pp.

Chouabbi A (1987) Etude géologique de la région de hammam N'Bails SE de Guelma, Constantinois, Algérie. PhD thesis, University Paul Sabatier de Toulouse, France, 123 pp.

David L (1956) Etude géologique de la haute Medjerda. Bulletin du Service de la carte géologique de l'Algérie 11:304

Djelit H (1987) Evolution tectono-métamorphique du socle Kabyle et polarité de mise en place des nappes de flysch en petite Kabylie occidentale (Algérie). PhD thesis, University Paris Sud, 180 pp

Durand-Delga M (1969) Mise au point sur la structure de l'Algérie septentrionale. Bulletin du Service de la Carte Géologique de l'algérie 39:89–131

Durand-Delga M (1980) La Méditerranée occidentale : étapes de sa genèse et problèmes structuraux liés à celle-ci. Mémoires du Service Géologique de France 10:203–224

Feinberg H, Hoyez B, Lahondère JC (1981) Nouvelles données biostratigraphiques sur le Numidien de l'Algérie et du Maroc. Cahier Micropal Paris 3:93–99

Frizon De Lamotte D, Saint-Bezar B, Bracène R, Mercier E (2000) The two main steps of the Atlas building and geodynamics of the western Mediterranean. Tectonics 19(4):740–761

Frizon De Lamotte D, Michard A, Saddiqui O (2006) Quelques développements récents sur la géodynamique du Maghreb. C R Géosci 338:1–10

Frizon de Lamote D, Leturmy P, Missenard Y, Khomsi S, Ruiz G, Saddiqi O, Guillocheau F, Michard A (2009) Mesozoic and Cenozoic vertical movements in the Atlas system (Algeria, Morocco, Tunisia): an overview. Tectonophysics 475:9–28

Gélard JP (1979) Géologie du NE de la grande Kabylie (un segment interne de l'orogène littoral Nord africain). PhD thesis, University of Dijon, 326 pp.

Guerrera F, Loiacono F, Puglisi D, Moretti E (1992) The Numidian nappe in the Maghrebian chain. State of the art. Bull Soc Geol It 11:217–253

Guerrera F, Martin-Algarra A, Perrone V (1993) Late Oligocene - Miocene syn/late-orogenic successions in western and central Mediterranean chains from the Betic cordilleras to the southern Apennines. Terra Nova 5:525–544

Guiraud R (1975) L'évolution post-triasique de l'avant-pays de la chaîne alpine en Algérie, d'après l'étude du bassin du Hodna et des régions voisines. Rev Geogr Phys Geol Dyn 17:427–446

Hoyez B (1989) Le Numidien et les flyschs oligo-miocènes de la bordure sud de la Méditerranée occidentale. PhD thesis, University of Lille, 459pp.

Khelil M, Khomsi S, Roure F et al (2021) Late Miocene-Quaternary thrusting in the Utique-Kechabta foreland basin of the Tell, Northern Tunisia. Arab J Geosci 14:108. https://doi.org/10.1007/s12517-020-06388-2

Khomsi S, Bédir M, Soussi M, Ben MG, Ismail B, Lattrache K (2006) Mise en évidence en sub-surface d'événements compressifs Éocène moyen-supérieur en Tunisie orientale: généralité de la phase atlasique en Afrique du Nord. C R Geosci 338:41–49

Khomsi S, Ben Jemia MG, Frizon de Lamotte D, Maherssi C, Echihi O, Mezni R (2009) An overview of the Late Cretaceous-Eocene positive inversions and Oligo-Miocene subsidence events in the foreland of the Tunisian Atlas: structural style and implications for the tectonic agenda of the Maghrebian Atlas system. Tectonophysics 475:38–582. https://doi.org/10.1016/j.tecto.2009.02.027

Khomsi S, Frizon de Lamotte D, Bédir M, Echihi O (2016) The Late Eocene and Late Miocene fronts of the Atlas Belt in eastern Maghreb: integration in the geodynamic evolution of the Mediterranean Domain. Arab J Geosci 9:650–670. https://doi.org/10.1007/s12517-016-2609-1

Khomsi S, Roure F, Khelil M, Mezni R, Echihi O (2019) A review of the crustal architecture and related pre-salt oil/gas objectives of the eastern Maghreb Atlas and Tell: need for deep seismic reflection profiling. Tectonophysics 766:232–248. https://doi.org/10.1016/j.tecto.2019.06.009

Khomsi S, Khelil M, Roure F et al (2021) Surface and subsurface architecture of the Kasseb structures: implications for petroleum exploration beneath the Tellian allochthon, the easternmost portion of the Maghrebides. Arab J Geosci 14:189. https://doi.org/10.1007/s12517-020-06401-8

Kriviakine B, Kovalenko E, Vnouchkov V (1989) Geologic maps of Souk Ahras N° 77 and Oued Mougras N° 78, 1/ 50 000 scale: national office of geology, Algeria

Lahondère JC (1987) Les séries ultra-telliennes d'Algérie Nord oriental et les formations environnantes dans leur cadre structural : PhD thesis, University of Paul Sabatier Toulouse, 242 pp.

Lahondère JC, Feinberg H, Haq BU (1979) Datation des grès numidien d'Algérie orientale : conséquences structurales. C R Acad Sci Paris 289(D):383–386

Leikine M (1971) Etude géologique des Babors occidentaux (Algerie). PhD thesis, University of Paris, 242 pp.

Leprêtre R, Frizon de Lamotte D, Combier V, Gimeno-Vive O, Mohn G, Eschard R (2018) The tell-Riff orogenic system (Morocco, Algeria, Tunisia) and the structural heritage of the southern Tethys margin. BSGF 189:10

Marmi R, Guiraud R (2006) End Cretaceous to recent polyphased compressive tectonics along the "Mole Constantinois" and foreland (NE Algeria). J Afr Earth Sci 45:123–136

Obert D (1981) Etude géologique des Babors orientaux (Domaine tellien. Algérie). PhD thesis, University of Paris, 635 pp.

Peybernès B, Chouabbi A, Vila JM (2002) Reconstitution du substratum Jurassique- Crétacé basal du domaine des Sellaoua (Marge téthysienne sud, Algérie nord oriental) à partir des galets du bassin Mio-Pliocène de Hammam N'Bails. C R Pale 1:681–688

Raymond D (1976) Evolution sédimentaire et tectonique du NO de la Grande Kabylie (Algérie) au cours du cycle alpin. PhD Thesis, Paris, 152 pp.

Raoult JF (1974) Géologie du centre de la chaîne numidienne (nord du Constantinois, Algérie). PhD thesis, University of Paris, 163 pp.

Riahi S, Soussi M, Kamel B, Ben Ismail Lattrache K, Stow D, Sami K, Mourad B (2010) Stratigraphy, sedimentology and structure of the Numidian Flysch thrust belt in northern Tunisia. J Afr Earth Sci 57 (1–2):109–126

Roure F, Casero P, Addoum B (2012) Alpine inversion of the North African margin and delamination of its continental lithosphere. Tectonics 31:TC3006

Rouvier, H., 1977, Géologie de l'extrême nord tunisien: Tectonique et paléogéographies superposées à l'extrémité orientale de la chaîne nord-maghrébine. PhD thesis, Paris VI, 703 pp.

Rudis (1985) Geologic map of Souk Ahras 1/200 000 scale, national office of la Geology, Algeria

Talbi F, Melki FK, Ben Ismail-Lattrache K, Alouani R, Tlig S (2008) Le Numidien de la Tunisie septentrionale: données stratigraphiques et interprétation géodynamique. Estudios Geol 64(1):3–44

Vila JM (1978) La chaine alpine de l'Algérie orientale et des confins Algéro-tunisiens, Structural map, 1/500000 scale, CNRS et BEICIP

Vila JM (1980) La chaîne alpine d'Algérie nord-oriental et des confins Algéro-tunisiens. PhD thesis, University of Pierre et Marie Curie, France. 665pp.

Voüte C (1967) Essai de synthèse de l'histoire géologique des environ d'Ain Fakroune, Ain Babouche et des régions limitrophes. P S C G Algérie N S Bull 3:36; Algiers

Wezel FC (1970) Numidian flyschs: an Oligocene-early Miocene continental rise deposit off the African platform. Nature 228:275–276

Wildi W (1983) La chaîne tello-rifaine (Algérie, Maroc, Tunisie): structure stratigraphique et évolution du Trias au Miocène. Rev Géol Dyn Géogr Phys 24:201–297

4 Facies Analyses and Basin Evolution of the Cretaceous-Tertiary Rift-Related Sedimentary Succession of Haddat Ash Sham Area, West Central Arabian Shield, Saudi Arabia

Ali Abdullatif Mesaed, Rushdi Jamel Taj, and Sami Khomsi

Abstract

This study aims to describe the facies architectures and basin evolution of the rift-related Cretaceous-Tertiary succession of Haddat Ash Sham area, Saudi Arabia. It is based mainly upon detailed field works and measurements augmented with lab techniques. The succession of the study area is subdivided into three members, i.e. (1) a lower fluvio-lacustrine member which is composed of conglomerates, sandstones and mudstones arranged in fining-upward cycles. These lithologies represent deposition in mid to distal alluvial fan and lacustrine environments syncontemporaneous with reactivation and block faulting of the Precambrian Arabian Shield rocks, (2) a middle shallow marine member which is composed mainly of oolitic ironstones, ferruginous sandstones and siltstones, dolostones and mudstone arranged in shallowing-upward cycles. This member was deposited in tidally influenced shallow marine settings under short-lived periods of sea level fluctuation and (3) an upper fluvio-lacustrine member which is completely similar in lithologic characters and depositional environments to the lower member. This indicates the second regime of block faulting and reactivation and deposition of alluvial fan succession terminated by lacustrine regime. The lower and upper members contain volcanic ash beds which indicate subaerial to subaqueous volcanism contemporaneous with the sedimentation.

Keywords

Cenozoic of Saudi Arabia • Ash Shumaysi and Haddat Ash Sham formations • Fluvio-lacustrine rift-related succession • Ash Shumaysi ironstones • Jeddah-Makkah area

4.1 Introduction

4.1.1 General

Haddat Ash Sham area lies in the west-central part of the Arabian Shield NE of Jeddah city (Fig. 4.1). The area is occupied by Arabian Shield rocks that overlained by sedimentary succession (the aim of the present chapter). The area is cross-cutted by numerous NE and NW small wadies. The studied succession of Haddat Ash sham area is a part of the rift-related Tertiary sedimentary succession of the western part of Saudi Arabia. This succession extends from the extreme northwestern part of Saudi Arabia to the southern part along the Saudi Arabia–Yemen borders. The sedimentary succession of this area has been subdivided into different rock units, such as Ash Shumaysi and Usfan formations (Brown et al. 1963; Karpoff 1958). Ash Shumaysi Formation is subdivided into the Haddat Ash Sham, Shumaysi, Khulays and Buraykah formations (Spincer and Vincent 1984). The economic importance of this succession is related to the presence of oolitic iron ores, clays and carbonate deposits. These deposits have been studied by many authors (Al Shanti 1966, Spincer and Vincent 1984, Mesaed et al. 2011 and Taj 2011, Taj and Mesaed 2013). The main groundwater aquifer of Makkah district is present within this succession.

A. Abdullatif Mesaed (✉) · R. J. Taj · S. Khomsi
Faculty of Earth Sciences, King Abdulaziz University, Gepexploration Department, P.O. BOX 80206 Jeddah, 21589, Saudi Arabia
e-mail: amesaed@kau.edu.sa

R. J. Taj
e-mail: rtaj@Kau.edu.sa

A. Abdullatif Mesaed
Geology Department, Faculty of Sciences, Cairo University, Giza, Egypt

S. Khomsi
CERTE, Georesources, University of Carthage, Carthage, Tunisia

S. Khomsi and F. Roure (eds.), *Geology of North Africa and the Mediterranean: Sedimentary Basins and Georesources*, Regional Geology Reviews, https://doi.org/10.1007/978-3-031-18747-6_4

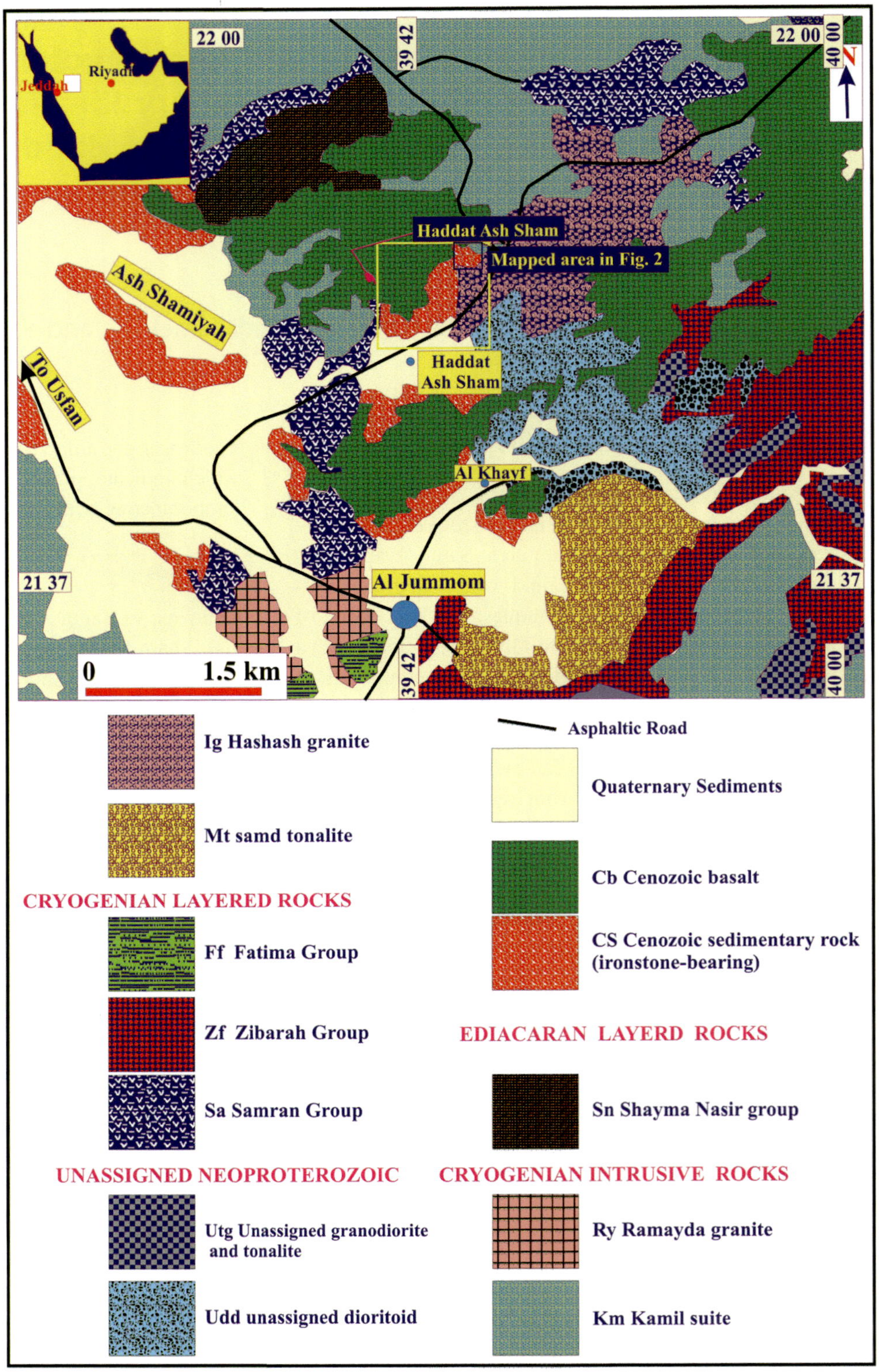

Fig. 4.1 Geologic map of Haddat Ash Sham area (Modified after Johnson 2006)

4.1.2 Aims and Methods of Study

The present study aims to clarify the facies analysis and basin evolution of the sedimentary succession of Haddat Ash Sham area. The study is based upon the following systematic methods of study: (1) Regional field works including correlation and selection of the best exposures for the different units of the studied section, (2) Measurement of stratigraphic sections; (3) drawing of the measured sections by computer programmes, i.e. Corel Draw 10; (4) Collection of hand samples of the different sedimentary facies, (5) microscopic description of the prepared thin sections of the different sedimentary facies and, (6) lumping the data and preparation of the manuscript by using different computer softwares.

4.2 Geologic Setting

The study area comprises the Arabian Shield Rocks, the Cretaceous-Tertiary sedimentary succession, the Tertiary-Quaternary volcanic (Harrat) and the Quaternary-Recent alluvial deposits (Fig. 4.1). The Arabian Shield rocks of the study area are previously studied by Al-Shanti (1966), Moore and Ar-Rehaili (1989) and Johnson (2006). According to Johnson (op. cit.), the Precambrian rock units of the study area (Fig. 4.1) are represented by: (1) Unassigned diorite and unassigned granodiorite and tonalite, (2) The Cryogenian layered rocks which are represented by the Samran (Sa), Zibarah (Zf) and Fatimah Groups (Ff) which are composed mainly of mafic to felsic lavas and the related volcaniclastic rocks, (3) The Cryogenian intrusive rocks which are represented by: A) the Sand tonalite (Mt), the Hashash granite (Ig), the Kamil suite (Km) and finally the Ramayda granite (Ry) and (4) The Edicarian layered rocks which area are represented by the Shyma Nasir group (Sn).

The Cenozoic succession is exposed below the flat-lying lavas and Quaternary deposits of the study area. Brown and others (1963) originated the names Shumaysi and Usfan formations to these sedimentary sequences after Karpoff (1958). Spincer and Vincent (1984) divided the Shumaysi formation into the Haddat Al-Sham, Shumaysi, Khulays and Buraykah formations. Haddat Al-Sham Formation has been divided into three members (Zeidan and Banat 1990): a lower, middle and upper member. A Middle Cretaceous age has been assigned to Haddat Al-Sham Formation. The Usfan Formation lies conformably on Haddat Al-Sham rocks and is characterized by its carbonate ledge. The formation has been divided into three members (Zidan and Banat 1990): a lower, middle and upper member. It has been dated as Eocene or Oligocene (Al-Shanti 1966). Mesaed et al. (2011) studied the different facies association of Al Shumaysi Formation in wadi Al Shumaysi area. They related the deposition of this formation to a fluvio-lacustrine depositional environment. The Khulays Formation is encountered in the Sugah trough and it is probably faulted against Precambrian rocks and the Usfan formation. According to Spincer and Vincent (1984), the Khulays formation is unconformable to the Shumaysi formation in Harrat Nuqrah but conformable on the same formation towards the southern end Sugah trough. The Khulays formation consists of alternating red and brown smectite clay, siltstone and fine-grained structureless sandstone with marly limestone.

The Oligo-Miocene lava fields occupy large surface areas in western Saudi Arabia, one of the largest lava fields occurs in Harrat Rahat, which extends from Al-Madinah Al-Munawarah southward for about 310 km into the study area. These basaltic rocks were mapped by Smith (1981 and 1982) and he assigned to them the Rahat group. These harrat rocks are mainly formed of flat-lying alkali basalts surrounded by a few pyroclastic cinder cones.

The Quaternary deposits cover large parts of the study area. They principally occur in the large drainage basins of Haddat Al-Sham, Wadi Al-Shumaysi, Al-Bayadi and wadi Al-Sugah. The principal units of the Quaternary rocks are the terrace gravel, alluvial fan deposits, tallus deposits, alluvial sands and gravels of wadi beds and some eolian edifices. The thickness of these deposits varies widely from one place to another.

4.3 Facies Analyses and Basin Evolution of the Sedimentary Succession of Haddat Ash Sham Area

The stratigraphic section of Haddat Ash Sham area is recently subdivided by Taj (2011) and Taj and Mesaed (2013) into three members: lower fluvio-lacustrine clastic member; middle shallow marine oolitic ironstones-carbonate member and upper tidal flat-fluvio-lacustrine clastic member. Each of these members was further subdivided into different units which were further subdivided into facies of characteristic sedimentary cycles (Figs. 4.2, 4.3A; Table 4.1).

4.3.1 Lower Fluvio-Lacustrine Clastic Member

This member forms a characteristic greyish-white horizon just overlying the weathered Precambrian igneous rocks with a characteristic quartz pebble conglomerate (Fig. 4.3B). The lower member ranges in lithology from conglomerates, coarse to fine-grained sandstones, siltstones and claystones. It is subdivided into the following units.

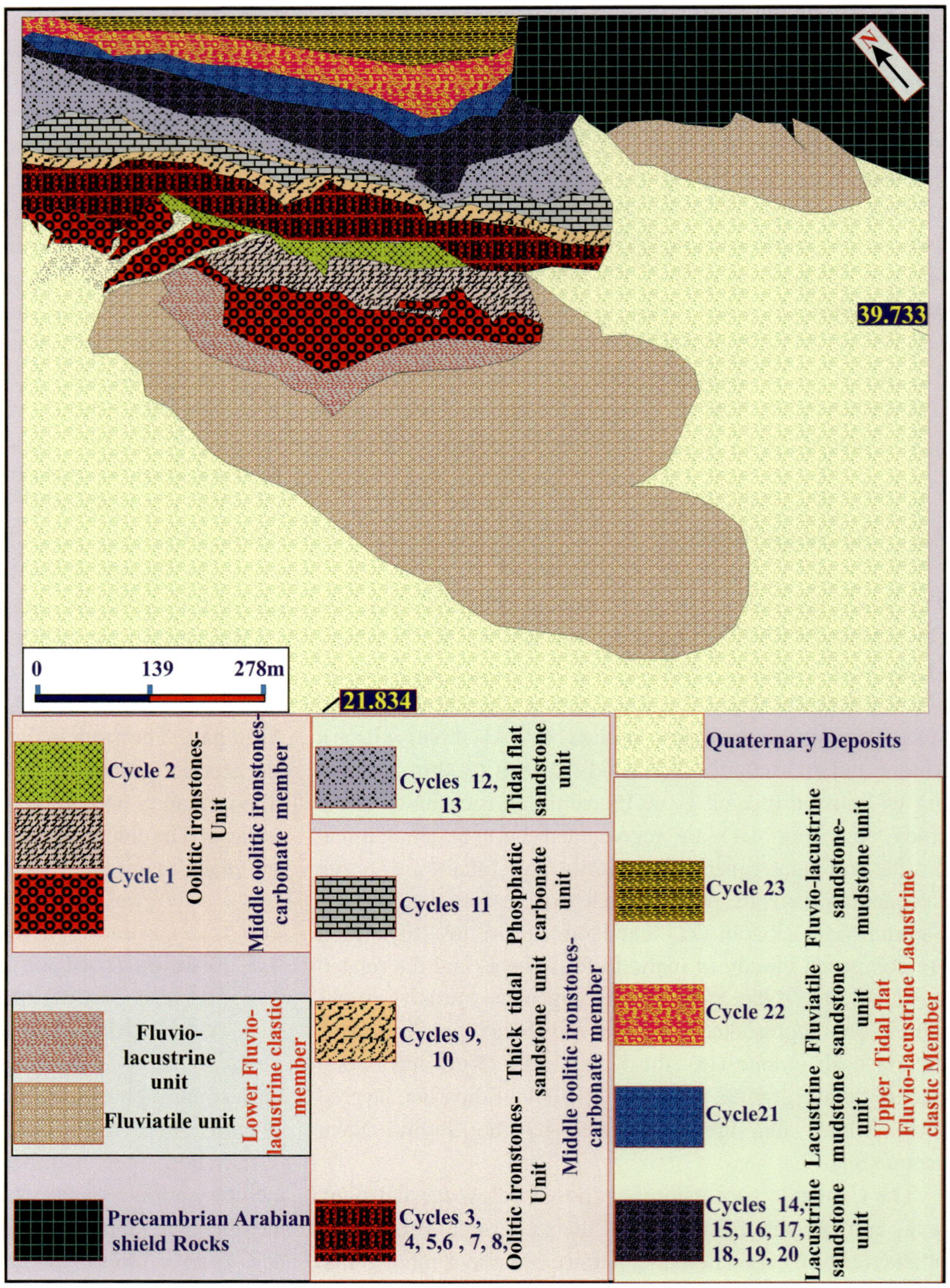

Fig. 4.2 Detailed geologic map showing the different members, units and cycles of the studied succession of Haddat Ash Sham area

4.3.1.1 Fluviatil Sandstone-Conglomerate Unit

This unit is present in the lowermost part of the measured section. It forms a characteristic kaolinitic tuffaceous horizon in the lower part of the section (Fig. 4.4, column A, B). It is composed of conglomerate, sandstone and siltstone and mudstone organized in multiple fining-upward cycles of braided stream system, being composed of four successive facies (F1, F2, F3, F4 Table 4.1), These facies are described here from bottom to top as follows:

Massive disorganized quartz pebble conglomerate Facie F1: The Massive diorganized tuffaceous kaolinitic conglomerate facies rests directly on lateritized Precambrian granites and ranges in thickness from 0.5 to 2 m and it represents channel lag deposits (Fig. 4.3B). The conglomerates are mostly friable, reddish brown and white, massive or crudely laminated (Fig. 4.3C). Rounded to subrounded red quartz pebbles and angular fragments of altered granites are the main framework components of these conglomerates (Fig. 4.3D). They set in a ferruginous and kaolinitic tuffaceous matrix.

Interpretation: This basal conglomerate unit is texturally like the alluvial facies “GM” of Mail (1977) which is related to deposition on low-relief longitudinal bars in the proximal (upper) reaches of braided streams. This facie is interpreted

Fig. 4.3 **A** = The complete stratigraphic succession of Haddat Ash Sham which consists of lower fluvio-lacustrine clastic member (1); middle shallow marine oolitic ironstones-carbonate member (2) and an upper tidal flat-fluvio-lacustrine clastic member (3); **B**, **C** = Massive disorganized quartz pebble conglomerate Facie F1 which composed mainly of quartz pebble conglomerates; **D** = The rounded to sub-rounded quartz pebbles of the quartz pebble conglomerates; **E**, **F** = Bedded quartz pebble conglomerates (1) and fine conglomerates (2) facie F2 in successive fining-upward cycles

as bedload gravel that was deposited from clast-by-clast accretion during higher discharges. This is supported by the presence of thin bedding and a clast-supported framework with no internal organization (Smith 1974; Hein and Walker 1977; Miall 1977; Rust 1978; Karpeta 1993a, b).

Bedded quartz pebble conglomerates and fine conglomerates F2: This facie is present just overlying and laterally changed to the aforementioned facies. It consists of successive fining-upward cycles. Each of these cycles begins with quartz pebble conglomerates and terminated with white kaolinitic conglomerates (Fig. 4.3E, F). The lower quartz pebble conglomerate unit consists mainly of rounded to subrounded quartz pebbles (clast-supported) embedded in friable white kaolinitic conglomerates (Fig. 4.5A, B). The upper kaolinitic conglomerates horizons of these cycles are cross-bedded and white small quartz granules are arranged along the bottom sets of these cross-beds. It is observed that there is an upward increase in the thickness of the white conglomerate horizons and a decrease in the thickness of the quartz pebble conglomerates from the lower to the upper part of this facie (Fig. 4.5C, D).

Interpretation: The quartz pebble conglomerate of the lower parts of these cycles represents deposition during high-energy stream flows while the kaolinitic conglomerate beds indicate deposition in low-energy distal parts of braided streams. The multistory character of the conglomerate bodies formed of superposed distinct gravel lithofacies (Gm, Gp, Gc; Miall 1978) is indicative of episodic ephemeral streams. A depositional model of facies F1 and F2 can be described as proximal coarse-unstratified conglomerates (debris flows), passing down fan into finely stratified conglomerates (turbulent stream flows).

Cross-bedded pebbly sandstone-conglomerate Facie F3: this facie consists of successive lenticular sand bodies, 1–3 m in diameter, displaying horizontal and low angle cross-bedding and lamination as well as fining-upward pattern (Fig. 4.5E, F). The lateral persistence and thickness of these lenticular bodies and their internal structures vary discriminately from place to place. Each fining-upward sequence has a basal erosion surface, along which pebbles and cobbles of quartz grains are distributed. These are followed by trough and tabular cross-stratified coarse to medium-grained kaolinitic sandstones (Figs. 4.5F, 4.6A). These sandstones are weakly consolidated and often penetrated by short and thin ferruginated root moulds and tubes giving a characteristic mottling. Microscopically, the sandstone consists of mono and polycrystalline quartz grains embedded in kaolinitic matrix (Fig. 4.6B, C).

Interpretation: The composition and sedimentological characters of this middle unit may reflect accumulation in distal reaches of a braided stream (Reading 1978). Both grain supported and matrix supported conglomerates form recognizable beds at the base of distinct cycles at the outcrop. The sandstone units are frequently cross stratified, generally moderately to well sorted and composed mainly of quartz and are thus texturally and mineralogically mature. The general characteristics of this sequence especially the fining-upward character, compositional and textural maturity and unidirectional paleo-current trends, suggest a fluvial depositional environment dominated by braided streams with sands deposited as channel bars consequent to fluctuating flow velocity. The stacked, amalgamated trough-cross beds in the upper part of this facie are interpreted as individual channel forms, and the upward fining-upward cycles of metre-scale are inferred to reflect waning flows (Leeder 1999; Allen 1985).

Parallel bedded tuffaceous sandstones-mudstone Facie F4: This facie consists of parallel bedded greyish white tuffaceous mudstone-siltstone and sandstone (Fig. 4.6D, E). The tuffaceous sandstone beds are hard and contain numerous black glass shreds. Microscopically, the tuffaceous sandstone consists mainly of quartz grains within microcrystalline devitrified volcanic glass (Figs. 4.6F, 4.7A).

Table 4.1 The different members, units and facies (F1-F25) of the studied sedimentary succession of Haddat Ash Sham area

	Member	Unit	Facies
Cretaceous-Tertiary Sedimentary Succession	Upper Tidal Flat-Fluvio-Lacustrine Clastic Member	Fluvio-lacustrine sandstone-mudstone unit (Cycle 23)	Tuffaceous sandstone Facies F 25 Tuffaceous Mudstone Facies F24
		Fluviatile sandstone unit (Cycle 22)	Tabular cross-bedded conglomerate-pebbly sandstone facies F23 Trough and tabular cross-bedded conglomerate facies F22 Disorganized corglomerate facies F21
		Lacustrine Mudstone unit (Cycles 21)	Cross-laminated light brown sandstone facies F20 Tuffaceous kaolinitic mudstone facies F19
		Lacustrine Sandstone unit (Cycles 14, 15, 16, 17, 18, 19, 20)	Cross-bedded sandstone Facie F18 Cross-laminated kaolinitic siltstone Facies F17 Ferruginous mudstone-sandstone Facies F16
		Tidal Flat sandstone Unit (Cycles 12, 13)	Ferruginous mudstone-sandstone Facies F15
	Middle Oolitic Ironstones-Carbonate Member	Phosphatic Carbonate Unit (Cycle 11)	Phosphatic dolostone facies F14 Thinly bedded carbonate facies F13 Ferruginous mudstone-sandstone Facies F12
		Thick Tidal Flat Sandstone Unit (Cycles 9, 10)	Thinly bedded ferruginous sandstone-sandy ironstone Facies F11 Thickly cross-bedded sandstone Facies F10
		Oolitic Ironstone Unit (Cycles 1, 2, 3, 4, 5, 6, 7, 8)	Mudstone- oolitic ironstone-sandstone Facies F9 Mudstone-silty ironstone-sandstone Facies F8 Grey and green chamositic mudstone-oolitic ironstones-sandstone Facies F7
	Lower Fluvio-Lacustrine Clastic Member	Lacustrine Mudstone-Sandstone Unit	Laminated mudstone Facie F6 Parallel bedded ferruginous sandstone-mudstone Facies F5
		Fluviatile Sandstone-conglomerate Unit	Parallel bedded tuffaceous sandstones-mudstone Facies F4 Cross-bedded pebbly sandstone-conglomerate Facies F3 Bedded quartz pebble conglomerates and fine conglomerates F2 Massive disorganized quartz pebble conglomerate Facies F1
Precambrian	Arabian Shield Rocks		

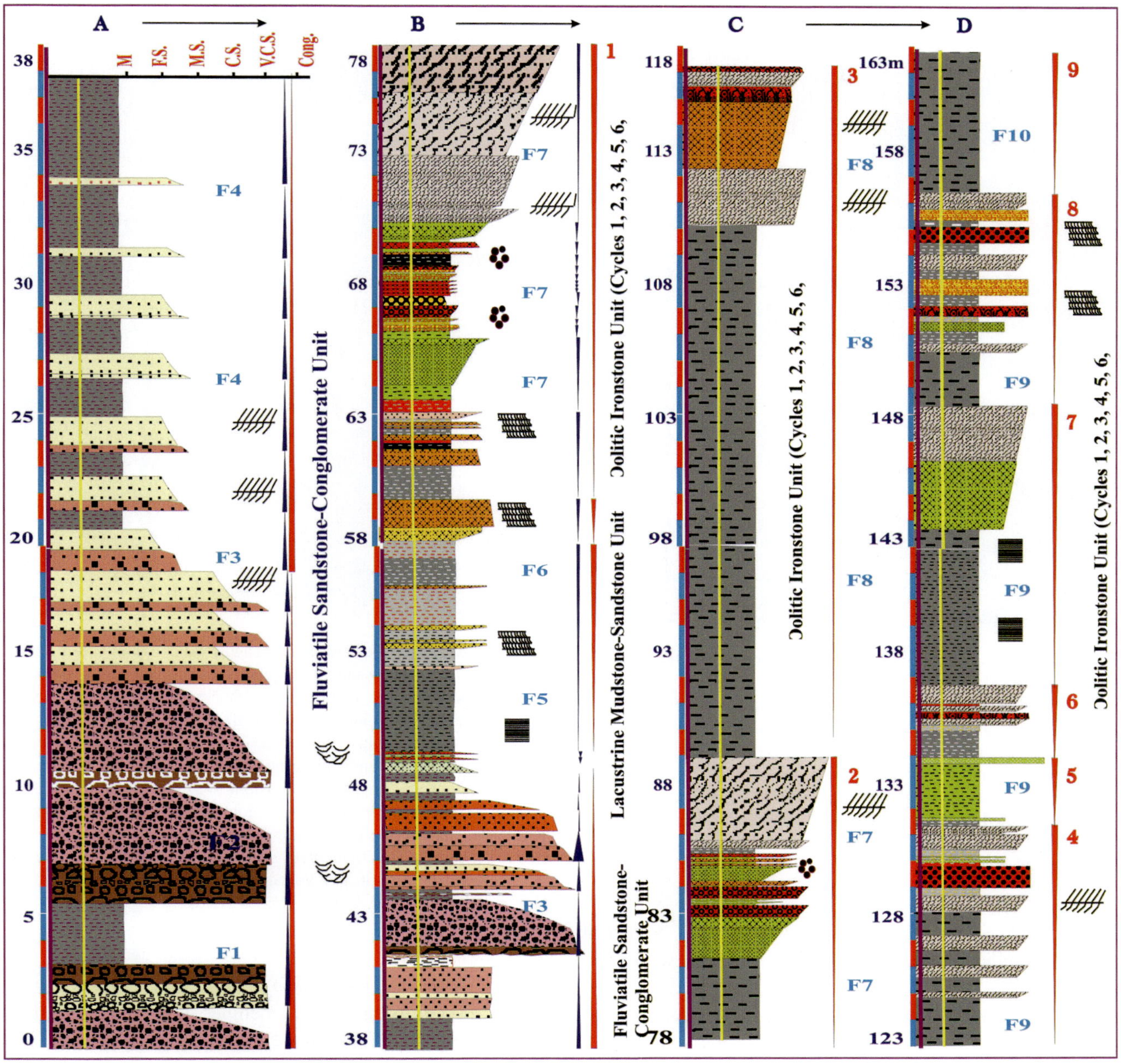

Fig. 4.4 Detailed Stratigraphic section of the lower part of the studied sedimentary succession of Haddat Ash Sham area (Modified after Taj and Mesaed 2012)

Interpretation: The parallel bedding of tuffaceous sandstone and mudstone indicates deposition in a quite depositional environment of fluctuation strength in the distal part of braided streams post the deposition of facie F3. The tuffaceous materials of this facie indicate the ash fall out and reworking of volcanic ashes of syncontemporanous volcanic activities during the deposition of this facie.

4.3.1.2 Lacustrine Mudstone-Sandstone Unit

Parallel bedded ferruginous sandstone-mudstone Facie F5: This facie consists of parallel white kaolinitic mudstone and reddish ferruginous sandstone (Fig. 4.7B). This unit attains up to 10 m thickness and it consists of successive 8 small-scale cycles each one began by mudstone and terminated by ledge-forming ferruginous sandstone/sandy ironstone.

Interpretation: The parallel bedded sandstone and mudstone are arranged in small-scale shallowing-upward cycles which indicates deposition in marginal marine lakes of saline affinities. The presence of green chamositic clays and the coarsening and thickening-upward characters indicate the general shoaling and shallowing of these lakes.

Laminated mudstone Facie F6: This facie is composed of thinly bedded greyish-white kaolinitic mudstone and

Fig. 4.5 **A, B** = Bedded quartz pebble conglomerates and fine conglomerates F2 which consists of successive fining-upward cycles of quartz pebble conglomerates (1) and terminated with white kaolinitic conglomerates (2); **C, D** = An upward increase in the thickness of the white conglomerate horizons (2) and decrease in the thickness of the quartz pebble conglomerates (1) from the lower to the upper part of the bedded quartz pebble conglomerates and fine conglomerates F2; **E, F** = Cross-bedded pebbly sandstone-conglomerate Facie F3 which consists of horizontal and low angle cross-bedded conglomerates of fining-upward pattern

Fig. 4.6 **A** = Cross-bedded pebbly sandstone-conglomerate Facie F3 which consists of fining-upward sequences of basal pebbles and cobbles of quartz followed by a trough and tabular cross-stratified coarse to medium-grained kaolinitic sandstones, **B, C** = Mono (1) and polycrystalline (2) quartz grains embedded in kaolinitic matrix (3) in the cross-bedded pebbly sandstone-conglomerate Facie F3; **D, E** = Parallel bedded tuffaceous sandstones-mudstone Facie F4 that consists of parallel bedded greyish white tuffaceous mudstone-siltstone (1) and sandstone (2); **F** = The tuffaceous sandstone of facie F4 of figures D, E consists mainly of fine- to medium-grained quartz grains (qz) embedded in microcrystalline devitrified volcanic glass (vg). T.S. = Thin Section, O.L. = Ordinary Light, C.N. = Crossed Nicols

fine-grained sandstone/siltstone (Fig. 4.7C). Numerous gypsum veinlets and streaks are predominated within this facie.

Interpretation: This facie is generally laminated, and this indicates deposition in quiet environments with slight fluctuation in the water depth. The predominance of gypsum veinlets and streaks within this facie indicates deposition in saline-restricted distal lakes in the distal alluvial fan depositional environments.

4.3.2 Shallow Marine Middle Oolitic Ironstones-Carbonate Member

This member attains up to 160 m thickness and it is composed of 11 shallowing-upward cycles (Fig. 4.4 Column B, C, D and Fig. 4.8E and upper part of column F). This unit forms a characteristic slope-forming unit of predominant mudstone lithology with characteristic protruded ferruginous sandstone and/or oolitic ironstone ledges demarcating the topmost parts of the shallowing-upward cycles (Fig. 4.7D). The ironstones of Haddat Ash Sham area are present within the succession of this middle member. It is composed of ten successive shallowing-upward cycles.

4.3.2.1 Oolitic Ironstone Unit (Cycles 1, 2, 3, 4, 5, 6 and 8)

Grey and green chamositic mudstone-oolitic ironstones-sandstone Facie F7 (Cycles 1, 2): This facie forms more than ¾ of the total thickness of the measured sedimentary succession in the study area. It is composed mainly of numerous large-scale shallowing-upward cycles. Each of these large cycles began with a characteristic slope-forming thick basal mudstone unit which grades upward into a middle rhythmic unit of parallel mudstone, siltstone, oolitic and non-oolitic ironstone and terminated by sheet-like cross-bedded tidal flat sandstone (Fig. 4.7E). They are generally showing evidence to support their deposition under upward increase in the intensity of the waves and current activities. This first ironstone-bearing cycle attains up to 17 m thickness

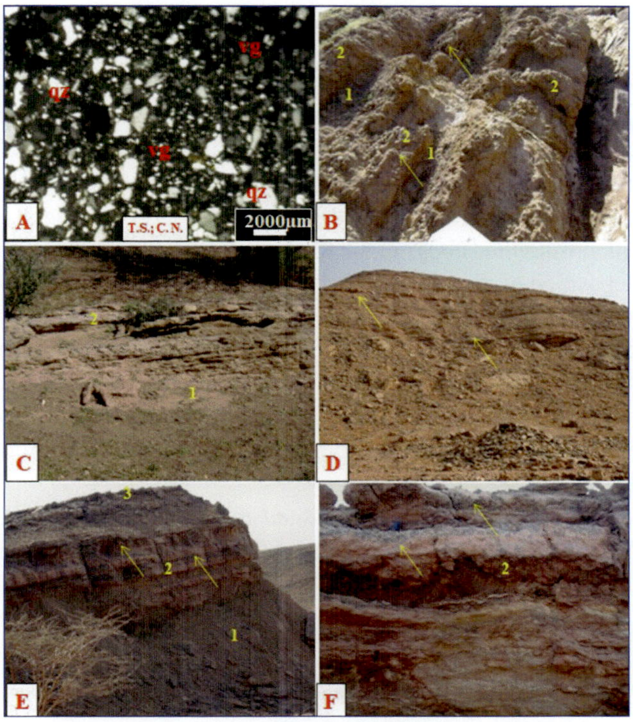

Fig. 4.7 **A** = The tuffaceous sandstone of facie F4 of the above-described plate 3F consists mainly of fine- to medium-grained quartz grains (qz) embedded in microcrystalline devitrified volcanic glass (vg); **B** = Parallel bedded ferruginous sandstone-mudstone Facie F5 that consists of parallel white kaolinitic mudstone (1) and reddish ferruginous sandstone (2); **C** = Laminated mudstone Facie F6 which composed of thinly bedded greyish white kaolinitic mudstone (1) and fine-grained sandstone/siltstone (2); **D** (From Taj and Mesaed 2012) = Complete succession of the shallow marine middle oolitic ironstones-carbonate member which is composed of successive shallowing-upward cycles with a characteristic protruded ferruginous sandstone and/or oolitic ironstone ledges (arrows) demarcating the topmost parts of the shallowing-upward cycles; **E; F** = The first ironstone-bearing cycle of the grey and green chamositic mudstone-oolitic ironstones-sandstone Facie F7 which begins with a characteristic mudstone unit (1) which grades upward into the middle rhythmic unit (2) of parallel mudstone, siltstone, oolitic and non-oolitic ironstone (arrows) and terminated by sheet-like cross-bedded tidal flat sandstone (3)

and forms a characteristic unit in the middle part of the sedimentary succession of the study area (Figs. 4.7E, F, 4.8A). This cycle comprises two medium-scale cycles each one began with a characteristic mudstone overlained by a rhythmic unit of chamositic mudstone, chamositic siltstone /sandstone and ironstones and terminated by sandstone (Fig. 4.8B, C). The oolitic ironstone beds of this first cycle are composed of rounded to subrounded goethite and hematite ooids embedded in goethitic groundmass (Fig. 4.8D, E). The second ironstone-bearing cycle is also remarked by the thick grey mudstones in its lower part just overlying the tabular cross-bedded sandstones terminating the first large cycle (Fig. 4.8B). It attains up to 15 m thickness and it is similar to the aforementioned one where it begins with a thick basal mudstone (3 m thick) grading upward into a middle oolitic ironstone-bearing unit (5 m thick) containing characteristic two reddish bench-like oolitic ironstone beds similar to those described in the aforementioned first large cycle (Fig. 4.8F, G).

Fig. 4.8 **A** = The first ironstone-bearing cycle of Facie F7 which consists of a characteristic mudstone unit (1) which grades upward into middle rhythmic unit (2) of parallel mudstone, siltstone, oolitic and non oolitic ironstone (arrows) and terminated by sheet-like cross-bedded tidal flat sandstone (3); **B, C** = The two medium-scale cycles of the first cycle of figure. A which began by a characteristic mudstone overlained by rhythmic unit of chamositic mudstone, chamositic siltstone/ sandstone and ironstones and terminated by sandstone (arrows); **D, E** = The petrographic composition of the oolitic ironstone beds of the first cycle shows the presence of rounded to subrounded goethite and hematite ooids embedded in goethitic groundmass; **F, G** = The second ironstone-bearing cycle of facie F7 which is similar to the aforementioned one where it begins with a thick basal mudstone grading upward into a middle oolitic ironstone-bearing unit contains characteristic two reddish bench-like oolitic ironstone beds (arrows); **H** = The bioturbated silty ironstone beds (arrows) of the mudstone-silty ironstone-sandstone facie F8 (Cycle 3) Fig. 4.8 C from Taj (2011)

Interpretation: The development of coarsening-upward packages within the middle member compared to the lower member indicates a distinct change to the depositional environment (Miall 1990; Plint and Walker 1992). The mudstone units represent periods of sea transgression (Hallam and Bradshaw 1979; Van Houten and Bhattacharyya 1982; Van Houten et al. 1984; Guerrak 1987; Van Houten and Arthur 1989; Young 1989, Cotter and Link 1993 and Mesaed 1995). Most of these authors used this green or dark mudstone as a good indication of new transgressive cycles. The coarsening-upward tendency of the studied ironstone-bearing succession is similar to that described in the Coniacian-Santonian oolitic ironstones of Aswan Area, Egypt, by Van Houten and Bhattacharyya (1982), Mesaed (1995), El Sharkawi et al. (1996) and El Aref et al. (1996), where they related these cycles to shoaling. The studied ironstone-bearing coarsening-upward cycles are characterized by a sharp contact against the underlying one and start with mudstone dominant unit, being formed of laminated green mudstone representing deposition in low-energy environment within inter-bar areas (troughs) of migrated sand bars. According to Taj (2011) and Taj and Mesaed (2013), the oolitic ironstone beds and the associated sandstone and mudstone interbeds of the described large-scale coarsening-upward cycles seem to be deposited in a shallow marine moderately to highly agitated environment. This is supported by: (a) the rippled cross-bedded, frequently burrowed and texturally mature sands of this unit which reflect a high-energy depositional environment of probably wave-swept subaqueous coastal barrier or swept and (b) the abundance of horizontal and ripple cross-laminations and also symmetrical ripples in the sandstone beds and the associated oolitic ironstones which reflect the importance of upper flow regime processes, producing the horizontal laminations and the role of wave-generated oscillation, producing symmetrical ripples in the lower foreshore or upper shoreface (Cotter and Link 1993).

Mudstone-silty ironstone-sandstone Facie F8 (Cycle 3): This facie is represented by the 3rd ironstone-bearing cycle (Fig. 4.4, column C). It began with thick basal mudstone which grades upward green chamositic mudstone and contains less frequent chamositic ironstone patches and domains. This facie is terminated by a characteristic bench-like protruded ferruginous siltstone/silty ironstones (Fig. 4.8H). This upper part is composed of successive 6 to 7 small-scale cycles. Each of these cycles began with fine-grained light colour sandstone/sandy sandstone. In the uppermost part of this large cycle, characteristic three sandstone-bearing cycles are present. These cycles began with friable ferruginous mudstone terminated by ripple cross-laminated to cross-bedded hematitic sandstone/sandy ironstone. The microscopic description of the first bioturbated hematitic sandstone/sandy ironstone beds revealed its composition from moderately to well-sorted, angular to subrounded fine- to medium-quartz grains. These grains are embedded within a green chamositic-kaolinitic clayey matrix. This matrix shows progressive stages of hematitization leading to the formation of dark hematitic domains in association with light kaolinitic unhematitic domains (Fig. 4.9A). In the dark hematitic patches and domains, the quartz grains show progressive and subsequent stages of corrosion and embayment by enclosing hematite cement.

Interpretation: The predominance of grey and green chamositic mudstones in the lower part of this facie indicates deposition during transgressive periods. The thinly laminated ferruginous siltstone/silty ironstones of the upper parts of the shallowing-upward cycles indicate deposition in slightly agitated environments. Ripple cross-lamination indicates that this lithofacie resulted from migrating current ripples (Harms and others 1982; Miall 1996; Nichols 1999).

Mudstone-oolitic ironstone-sandstone Facie F9 (Cycle 4, 5, 6, 7, 8): This facie is present in the ironstone-bearing

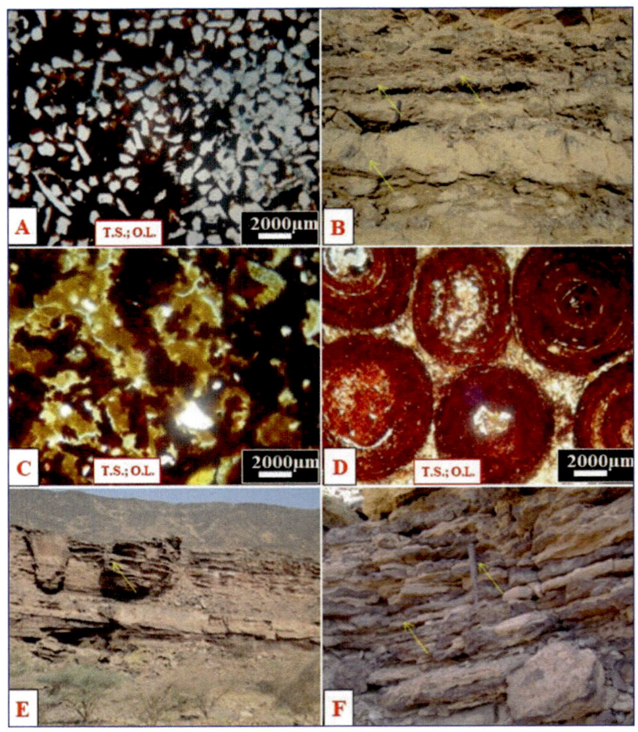

Fig. 4.9 **A** = The bioturbated hematitic sandstone/sandy ironstone beds are composed of moderately to well sorted, angular to subrounded fine- to medium-quartz grains (white) embedded within the green chamositic-kaolinitic clayey matrix; **B** = cycle no 4 of the mudstone-oolitic ironstone-sandstone Facie F9 which terminated red oolitic ironstone beds (arrows); **C** = The lower horizon of the ironstone beds of cycle 4 is composed of thinly laminated chamositic ironstone which is composed mainly of green chamositic clays (green); **D** (From Taj 2011) = The oolitic ironstone beds of cycle 4 are composed of matrix- to grain-goethite (blood red) and hematite ooids (black); **E** = The thickly cross-bedded sandstone Facie F10 (Cycles 9) which consists mainly of cross-bedded sandstone; **F** = The thinly bedded ferruginous sandstone-sandy ironstone Facie F11 (Cycles 10) which consists of rhythmically bedded yellow and red ferruginous sandstone and reddish break to black sandy ironstone beds

cycles no. 4, 5, 6, 7 and 8 (Fig. 4.4, Column D). Cycle no. 4 attains up to 10 m thickness and it began with successive 4 small cycles of grey mudstone and cross-laminated to cross-bedded sandstone. These four cycles are followed upward by bench-like characteristic red oolitic ironstone bed which contain thin ferruginous mudstone horizon in the middle part (Fig. 4.9B). The lower horizon of this ironstone bed is of silty type, and it is composed of thinly laminated red goethitic silty ironstone and black hematitic silty ironstone (Fig. 4.9C). The upper part of this ironstone bed is oolitic in composition and it is composed mainly of matrix- to grain-supported goethitic and hematitic ooidal ironstone (Fig. 4.9D). The ooids are hematitic, rounded to subrounded, moderately to well sorted and are embedded within hematitized goethite cement contains some white kaolinitic patches and domains (Fig. 4.9D). Cycle no. 5 is composed mainly of thick green chamositic mudstone and terminated by green chamositic fine-grained ironstone (cycle No. 5, middle part of column D, Fig. 4.4). Cycle no. 6: attains up to 3 m thickness and it begins with a characteristic slope-forming grey mudstone unit (1 m thick) which grades upward into bedded unit of thin mudstone and chamositic siltstone/chamositic silty ironstone (middle part of column D, Fig. 4.4). Cycle no. 7 attains up to 11 m thickness and it is composed of basal grey mudstone (6 m thick) grading upward into green chamositic mudstone and terminated by ledge-forming cross-bedded kaolinitic ironstone (Fig. 4.4 Column D). Cycle no. 8 attains up to 14 m thickness and it begins with grey mudstone which grades upward into interbedded unit of kaolinitic sandstone, chamositic ironstone, red oolitic, sandy oolitic and sandy ironstone (Fig. 4.4, upper part of Column D).

Interpretation: According to Taj (2011) and Taj and Mesaed (2013), the stratigraphic setting of Haddat Ash Sham oolitic ironstones, their composition, textural maturity, internal sedimentary structures and lateral facies changes strongly support the concentration of the ferruginous ooids in highly agitated conditions along bar flanks and bar crests during the regressive events which terminate the short-lived (small-scale) prograding regimes. During the regressive events, winnowing, transportation and redeposition of the formed ooids take place, and the coarse sediments including quartz grains and/or iron ooids prograde gradually into the interbars shelf muds. This is indicated by the graded bedding, rippling and burrowing of the oolitic ironstone beds. The mudstones of the lower parts of the shallowing-upward can be also correlated with the shale-dominated distal facies described by Jennette and Pryor (1993) and the shelf prodelta shales and siltstones described by Tankard and Barwis. Similar muddy facies are described by Teyssen (1989) in the coarsening-upward depositional cycles of the Minette oolitic ironstones of Luxembourge and Lorraine, France.

4.3.2.2 Thick Tidal Flat Sandstone Unit (Cycles 9, 10)

Thickly cross-bedded sandstone Facie F10 (Cycles 9): This facie attains up to 25 m thickness. It begins with 7 m grey mudstone which grades upward into interbedded kaolinitic sandstone, green chamositic sandstone and red sandy ironstone (Fig. 4.10, Column E, Fig. 4.9E). This facie is terminated by thickly bedded cross-bedded sandstone. The sandstone shows an upward increase in the angle of the cross beds and grain size. It is composed mainly of rounded to subrounded, moderately sorted fine- to medium-quartz grains.

Interpretation: The uppermost sheeted sandstone units terminating these large-scale coarsening-upward cycles are suggested to be deposited during regressive phases. Parallel laminations have been formed as sand falls out from suspension. Waning current conditions prevailed during the deposition of these sandstone beds is evidenced by the occurrence of lower flow regime structures, such as current and wave-ripple cross-laminations, immediately above the parallel laminated or cross-stratified division.

Thinly bedded ferruginous sandstone-sandy ironstone Facie F11 (Cycles 10): This facie represents the uppermost part of this unit. It consists in the western part of the study area from about 5 m of rhythmically bedded units of yellow and red ferruginous sandstone and reddish break to black sandy ironstone beds (Fig. 4.9F). In the eastern part of the study area, this cycle is mudstone dominated where it consists mainly from two small cycles each one begins with thick grey mudstone and terminated by ripple cross-laminated to cross-bedded fine-grained sandstone (Fig. 4.10; Column E, Cycle 10; Fig. 4.11A). The cross-laminated to cross-bedded fine-grained sandstone terminating the upper part of this cycle attains up to 4 m thickness.

Interpretation: The rhythmic nature of this facie reflects deposition under fluctuating water tables in marginal lagoons. The presence of silty and sandy ironstones in the topmost parts of the shallowing-upward cycles reflects upward shoaling and reduction in the clastic input which favour the Fe-oxyhydroxides accumulation.

4.3.2.3 Phosphatic Carbonate Unit (Cycle 11)

This unit attains up to 25 m thickness (Fig. 4.10, upper part of column E and lower and middle part of column F).

Ferruginous mudstone-sandstone Facie F12: This facie consists of interbedded grey mudstone, cross-laminated siltstone, fine sandstone, red sandy and silty ironstone (Fig. 4.11B, C). It is terminated by green phosphatic chamositic sandstone bed (60 cm thick lower part of column F, Fig. 4.4, Pl. 7D) just underlying the egg yellow dolostone beds of the next facies. This green chamositic phosphatic sandstone bed has a great lateral extension and can be used for stratigraphic correlation. The microscopic description of

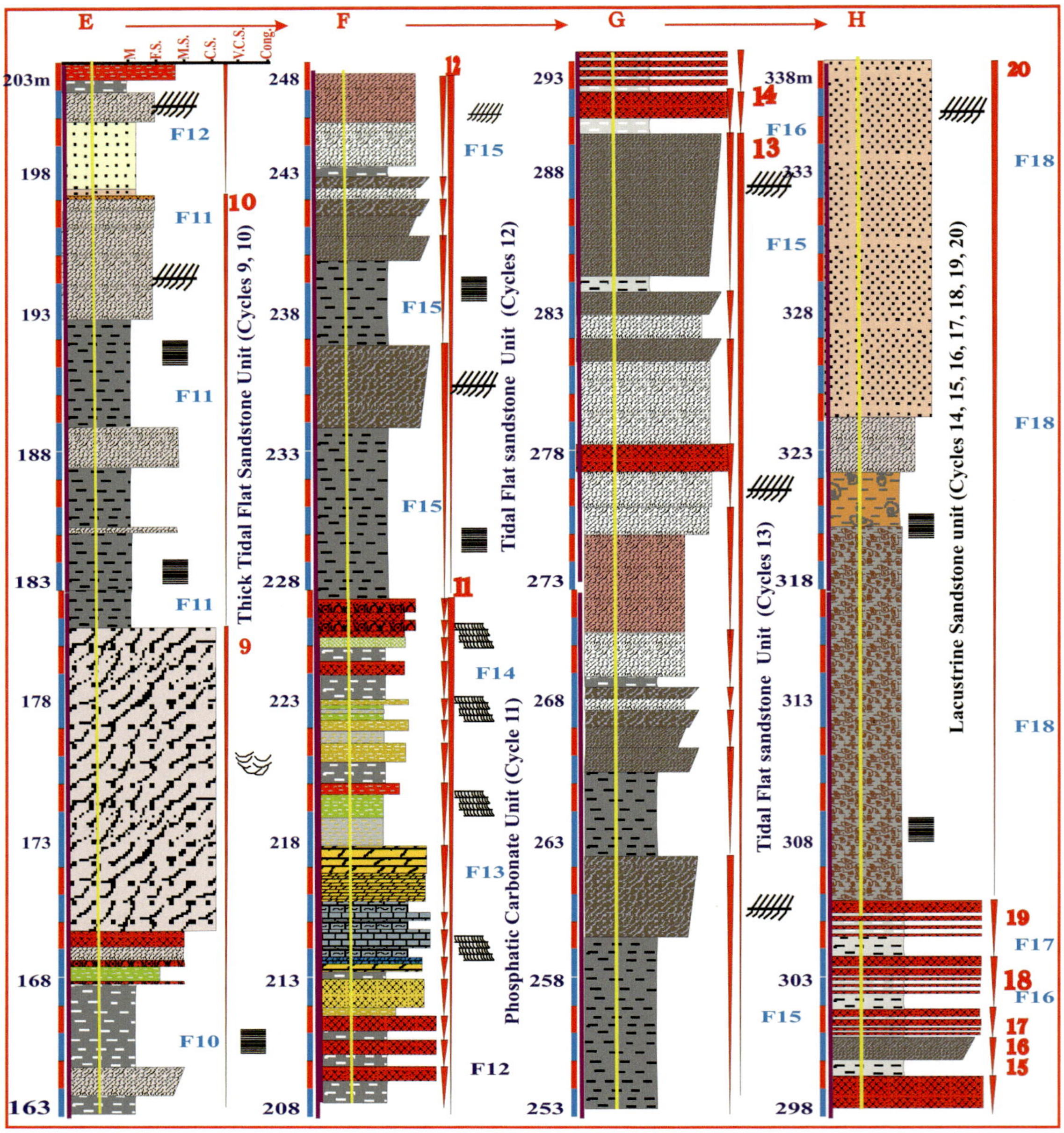

Fig. 4.10 Detailed Stratigraphic section of the studied succession of Haddat Ash Sham area (Modified after Taj and Mesaed 2012)

this bed revealed its composition from angular to subrounded quartz grains, green chamositic peloids, light yellowish-white dahlitic skeletal phosphatic grains (bones and teeth). These components are embedded within a slightly, hematitized green chamositic clay matrix (Fig. 4.11E).

Interpretation: The interbedded green mudstone, siltstone and sandstone in shallowing-upward cycles reflect fluctuation in water tables of marginal lagoons. The presence of this green chamositic phosphatic sandstone unit in the uppermost part of ironstone-bearing horizon just below the middle carbonate horizon confirms the green clays (glauconitic)-phosphatization model suggested by many authors and confirmed by Mesaed (2004a) in the topmost part of Bartonian glauconitic iron ore of El Gedida mine western Desert, Egypt. This model is based on the releasing of phosphorous from the clays in the reducing dysaerobic condition during the conversion of Fe^{3+} to Fe^{2+} and the formation of green clays. During the progressive shallowing, the level and concentration of P in the marine water increases which led to the formation of the different phosphatic components within the marine environments.

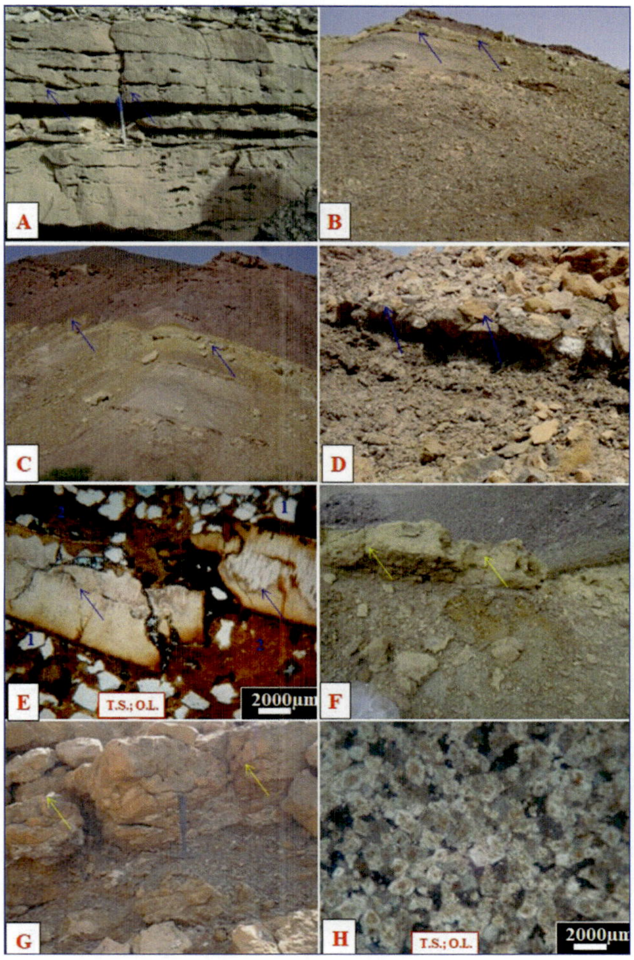

Fig. 4.11 **A** = The thickly cross-bedded sandstone Facie F10 (Cycles 9) in the eastern part of the study area, consists mainly of two small cycles each one begins with thick grey mudstone and terminated by ripple cross-laminated to cross-bedded fine-grained sandstone (arrows); **B, C** = The ferruginous mudstone-sandstone facie F12 which consists of interbedded grey mudstone, cross-laminated siltstone, fine-sandstone, red sandy and silty ironstone; **D** = Green phosphatic chamositic sandstone bed (arrows) terminating the ferruginous mudstone-sandstone facie F12 of figures. **B, C; E** = The green phosphatic chamositic sandstone bed terminating F12 consists of angular to subrounded quartz grains (1), green chamositic peloids, light yellowish white dahlitic skeletal phosphatic grains (bones and teeth, arrows) embedded within slightly, hematitized green chamositic clay matrix (2); **F, G** = Thinly bedded carbonate facie F13 which consists of interbedded marl, grey fossiliferous limestone and egg yellow dolostone (arrows); **H** = The dolostone beds of F13 are composed mainly of zoned idiotopic dolomite rhombs. Fig. 4.11C, F, H from Taj (2013)

Fig. 4.12 **A** = The upper part of facie F13 consists mainly of interbedded variegated mudstone, siltstone and terminated with a characteristic red silty ironstone beds (arrows); **B** = The phosphatic dolostone facie F14 which is represented by light colour phosphatic discontinuous thin beds as well as large patches and domains (arrows); **C** = Petrographic composition of the phosphatic dolostone from fine crystalline dolomite rhombs with less frequent yellow colour phosphatic grains (arrow); **D, E** = the ferruginous mudstone-sandstone facie F15 which consists of 5 small shallowing-upward cycles of grey mudstone which grades upward into rippled cross-laminated to cross-bedded kaolinitic fine-grained ironstone (arrows); **F** = Cycle 13 of facie F15 attains which consists of lower grey fissile mudstone (1) which grades upwards into cross-laminated to cross-bedded red ferruginous sandstone (2); **G** = Cross-bedded sandstone (arrows) in the upper part of facie F15; **H** = The ferruginous mudstone-sandstone facie F16 and cross-laminated kaolinitic siltstone facie F17 which are composed of white kaolinitic mudstone/siltstone (white) which grades upward into reddish ferruginous and kaolinitic fine-grained sandstone/siltstone (reddish)

Thinly bedded carbonate facie F13: This facie consists of interbedded marl, grey fossiliferous limestone and egg yellow dolostone (Fig. 4.11F). This facie begins with yellow colour dolomitic mud which grades upward into successive cycles of light greyish white limestone and terminated with yellow dolostone (Fig. 4.11G). Microscopically, the dolostone beds are composed mainly of zoned idiotopic dolomite rhombs showing some signs of hematitization and formation of hematite patches and domains (Fig. 4.11H). The upper horizon of this unit (Cycle II) is about 10 m thick and it consists mainly of interbedded variegated mudstone and siltstone interbeds and terminated with a characteristic red silty ironstone bed in shallowing-upward cycles (middle part of column F, Figs. 4.10, 4.12A).

Interpretation: The interbedded mudstone with very thin horizontally extended siltstone and ironstone interbeds underlying the studied carbonate sections suggests deposition from suspension within shallow protected water of a low-energy regime (Pettijohn et al. 1987). The studied carbonates are present overlying ferruginous mudstone-sandstone facie F11 which indicates their formation in protected (restricted) lagoons formed during ultimate stages of shoaling post-date the agitated shallow marine environments that predominated during the deposition of the oolitic ironstones underlying the studied carbonates (Taj 2011, 2013).

Phosphatic dolostone facie F14: This facie is represented by light colour phosphatic discontinuous thin beds as well as large patches and domains (Fig. 4.12B). Under the microscope, this facie consists mainly of fine crystalline dolomite rhombs with less frequent yellow colour phosphatic grains (Fig. 4.12C).

Interpretation: The presence of phosphatic grains indicates deposition in starved-protected lagoonal conditions that led to the formation of some authigenic phosphatic components. This facie is similar to the phosphatic dolostones and dolomitic phosphorite facies described by Mesaed (2004b) in Abu Tartour phosphorites of Egypt. He related this facie to deposition in restricted shallow subtidal lagoonal environment of general low clastic input and high Ca, Mg and Fe content.

4.3.3 Upper Tidal Flat-Fluvio-Lacustrine Clastic Member

4.3.3.1 Tidal Flat Sandstone Unit (Cycles 12, 13)

This unit is represented by the upper part of column F and the lower 95% of column G (Fig. 4.10). It is represented by cycle No. 12, 13.

Ferruginous mudstone-sandstone Facie F15: This facie comprises cycles no.12, 13. Cycle 12 attains up to 20 m thickness. It consists of 5 small shallowing-upward cycles. Each of these cycles begins with grey mudstone which grades upward into rippled cross-laminated to cross-bedded kaolinitic fine-grained ironstone (Fig. 4.12D, E). This cycle is terminated by ledge-forming cross-bedded sandstone. Cycle 13 attains up to 35 m thickness. It is represented by column G (Fig. 4.10) and it consists mainly of shallowing-upward cycles of lower grey fissile mudstone which grades upwards into cross-laminated to cross-bedded red ferruginous sandstone (Fig. 4.12F). This sandstone shows upward increase in the grain size and thickness of beds and cross-beds thickness (Fig. 4.12G).

Interpretation: The cyclic nature of this facie and its composition from interbedded mudstone and fine-grained sandstone indicate their deposition in coastal tidal-dominated lagoons of fluctuated water levels. The upward coarsening and thickening characters indicate the general shoaling and infilling of these lagoons. The cross-bedded dominated uppermost part of this facie represents deposition in highly agitated tidal sand bars.

4.3.3.2 Lacustrine Sandstone Unit (Cycles 14, 15, 16, 17, 18, 19, 20)

This unit attains up to 60 m thickness. It is represented in the field by two main horizons, i.e. the first horizon includes cycles 14, 15, 16, 17, 18, 19 (Column G, H, Fig. 4.10) while the second horizon includes cycle 20 (Fig. 4.10, Column H and Fig. 4.14, Column I). Figure 4.13 shows the symbols used in the drawing of the stratigraphic sections and the sedimentologic model.

Ferruginous mudstone-sandstone Facie F16 and Cross-laminated kaolinitic siltstone Facie F17: These two facies are present together and are represented by Cycles 14, 15, 16, 17, 18, 19. These cycles are relatively thin and present in one ledge-forming cliff. They are composed of rhythmic bedded unit of lower white kaolinitic mudstone/siltstone which grades upward into reddish ferruginous and kaolinitic fine-grained sandstone/siltstone (Figs. 4.12H, 4.15A). These alternating beds are organized in small-scale shallowing-upward cycles showing an upward increase in the thickness of the sandstone beds of the upper parts of the shallowing-upward cycles and decreasing of the white kaolinitic mudstone/siltstone beds of the lower parts of the shallowing-upward cycles (upward coarsening and thickening). This unit is affected by a series of normal faults (Fig. 4.15A).

Interpretation: The lower deeper part of the shallowing-upward cycles consists mainly of grey and white kaolinitic mudstone/kaolinite which indicates deposition by suspension from slightly saline to fresh water. The cross-laminated to cross-bedded siltstone/fine sandstones of the upper parts of these cycles indicate deposition in agitated conditions. The predominance of kaolinite advocates the continental lake depositional environments (lacustrine) instead of marginal lagoons. This is also supported by the absence of green chamositic clays and the associated chamositic ironstones as described in the aforementioned marginal lagoons facies.

Cross-bedded sandstone Facie F18: This facie is represented by Cycle 20. It attains up to 47 m thickness (Column H, Fig. 4.4). It begins with a characteristic 17-m thick reddish mudstone contains mainly reddish-brown ferruginous reworked mudstone clasts (Fig. 4.15B). This mudstone grades upward into ripple-cross-laminated siltstone (5 m thick). The upper part of this cycle is composed of successive small-scale cycles of lower cross-laminated siltstone (Fig. 4.15C) which grades upward into ledge-forming cross-bedded sandstone in the uppermost parts of these cycles (Fig. 4.15D), Column I (Fig. 4.14).

Fig. 4.13 Legend of the symbols used in the drawing of the stratigraphic sections of Figs. 4.2, 4.4, 4.10, 4.14, and the block diagrams of Figs. 4.17 and 4.18

Interpretation: This facie represents a general continuation of the above-described lacustrine facies (F 17). This is supported by the general thickening and coarsening-upward character.

4.3.3.3 Lacustrine Mudstone Unite (Cycle 21)

This unit attains up to 40 m thickness. It forms a characteristic slope-forming unit just overlying the sandstone of facie F18 and underlying the thick cliff-forming fluviatil sandstone of cycle 22 (Fig. 4.14, Column I, J, Fig. 4.15E). This unit contains two main facies:

Tuffaceous kaolinitic mudstone facie F19: This facie attains up to 25 m thickness (Fig. 4.15F). It consists of successive shallowing-upward cycles of lower thick mudstones and upper thin sandstones terminating these shallowing-upward cycles (Fig. 4.15F, column I, Fig. 4.14).

Interpretation: The rhythmic nature of this facie as well as the shallowing-upward cycles of mudstone and the overlying cross-laminated to cross-bedded fine sandstone represent a typical lacustrine depositional environment.

Cross-laminated light brown sandstone facie F20: This facie represents the uppermost part of cycle 21 (column, J,

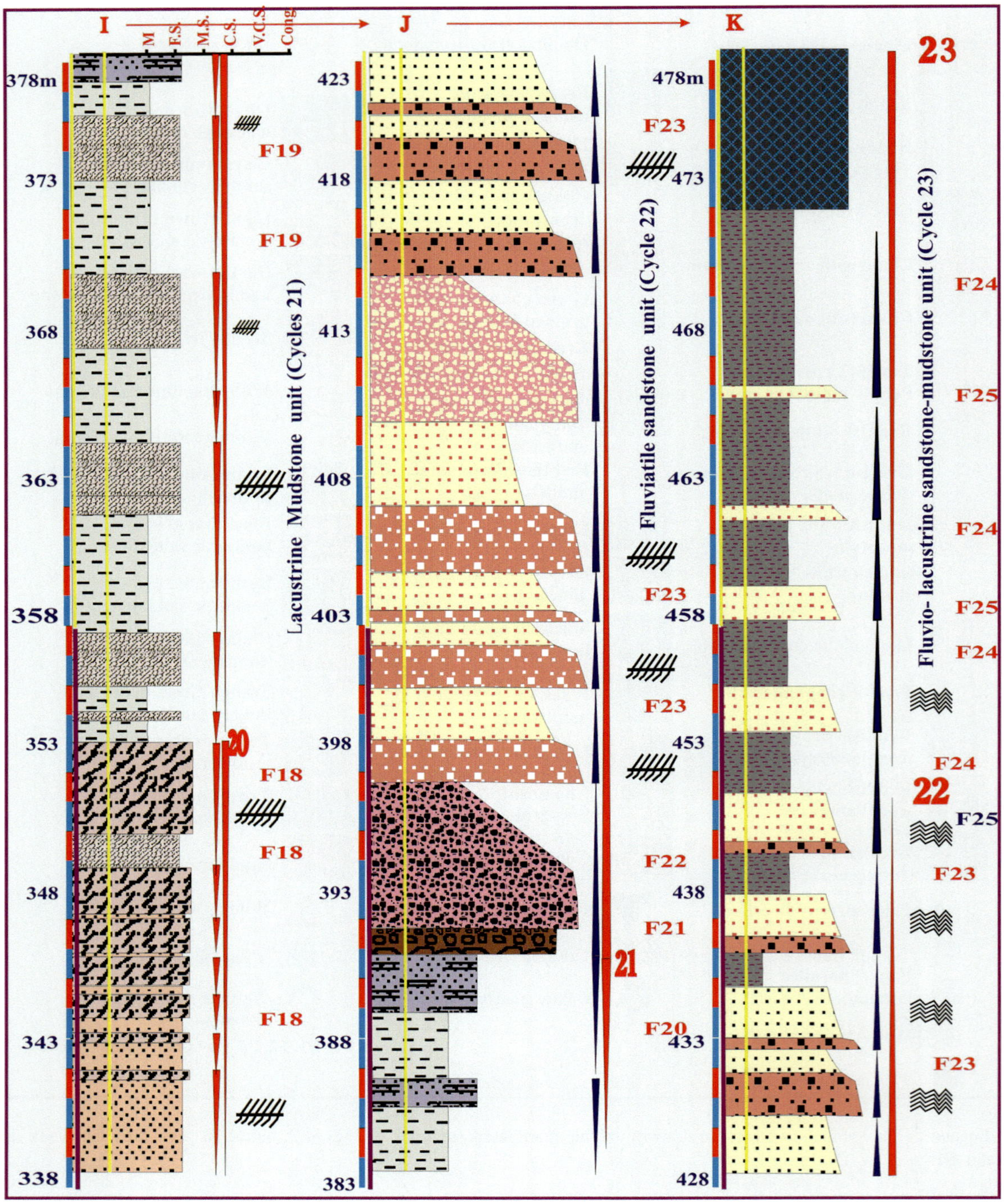

Fig. 4.14 Detailed Stratigraphic section of the upper part of the studied sedimentary succession of Haddat Ash Sham area

Fig. 4.14). It consists of successive shallowing-upward cycles of relatively thin lower mudstone and thicker cross-laminated to cross-bedded fine-grained sandstone. The uppermost cycle of these facies is terminated by white calcareous sandstone/ pebbly sandstone (Fig. 4.16A).

Interpretation: This facie represents the upper part of progressively shoaled lakes. This is indicated by the upward increase in the thickness and number of sandstone beds relative to the alternating mudstone beds. This is also supported by the presence of pebbly calcareous sandstone bed in the uppermost part of this facie which indicates the complete filling of the lakes and deposition of channel fluviatil calcareous sandstone.

4.3.3.4 Fluviatil Sandstone Unit (Cylce 22)

This unit forms the last characteristic cliff-forming horizon in the uppermost part of the studied succession (Fig. 4.16B, column J, K; Fig. 4.14). It attains up to 45 m thickness and it

Fig. 4.15 **A** = Series of normal faults (arrows) affecting the ferruginous mudstone-sandstone facie F16; **B** = Cycle 20 of the cross-bedded sandstone facie F18 which consists of thick reddish mudstone contains mainly reddish brown ferruginous reworked mudstone clasts (arrows); **C** = The ripple-cross-laminated siltstone (arrows) terminated the cycles of the middle part of facie F18; **D** = The ledge-forming cross-bedded sandstone (arrows) in the uppermost parts of facie F18, **E** = The lacustrine mudstone unite (Cycle 21) which forms a characteristic slope-forming part just overlying the sandstone of facie F18 and underlying the thick cliff-forming fluviatil sandstone of cycle 22; **F** = The tuffaceous kaolinitic mudstone facie F19 which consists of successive shallowing-upward cycles (arrows) of mudstones and thin sandstones

Fig. 4.16 **A** = The white calcareous sandstone/pebbly sandstone (arrows) in the upper parts of the Cross-laminated light brown sandstone facie F20*;* **B** = The cliff-forming horizon of the Fluviatil sandstone unit (arrows, Cylce 22); **C** = The reddish white and milky quartz pebbles at the base of the disorganized conglomerate facie F21; **D** = Successive fining-upward cycles (arrows) of quartz pebble conglomerate and tabular cross-bedded conglomerates, trough and tabular cross-bedded conglomerate facie F22; **E** = Successive fining-upward cycles of trough cross-bedded conglomerates and tabular cross-bedded sandstone (arrows) in the tabular cross-bedded conglomerate, pebbly sandstone facie F23; **F** = Cross-laminated to cross-bedded fine-grained sandstone and terminated with thick mudstone interbeds (arrows) in the upper parts of the tabular cross-bedded conglomerate, pebbly sandstone facie F23; **G** = The successive small-scale fining-upward cycles (arrows) of the tuffaceous mudstone facie F24 and tuffaceous sandstone facie F 25; **H** = The grey tuffaceous ferruginous pebbly sandstone of the lower parts of the fining-upward cycles of the tuffaceous sandstone facie F 25

includes the following facies which are arranged from the base to the top as follows:

Disorganized corglomerate facie F21: This facie (5-m thick) is present in the lowermost part of this unit. It attains up to 5 m thickness and it consists of intermixed reddish white quartz pebbles (2–4 mm in diameter) and milky white quartz pebbles embedded within ferruginous kaolinitic clay matrix (Fig. 4.16C).

Interpretation: This facie represents the proximal reaches of braided stream system drained the study area after the infilling of the above-mentioned lakes of facie F19, F20. It is completely similar to F1 of the lowermost part of the studied sedimentary succession of the study area. This facie is very thin when compared with facie F1 which indicates the absence of weathered uncovered Precambrian Arabian Shield rocks. The facies was formed by sheetfloods in the mid to distal alluvial fan. This interpretation is supported by the presence of stratified gravels and sands, the clasts parallel to bedding planes, the moderate sorting of clast-supported pebbles and cobbles and the absence of cross-bedding (Gloppen and Steel 1981; Nilsen 1985; Blair and McPherson 1994; Nichols and Hirst 1998; Blair 1999a,b).

Trough and tabular cross-bedded conglomerate facie F22: This facie is present in the lowermost part of the successive fining-upward cycles of this unit (column J, Figs. 4.14, 4.16D). It consists mainly of successive fining-upward cycles. These cycles begin with very thin quartz pebble conglomerate and terminated by tabular cross-bedded conglomerates.

Interpretation: This facie represents deposition in the middle parts of alluvial fan by braided stream system. This is indicated by the absence of flood plain mudstones which indicated deposition by strong stream current. This is also supported by the rarity of clay matrix in the conglomerate beds. Planar cross-stratification in the clast-supported conglomerate indicates that facies Gp was deposited by the migration of large, isolated, gravelly, straight-crested, transverse bedforms (Miall 1977; Hein and Walker 1977; Smith 1990; Karpeta 1993a b).

Tabular cross-bedded conglomerate-pebbly sandstone facie F23: This facie is present just above the aforementioned facie F22. It consists mainly of successive fining-upward cycles. Each of these cycles begins with trough cross-bedded conglomerates and terminated with tabular cross-bedded sandstone (Fig. 4.16E). Towards the upper parts of this unit, the fining-upward cycles become composed of cross-laminated to cross-bedded fine-grained sandstone and terminated with thick mudstone interbeds (Fig. 4.16F).

Interpretation: The fining-upward characters of this facie indicate deposition during a progressive decrease in the strength of braided stream. It is similar to the aforementioned facie F3 in the lower member of the studied sedimentary succession.

4.3.3.5 Fluvio-Lacustrine Tuffaceous Sandstone-Mudstone Unit (Cycle 23)

Tuffaceous Mudstone Facie F24 and Tuffaceous sandstone Facie F25: These two facies are represented by cycle no. 23 (column K, Fig. 4.14). They are present just underlying the Oligo-Miocene basaltic sheet. It forms a characteristic slope unit and attains up to 22 m thickness and consists of 4 successive small-scale fining-upward cycles (Fig. 4.16G). Each of these cycles is composed of grey tuffaceous ferruginous pebbly sandstone and terminated by mottled ferruginous siltstone (upper part of column K, Figs. 4.14, 4.16H).

Interpretation: This facie represents deposition in a slightly quite depositional environment in the extreme distal reaches of braided stream. The increasing thickness and number of the mudstone beds of the upper parts of this facie indicate deposition in flood plain lakes of meander streams. The mudstone beds are tuffaceous in composition which indicates the deposition of this facie during contemporaneous volcanic activities. This is also supported by the presence of glass shreds. Similar tuffaceous facies is also described in the uppermost part of Ash Shumaysi Formation in wadi Ash Shumaysi area (Taj and Mesaed 2013).

4.4 Depositional History of the Study Area

From the sedimentologic point of view, the sedimentary succession of Haddat Ash Sham area represents an ideal case of transition from the fluvio-lacustrine continental sedimentation into lacustrine delta and shallow marine succession (upper to lower shoreface) which is again grades upward into fluvio-lacustrine facies formed general uplifting, river sedimentation and the associated volcanic deposits.

The lower fluviatil succession was deposited during general uplifting along fault scarps of the Arabian shield rocks. This led to the formation of alluvial fan succession with characteristic proximal and distal parts (Fig. 4.17A). The presence of the quartz pebble conglomerate facies supports the presence of remarkable subaerial sheet flood sedimentation. It is only dominated by proximal reaches of braided stream sedimentation with sporadic gravel bars (Fig. 4.17A). In the distal reaches of these braided streams, bedded pebbly sandstones, siltstones and mudstones of general fining-upward characters were deposited. Thinly bedded to laminate lacustrine facies was deposited in the extreme distal parts of these braided streams.

With the general quiescence of uplifting, new marine tongues transgress Haddat Ash Sham area (Fig. 4.17B). During this time, shallow marine to lagoonal oolitic ironstone-bearing succession was deposited. This succession is of general shallowing-upward character and it consists of repetitive small-scale cycles representing short-lived periods of sea level fluctuation (regressive-transgressive cycles). The oolitic ironstone beds of the middle parts of these cycles were deposited during periods of low clastic input (starved time) and slightly reducing conditions (Dysaerobic zone of Berner 1981). During these time periods, the basin becomes of high Fe^{2+} activities and organic matter activities which led to the authigenesis of green chamositic clays in the middle parts of the shallowing-upward cycles. The diagenetic hematitization of these clays led to the formation of different types of muddy, silty, sandy and oolitic ironstones. There is a reverse relationship between the percentage of clastic input and the formation of the green chamositic clays and hence the formation of the different types of ironstone. During the progressive shoaling, the depositional environments become shallower and of high clastic input which resulted in the deposition of the thick tidal sand bar facies (top of cycle No. 9 column A, Figs. 4.10, 4.17C). After, this part of the section, the depositional environments become converted into protected lagoons of very low clastic input which led to the deposition of thick greyish black mudstone contains very thin siltstone interbeds and terminated by egg yellow dolostone facies representing deposition in protected lagoonal conditions (Fig. 4.17C, D). After, the deposition of the lagoonal dolostone facies, the freshwater influxes increase

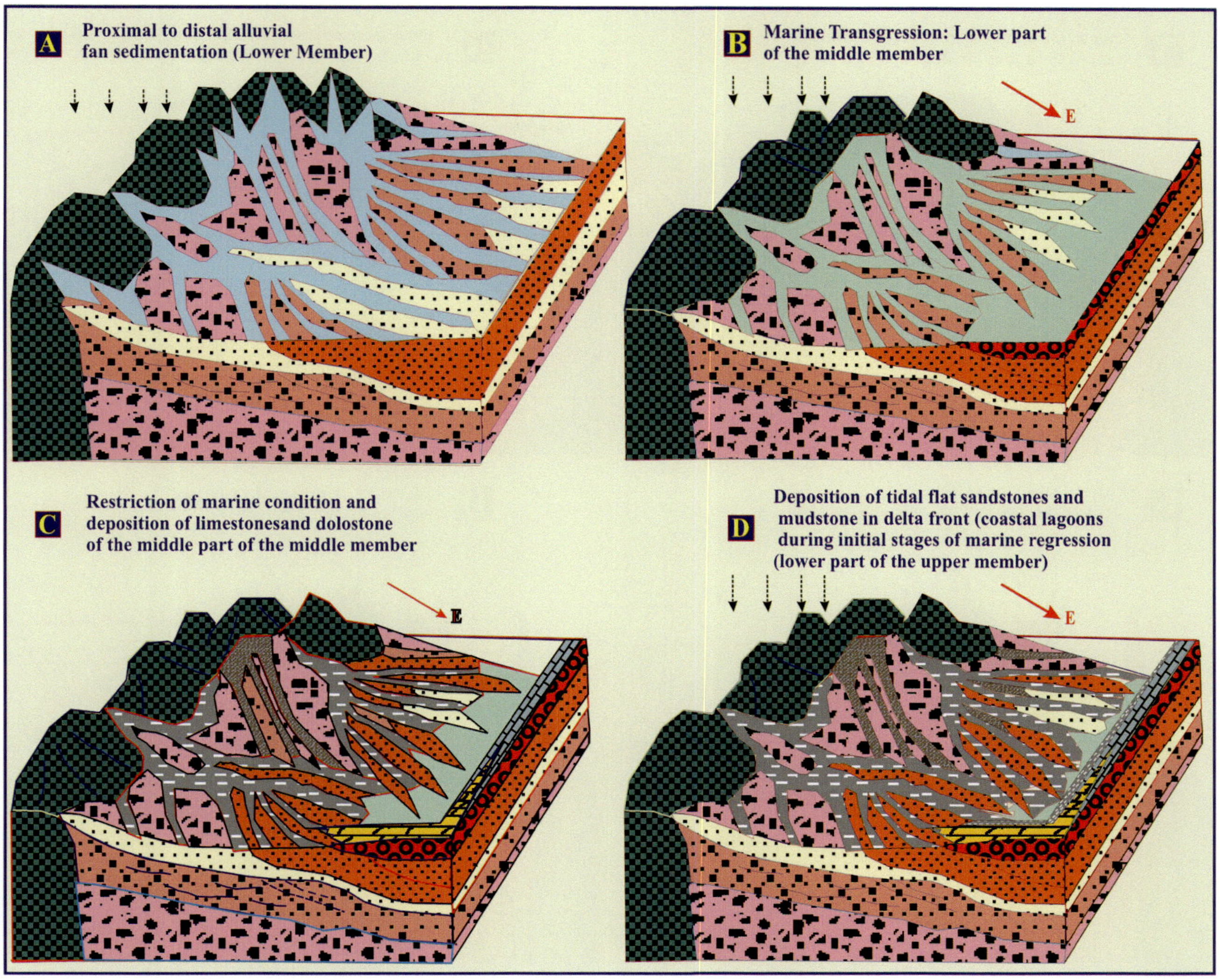

Fig. 4.17 Block diagram showing the depositional model of the lower and middle parts of the studied sedimentary succession of Haddat Ash Sham area

and the conditions become freshwater lakes depositing rhythmic alternations of white kaolinitic mudstone and red ferruginous sandstone, siltstone (cycles 15, 16, 17, 18, 19, column H, Figs. 4.10, 4.17D, 4.18A). These cycles grades upward into interbedded mudstone and sandstone facies (cycles 20, 21, Column I, J, Fig. 4.14) represent which represent shallowing-upward cycles in regressive lakes (Fig. 4.18A).

After, the deposition of this thick lacustrine succession, the area was subjected to uplifting regime accompanied by the deposition of another fluviatile facies in the upper part of the succession (Fig. 4.18B). This part shows an upward increase in the flood plain mudstones and becomes completely composed of mudstone with less frequent siltstone, fine-sandstone interbeds which support deposition in lacustrine conditions (Fig. 4.18C). In the uppermost part of the succession, volcaniclastic facies of dark grey to black tuffaceous, mudstone and tuffaceous sandstone are present (Fig. 4.18D). This indicates the intimate association between the subaerial volcanics and lacustrine sedimentation (tephra facies). Similar conditions of volcanism and sedimentation have been previously described by Mesaed (2004c) in the Oligocene volcaniclastic deposits of Abu Roash area, West Greater Cairo, Egypt.

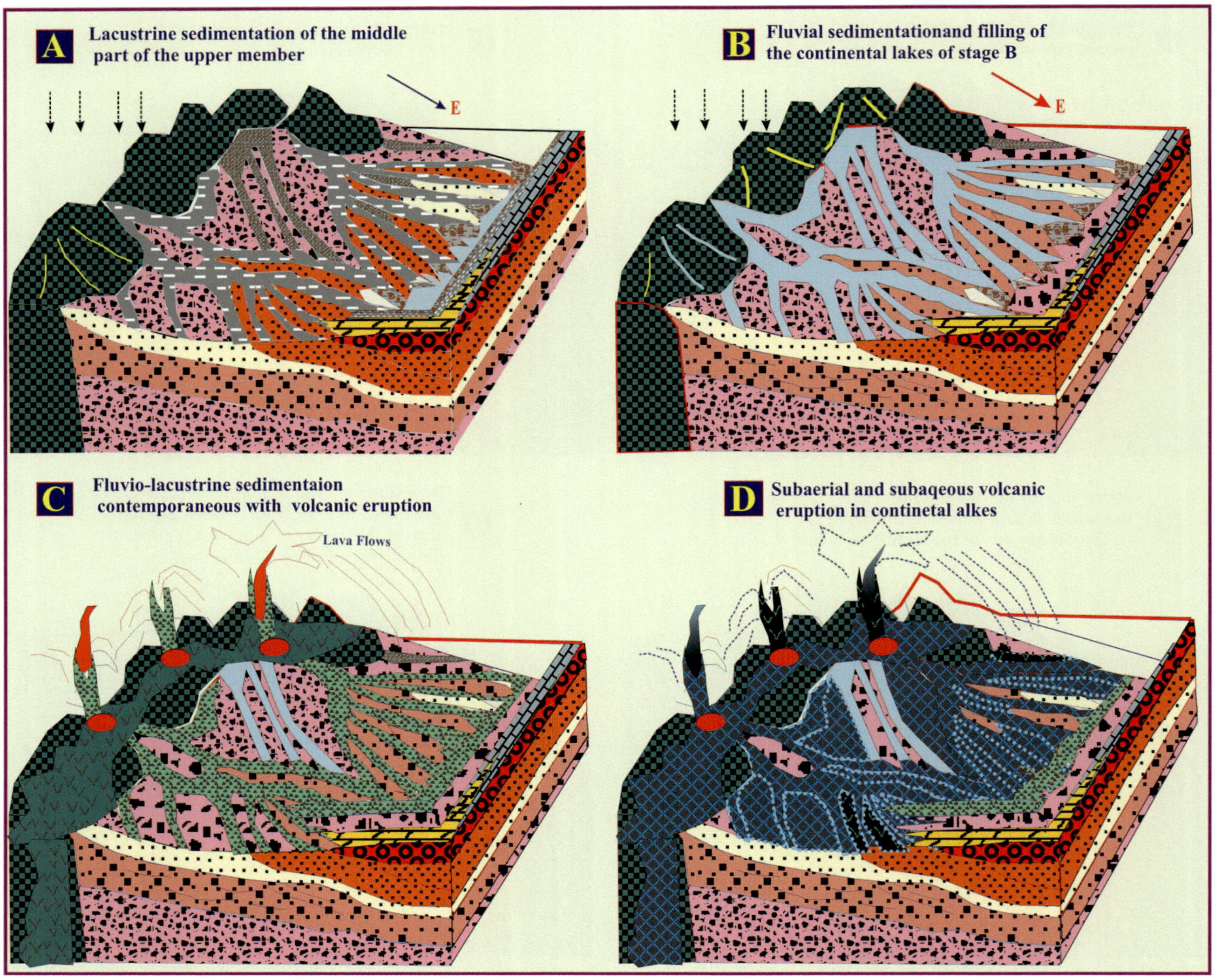

Fig. 4.18 Block diagram showing the depositional model of the upper part of the studied sedimentary succession of Haddat Ash Sham area

4.5 Discussion and Conclusions

The studied succession of the Cretaceous-Oligocene sedimentary succession of Haddat Ash Sham area represents fluvio-, fluvio-lacustrine and shallow marine sedimentation postdate the formation of Arabian Shield Rocks. The succession began with characteristic quartz pebble lag deposits representing deposition in the meander streams. The meander streams become more mature which resulted in the formation of bedded mudstone-sandstone facies of meander steams. Thin lacustrine/fluvio-lacustrine facies represents deposition in fresh to mixed water coastal lakes just underlying the sedimentary tidal flat oolitic ironstone faces.

This tidal flat facies forms the middle part of the succession. It is composed mainly of successive four large shallowing-upward cycles. These cycles represent deposition during progressive shallowing accompanied by an upward increase in the current and wave activities. The different types of ironstones are recorded within the middle parts of these large cycles.

The position of the chamositic ooids within the middle parts of the small-scale coarsening-upward cycles between the bioturbated and thinly laminated basal, mudstone below and increasing higher energy hematitic ooid above, may indicate their formation in the reducing dysaerobic transition zone (Cotter and Link 1993) along the lower and upper flanks of sand bars. On the other hand, the abundance of hematitic, kaolinitic-hematitic and kaolinitic ooids within the upper parts of the small-scale coarsening-upward cycles indicates their formation under somehow agitated oxidized conditions.

The stratigraphic positions of the oolitic ironstones reflect their intimate genetic relation with short-lived transgressive–regressive events which are expressed by the small- and medium-scale coarsening-upward cycles, developed during the long-lived progradation or coarsening-upward cycles. The uppermost parts of these cycles are composed of cross-laminated to cross-bedded tidal flat sandstone reflecting deposition in highly agitated conditions.

Acknowledgements The authors would like to thank the Deanship of Scientific Research (DSR), where this paper is from the data of the Project funded by the DSR, King Abdulaziz University, Jeddah, under grant no. 472/145/1431. The authors, therefore, acknowledge with thank to DSR technical and financial support.

References

Al Shanti AMS (1966) Oolitic iron ore deposits in Wadi Fatima between Jeddah and Mecca, Saudi Arabia. Saudi Arabian Directorate Gen Miner Resour Bull 2:51p

Allen JR (1985) Principles of physical sedimentology. Blackwell, London

Berner RA (1981) A new geochemical classification of sedimentary environments. J Sed Pet 51:359–365

Blair TB, McPherson JG (1994) Alluvial fans and their natural distinction from rivers based on morphology, hydraulic processes, sedimentary processes and facies assemblage. J Sediment Res 64:450–489

Blair TC (1999a) Cause of dominance by sheetfloods vs. Debris flow processes on two adjoining alluvia fans, Death Valley, California. Sedimentology 46:1015–1028

Blair TC (1999b) Sedimentary processes and facies of the waterlaid Anvil Spring Canyon alluvia fan, Death Valley, California. Sedimentology 46:913–940

Brown GF, Jackson RO, Bogue RG, MacLean WH (1963) Geology of the Southern Hijaz quadrangle, Kingdom of Saudi Arabia: Saudi Arabian directorate general of mineral resources, miscellaneous geologic investigations MAP I -210 A, 1: 500,000 scale

Cotter E, Link JE (1993) Deposition and diagenesis of Clinton ironstones (Silurian) in the Appalachian Foreland Basin of Pennsylvania. Geol Soc Am Bull 105:911–922

El Aref MM, El Sharkawi MM, Mesaed AA (1996a) Depositional and diagenetic microfabric evolution of the Cretaceous oolitic ironstones of Aswan, Egypt. Geol Soc Egypt Spec Publ 2: 279–312

El Sharkawi MM, El Aref MM, Mesaed AA (1996b) Stratigraphic setting and paleoenvironment of the Coniacian-Santonian ironstones of Aswan, South Egypt. Geol Soc Egypt Spec Publ 2: 243–278

Gloppen TG, Steel RJ (1981) The deposits, internal structure and geometry in six alluvial fan-delta bodies (Devonian-Norway-A) study in the significance of bedding sequence in conglomerate. In: Ethridge FG, Flores RM (eds), Recent and ancient non marine depositional environments, Soc. Eco. Pal. Miner. Spec. Pub., vol 31, pp 49–69

Guerrak S (1987) Metallogenesis of cratonic oolitic ironstone deposits in the Bled el Mass, Azzel Matti, Ahnet and Mouydir basins, Central Sahara, Algeria Geol Rundsch 76(3): 903–922

Hallam A, Bradshaw MJ (1979) Bituminous shales and oolitic ironstones as indicators of transgressions and regressions. J Geol Soc Lond 136:157–164

Harms JC, Southard JB, Walker RG (1982) Structures and sequences in clastic rocks: society of economic paleontologists and mineralogists short course 9

Hein FJ, Walker RG (1977) Bar evolution and development of stratification in the gravelly, braided, Kicking Uorse River, British Columbia. Can J Earth Sci 14:562–570

Jennette DC, Pryor WA (1993) Cyclic alternation of proximal and distal storm facies: Kope and Fairview Formations (Upper Ordovician), Ohio and Kentucky. SJ Sed Petrol 63:183–203

Johnson PR (2006) Explanatory notes to the map of Proterozoic geology of western saudi Arabia, technical report SGS-Tr-2006-4

Karpeta WP (1993a) Sedimentology and gravel bar morphology in an Archean braided river equence; the Witpan Conglomerate Member (Witwatersrand Supergroup) in the Welkom Goldfield, South Africa. In: Best JL, Bristow CS (eds) Braided rivers: geological society of London special publication, vol 75, pp 369–388

Karpeta WP (1993b) Sedimentology and gravel bar morphology in an Archean braided river sequence; the Witpan Conglomerate Member (Witwatersrand Supergroup) in the Welkom Goldfield, South Africa. In: Best JL, Bristow CS (eds) Braided rivers: geological society of London special publication,vol 75, pp 369–388

Karpoff R (1958) Esquisse geologiques de I, Arabie Saoudite: Bulletin Societe Geologique de France, 6 ser. VIIm, pp 653–696

Leeder M (1999) Sedimentology and sedimentary basins: from turbulence to tectonics. In: Blackwell science, 608 pp.

Mesaed AA (2004a) Mechanism of formation of the upper Eocene glauconitic ironstones and red beds of Gabal Qalamoon area, Western Desert, Egypt. Egypt J Geol 48:17–44

Mesaed AA (2004b) Facies types and mechanism of formation of Abu Tartour phosphorites, Western desert, Egypt. An example of heterogeneous Glaucony-bearing phosphorite type. (Abstract 42th Annual meeting of the Geologic Soc. of Egypt)

Mesaed AA (2004c) Depositional and diagenetic history of the fluvio - lacustrine ironstones and red beds of the Oligocene Qatrani formation, North Abu Roash Area, Egypt. In: Egyptian Mineralogist, vol 16 (accepted for publication in V.16).

Mesaed AA, Taj RJ, Harbi H (2011) Stratigraphic setting, facies types, depositional environments and mechanism of formation of the Ash Shumaysi ironstones, Wadi Ash Shumaysi, Jeddah district, West Central Saudi Arabia (Accepted for publication in the Arabian Journal of Geosciences)

Mesaed AA (1995) Geological, mineralogical and geochemical studies on the ironstones and related lateritic products of Aswan region, Egypt. Ph. D. Thesis, Faculty of Science, Geology Department, Cairo University, 289p

Miall A (1990) Principles of sedimentary basin analysis. Springer, New York, p 616

Miall AD (1978) Lithofacies types and vertical profile models in braided river deposits: a summary, In: Miall AD (ed) Fluvial sedimentology. Canadian society of petroleum geologists, Calgary, Memoir,vol 5, pp 578–586

Miall AD (1977) A review of the braided river depositional environment. Earth Sci Rev 13:1–62

Miall AD (1996) The geology of fluvial deposits: sedimentary facies, basin analysis, and petroleum geology, Springer, New York, 582 p.

Nichols GJ (1999) Sedimentology and Stratigraphy: Oxford. Blackwell Science, United Kingdom, p 355

Nichols GJ, Hirst JP (1998) Alluvial fans and fluvial distributary systems, oligo-miocene, Northern Spain: contrasting processes and products. J Sediment Res 68:879–889

Nilsen TH (1985) Modern and ancient alluvial fan deposits, editorial introduction. Van Nostrand Reinhold Company, New York, 372 pp.

Pettijohn FJ, Potter PE, Siever R (1987) Sand and Sandstone, 2nd edn. Springer, New York, p 553

Plint AG, Walker RG (1992) Tidal deposits. In: Walker RG, James NP (eds), Facies models: a response to sea level change. Geological Association of Canada, pp 220–230

Reading HG (1978) Sedimentary environments and facies. Blackwell Scientific Publications, xi+557 pp

Rust BR (1978) Depositional models for braided alluvium. In: Miall AD (ed) Fluvial sedimentology: canadian society of petroleum geology memoir, vol 5, pp 605–625

Smith (1982) Reconnaissannce geologic map of the Wadi Hammah quadrangle, sheet22/40C, Kingdom of Saudi Arabia: Saudi Arabian deputy ministry for mineral resources geologic map GM-65, 1:100,000 scale, with text, 19 p.

Smith ND (1974) Sedimentology and bar formation in the upper Kicking Uorse River, a braided outwash stream. J Geol 82:205–224

Smith SA (1990) The sedimentology and accretionary styles of an ancient gravel-bed stream; the Budleigh Salterton pebble beds (Lower Triassic). Southwest Engl: Sediment Geol 67(3–4):199–219

Spincer CH, Vincent PL (1984) Bentonite resource potential and geology of the Cenozoic sediments, Jeddah region: Saudi Arabian deputy ministry for mineral resources, open-file report BRGM-O-F-02-34, 34 p

Taj RJ (2011) Stratigraphic setting, facies types and depositional environments of Haddat Ash Sham ironstones, Western Arabian Shield, Saudi Arabia. Asian Trans Basic Appl Sci 1(2). (ATBAS ISSN: 2221–4293)

Taj RJ (2013) Microfacies, diagenesis, and depositional environments of the Tertiary carbonates of Usfan Formation in Haddat Ash Sham area, Western Arabian Shield, Saudi Arabia. Arab J Geosci 6:1011–1031. https://doi.org/10.1007/s12517-011-0410-8

Taj RJ, Mesaed AA (2012) Mechanism of formation of Haddat Ash Sham ironstones (Oligo-Miocene), Makkah Al Mokaramah District, West Central Arabian Shield, Saudi Arabia. Arab J Geosci 6:4299–4321. https://doi.org/10.1007/s12517-012-0694-3

Teyssen TAL (1989) A depositional model for the Liassic Minette ironstones (Luxembourg and France), in comparison with other Phanerozoic oolitic ironstones. In: Young TP, Taylor WEG (eds) Phanerozoic ironstones, geological society, London, special publication, vol 46, pp 79–92

Van Houten FB, Arthur MA (1989) Phanerozoic Ironstones, Temporal patterns among Phanerozoic oolitic ironstones and oceanic anoxia. In: Young TP, Taylor WEG (eds) Geological society, London, Special Publication, vol 46, pp 33–50

Van Houten FB, Bhattacharyya DP (1982) Phanerozoic oolitic ironstones—geologic records and facies model. Annu Rev Earth Planet Sci 10:441–457

Van Houten FB, Bhattacharyya DPC, Mansour SEI (1984) Cretaceous Nubia Formation and correlative deposits, Eastern Egypt. Major regressive-transgressive complex. Bull Geol Soc Am 95:397–405

Young TP (1989) Phanerozoic ironstones: an introduction and review. In: Young TP, Taylor WEG (eds) Phanerozoic ironstones. Geological society, London, special publication, vol 46, pp19–30

Zeidan R, Banat K (1990) Petrology, mineralogy and geochemistry of the sedimentary formations in Usfan, Haddat Ash-Sham and Shumaysi Areas, and the associated oolitic ironstone interbeds, North East and East of Jeddah, Saudi Arabia, King Abdulaziz University, Directorate of Research Projects 035/406, 264 p.

Part III

Third Thematic: Tectonic Evolution, Structural Styles, and Oil/Gas Traps of the Northern Africa Foreland Basins

5 Petroleum System Evaluation of Upper Cretaceous and Eocene Plays, Offshore and Onshore Southern Pelagian Basin, Tunisia

Karin Göttlich, Jean Rodriguez, Sabrine Mnii, Wala Mzoughi, Tam Lovett, and Gabor Tari

Abstract

Oil and gas exploration has been active in the southern Pelagian Basin since 1949. The aim of this study was to review Upper Cretaceous and Eocene plays which are found to be the most commercially significant. The Upper Cretaceous plays comprise the carbonate reservoirs of the Bireno (Turonian) and Douleb Members (Coniacian) and the Abiod Formation (Campanian-Maastrichtian). The Eocene plays are centred on the carbonate reservoirs of the El Garia Formation (Ypresian) and the Reineche Member (Lutetian-Bartonian). Reservoir, seal and charge were separately evaluated by refining depositional environment and maturity maps using well and 2D seismic data. Good reservoir properties are observed in reservoirs mainly deposited along a NW-SE trend, following the inner-ramp settings of the Bireno and Douleb rimmed carbonate platform. Abiod chalky limestones are present in the north eastern area and highly depend on the enhancement of reservoir quality by local fracturing, karsting and/or dolomitization. This trend has also impacted the development of Eocene El Garia and Reineche nummulitic carbonate reservoirs which lie along a shallow marine edge and are found in both the offshore and onshore areas of the southern Pelagian Basin. The nummulitic facies zone of the Reineche Member appears to be much narrower than that of the underlying El Garia Formation. Over the southern and southwestern parts of the study area, the main source rocks for the Upper Cretaceous plays, the Lower Fahdene (Albian) and Bahloul (Cenomanian-Turonian) Formations, are found to be absent and, thus, hydrocarbon accumulations in these areas are considered to be of higher risk. For the Upper Cretaceous plays, charge is considered to be a low risk where proven source kitchens lie directly underneath. Therefore, reservoir charge is dependent on vertical migration or on short distance lateral migration. For the Eocene plays, sourcing will be through lateral migration from the Ypresian Bou Dabbous Formation organic rich facies into the El Garia reservoir facies, with remigration via faults into the overlying Reineche reservoir rocks. Various seals are considered to be effective over the area, except for areas of non-deposition and erosion. Most of the small hydrocarbon accumulations appear restricted to settings near fault zones generally affected by fracturing. Within the framework of this study, predictability for prospect evaluation appears low due to the lack of data to delineate characteristic fault block patterns or identify areas of local uplift which can support dolomitization and/or karsting. The aim of the evaluation is to gain a further understanding of the risks on key elements for hydrocarbon prospectivity in the basin in an attempt to highlight areas of low risk for individual plays and not to identify specific prospects for drilling.

K. Göttlich (✉) · S. Mnii · T. Lovett · G. Tari
OMV Exploration and Production, Vienna, Austria
e-mail: karin.goettlich@outlook.com; kgruber13@gmail.com

J. Rodriguez · S. Mnii
OMV Tunisie Production GmbH, Tunis, Tunisia

W. Mzoughi
STUDI International, Tunis, Tunisia

5.1 Introduction

The Pelagian Basin lies in the eastern onshore area of Tunisia, extending into the offshore area and into the southern part of offshore Libya where it progressively merges into the Sirte Basin. Towards the north lies Malta and the offshore areas of Italy. For the purpose of this study only the southern Pelagian Basin in Tunisia was evaluated to further understand the petroleum systems and focus on the main play fairways. The area of interest (Fig. 5.1) has been explored for oil and gas since 1949 with many evaluations of the Pelagian Basin and its surrounding geological provinces previously undertaken to better define its petroleum systems

S. Khomsi and F. Roure (eds.), *Geology of North Africa and the Mediterranean: Sedimentary Basins and Georesources*, Regional Geology Reviews, https://doi.org/10.1007/978-3-031-18747-6_5

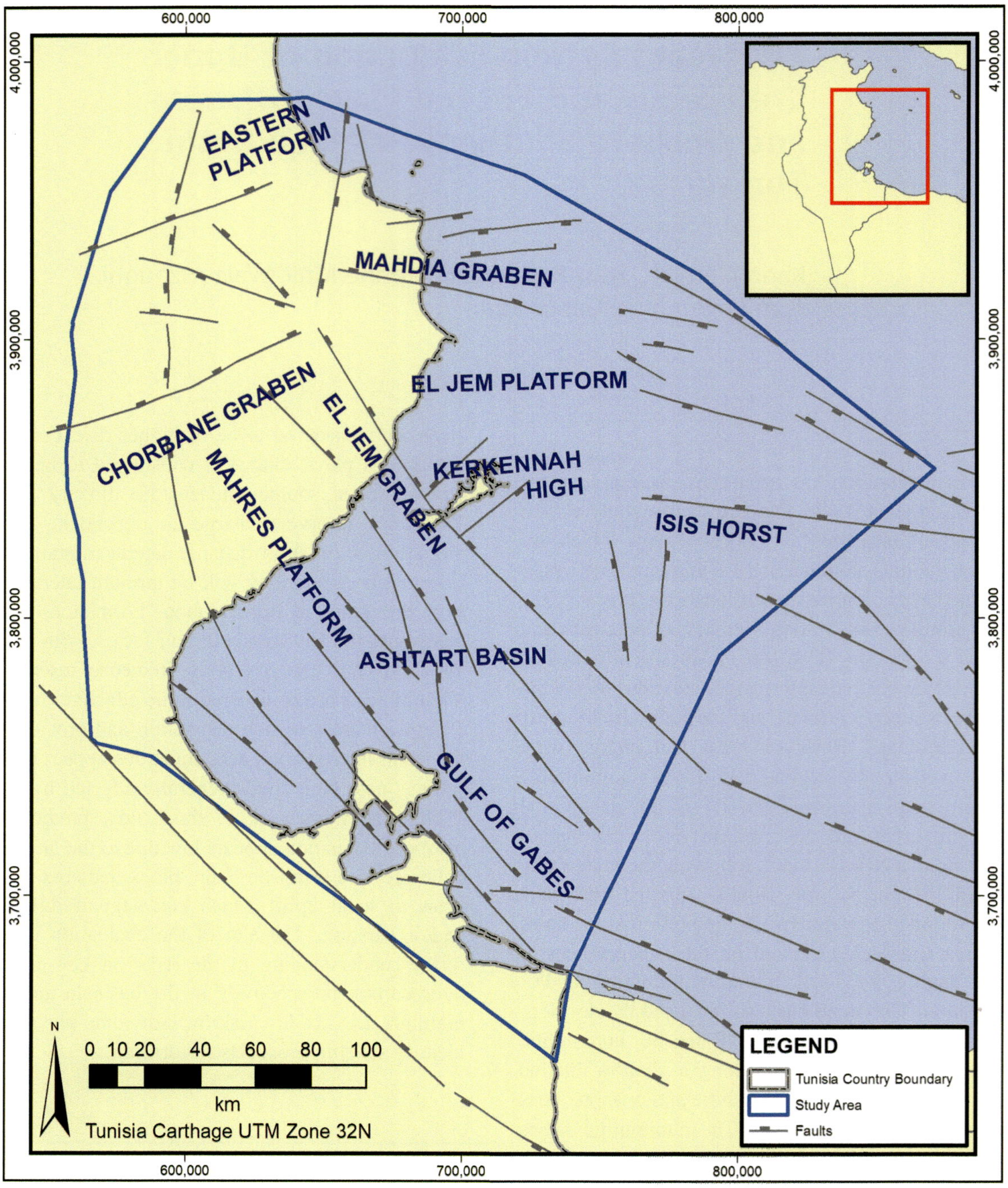

Fig. 5.1 Location of study area (blue outline) and structural elements (modified after Bédir 1995; Khomsi 2004; de Lamotte et al. 2009; Khomsi et al. 2012, 2016, 2019a; Bédir et al. 2018; Mezni et al. 2019) in the southern Pelagian Basin, onshore and offshore Tunisia

(i.e. Burollet 1956; Bishop 1975; Burollet et al. 1978; M'Rabet 1987; Bishop 1988; Ferjani et al 1990; Bédir et al. 1992; Bédir 1995; Bédir et al. 1996; Bishop and Debono 1996; Gumati et al. 1996; MacGregor et al. 1998; Troudi 1998; Ziegler and Roure 1999; Bédir et al. 2001; Klett 2001; Fourati et al. 2002; Hallett 2002; Acheche et al. 2003; Adouani et al. 2003; Brahim et al 2003; Fourati et al. 2003; El Euchi et al. 2004; Khomsi 2004; Mejri et al. 2006; de Lamotte et al. 2009; Fiduk 2009; Tlig et al. 2010; Brahim et al. 2012; Khomsi et al. 2012; Haddad et al. 2013; Abidi et al. 2015; Dkhaili et al 2015; Khomsi et al. 2016; Elfessi 2017; Bédir et al. 2018; Chalwati et al. 2018; Khomsi et al. 2019a, b; Lučić and Bosworth 2019; Mezni et al. 2019).

Several plays within Jurassic, Cretaceous and Palaeogene sections exist in the southern Pelagian Basin, but only five plays within the Upper Cretaceous and Eocene epochs are investigated in more detail in this study. A total of 348 wells and 194 2D reflection seismic lines, covering an area of approximately 73,500 km^2, were used in this evaluation. Due to the paucity of data and insufficient data quality for the deeper plays, it was decided, for the purpose of this study, that an adequate determination of the individual petroleum system elements of the Jurassic and Lower Cretaceous plays was not feasible.

The review focused on Upper Cretaceous and Eocene plays which are found to have economic significance as well as those which may also have some potential for oil and gas. A play is defined as a stratigraphic interval, which may have already been tested, is sharing a common geological history and is separated by regional or semi-regional seals. The Upper Cretaceous plays comprise the carbonate reservoirs of the Bireno (Turonian) and Douleb Members (Coniacian) and the Abiod Formation (Campanian-Maastrichtian). The Eocene plays are centred on the carbonate reservoirs of the El Garia Formation (Ypresian) and the Reineche Member (Lutetian-Bartonian). This study aimed at refining previously published depositional environment and maturity maps (Bishop 1988; Saïdi 1993; Saïdi and Belayouni 1994; Saïdi and Philip 1997; Troudi and M'Rabet 1998; Touir and Soussi 2003; Troudi et al. 2000; Racey et al. 2001; El Euchi et al. 2004; Lüning et al. 2004; Mejri et al. 2006) for the individual petroleum system elements by using well and 2D seismic data. Depositional Environment maps were produced for the main known hydrocarbon producing reservoirs (Bireno, Douleb, Abiod, El Garia and Reineche). For each of these reservoirs, lithology, biostratigraphy, depositional environment, reservoir facies and quality, presence and level of risk were noted. Similarly, seal depositional environment maps were produced for each of the above plays, with lithology, quality/effectiveness and presence defining the main risks. For the Upper Cretaceous plays, the charge was based on the presence, quality and maturity of the two main source rocks, the organic rich shales of the Lower Fahdene (Albian) and the Bahloul (Cenomanian-Turonian) Formations, with migration assumed to be predominantly through faulting. Maturity was estimated by mapping near top Lower Fahdene and Bahloul events which were depth converted using well data only and the maturity range calibrated against geochemical data. For the Eocene plays the main source rock is the Bou Dabbous Formation of Ypresian age. Mapping of this event was based on a regional interpretation of the Near Top El Haria shale (Uppermost Maastrichtian-Lower Thanetian), which is the top seal for the Cretaceous plays, especially the Abiod, lying approximately 100–150 m below the Top Bou Dabbous. This event was also depth converted by using well data only and the maturity map was calibrated with geochemical data. Migration into the El Garia reservoir is predominantly lateral and into the overlying Reineche Member vertically via faults. Faults may also provide the potential for more mature hydrocarbon migration from the underlying Cretaceous into these younger Palaeogene reservoirs.

Ideally, a regional 3D seismic coverage over the basin would aid in improving the delineation of individual plays by refining the regional depth surfaces and allowing the evaluation of deeper intervals. For better delineation of the charging mechanisms for all play levels, it is recommended that a fluid to source correlation study in combination with a regional 3D Basin Model be undertaken. Hence, the maturity maps produced in this study are an approximation due to the various 2D seismic vintages used and, therefore, will be subject to future modifications as more data become available.

5.2 Petroleum Geology of the Southern Pelagian Basin

In the offshore area of the Pelagian Basin, the Gulf of Gabes was heavily subsiding during the Triassic and Jurassic, becoming more intermittent throughout Cretaceous times (Burollet 1956; Baird et al. 1967; Bishop 1975; Burollet et al. 1978; Dercourt et al. 1993; Morgan et al. 1998; MacGregor et al. 1998; Stampfli 2000; Stampfli and Borel 2004; Mejri et al. 2006; Stampfli and Kozur 2006; Scotese and Schettino 2017). The major structural style in the Pelagian Basin is composed predominantly of alternating NW-SE trending horsts and grabens separated by normal faults (Castany 1947, 1951; Burollet 1991; Bédir 1995; Zourai 1995; Chikhaoui 1988; Badalini et al. 2002; Adouani et al. 2003; Dey and Kilani 2003; Saïd et al. 2011; Melki et al 2012; Roure et al. 2012; Gharbi et al. 2013; Abidi et al. 2015; El Rabia et al. 2018). From Albian to Maastrichtian times, the Sirte Basin rifting is responsible for the major structures observed today in the Pelagian Basin (Ben Ayed 1986; Chihi 1995; Guiraud 1998; Grasso et al 1999). Associated fault displacements and uplift of horst blocks controlled sedimentation (Morgan et al. 1998). Between Late Cretaceous and Early Eocene times, Africa drifted northwards, whilst continuing to rotate slightly anticlockwise (Stampfli and Kozur 2006). Due to the closure of Tethys, short compressional events during the Palaeocene and Eocene caused uplift and erosion of the southern part of the Pelagian Platform. The subsequent rifting engendered volcanic activity in different parts of the basin and many rifts underwent a period of reactivation (MacGregor et al. 1998; Mejri et al. 2006; Kort et al. 2009). Early Eocene facies belts generally followed similar NW-SE trends to those of the Late Cretaceous (Bédir 1995; Bédir, et al. 1992, 1996, 2001). Sea levels remained relatively high at this time, with flooding consequently occurring in many of the African and Arabian cratons. In the Tunisian and Libyan areas, shallower, restricted marine gulfs accumulated organic and phosphate rich carbonates. Towards land, the presence of evaporites reflects restricted connections to open sea due to lower sea levels and the formation of narrow straits between exposed coastal islands. Offshore, deeper shelf and deep marine facies accumulated argillaceous limestones and calcareous shales, which formed source rock facies in the Pelagian Basin (Mejri et al. 2006). The African Plate continued to drift northwards with rifting occurring along the northern margin (Stampfli and Kozur 2006). This resulted in the development of a complex horst and graben system in the Pelagian Basin, such as the Ashtart Basin (Fig. 5.1). Tectonic activity promoted major diapirism of Triassic salt (El Rabia et al. 2018). The area was lagoonal to the southwest and became increasingly more basinal to the northeast throughout much of Mesozoic and Palaeogene times. A transpressional regime occurred post-Miocene to Recent times. Halokinesis driven by tectonism was very active throughout this period with occasional diapirs piercing through the Mesozoic and the Cenozoic series, and often forming long submarine salt walls. Halokinesis had an impact on the evolution of local paleohighs and resulted in the growth and development of reefal build-ups (Mejri et al. 2006; El Rabia et al. 2018). Episodic volcanic activity is recorded throughout the Cretaceous (Wilson and Guiraud 1998; Kort et al. 2009).

The onshore area is located in central eastern onshore Tunisia, limited to the West by the N-S axis (Fig. 5.1) (Burollet et al. 1978). The region comprises large and shallow continental shelf sequences. During Mesozoic and Paleogene times, it was a stable platform with a dominance of shallow marine carbonates to the southwest grading to open marine shaly facies to the northeast. This limit was controlled by active E-W and NW-SE fault systems (Castany 1947; 1951; Bédir 1995; Zourai 1995; Chikhaoui 1988; Badalini et al. 2002; Adouani et al. 2003; Dey and Kilani 2003; Abidi et al. 2015). During Neogene times, an inversion with active subsidence occurred in most of the area caused by a NW-SE compressive phase. This is clearly observed all over Tunisia and is represented by the regional Late Miocene unconformity (Base Ségui Formation) (Fig. 5.2) (Burollet 1956; Bédir 1995; Bédir, et al. 1992, 1996, 2001; MacGregor et al. 1998; Mejri et al. 2006; Khomsi et al. 2009, 2016). The main geological features are fold amplification, faulting, local Triassic salt reactivation and basin inversion (Khomsi et al. 2009; El Rabia et al. 2018).

Although the main structural elements within the offshore Pelagian Basin are extensional, seismic data reveals the presence of tightly folded compressional structures that could have considerable exploration potential (Fiduk 2009; Khomsi et al. 2009, 2016). The tectonic evolution of the Pelagian Basin controlled the regional distribution of reservoir and source rocks, resulting in marked lateral variations in facies composition and thickness. Wells close to the southern margin of the basin (near the current shoreline) show, in general, a relatively thin succession of Palaeogene and Upper Cretaceous sediments, a highly truncated Lower Cretaceous and a significant thickness of Jurassic/Triassic sediments. Further offshore, within the grabens, there is a marked increase in the thickness of Oligocene-Miocene and Cretaceous sediments. In the northern part of the province Upper Cretaceous sediments are observed to lie directly below middle Eocene shales. Well data obtained in the western onshore area, for example in the Chorbane graben, show markedly the impact of halokinesis, with Triassic and Jurassic evaporites directly underlying Cretaceous (Fig. 5.3, Mezni et al. 2019) or Upper Palaeogene sediments.

There are several source rocks observed in the area (mostly Type II) but the main ones are to be found within the Albian-Cenomanian, Cenomanian-Turonian and the Ypresian (Mejri et al. 2006). There are two main sub-basins separated by the Kerkennah High (Fig. 5.1) where gas and oil are currently being generated today and have been, for the older source rocks, generating since early Palaeogene times (Fig. 5.4). The various evaluated play types are displayed in Fig. 5.5. Hydrocarbon producing reservoirs are predominantly Jurassic, Cretaceous and Eocene carbonates. The former being either fractured tight limestones or rudist carbonates, exhibiting good reservoir matrix properties which may be enhanced through secondary dolomitization and dissolution (Troudi and M'Rabet 1998; Troudi et al. 2000; Mejri et al. 2006). The latter can include a nummulitic facies member that may exhibit a reasonable to good reservoir matrix quality, often enhanced by fractures (Racey 2001; Mejri et al. 2006). The main top seals are provided by various effective thick shale members found within Cretaceous, Palaeocene and Eocene aged sediments. A major seal for the Cretaceous at Palaeocene level, the El Haria Formation (Fig. 5.2), is often of sufficient thickness to provide an incompetent zone where dislocation of structuring between the Cretaceous and the overlying Palaeogene can be observed. Migration of hydrocarbons from source rock to reservoir is considered to occur predominantly through faulting (Mejri et al. 2006).

In the Gulf of Gabes area, the main producing reservoirs are the Upper Cretaceous carbonates of the Abiod Formation and the Douleb and Bireno Members of the Aleg Formation. These reservoirs produce gas from the Abiod and Miskar Formations (Fig. 5.2) and oil (of approximately 35° API) from the Douleb and Bireno Members from the onshore fields. However, they can also contain fluids with a high inert gas content, predominantly CO_2 (e.g., from 30–70% CO_2—Miskar and Ashtart fields) and minor N_2 (Isis field). H_2S is common, especially from the Bireno reservoir where it is believed to be mainly associated with the presence of anhydrite layers within the reservoir. The origin of these inert gases (especially CO_2) is most probably due to the heating of carbonates as a result of their depth of burial as well as from heat associated with volcanic activity (Gaaya and Kharbachi 1997; Mejri et al. 2006).

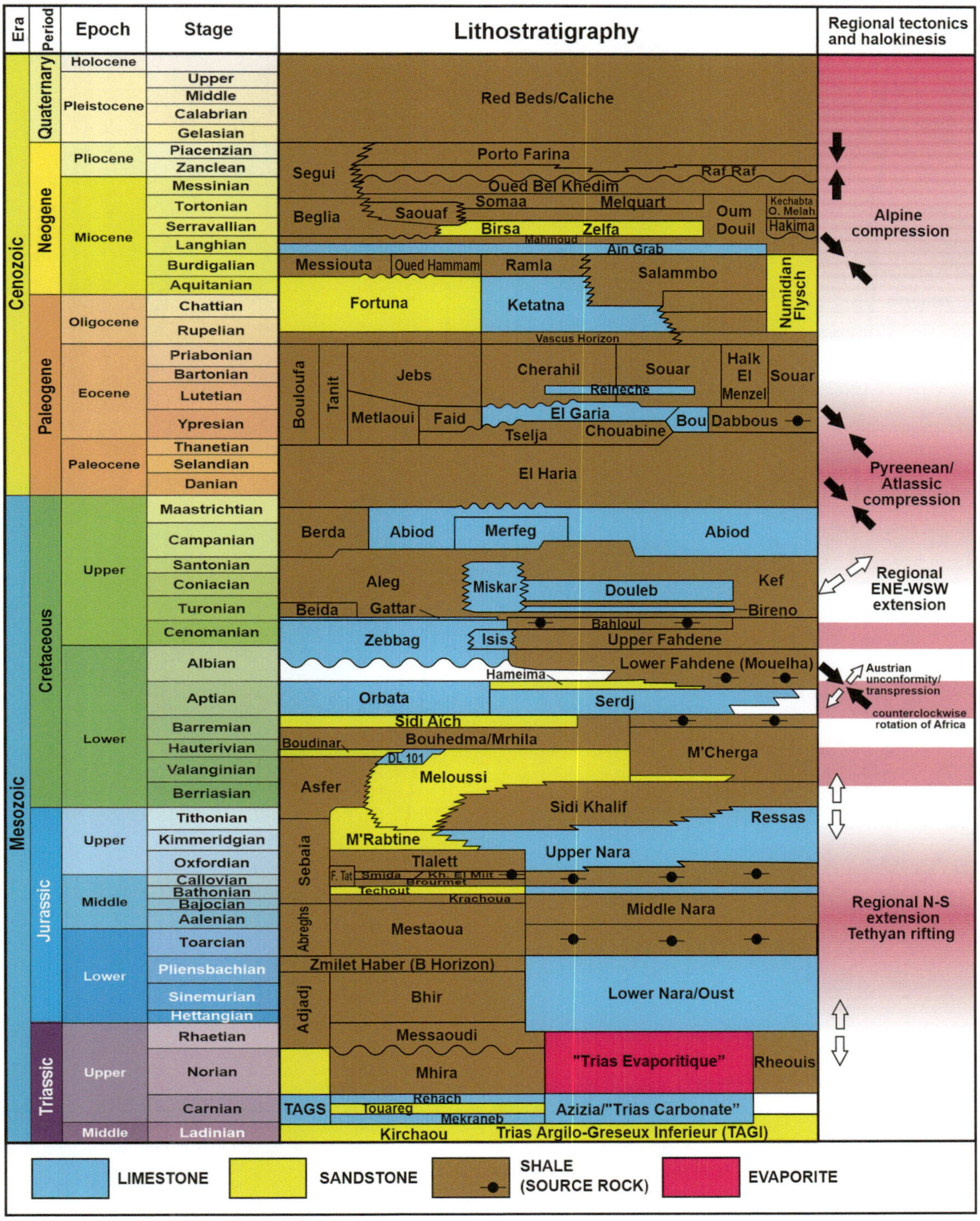

Fig. 5.2 Cenozoic and Mesozoic Stratigraphic chart, regional tectonics and halokinesis events for most areas of Tunisia, including the southern Pelagian Basin. Lithostratigraphy modified after Klett (2001), regional tectonics and halokinesis taken from Troudi et al. (2017), modified after Melki et al. (2012)

5.2.1 Source Rocks

Early Cretaceous (Albian): Lower Fahdene Formation

Dark shales and platy limestones cover the northern and north eastern part of the study area and were deposited during Albian times (Fig. 5.2). Along a NW-SE trend, the facies changes from dark shales and platy limestones of the Lower Fahdene Formation into interbedded limestones and shales of the Lower Zebbag Formation (El Euchi et al. 2004) (Fig. 5.6). The thickness of the Lower Fahdene Formation varies and can exceed up to 400 m in some depocenters. The

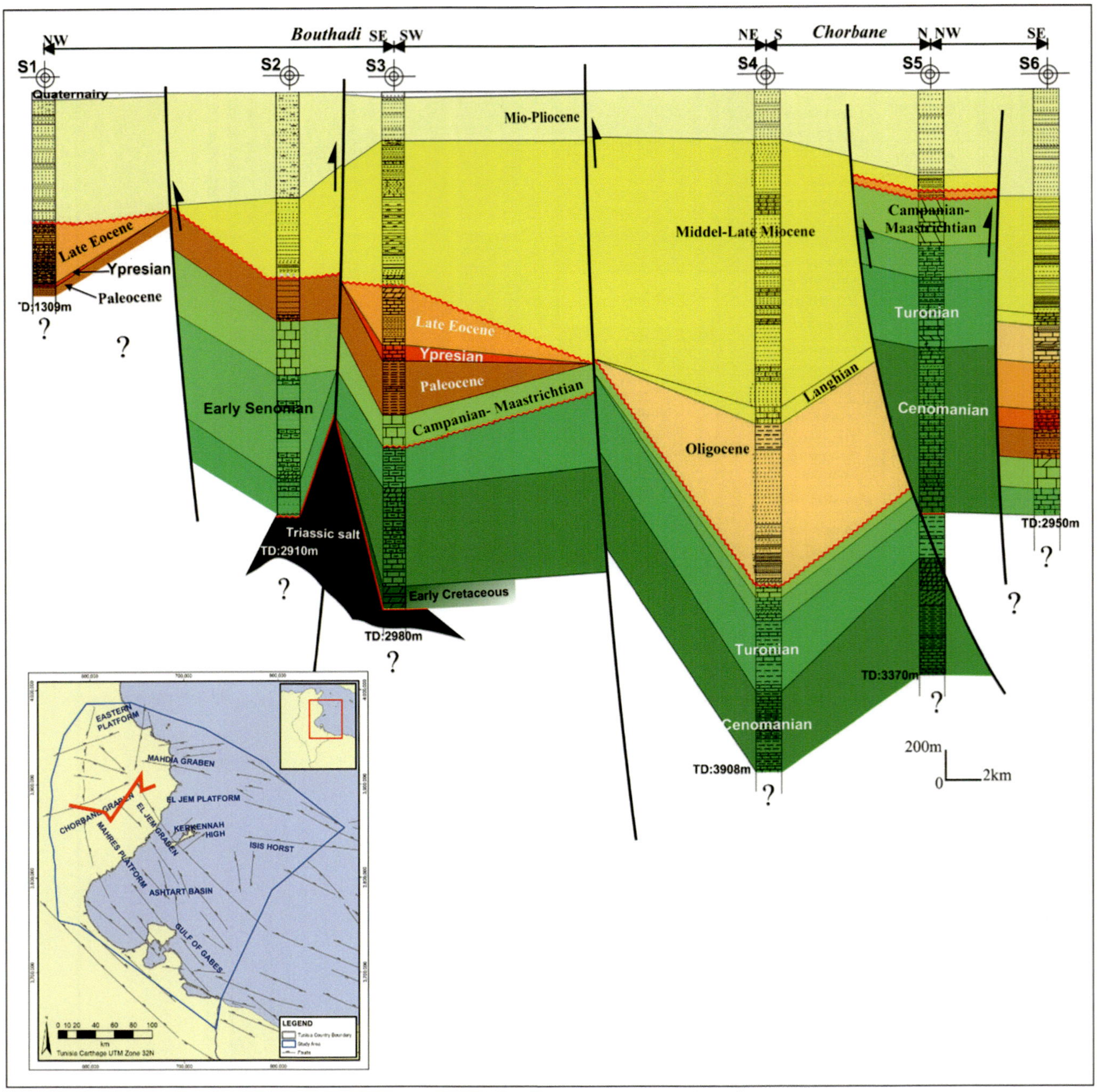

Fig. 5.3 Lithostratigraphic well correlation for wells S1 to S6, over Bouthadi and Chorbane, Chorbane Graben, red line on inset map indicates approximate location of well correlation, taken from Mezni et al. (2019)

geochemical analysis shows that the Lower Fahdene is found to be organic rich mainly in northwest Tunisia and the eastern part of the Pelagian sea. TOC (total organic content) values range from 0.65% to 4% and the hydrocarbon potential varies between 9 kg/t of rock to 17.1 kg/t of rock. The organic matter of the Lower Fahdene is predominantly type II/III kerogen (Saïdi 1993; Hughes and Reed 1995; Nacer Bey et al. 1995; Saïdi and Philip 1997; Klett 2001; Fourati et al. 2002, 2003; El Euchi et al. 2004; Mejri et al. 2006; Khalifa et al. 2018).

In the north eastern offshore part of the basin, the Lower Fahdene is considered to be immature. Much of the onshore area is considered to be immature or uncertain, as at the time of writing no data was available. In the

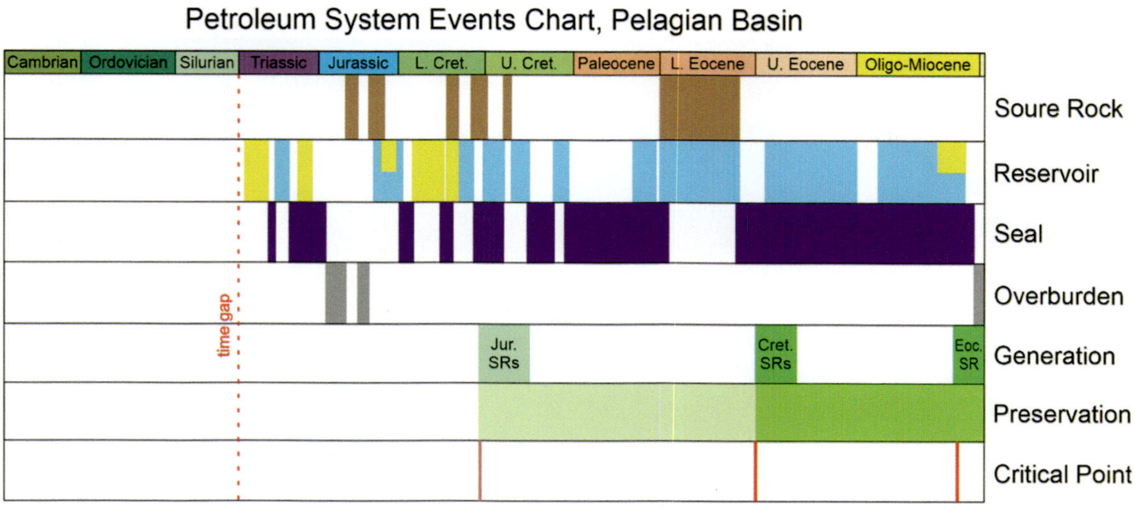

Fig. 5.4 Petroleum System Events Chart for the Pelagian Basin details the main elements of the petroleum system in the Pelagian Basin onshore and offshore. The various evaluated play types are displayed in Fig. 5.5. There are essentially five key plays observed in the area. These are to be found in the Late Jurassic, Early and Late Cretaceous, and Early and Late Eocene (Figs. 5.2 and 5.5); reservoir: blue = carbonate, yellow = sandstone

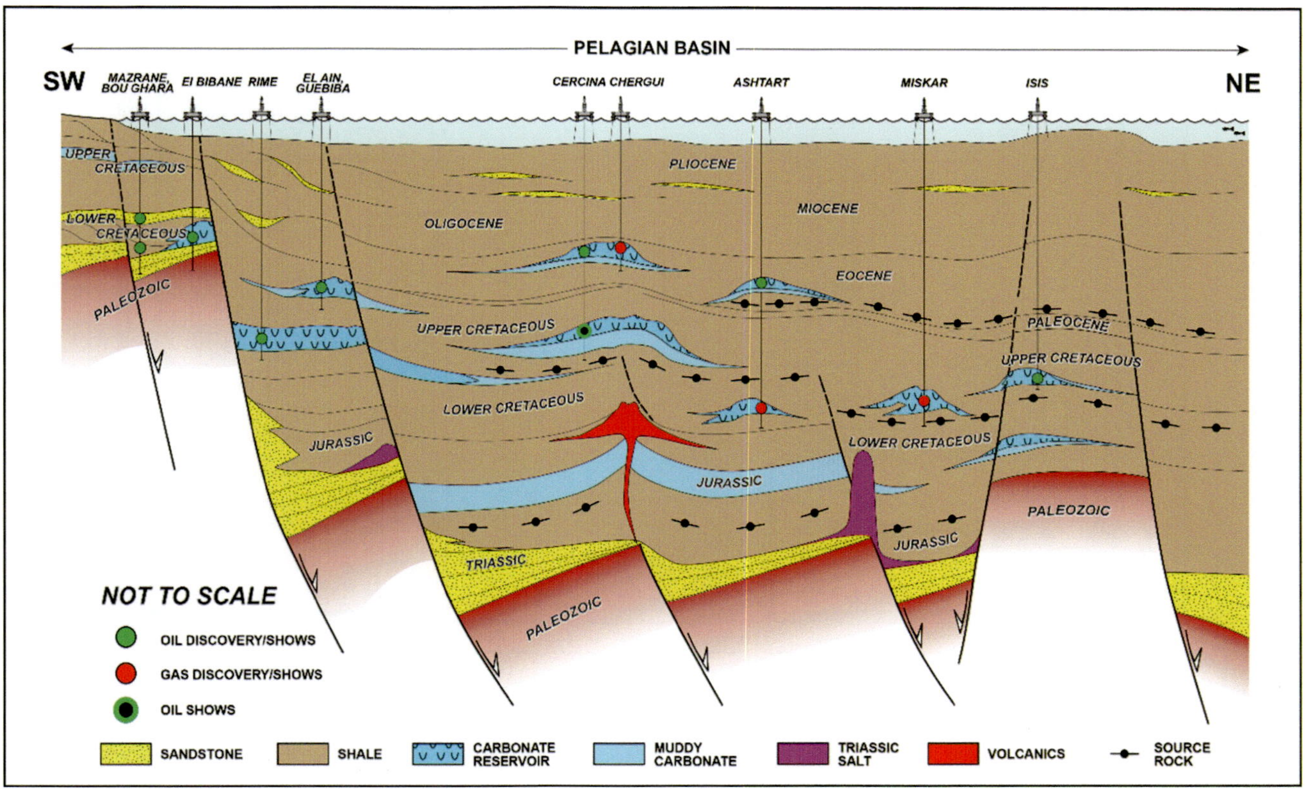

Fig. 5.5 Play sketch for the Pelagian Basin, schematically illustrating the individual reservoir distributions, seal and source presence and trapping styles for the different plays present in the Pelagian Basin. Drawn not to scale

southeastern area, the Lower Fahdene facies, where deposited, is considered to be gas and condensate prone with a narrow zone of peak oil trending NW-SE along the NW-SE trending shelf (Saïdi and Belayouni 1994; Mejri et al. 2006). It is assumed that the condensate-gas zone is a result of burial during Cenozoic times. Migration into reservoirs is considered to be predominantly via faulting.

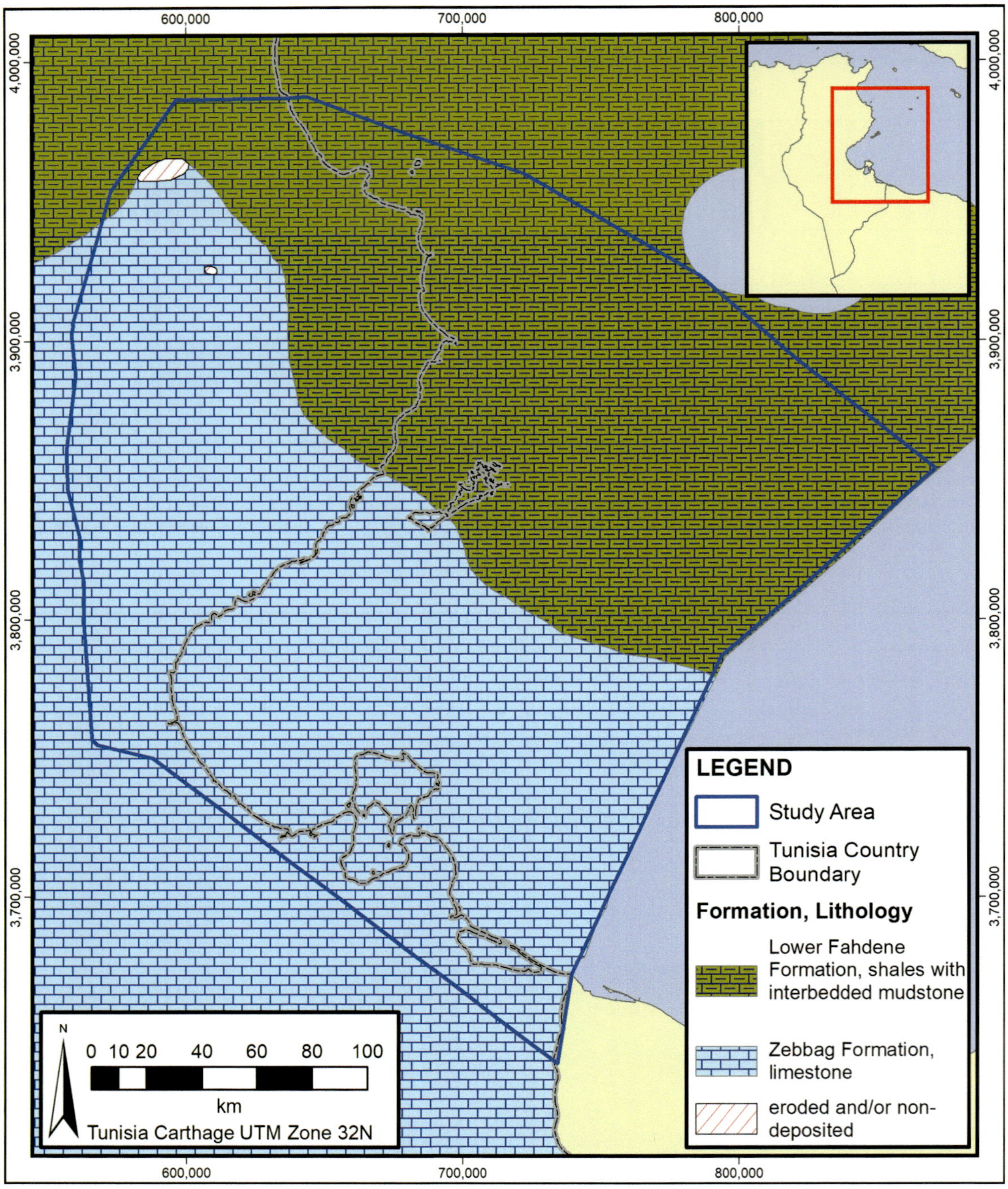

Fig. 5.6 Albian to Cenomanian gross depositional environment map, organic matter rich Lower Fahdene Formation and lateral equivalent Lower Zebbag Formation, redrawn after El Euchi et al. (2004)

Late Cretaceous (Late Cenomanian to Turonian): Bahloul Formation

The Bahloul source rock (Fig. 5.2) is present in the central and north eastern offshore area following the Late Cretaceous change from platform to basin configuration. In the southwestern and southern part offshore, the facies change into the lateral time equivalent Zebbag carbonate platform (Fig. 5.7) (Bishop 1988; Hallett 2002; El Euchi et al. 2004; Lüning et al. 2004; Mejri et al. 2006). Onshore, the Late Cenomanian to Turonian source rock is present in most of the area except in the northwestern most part where a local paleohigh has led to the truncation of the Bahloul Formation. In the westernmost part onshore, the succession changes into the lateral time equivalent Zebbag carbonate platform facies (Fig. 5.7). The occurrence of the source rock is governed by the Base Tertiary unconformity as well as the pre-Cenomanian paleohigh pattern (Affouri et al. 2013). The

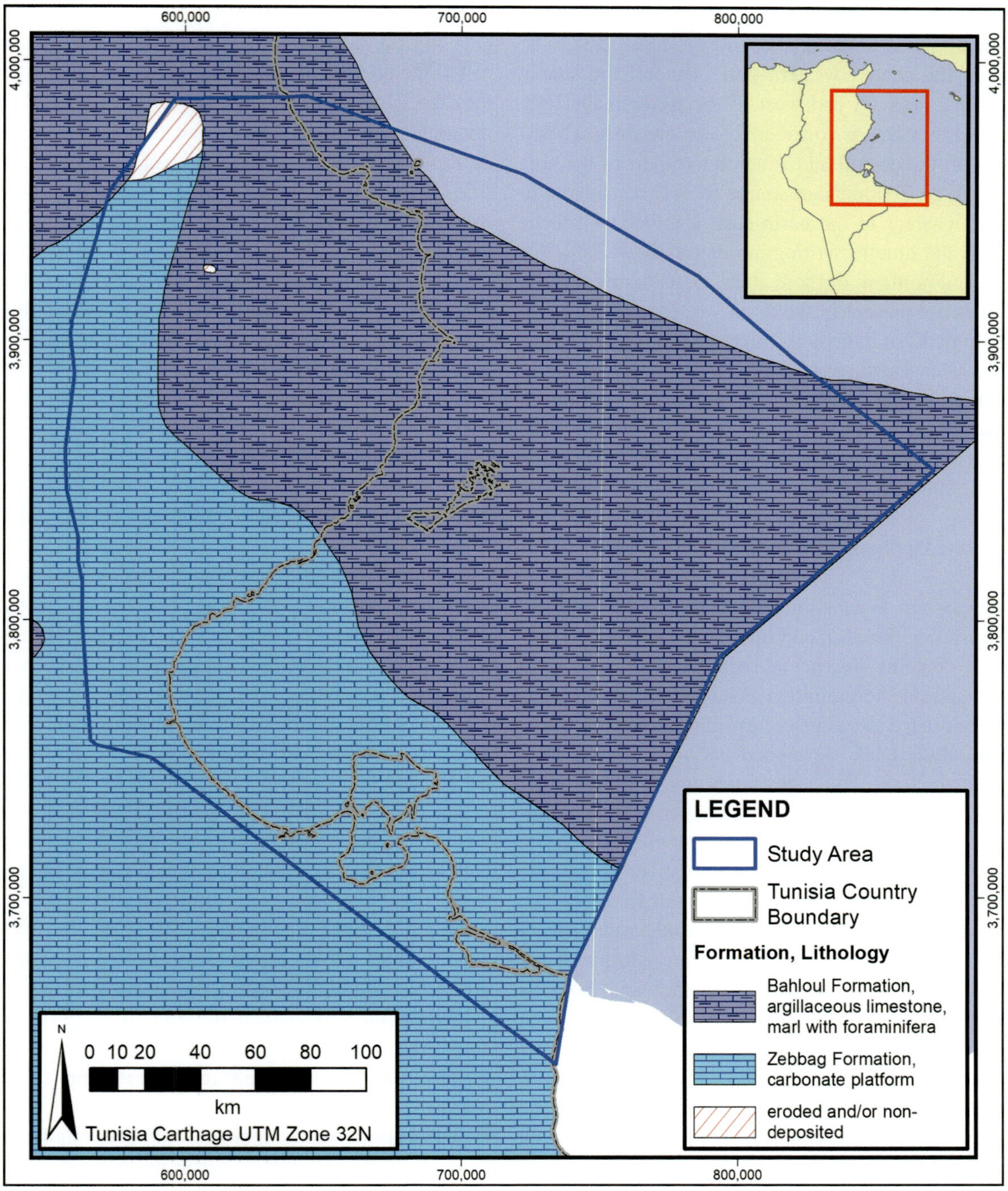

Fig. 5.7 Late Cenomanian to Turonian gross depositional environment map, source rock Bahloul and lateral equivalent reservoir Zebbag carbonate platform, redrawn after El Euchi et al. (2004)

Late Cenomanian to Turonian source rock consists of laminated dark marly limestone, deposited in a restricted marine shelf environment, with large Globigerinids of Late Cenomanian age. Its thickness ranges from 10 to 50 m (Robaszynski et al. 1990a, 1990b; Mejri et al. 2006). The Bahloul Formation is a type II/III kerogen carbonate source rock based on internal geochemical analysis. Geochemical data from surface sections and well samples provided a TOC range from 1.1% to 8%. The hydrocarbon potential of this horizon can reach 33 g/kg of rock (Saïdi 1993; Saïdi and Philip 1997; Mejri et al. 2006;Affouri et al. 2013). The Bahloul Formation is considered to be the main source rock for the Upper Cretaceous reservoirs of Cenomanian, Turonian and Maastrichtian age and may also contribute some hydrocarbon sourcing to the Ypresian reservoir. The latter can be observed in the case of the El Hajeb field, where the fault system in the area allowed Bahloul sourced oil to migrate upwards into the Ypresian reservoir (Klett 2001;

Fourati et al. 2002; El Euchi et al. 2004; Mejri et al. 2006). The north eastern part of the offshore area is considered to be immature. In the central western part offshore, the source kitchen reaches peak oil just west of the Kerkennah High. The central eastern area is considered to be gas prone (Mejri et al. 2006). In the central and southeastern part of the area onshore, the Late Cenomanian source rock is considered to be oil prone whereas to the north east, it is believed to be immature. At the time of writing, no data was available for the western and northern onshore areas. Considering that the source rock is not present in the north western part onshore, it can be assumed that there is no migration coming from this area. The condensate to gas window corresponds to those areas in which effective Tertiary burial causes continuous hydrocarbon maturation. Migration into reservoirs is considered to be predominantly via faulting (Bishop 1988; Lüning et al. 2004).

Ypresian: Bou Dabbous Formation

The Bou Dabbous Formation (Fig. 5.2) constitutes open marine deep-water facies, made up of a well bedded mudstone which contains micrite and globigerinid marl with abundant planktonic foraminifera (Mejri et al. 2006). It is believed to be related to a zone of upwelling in front of the El Garia nummulitic bank developed on the shelf margin (Berrocoso et al. 2013). The depositional environment for the Bou Dabbous is described as having an open marine, hemipelagic, outer shelf or ramp position (Mejri et al. 2006; Affouri and Montacer 2014). This formation covers the central part offshore with a depositional trend from NW-SE along the Ypresian shelf (Fig. 5.8). In the northeastern part of the offshore area, the source rock is not present due to the presence of the El Jem platform (Figs. 5.2 and 5.8). Its thickness varies between 50 and 350 m (Saïdi 1993; Saïdi and Philip 1997; Mejri et al. 2006). The formation also covers the northern and eastern part onshore with a NW-SE depositional trend along the Ypresian shelf. Along this trend the transition into the El Garia reservoir facies is confirmed by well data (Saïdi 1993; Saïdi and Philip 1997; Mejri et al. 2006). This globigerinid marl is a predominantly type II carbonate and black shale source rock with TOC content ranging from 0.5% to 8% with an average of 2% (Affouri and Montacer 2014). The geochemical results indicate a type II to type I kerogen with a slight dominance of land-derived organic matter (Mejri et al. 2006; Affouri and Montacer 2014). The Bou Dabbous shale is considered to be the main oil source into its lateral time equivalent, the El Garia carbonate reservoir, as well as via faults into the younger Lutetian to Bartonian nummulitic Reineche carbonate reservoir (Mejri et al. 2006). It is also considered to be the major oil source for all Cenozoic reservoirs both onshore and offshore in the southern Pelagian Basin. The oil prone kitchen area lies on the southwestern part of the depositional area, just to the northeast and adjacent to where the facies changes into the El Garia Formation. Here, the Bou Dabbous Formation is observed to be buried to a depth of more than 2200 m below MSL (mean sea level) today (Fig. 5.9). Further to the northeast, the source rock is considered to be immature. Hydrocarbon generation started at the end of the Miocene and is still ongoing (Fig. 5.4). Over the Kerkennah High, the Bou Dabbous is currently immature but migration into Cercina, Chergui, Ashtart and Ras El Besh fields indicate that relatively long lateral and fault migration is effective from mature zones (Fig. 5.9).

5.2.2 Reservoir Rocks

Middle Turonian: Bireno Member (Aleg Formation)

The Bireno Member lies within the Aleg Formation of Middle Turonian age (Fig. 5.2) (Touir et al. 1989; Abdallah et al. 2000), below which lies the Bahloul Formation and above lies the base of the Campanian to Maastrichtian Abiod Formation. The base of the reservoir unit is underlain by the Annaba shales of the Aleg Formation (Fig. 5.2). Bireno dolomites and their deposition have been extensively studied (M'Rabet 1981; Camoin 1993; Troudi et al. 2000; Abdallah et al. 2000; Abdallah et al. 2003; Touir et al. 2009) and represent a known productive oil reservoir within the Pelagian Basin. Previously published models describe the Bireno ramp type platform as a rudist shoal-rimmed carbonate platform or rimmed carbonate shelf (Camoin 1989; Troudi and M'Rabet 1998; Troudi et al. 2002; Touir and Soussi 2003).

Within the study area, the Bireno platform comprises anhydrites with thin dolomitic intercalations in the southwestern part, transitioning into inner-ramp settings with bioclastic and derived reef deposits along a NW-SE direction in the central part. Interbeds of microbially laminated dolomites and anhydrites occur in the northwestern part onshore. These rudist bioherms develop in quiet environments where peritidal flat and lagoonal facies dominate. Limestone with mixed benthic and planktonic foraminifera and mudstone develop further to the northeast in a mid to outer ramp setting (Fig. 5.10) (Gili et al. 1995; Troudi and M'Rabet 1998; Touir and Soussi 2003). The deep marine Lower Aleg shale overlies the Bireno carbonate ramp type platform (Fig. 5.2) and indicates the drowning of the platform and filling of accommodation space with the deeper marine shale. The main producing Turonian Bireno carbonates fields are El Ain, Rhemoura, Mahares and Guebiba (Mejri et al. 2006).

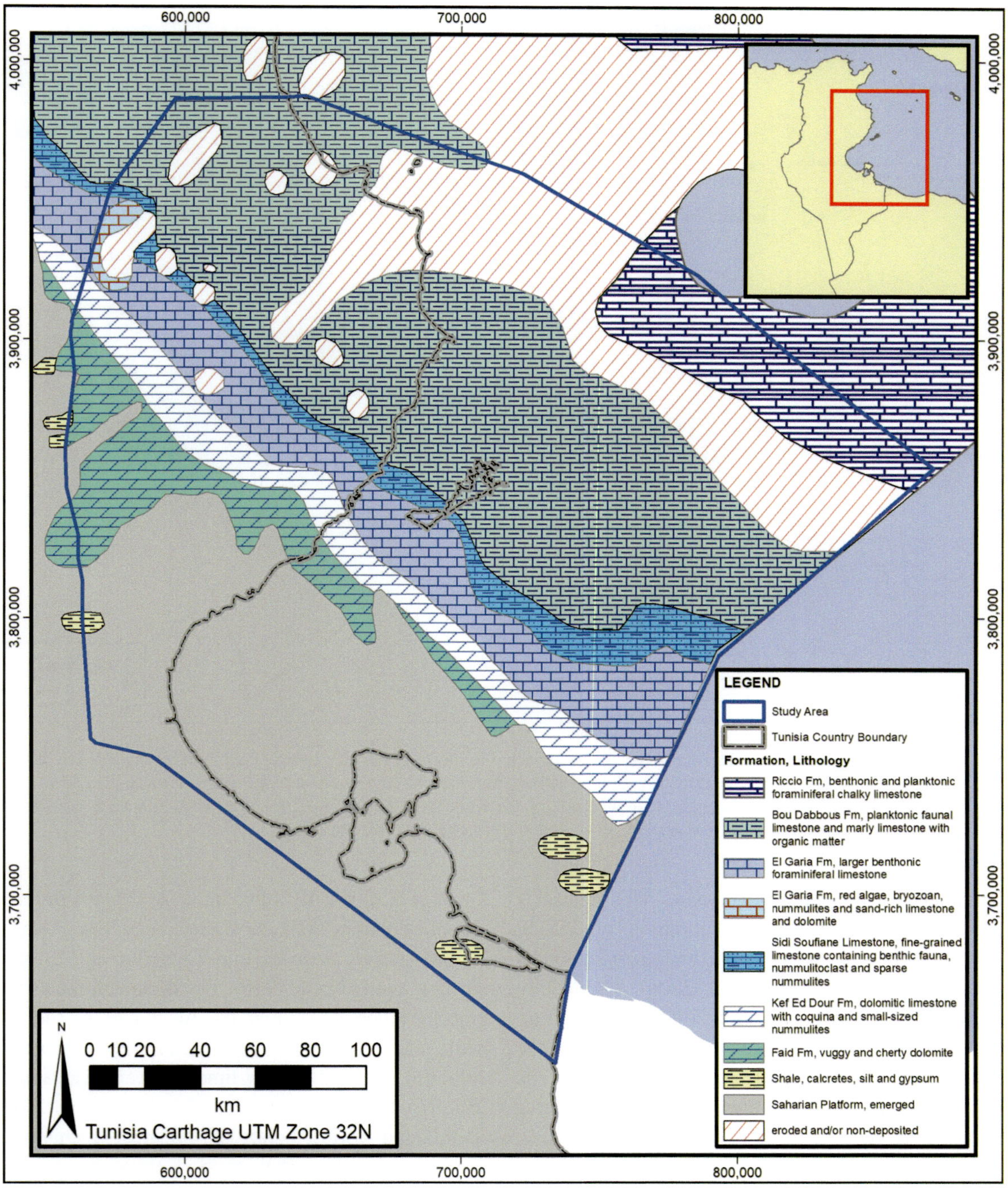

Fig. 5.8 Ypresian depositional environment map, organic rich facies of the Bou Dabbous Formation and lateral equivalent reservoir facies of the El Garia Formation, redrawn after Mejri et al. (2006). Note: El Garia reservoir comprises nummulitic limestones, packstones/grainstones, minor shales and dolomites

Turonian to Coniacian: Douleb Member (Aleg Formation)

The Douleb Member corresponds to another carbonate ramp Formation due to regression imbedded between the underlying Lower Aleg shales and overlying Upper Aleg shales (Fig. 5.2) (Klett 2001; MacGregor et al. 1998; Mejri et al. 2006). The Upper Aleg shales and overlying El Haria shales act as seal. The Douleb Member facies zones range from restricted shallow lagoon to inner ramp in the southwest, transitioning into a carbonate platform deposited along a NW-SE direction. Further to the northeast, the depositional environment changes to open shelf—lower slope and consequently to a deeper marine setting (Fig. 5.11) (Troudi and M'Rabet 1998). Good reservoir properties are developed in the central part of the study area where the carbonate

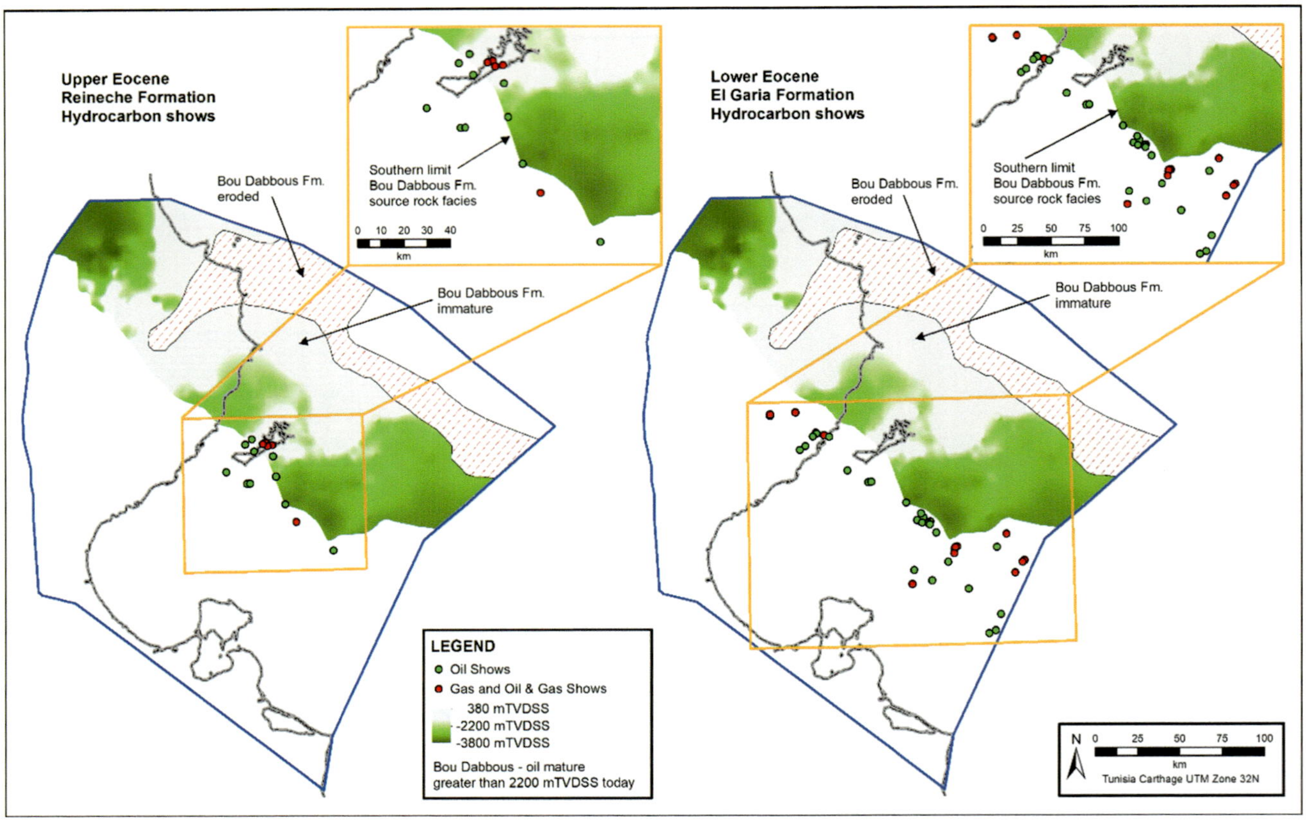

Fig. 5.9 Ypresian Bou Dabbous Maturity map and shows in El Garia and Reineche reservoirs. Lateral migration southwards into the Ypresian El Garia reservoir, minor contribution coming vertically from Bahloul source rock. Migration into the Lutetian-Bartonian Reineche reservoir vertically via faults with focus from both kitchens on either side of the Kerkennah High (Fig. 5.1). Generally, migration centred on El Garia reservoir due to shorter migration pathways, remigration possible into Reineche reservoir via faults

platform with oolitic limestones dominate. This indicates reef development on local shoals (Fourati et al. 2002). Porosity ranges between 10 and 20% with permeabilities generally less than 10 mD (Mejri et al. 2006). The locally developed Miskar Member consists of rudist reefal build-up and detrital carbonates and is a lateral time equivalent to the Douleb Member and Aleg shales (Fig. 5.2). In the wells around the Miskar field, several intercalations of volcanics have been recorded (Knott et al. 1995).

Campanian to Maastrichtian: Abiod Formation

The Abiod Formation (Fig. 5.2) was deposited during a transgression on a widespread carbonate platform during the Campanian to Maastrichtian (Bishop 1975; El Asmi 2015). It consists of two hemipelagic chalky limestone members separated by marl and limestone layers (Fourati and Hamouda 1996; El Euchi et al. 2004; Mejri et al. 2006) with locally traces of volcanics and is overlain by the El Haria Formation which represents the deep marine shale facies at the end of the highstand (Burollet 1956; Bishop 1975; Burollet et al. 1978; Bishop and Debono 1996; MacGregor et al. 1998; Klett 2001; Hallett 2002; Ferjani et al. 2003; El Euchi et al. 2004; El Rabia et al. 2018). Unconformities are identified within the Late Cretaceous (Base Aleg and Base Abiod) as well as at the mid-Thanetian age 'Base Tertiary' (Bishop and Debono 1996). The dominant facies is a chalky limestone deposited over the main part of the study area with intercalations of partly calcareous shale, partially dolomitized, argillaceous limestone and calcareous shales (Burollet 1956; Negra et al. 1994; Abidi et al. 2015). In the central and northern part of the area, a chalky fractured limestone is deposited with a thin zone of partially dolomitized limestone along a NW-SE trend following the shallow marine shelf environment. In the southern and south western area, the Abiod Formation is eroded or non-deposited (Troudi et al. 2000) (Fig. 5.12). The chalky limestone is heavily recrystallized and has poor primary reservoir quality. Enhancement of the reservoir is dependent on local fracturing, karsting and/or dolomitization (Troudi et al. 2000; Fourati et al. 2002; Deng et al. 2012; Abidi et al. 2015).

Variation in thickness of the Campanian-Maastrichtian deposits is recorded by wells mainly drilled on structural highs. The thinning of the carbonate platform is either explained as a depositional condensed section or by severe

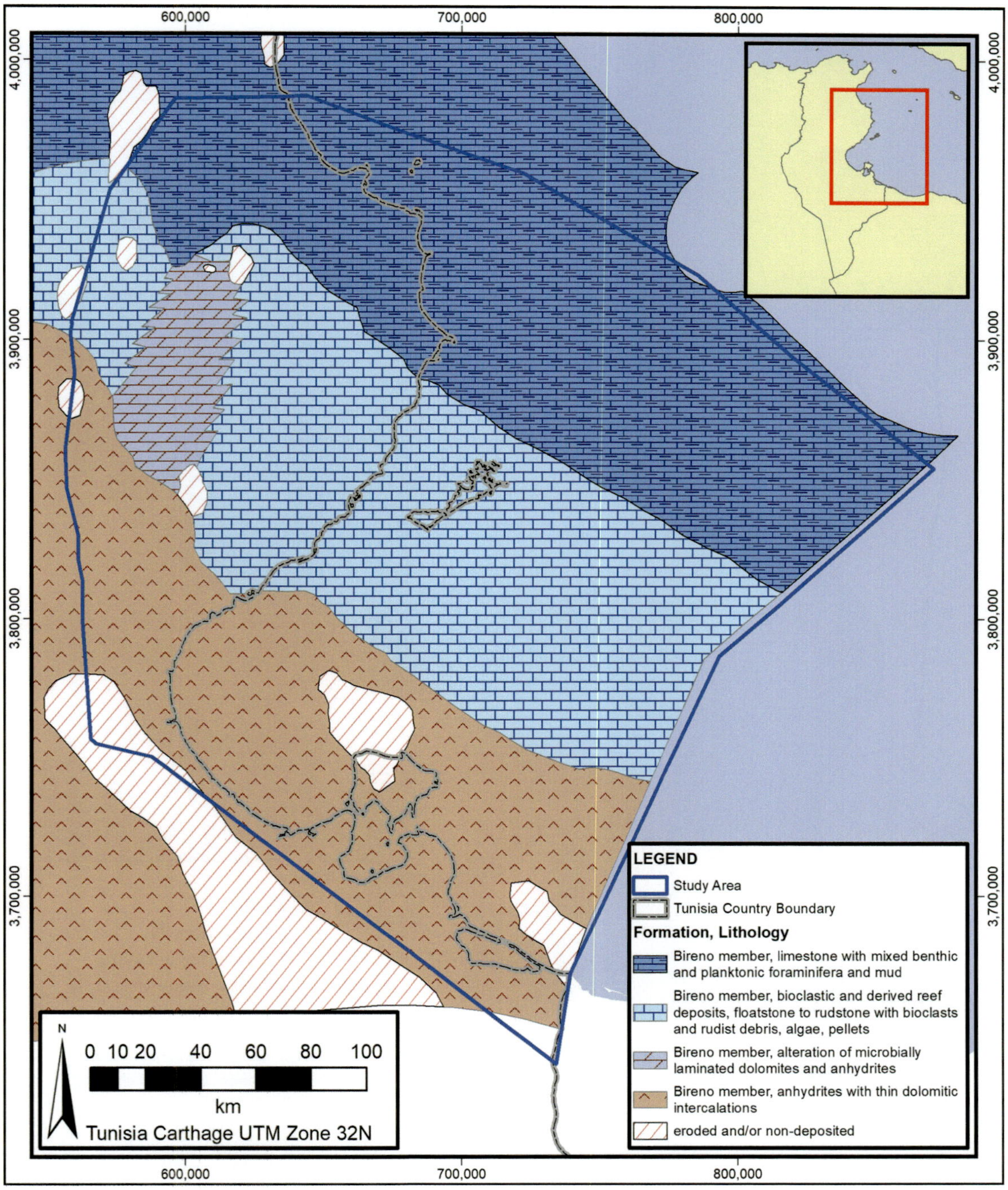

Fig. 5.10 Middle Turonian depositional environment map, Bireno Member (Aleg Formation), redrawn after Troudi and M'Rabet (1998); Touir and Soussi (2003)

erosion (Mabrouk et al. 2003a). The major mechanisms are suggested to be sea level fall and salt diapirism (Mabrouk et al. 2003b).

The shaly Middle Abiod Member has often been misinterpreted as Maastrichtian El Haria which has led to a reduction of the actual total thickness of the Abiod whilst adding to the thickness of the overlying El Haria. Large areas are affected by significant erosion and subsequent transgressive onlap of Palaeocene, often Thanetian aged El Haria shales. The preserved thickness of the Abiod ranges between 50 to 250 m over wide areas with larger local thicknesses of up to around 1500 m recorded (e.g., Sidi El Kilani 1562 m, El Jem 1482 m). The main fields producing from the Campanian-Maastrichtian Abiod fractured limestone is Sidi el Kilani (Mejri et al. 2006) and Miskar (Troudi et al 2000 and Brown 2003).

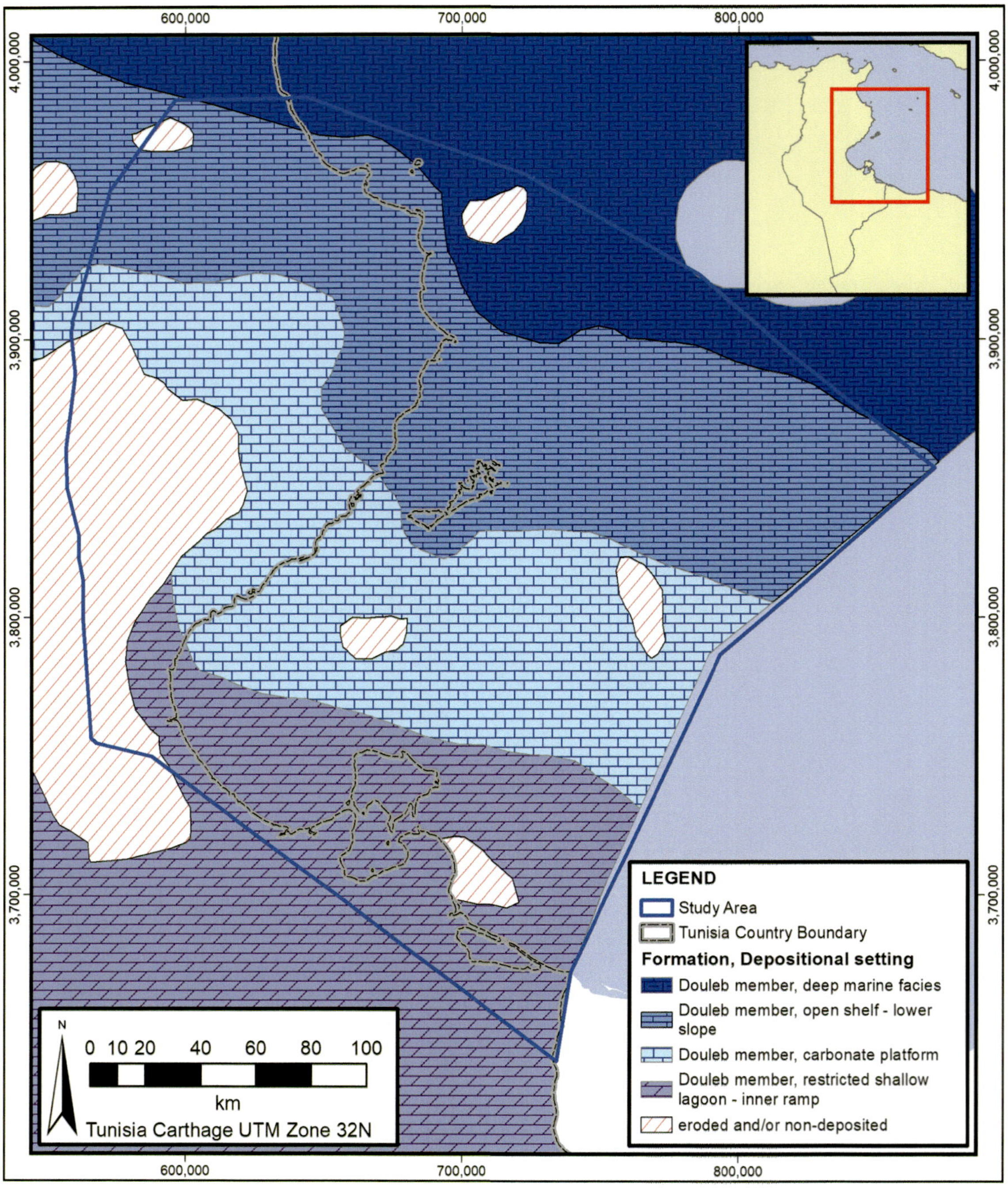

Fig. 5.11 Late Turonian to Coniacian depositional environment map; Douleb Member (Aleg Formation), redrawn after Troudi and M'Rabet (1998)

Ypresian: El Garia Formation

Two main facies zones of the Early Eocene Ypresian Metlaoui group can be differentiated: a broad basinal globigerinid limestone belt of the Bou Dabbous Formation and a narrow nummulitic limestone fairway representing the El Garia Formation (Fig. 5.2). The distinct facies contact between Bou Dabbous and El Garia follows a NW-SE trend turning into a W-E trend near the Libyan border (Fig. 5.8; Bishop 1988; Loucks et al. 1998; Anketell and Mriheel 2000; Klett 2001; Hallett 2002; Jorry et al. 2003; Vennin et al. 2003; Beavington-Penney and Racey 2004; Beavington-Penney et al. 2005; Mejri et al. 2006; Taktak et al. 2010; Tlig et al. 2010; Swei and Tucker 2012; Berrocoso et al. 2013; Affouri and Montacer 2014; Elfessi 2017). The El Garia consists of nummulitic limestones, packstones and grainstones, minor intercalations of shales, and locally dolomitized horizons. Depositional environments for the El Garia Formation are

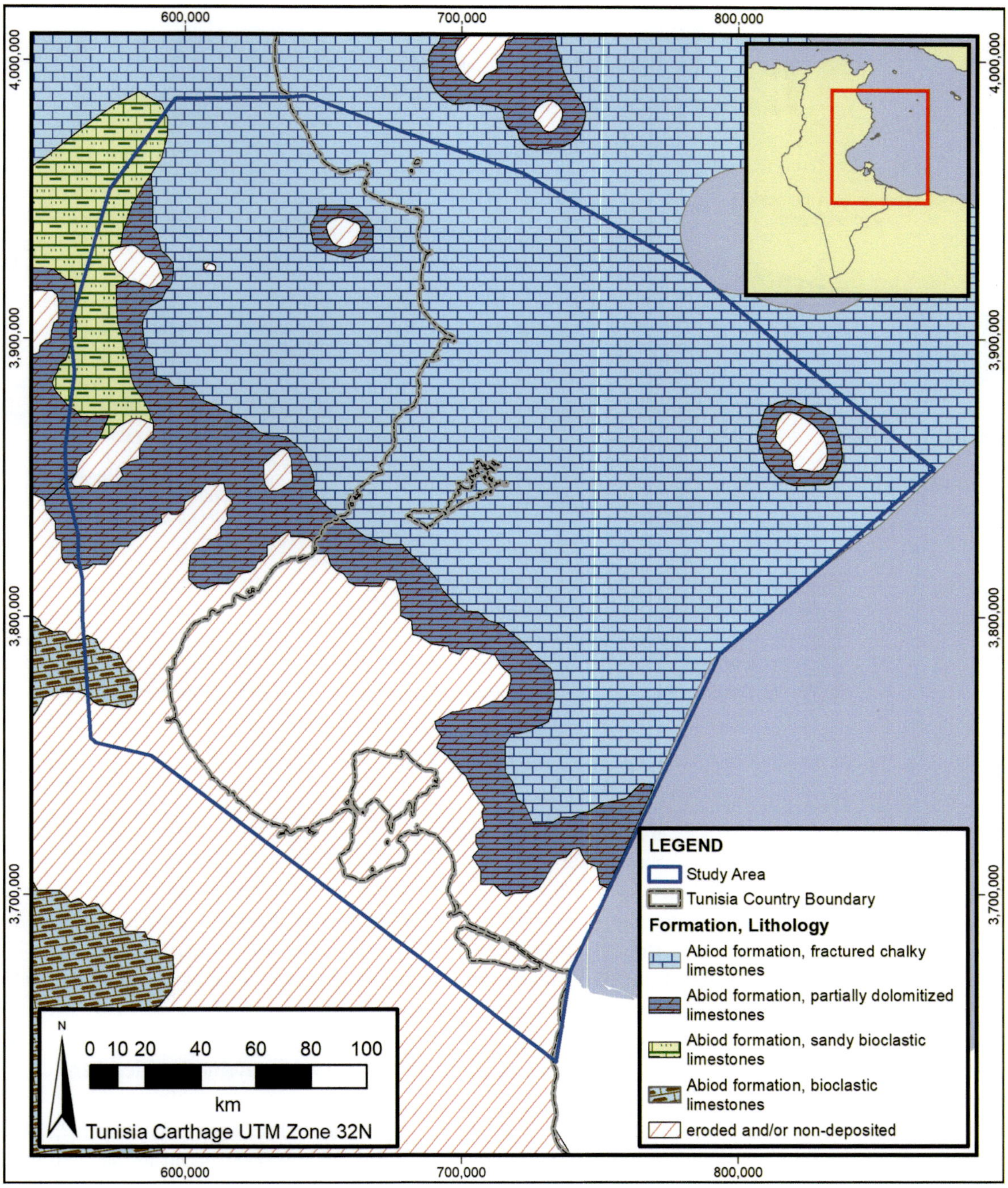

Fig. 5.12 Campanian to Maastrichtian depositional environment map, Abiod Formation, redrawn after Troudi et al. (2000)

described as an open marine, mid ramp position between fair weather and storm wave base (Racey et al 2001; Beavington-Penney et al. 2005; Nadhem et al. 2017). The thickness ranges from 0 to 220 m. El Garia nummulitic limestones are deposited along a 20–40 km broad NW-SE trend in the central part of the study area (Fig. 5.8). Porosities and permeabilities for the reservoir ranges from 10 to 26% with permeabilities from 40 to 100 mD (Ligtenberg and Wansink 2001; Mejri et al. 2006). Further south and southwest the Metlaoui group is absent due to emersion or replacement by lacustrine to continental units of the clayey Tanit Formation which is the lateral equivalent to the Cherahil Formation (Fig. 5.8). The El Garia is deposited during a transgressive period on an extensive carbonate platform within a widespread warm shallow marine water environment. A narrower and more linear facies belt indicates a steeper shelf/basin transition. A depositional model for the El Garia is illustrated in Fig. 5.13 based on studies undertaken

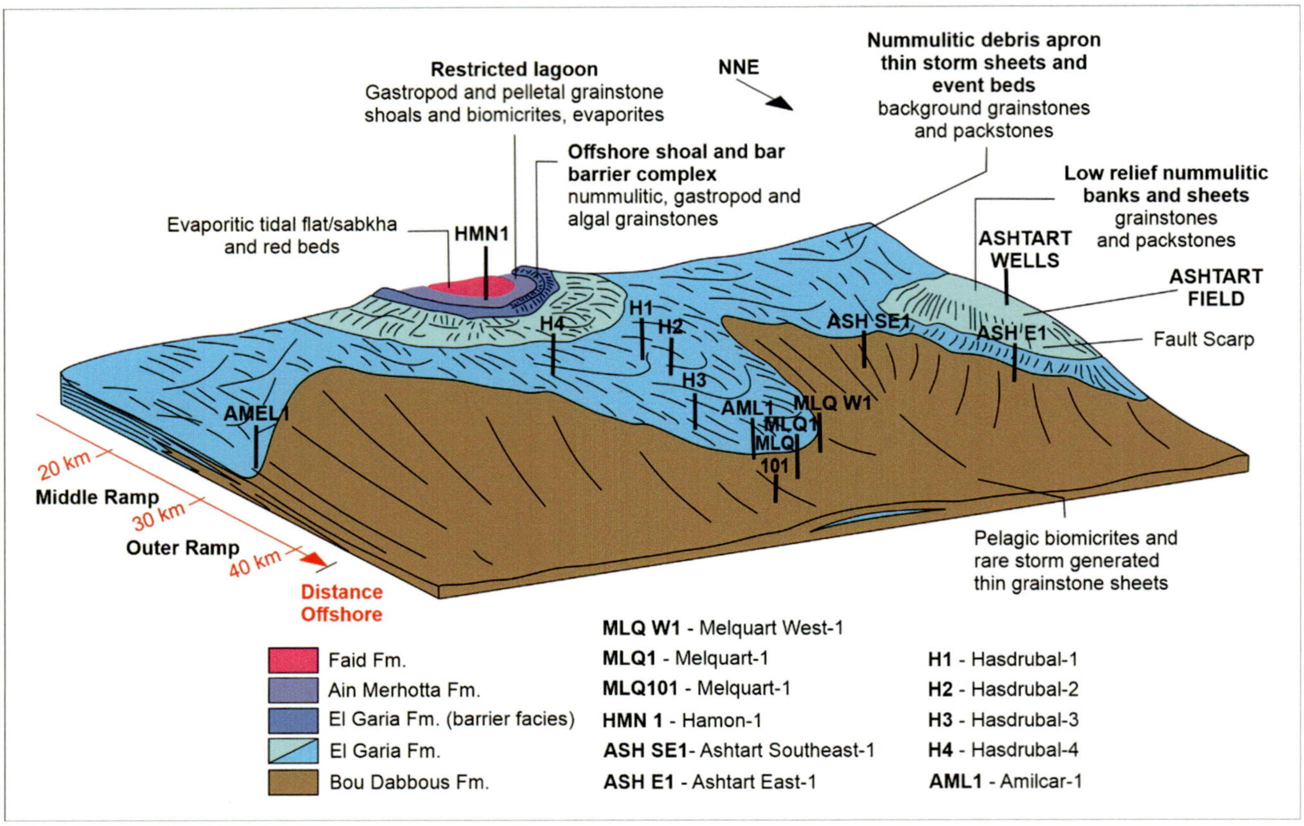

Fig. 5.13 Depositional Model for the El Garia over Ashtart and Hasdrubal fields after Racey et al. (2001)

over the Ashtart and Hasdrubal fields (Racey et al. 2001). The carbonate shelf progressively diminishes throughout the Early Eocene. The lower boundary is marked by a gradual transition from Chouabine Limestone into the El Garia nummulitic limestone and then into Bou Dabbous limestone/shale alternations. The transition from Chouabine into El Garia is marked by the disappearance of shale interbeds and shaly limestones. These Ypresian sediments are transgressively overlain by a well cemented, earliest Lutetian, locally silicified micritic limestone ("Compact Micrite"). Local evidence of pre-Lutetian erosion affecting the top of El Garia can be seen which are subsequently overlain by Lutetian Cherahil limestone/shale alternations and Souar shales respectively. Nummulitic limestones exhibit excellent primary porosities due to nummulite chambers. Permeabilities depend on the destruction of the nummulite test by either bioturbation, storm activity or gravity-triggered redeposition on a slope. However, the outcome of such processes is difficult to predict, because the destruction of the tests also supports subsequent cementation and loss of primary porosities. Enhancement of primary porosities by dolomitization has only been observed locally around small islands with exposure to meteoric waters. More commonly at Hasdrubal, porosities are significantly improved by acidic waters associated with early oil migration from the adjacent Bou Dabbous source rock (Racey 2001). Bou Dabbous micritic limestones and shales lack primary porosities generally. As with the Abiod, only fracturing and karsting can create a valid Bou Dabbous reservoir. Significant hydrocarbons are produced from the El Garia nummulitic limestones in Ashtart, Sidi El Itayem, Hasdrubal, Didon and Zarrat fields (Mejri et al. 2006).

Lutetian to Bartonian: Reineche Member

The Reineche nummulitic limestones (Fig. 5.2) are deposited along a 50 km wide NW-SE trend in the central part of the study area and are the last prominent culmination of a thin and narrow carbonate platform that was initiated during Mesozoic times (Fig. 5.14) (Burollet 1956; Bishop and Debono 1996; Hauptman et al. 2000; Klett 2001; Hallett 2002; Mejri et al. 2006; Taktak et al. 2010, 2012; Elfessi 2017; Njahi et al. 2017). Compared to the Metlaoui group, the platform—shelf margin—ramp boundaries are shifted further north due to a general relative sea level fall. Thickness of the gross unit ranges from 25 to 75 m with a net reservoir thickness of the nummulitic facies of 2–10 m. Porosities range from 20 to 30% with permeabilities poor to excellent dependent on secondary dissolution, dolomitization and fracturing (Elfessi 2017). A pronounced basinward pinch out

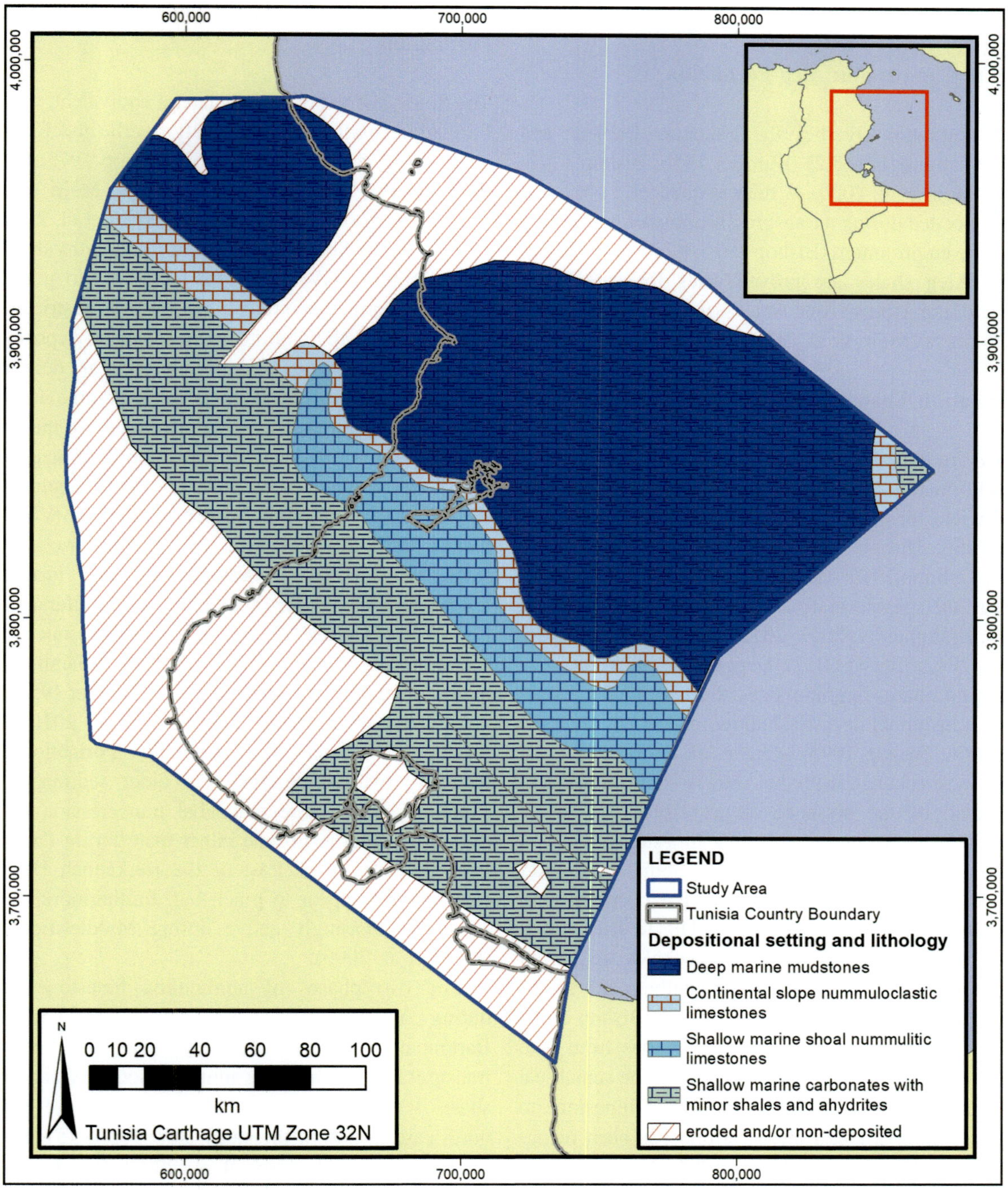

Fig. 5.14 Regional depositional trend of late Eocene Lutetian-Bartonian Reineche

to the north east can be seen in wells from the Chergui and Cercina fields. Lutetian-Bartonian deposits comprise a stack of partly nummulitic limestones and minor, partly calcareous shales. Landward Reineche nummulitic limestones disappear into intercalations of limestone and shale of the Cherahil Formation, whilst basinward a marked pinch out into Souar shales is observed. The Reineche limestones subdivide the Cherahil A-Souar A from the upper Cherahil B-Souar B. Depositional environments are open marine, mid ramp, between fair weather wave base and storm wave base and are considered to be comparable to that of the El Garia nummulitic limestones depositional regime (Racey 2001). Alternatively, it is also suggested that nummulites, during this time, could only develop on the shelf or flat ramp just below storm wave base where environmental conditions were preferable for survival (Hauptmann et al. 2000). The two main producing fields of the Mid-Eocene Reineche limestones are Cercina and Chergui (Mejri et al. 2006; Elfessi 2017).

5.2.3 Seal Rocks

Turonian to Campanian: Aleg Formation

The Aleg Formation overlies the Fahdene, Bahloul and Zebbag Formations (Fig. 5.2) (Burollet 1956; Bishop 1975; Klett 2001; Mejri et al. 2006). A thick sequence of shale and marl was deposited during a transgressive phase representing a deep marine environment (Bishop 1975; Abidi et al. 2015). The lower Aleg shales are proven seals for the Bireno Member and the Upper Aleg shales for the Douleb and Miskar Members (Mejri et al. 2006).

Maastrichtian to Thanetian: El Haria Formation

The Maastrichtian to Thanetian El Haria Formation overlies the Abiod Formation (Fig. 5.2), locally unconformably, and is the main seal for the Upper Cretaceous Abiod reservoir. (Burollet 1956; Bishop 1975; Zourai 1995; Bédir 1995; Bishop and Debono 1996; MacGregor et al. 1998; Zaïer et al. 1998; Klett 2001; Rabhi et al. 2001; Hallett 2002; Adouani et al. 2003; Brahim et al. 2003, 2012; Mabrouk et al. 2003a; Melki et al. 2012; Elfessi 2017). Deposition started during a long highstand phase, ongoing since the Late Upper Cretaceous, in a transitional deep to shallow marine environment (Burollet 1956; Bishop 1975; Mejri et al. 2006; Tlig et al. 2010). The facies change from the Abiod to the overlying El Haria is noted by the presence of grey, black or brown mudstone with intercalations of thin limestone beds, in combination with planktonic diversity and abundance versus benthonic diversity and abundance and the occurrence of ostracods (Burollet 1956; Bishop 1975; Donze et al. 1985; Mejri et al. 2006). In general, the El Haria shales are present in most of the study area, except in the southern and southwestern part due to non-deposition or erosion (Bishop 1975; Mejri et al. 2006). Thickness of the unit varies from 0 to 500 m (El Euchi et al. 2004; Mejri et al. 2006).The actual seal quality depends strongly on the proportion of limestone to shale. A local intra-El Haria unconformity has been recognised by previous interpreters (Haller 1983; Zargouni 1985; Zourai 1995; Mejri et al. 2006; Boussiga 2008), in the Gulf of Gabes within the southern edge of the Pelagian Basin (Taktak et al. 2010) based on well and seismic data. In addition, it is important to also consider the possible misinterpretation of the contact between the Abiod and El Haria, and the subsequent impact on the thickness variations observed from wells and seismic, as briefly discussed in Sect. 5.2.2.

Middle Eocene: Souar and Lateral Equivalent Cherahil Formations

The Souar Formation and its lateral equivalent, the Cherahil Formation, partly unconformably overlie the El Garia formation (Fig. 5.2). (Burollet 1956; Bishop 1975; Bishop and Debono 1996; MacGregor et al. 1998; Mejri et al. 2006; Baklouti et al. 2017; Elfessi 2017; Njahi et al. 2017). In the study area, the Souar and Cherahil Formations show varying facies trending NW-SE. From southwest to northeast, the Cherahil Formation transitions from a restricted marine environment with dolomite, marl, shale and gypsum deposits into a more open marine environment dominated by marls, shales and limestones rich in benthic foraminifera, mixed with coquina. Towards the northeast of the study area, the Cherahil Formation transitions into the Souar Formation when the percentage of limestone and coquina becomes smaller than 50%. The deeper marine facies consists of planktonic and benthonic foraminifera rich marl and shale, with nodular limestone beds and further to the northeast with larger benthonic foraminiferal limestone (Fig. 5.15). During Eocene times, subsidence was more active leading to a transgressive phase, explaining the wider extension of deeper marine deposits (Burollet 1956; El Euchi et al. 2004; Mejri et al. 2006; Tlig et al. 2010). In many places, the Souar and Cherahil Formations unconformably overlay progressively older sediments. This is explained by a more extensive transgression of the sea during the Middle Eocene rather than during Palaeocene or Early Eocene times. East of the Kerkennah High, Lower Eocene beds appear to pinch out. Furthermore, subsidence became increasingly active during Middle Eocene times (Mejri et al. 2006).

The two phases of nummulitic limestone deposition during Ypresian (El Garia Formation) and Lutetian-Bartonian (Reineche Member) times are separated by transgressive diachronous units of Souar A and B pelagic shales which act as seals over much of the central and north eastern part of the study area (Fig. 5.15) (Klett 2001; Racey et al. 2001; Jorry et al. 2003; Taktak et al. 2012; Mejri et al. 2006; Chalwati et al. 2018). Alternatively, if unfractured, the El Garia reservoir can also be sealed by the overlying late Ypresian Compact Micrite and by the early Lutetian Cherahil A limestone and shale intercalations that are present narrowly trending NW-SE in the central part of the study area (Racey et al. 2001; Mejri et al. 2006; Chalwati et al. 2018). For seal integrity, the proportion of

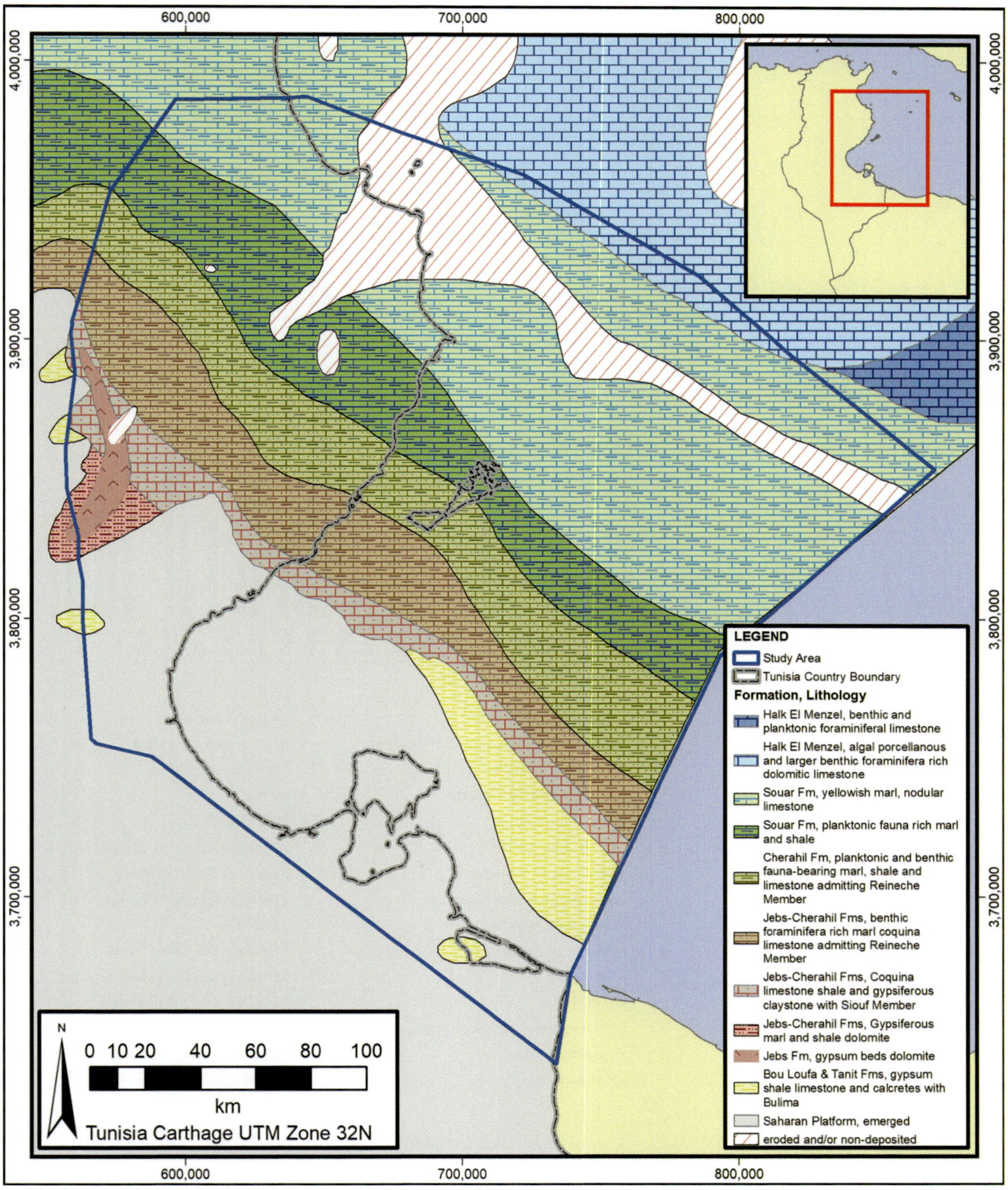

Fig. 5.15 Middle Eocene: Souar and lateral equivalent Cherahil Formations, modified after El Euchi et al. (2004)

limestone interbeds becomes critical for the Lutetian reservoir. In the south and southwestern area, the Middle Eocene deposits are either eroded or not deposited (Fig. 5.15).

By Bartonian times, the Cherahil B–Souar B transition is located basinward compared to the situation by the end of the Ypresian, indicating that the Reineche transgressive pulse is less pronounced. During the transgressive period, the carbonate shelf diminishes throughout the Early Eocene (Mejri et al. 2006).

5.3 Data Sets, Methodology and Observations

372 wells have been drilled in total in the study area since 1949 with 136 exploration wells drilled offshore and 75 exploration wells drilled onshore. From these wells, 26 fields were discovered offshore and 16 fields onshore.

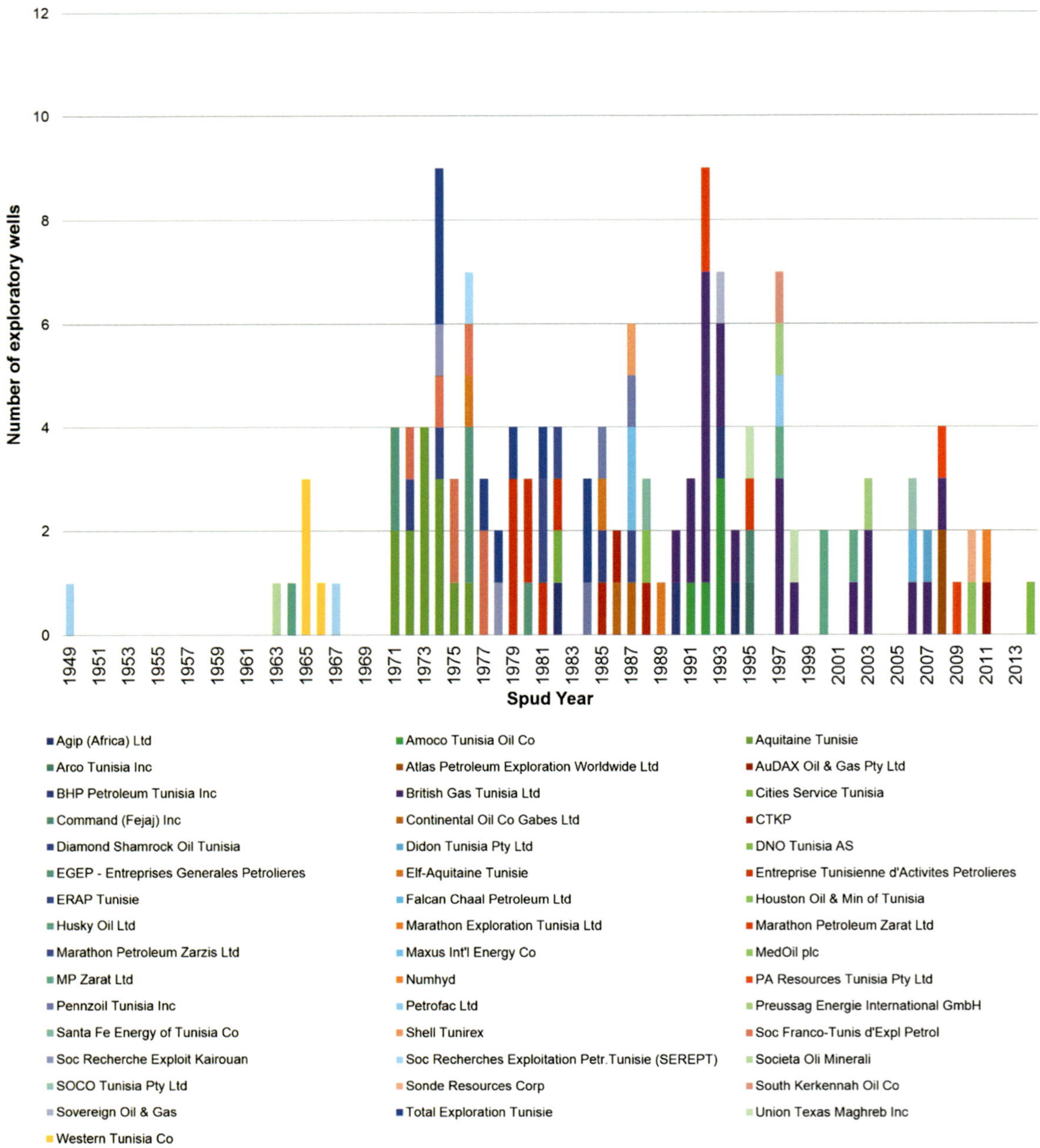

Fig. 5.16 Pelagian Basin Exploration history offshore, drilling activity from 1949 until 2015

Figures 5.16 and 5.17 show the drilling activity offshore versus onshore from 1949 until 2015. Many of the earlier wells drilled in the area were drilled on wide spaced 2D seismic datasets and as such definition of structure is considered to be the main cause of failure. A secondary cause is the poor quality of the targeted reservoir as many of the carbonate reservoirs require either the presence of a significant fracture system—related to fault activity and halokinesis—or enhancement of matrix reservoir properties—through dolomitization, dissolution, early hydrocarbon migration or facies quality, such as presence of nummulites in the Ypresian El Garia Formation and Lutetian-Bartonian Reineche Member (Racey et al 2001; Elfessi 2017).

The data set used for this evaluation is composed of well data for 348 wells and 194 selected key 2D seismic lines covering an area of approximately 73,500 km^2. The 2D seismic data is of varying quality and different vintages. However, for regional work where the seismic quality was reasonable, these lines were used for mapping Cretaceous and Palaeocene horizons. Well Statistics, Primary Targets and Success Rate for Cretaceous, Jurassic and Palaeogene horizons of Exploration wells are shown in Fig. 5.18.

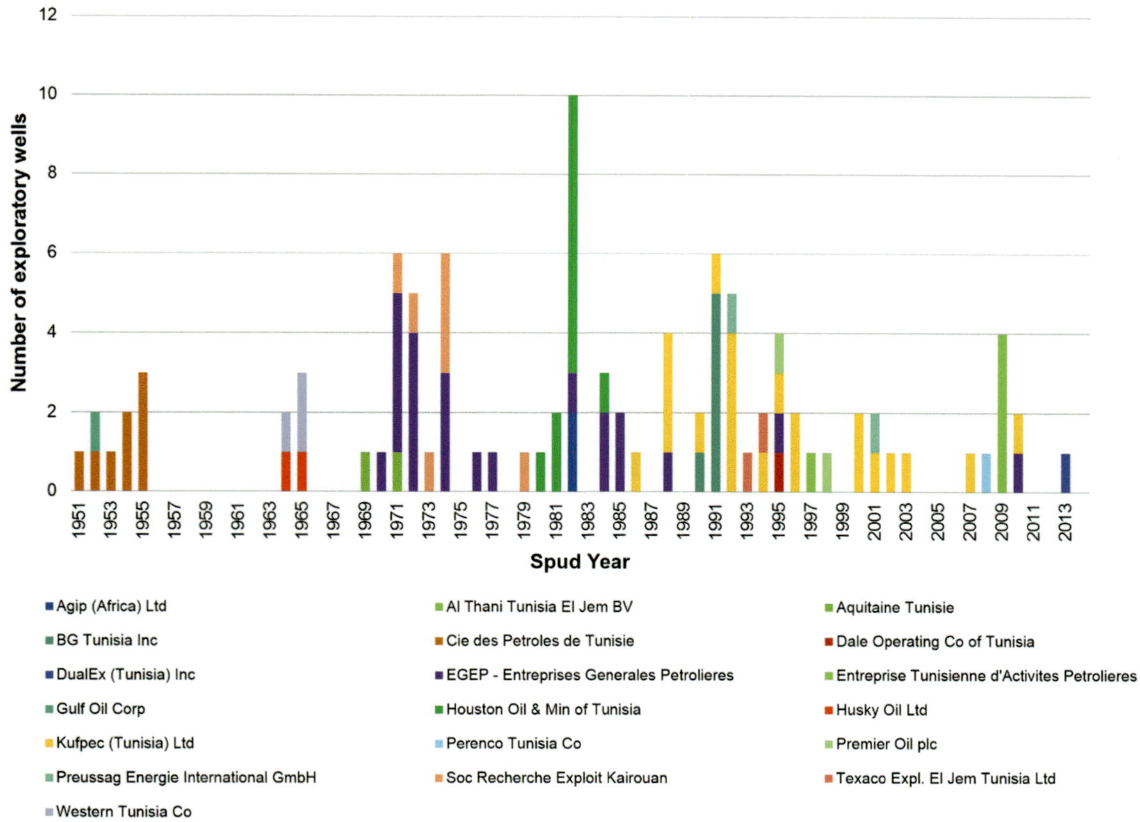

Fig. 5.17 Pelagian Basin Exploration history onshore, drilling activity from 1949 until 2015

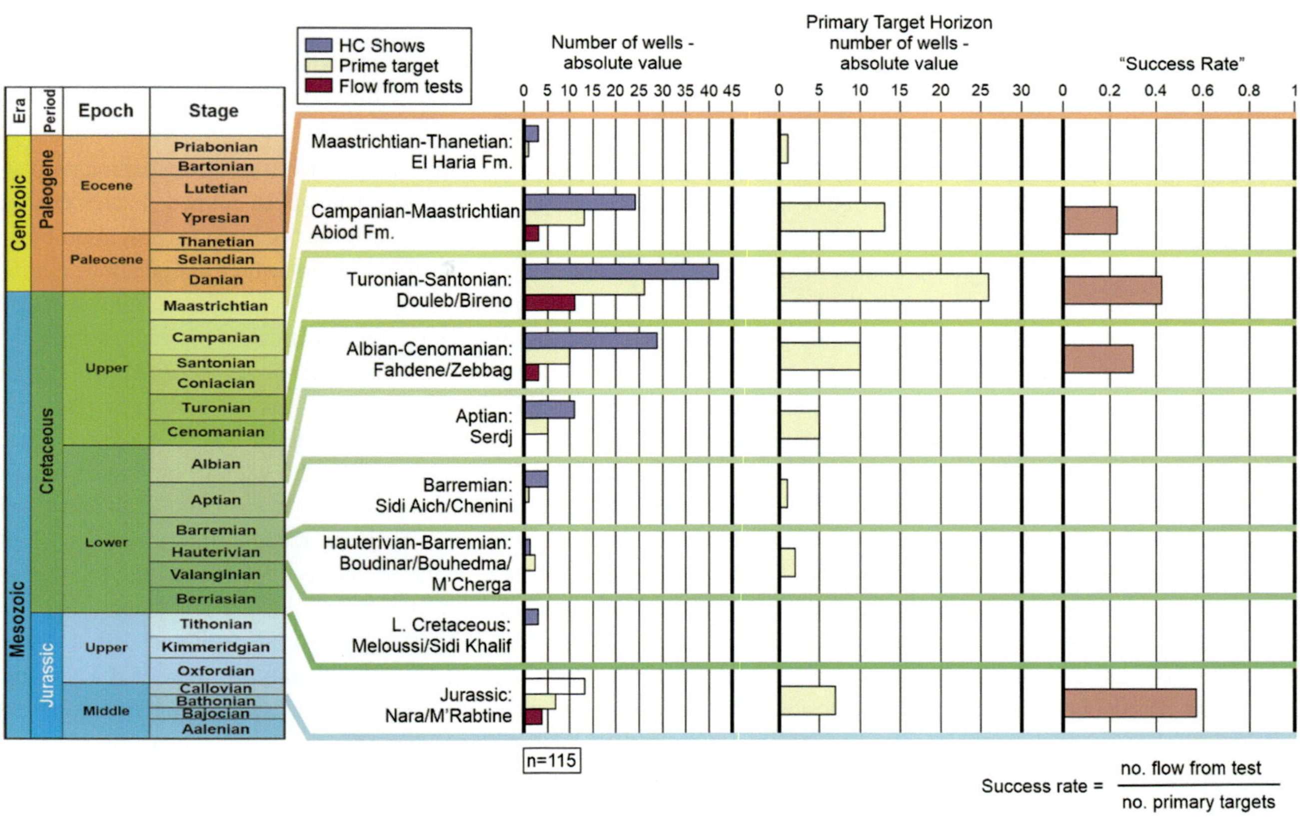

Fig. 5.18 Well Statistics, Primary Targets and Success Rate for Cretaceous, Jurassic and Paleocene horizons of Exploration wells (excludes Eocene)

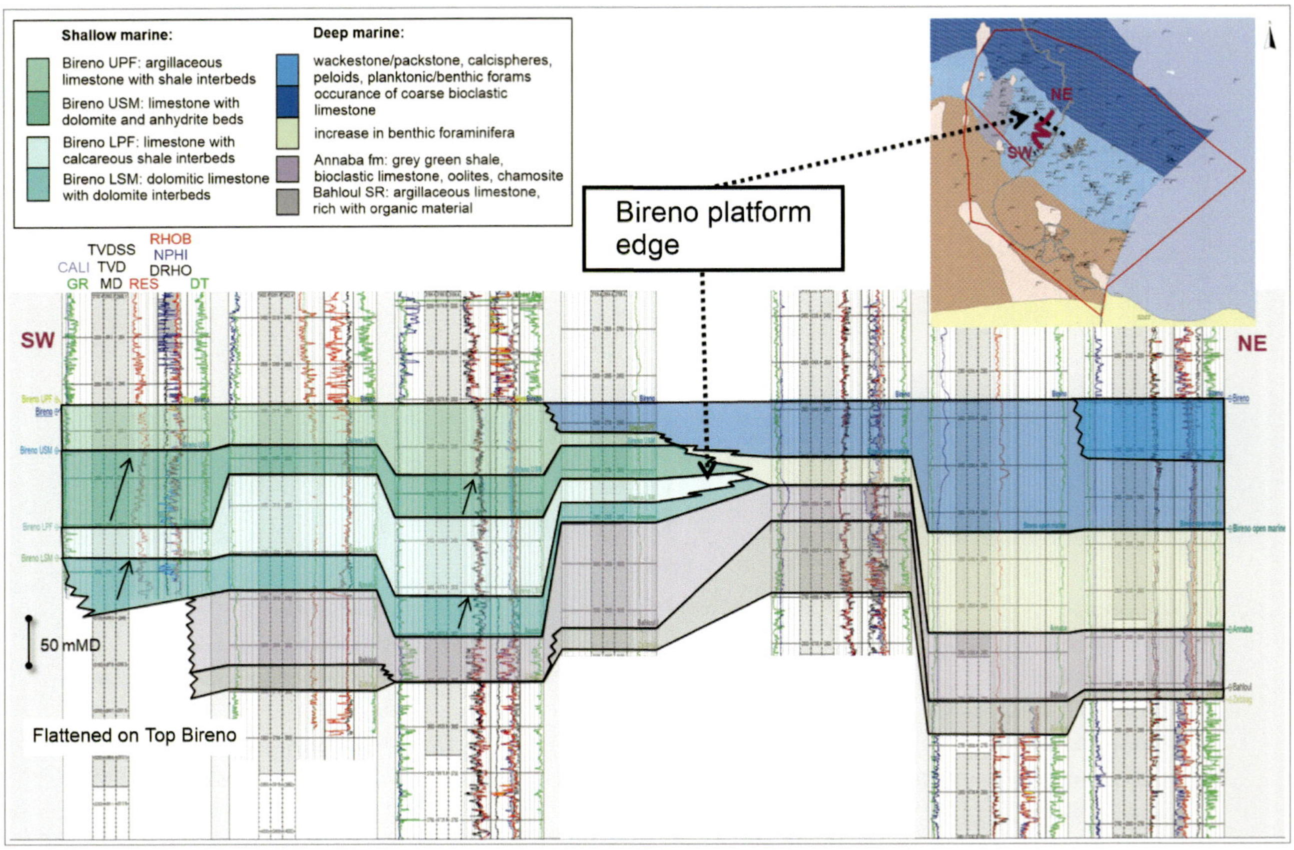

Fig. 5.19 SW-NE well correlation of the Early to Middle Turonian, focusing on the reservoir within the Bireno Member, showing a facies change from the SW towards the NE is recognised by an increase in benthic foraminifera where the shallow marine carbonate platform environment transforms into an open marine environment. Inset map on the upper right-hand side Middle Turonian Reservoir Bireno depositional environment map Fig. 5.10, modified after Troudi & M'Rabet (1998); Touir & Soussi (2003); UPF = upper platform, USM = upper shallow marine, LPF = lower platform, LSM = lower shallow marine

For the Cretaceous plays a selection of wells was chosen for the regional interpretation. The selection was based on log data quality and available runs in target horizons as well as the location on selected 2D seismic lines of the regional interpretation grid. 26 wells were selected offshore and 23 wells were selected onshore for the regional interpretation of Cretaceous horizons.

Over the study area there are various facies changes within the Cretaceous section. In the frame of this study previously published maps of the depositional environment were reviewed (see Figs. 5.6, 5.8, 5.10, 5.11, 5.12, and 5.15) (Troudi and M'Rabet 1998; Troudi et al. 2000; Touir and Soussi 2003; El Euchi et al. 2004; Mejri et al. 2006) and updated. The depositional environment boundaries were refined by using well correlations and lithostratigraphic and biostratigraphic reviews of selected wells using cuttings, sidewall core and barrel core descriptions based on availability and composite logs. The wells were selected based on their well class with the main focus on exploration wells, their locations and the necessity to target the Cretaceous formations.

Figure 5.19 shows a southwest-northeast carbonate platform edge location which has been refined. Facies change from the southwest towards the northeast is recognised by an increase in benthic foraminifera where the shallow marine carbonate platform environment transforms into an open marine environment. The best quality reservoir within the Bireno Member is within the lower peritidal and upper peritidal facies of the shallow marine carbonate platform. The peritidal facies are interbedded with claystones. These thin layers are visible within the gamma ray log response. The peritidal facies overlay the lower shelf margin and upper shelf margin facies. They are recognisable in the well logs by an increase of the resistivity log towards the top due to facies change from dolomite to anhydrite layers.

Figure 5.20 shows a WSW-ENE well correlation where the main focus was on identifying the Mid-Thanetian Unconformity as a support for seismic interpretation. In the southwestern part of the study area the Campanian to Maastrichtian succession is absent, which is well documented in the wells penetrating this area.

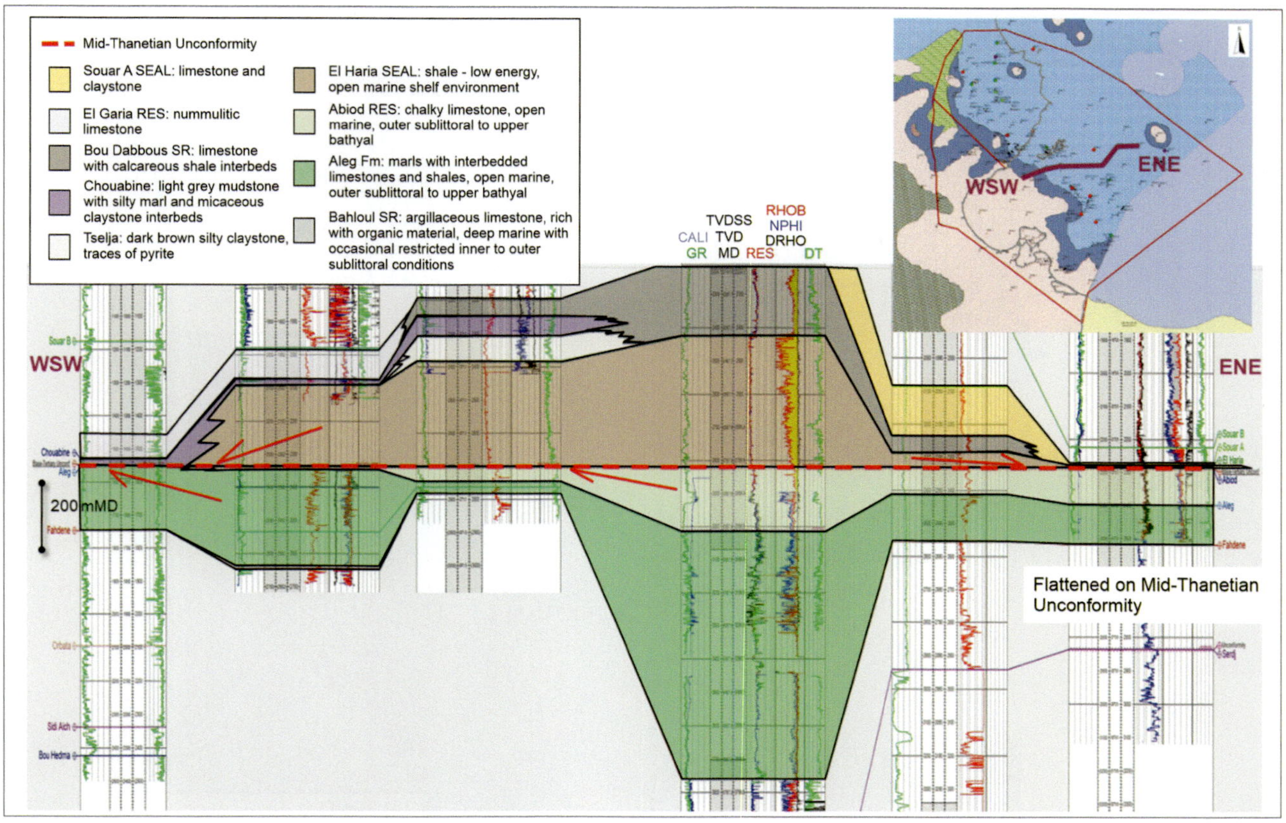

Fig. 5.20 WSW-ENE well correlation of the Mid-Thanetian Unconformity, showing toplaps of the Campanian-Maastrichtian Abiod Formation and the Santonian Aleg shales against the Mid-Thanetian Unconformity and downlaps of the Thanetian El Haria shales against the Mid-Thanetian surface in seismic sections, recognised by sharp contact in gamma ray response log response. Inset map on the upper right-hand side Campanian—Maastrichtian Reservoir Abiod depositional environment map Fig. 5.12, modified after Troudi et al. (2000)

In addition, a schematic map of hydrocarbon discoveries and shows were plotted for the Middle Cretaceous and Late Cretaceous (Fig. 5.21) overlying the regional source rock lithofacies zones of the Albian Lower Fahdene and the Turonian-Cenomanian Bahloul Formations. The data for these plots were taken from the final well reports. These maps show hydrocarbon indications and successful tests grouped by Epoch. The hydrocarbon indications and discoveries are plotted against the facies distributions of the two main source rocks in the study area, the Lower Fahdene and Bahloul Formations.

Figure 5.21 displays hydrocarbon indications within the Maastrichtian El Haria shale Formation, the Campanian-Maastrichtian Abiod Formation and the Turonian Bireno and Coniacian-Santonian Douleb Members. The Late Cretaceous has the most prolific hydrocarbon indications and plotted against the major structural elements, it is interesting to observe that most of the successfully tested wells are located southwest of a lineament trending NW-SE in the central part in the area. Magmatic activity is noted to have occurred extensively, much of it during the Aptian to Palaeocene rifting phase of the Pelagian Shelf and in the Neogene to Quaternary due to the Alpine (Atlassic) orogeny and subsequent opening of the western Mediterranean (Finetti 1982; Wilson and Guiraud 1998; Guiraud 1998; Guiraud et al. 2005).

348 wells in the area were used in the evaluation of the two Palaeogene plays focused on the Eocene Ypresian El Garia and Lutetian-Bartonian Reineche reservoirs. In addition to all these wells which also included relevant information regarding either missing section or non-reservoir facies, 54 wells were studied in detail for the El Garia play and 57 wells for the Reineche. Many of these wells contained wireline data which allowed for a detailed analysis of how the lithostratigraphic units changed over the basin. Figure 5.9 shows the maturation map for the Bou Dabbous source rock and the hydrocarbon shows observed in wells from the El Garia and the Reineche nummulitic limestones.

The El Garia nummulitic limestone is one of the most prolific hydrocarbon producing reservoirs in the Pelagian Basin. It is the reservoir for the largest oil field, Ashtart, and the gas and condensate field, Hasdrubal (Racey et al 2001). The trend of the main nummulitic reservoir bank (Fig. 5.8) shows the continuation of the NW-SE lineament that was

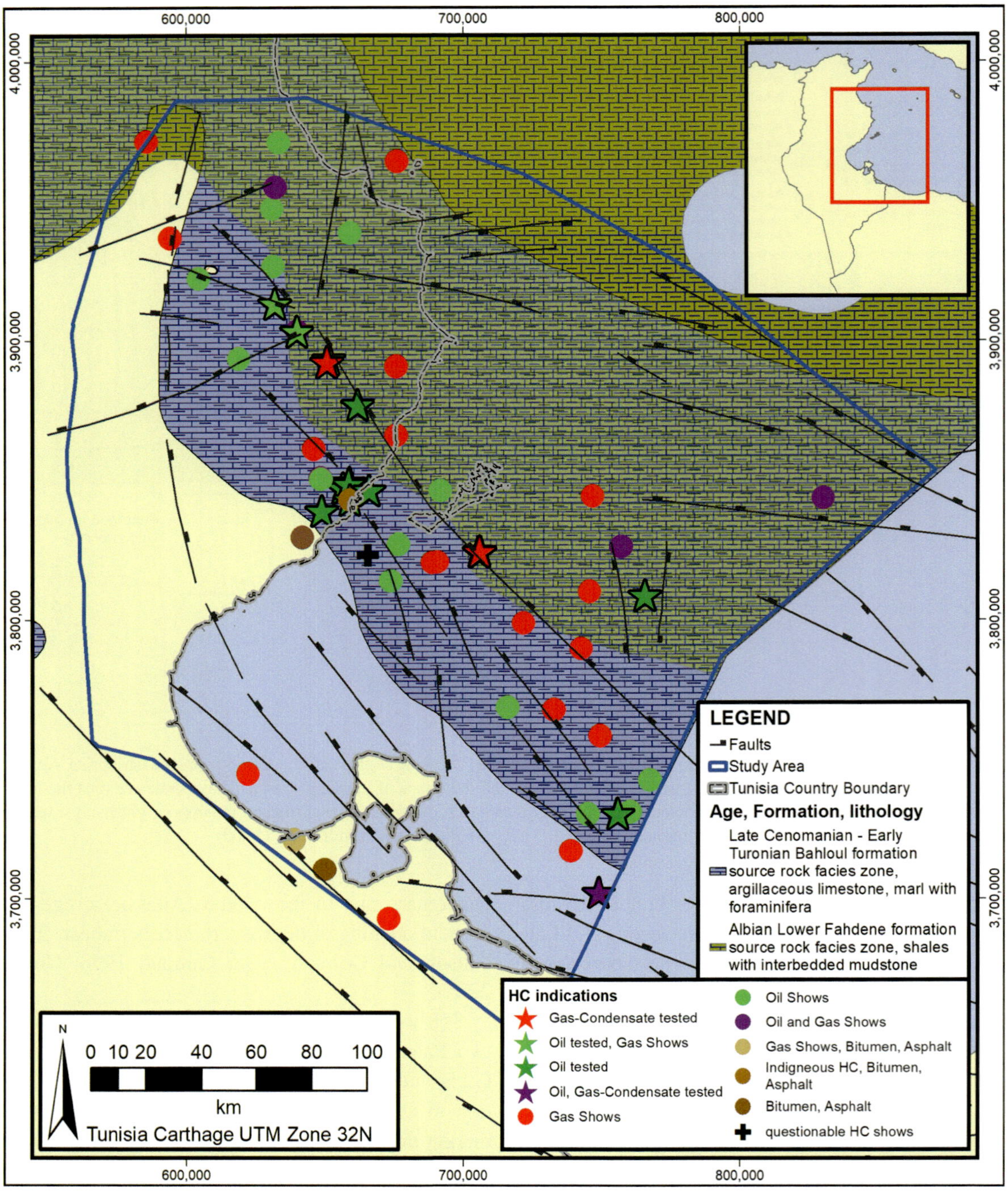

Fig. 5.21 Schematic hydrocarbon discoveries and shows map for Middle—Late Cretaceous. Hydrocarbon indications within Late Maastrichtian—Palaeocene El Haria Formation, Campanian—Maastrichtian Abiod Formation, Coniacian—Santonian Douleb Member and Turonian Bireno Member are plotted together with structural elements (modified after Bédir 1995; Khomsi 2004; de Lamotte et al. 2009; Khomsi et al. 2012, 2016, 2019a; Bédir et al. 2018; Mezni et al. 2019) and source rock facies zones of the Albian Lower Fahdene Formation and Late Cenomanian—Early Turonian Bahloul Formation

key to the development of the earlier Pre-Tertiary reservoirs and reflects the importance of the underlying structural configuration within the basin. Both Ashtart and Hasdrubal (Macaulay et al. 2001) are situated on underlying highs and are defined to the north by pinch out of the reservoir facies (Fig. 5.13). Hydrocarbon charging is predominantly from lateral migration from the Bou Dabbous Formation and is confirmed by oil typing (Racey et al 2001) (Fig. 5.22). There is a trend for increasing hydrocarbon maturity, from west to east, closest to the Bou Dabbous kitchen e.g., Ashtart, oil and Hasdrubal and Zarrat, gas/condensate and gas. Further south, the trend is for oil which could indicate that hydrocarbon charging was less mature due to the time taken to travel the greater distance from the Bou Dabbous kitchen. Another trend which is observed is the increasing levels of inert gas identified towards the east, e.g., Hasdrubal with approximately 4% CO_2 (Macaulay et al. 2001) as opposed to Zarrat further to the east which contains a wide range from 12 to 58% CO_2. H_2S levels are also noted to increase from west to east. It is likely that the contribution of inerts is coming from the underlying sediments via faulting associated with interbedded volcanics and evaporites.

The Lutetian-Bartonian Reineche carbonate Member is observed following a NW-SE trend similar to that of the underlying El Garia (Fig. 5.22). Two fields in the onshore part of the Pelagian Basin are commercially producing hydrocarbons from the Lower Reineche nummulitic carbonate facies. Both are located around the Kerkennah High (Figs. 5.1 and 5.28), with the Cercina oil field to the west and the Chergui gas field to the east (Elfessi 2017). Hydrocarbon charging is from the Bou Dabbous, located either side of the Kerkennah High, through a network of faults linking these fields with two kitchens that may have marginally different maturities (Fig. 5.9). This is observed by the differences in fluid type from both fields. However, in addition, Chergui is structurally higher than Cercina and, therefore, could just reflect the presence of a common gas cap over the whole Kerkennah High complex. The fields have a structural/stratigraphic trap configuration with pinch out of the reservoir facies apparent to the north and east. The best reservoir quality is from the nummulitic facies which has a net range of 2 m predominantly in the onshore area to approximately 10 m (net) in the west and central areas offshore. In addition, the units themselves vary over the nummulitic bank from northwest to southeast.

Offshore in the study area, three domains are observed for the whole of the Reineche Member (Fig. 5.24). In the west there are four units, in the central part there are five and in the east, there are two. In the west and central areas, it appears that the Upper Reineche does not vary. However, in the west, the lowermost unit of the Lower Reineche (which is observed in the central area) is absent. The east domain is significantly different in that the Upper Reineche appears to be missing. The relatively thin nummulitic facies appears to be present in all domains. Although not fully understood, one explanation is that these three areas were experiencing differing tectonic settings during this time, resulting in the subsequent variation in the Reineche unit itself.

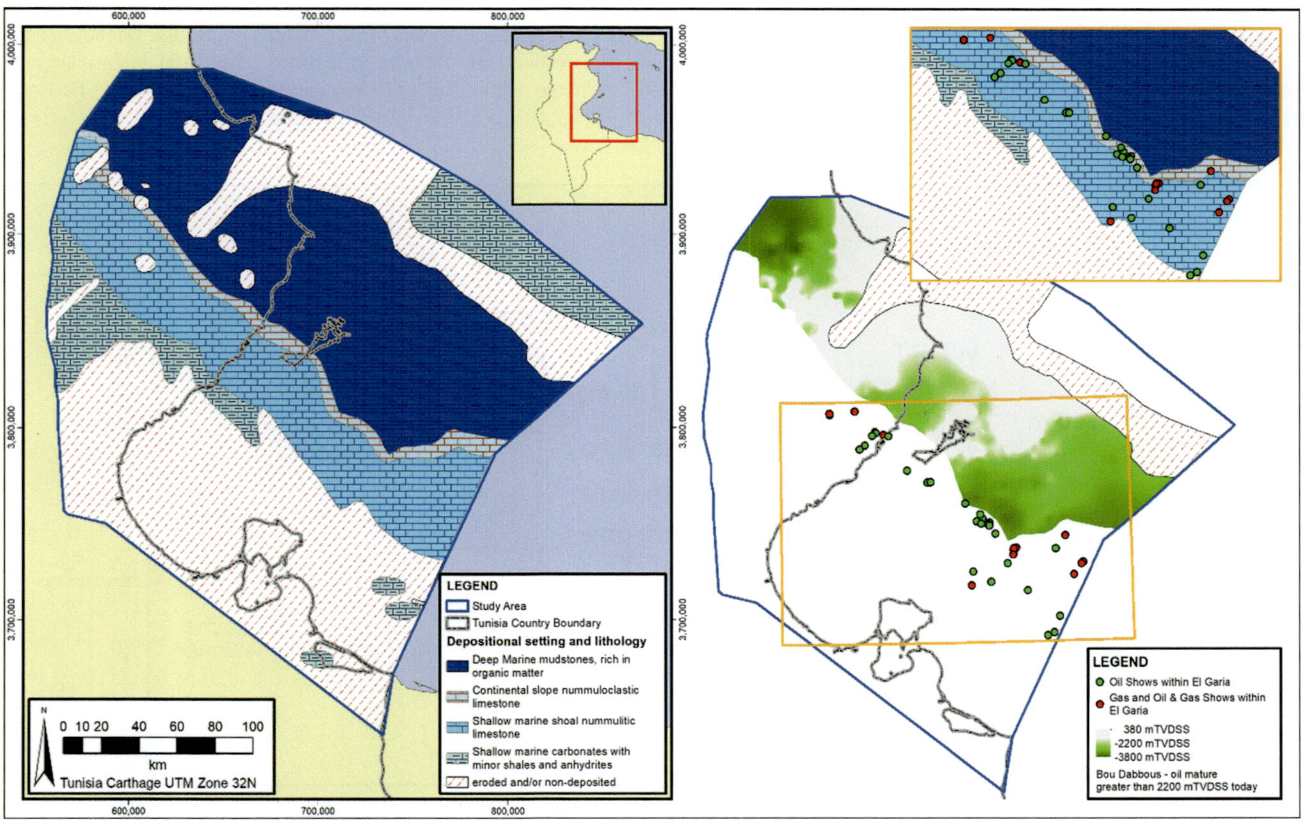

Fig. 5.22 Ypresian depositional environment map of El Garia reservoir and Bou Dabbous source rock and inset showing maturity map of the Bou Dabbous source rock inferred on depositional environment map, depth map calibrated against geochemical data, Vitrinite Reflectance values from final well reports, geochemical reports and field reports by hydrocarbon type. Hydrocarbon indications within El Garia Formation plotted on Ypresian depositional environment map

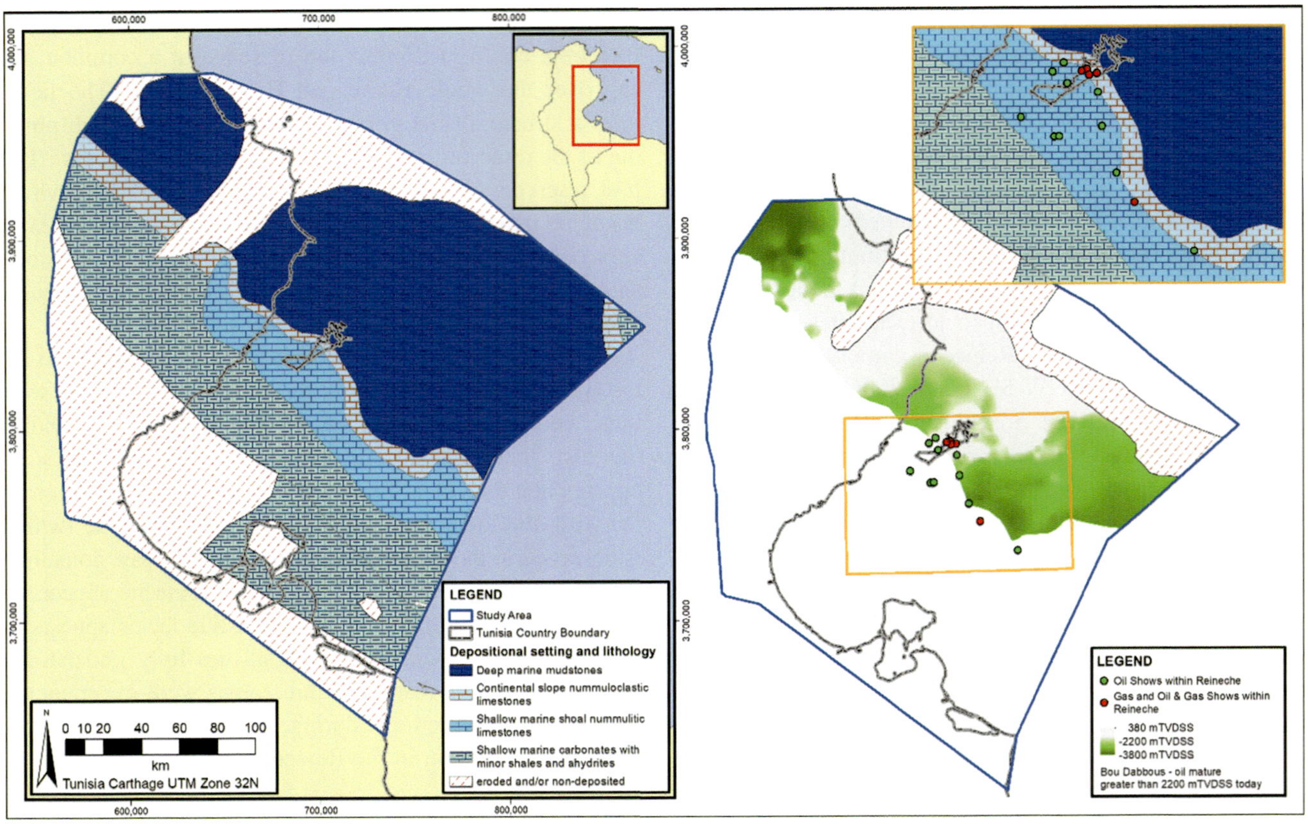

Fig. 5.23 Depositional environment map of the Lutetian-Bartonian Reineche Member. Inset showing maturity map of the Bou Dabbous source rock inferred on depositional environment map, depth map calibrated against geochemical data, Vitrinite Reflectance values from final well reports, geochemical reports and field reports by hydrocarbon type. Hydrocarbon indications within Reineche Member plotted on Lutetian—Bartonian depositional environment map

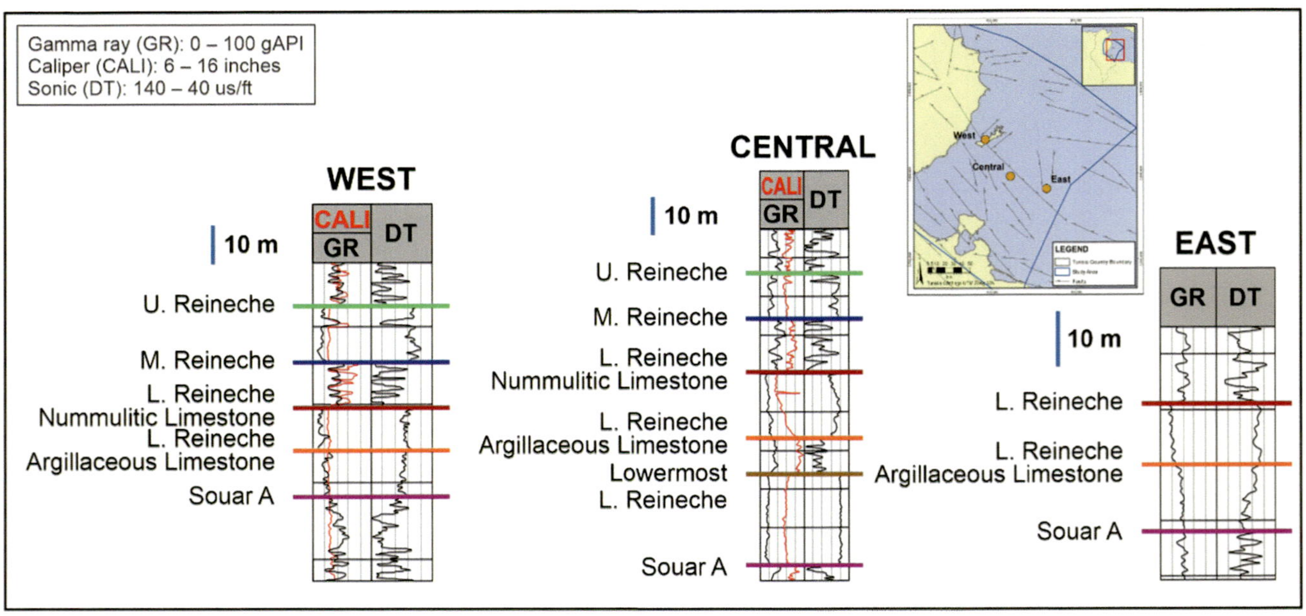

Fig. 5.24 W-E well log correlation of the Reineche Member. Three domains of the Reineche Member are observed offshore, with four units being present in the west, five units in the central part and two units in the east

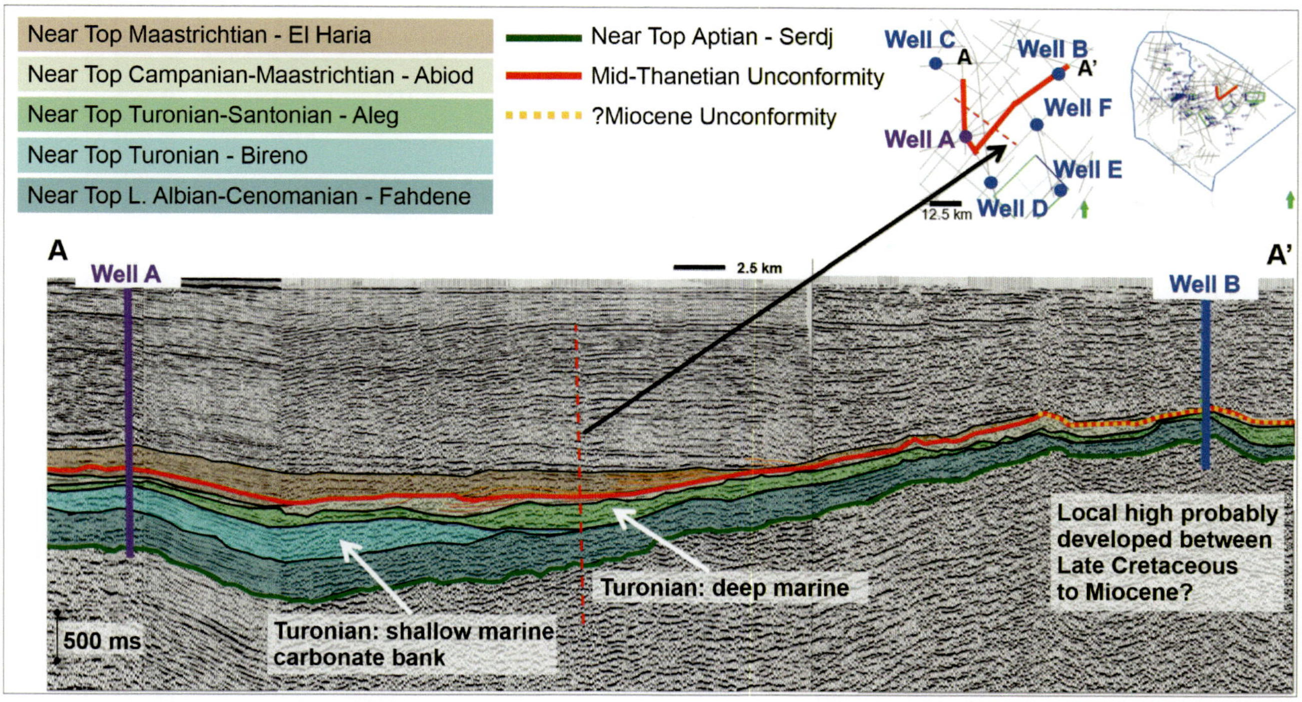

Fig. 5.25 Regional seismic line on poor quality 2D data, seismic interpretation to identify six key seismic markers in the Upper Cretaceous and Palaeocene

A regional seismic interpretation was undertaken for six key seismic markers in the Late Cretaceous and the Palaeocene in combination with well-log interpretation and a lithostratigraphic and biostratigraphic review. The 2D surveys were acquired between 1970 and 2009, resulting in varying quality over the study area. Some surveys are only available as scanned versions, where there is no seabed clearly visible and the lines themselves look cut, tilted or squeezed. Furthermore, the different surveys vary in phase and polarity. The prominent Mahmoud shale event (Langhian to Serravallian) was therefore used as the reference horizon. The lower part of the Formation has interbedded layers of limestone (Mejri et al. 2006), which are visible as strong reflectors. For the purpose of regional mapping the available data is suitable, but for detailed prospect evaluation more effort needs to be put into improving the data quality, with the main focus on improving the time-depth relationship and the seismic to well ties for wells where digital data is available.

Due to a lack of data and quality checks the horizons are only mapped as near top events. The purpose of the regional mapping was to understand the regional topography of the near top Maastrichtian–Palaeocene El Haria shale, the near Top Lower Albian–Cenomanian Fahdene Formation and the near Top Aptian Serdj Formation. These near top events could be used as near base surfaces for the three major source rock intervals investigated during this study, which are near base Ypresian Bou Dabbous, near base Cenomanian Bahloul and near base Albian Lower Fahdene (Fig. 5.25). The available poor seismic quality limited a seismic facies evaluation. Mapping of the Mid-Thanetian unconformity was undertaken to understand the distribution of the major seal for the Cretaceous petroleum system, the El Haria shale. To enhance the visibility of onlaps and top laps onto the unconformity a wiggle trace display of the 2D seismic lines was used (Fig. 5.25). Through mapping of the Mid-Thanetian Unconformity, an attempt was made to further identify the presence of Cretaceous paleohighs. These highs could indicate possible leads for further prospectivity investigation.

Based on depth-converted maps, a regional maturity trend could be established for the three major source rocks: the Albian Lower Fahdene (Fig. 5.26), the Cenomanian Bahloul (Fig. 5.27) and the Ypresian Bou Dabbous (Fig. 5.9) within the basin, with the Top Oil Window approximately estimated at 2800 mTVDSS (true vertical depth subsea) and the Top Gas window approximately estimated at 3800 mTVDSS. The near top picked horizons are effectively the base of the source rocks. For generating depth maps a simplified approach was taken. The average velocity from SRD or 0 mMSL (mean sea level) to each horizon was calculated from the well top information [m] and the horizon seeds TWT (two-way time) pick [ms]. These average velocity trends are suitable for an estimated regional maturity trend evaluation, but more effort will need to be put into understanding the regional average velocity trend as well as

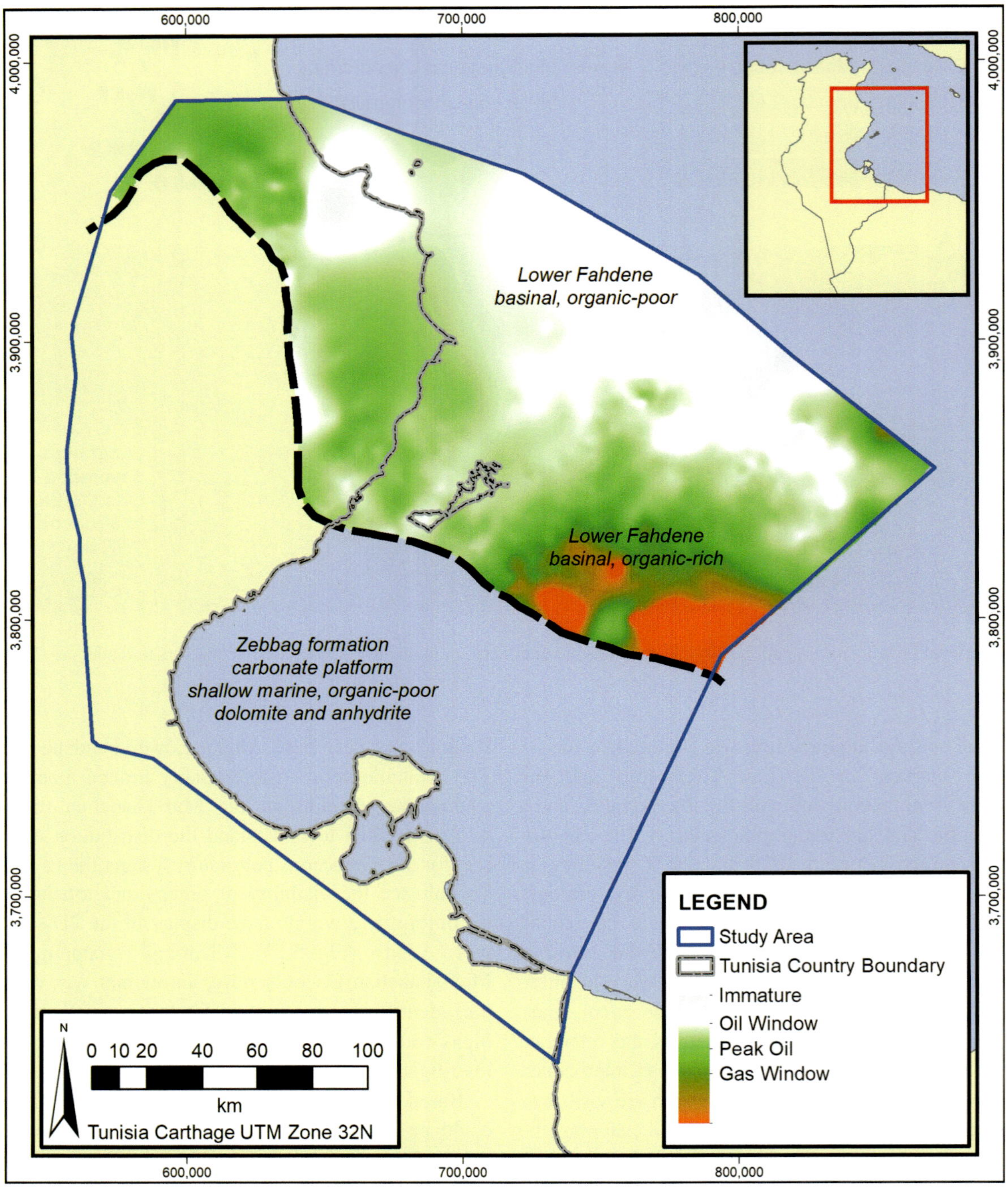

Fig. 5.26 Maturity Map of the Albian Lower Fahdene source rock inferred on depositional environment map, depth map calibrated against geochemical data, Vitrinite Reflectance values from final well reports, geochemical reports and field reports by hydrocarbon type

building a regional 3D Basin Model if any further investigation of the area for lead or prospect scale evaluation is undertaken. The regional depth maps honour the well data but there is uncertainty in the accuracy of the maps away from the wells. Figure 5.23 shows hydrocarbon indications in the Middle–Late Eocene Reineche reservoir plotted on a depositional environment map.

Structural observations during the regional mapping show generally steep, normal faults with major ones striking NW-SE and minor ones NE and N. Two separated fault systems are visible. In the Tertiary, faults appear to sole out in the El Haria shale with most reaching the surface. The El Haria shale seems to act as a detachment layer for the two component systems. The Maastrichtian to Palaeocene shale

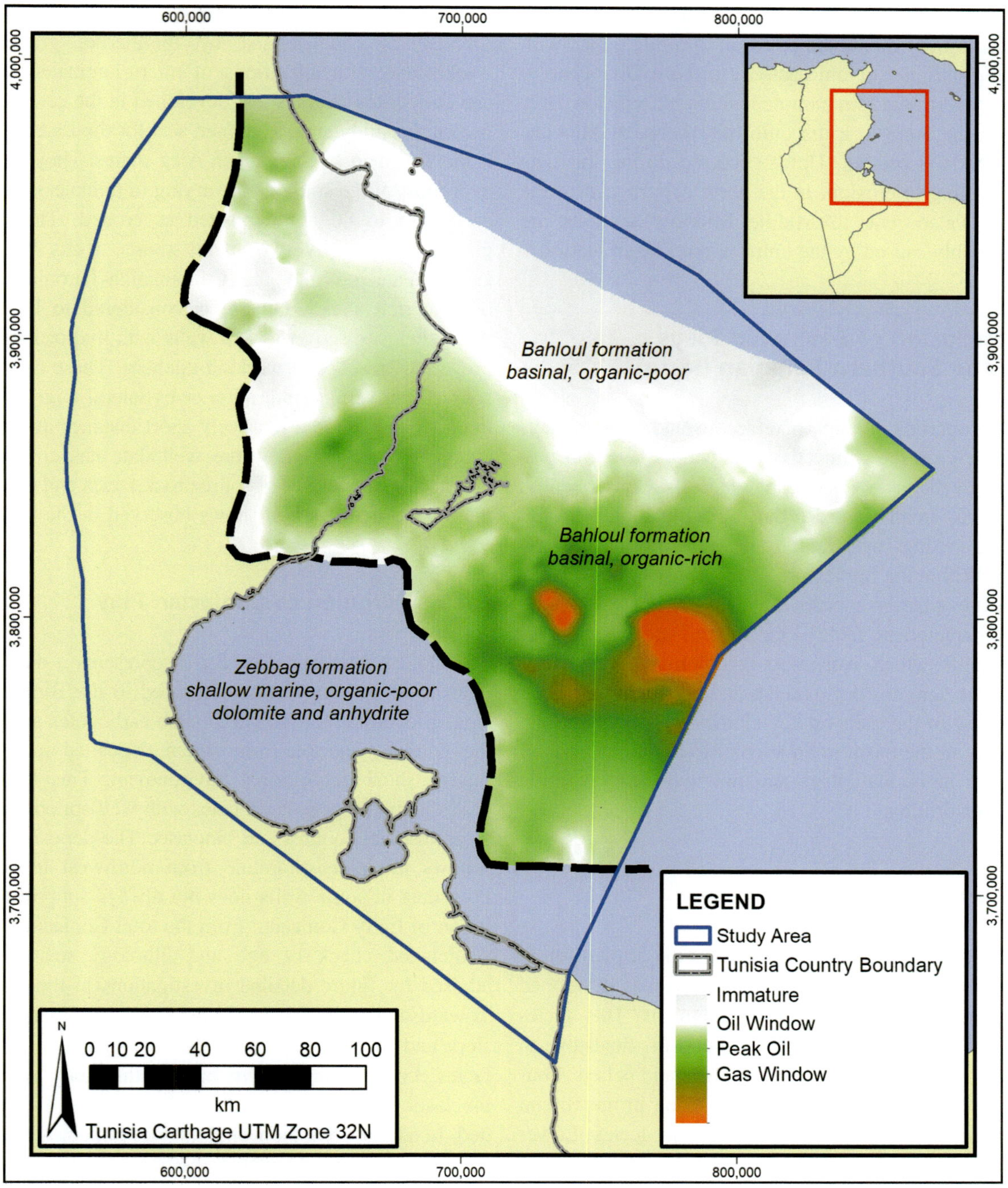

Fig. 5.27 Maturity Map of the Cenomanian Bahloul source rock inferred on depositional environment map, depth map calibrated against geochemical data, Vitrinite Reflectance values from final well reports, geochemical reports and field reports by hydrocarbon type

mechanically decouples the Cretaceous and older layers from the Tertiary system. Horst-graben structures and rollover-anticlines are the dominant structural features in the Tertiary system where fault displacements are very small. The Cretaceous appears to be a paleo-relief influenced by older tectonic phases, volcanism and halokinesis. Most of the discoveries in the study area are fault-related structures. During the Late Cretaceous, transtensional to transpressional stress fields affected the paleogeography, facies distribution, thickness and later the trapping. The faults were reactivated during the alpine orogeny (Miocene–Pliocene) with associated folding, wrenching and inversion (Troudi et al. 2000).

In the early Eocene, W-E or WNW-ESE transfer faults were reactivated. Sedimentation was controlled by fault activity and facies boundaries of Lower Eocene rocks reflect the orientation of these strike-slip faults (Bishop 1988; Morgan et al 1998). This can be observed in many of the Tertiary fields which contain an element of stratigraphic pinchout to determine the trapping mechanism. The Oligocene Unconformity marks the collision of a microplate with the African margin. Tectonic activity in Late Oligocene to Eocene times resulted in non-deposition or erosion over some areas. In Tunisia, a disconformity between Miocene and older beds is present. Uplift occurred during the Late Miocene leading to erosion. In the north eastern part of the study area Palaeocene to Middle Miocene sections are absent, probably caused by the uplift of the El Jem platform.

5.4 Conclusion of Evaluated Plays in the Southern Pelagian Basin

Reservoir properties in Upper Cretaceous and Eocene plays depend largely on the connectivity of porous layers in predominantly cyclic or low permeable deposits. Consequently, most of the small hydrocarbon accumulations appear restricted to settings near fault zones generally affected by fracturing. Within the framework of this study, predictability for prospect evaluation appears low due to the lack of data to delineate characteristic fault block patterns or identify areas of local uplift which can support dolomitization and/or karsting. The aim of the evaluation is to gain a further understanding of the risks on key elements for hydrocarbon prospectivity in the basin in an attempt to highlight areas of low risk for individual plays and not to identify specific prospects for drilling.

5.4.1 Turonian Play

Turonian sedimentation is dominated by a major platform in the study area (Fig. 5.28). The main reservoir is developed in an organic reef depositional environment. This facies comprises bioclastic and derived reef deposits, floatstone to rudstone with bioclasts, rudist debris, algae and pellets. Four cycles are described in the organic reef zone in the Bireno Member based on well data: Lower peritidal facies, Lower shelf margin, Upper peritidal facies and Upper shelf margin which are cyclic alternations of carbonates, shales and evaporites with the peritidal facies having the most prospective reservoir quality. Due to the varying age of the information and poor well data, a detailed analysis of the four known cycles within the Bireno succession was not included in the study. Detailed prospect analysis requires a revised check of age and lithology relationships. This narrow facies zone has a NW-SE trend, transitioning into a continental slope facies and deep marine facies towards the northeast. The continental shelf facies is characterised by peritidal packstone and wackestone which are affected by dolomitization. The deep marine facies is dominated by limestone with mixed benthic and planktonic foraminifera and mud. Towards the southwest, the shallow marine environment is characterised by biomicrite, grainstone to wackestone with alterations of micro-laminated dolomites and anhydrites with a shoal developed in the central western part of the study area. The main seal for the Late Cretaceous is the Turonian to Campanian Aleg shales. They are present over most of the study area, varying in thickness and locally are found to be non-deposited or eroded. The Turonian play is charged by two major source rocks, the Albian Lower Fahdene and the Cenomanian-Turonian Bahloul organic rich shales. Charge is considered to be low risk where fields, discoveries and wells with hydrocarbon shows prove the presence of working kitchens. These can be either sourced via vertical migration or by lateral migration where the kitchen is sited a relatively short distance away. Higher charge risk is present where well data has confirmed the absence or immaturity of source rock facies and areas where there is no hydrocarbon shows observed in the play.

5.4.2 Turonian to Coniacian Play

The Douleb Member (Fig. 5.29) represents a deeper water outer shelf or ramp facies compared to the Bireno succession. Biodetrital and ooidal Douleb carbonates at the top of the Douleb sequence were again deposited in a shallow marine shoal depositional environment. This facies comprises high energetic carbonates with well sorted packstones to grainstones, oolites and bioclasts. The depositional trend follows the paleo-shoreline from northwest to southeast. Poor data in some wells does not always support the separation of Early Coniacian from the total Coniacian package. A detailed check of age and lithology relationships is advised for future detailed investigations in the study area. Towards the northeast, the deeper basinal geometry develops slope and basin facies without the development of a reservoir facies. In the southwestern part of the study area, oolitic bioclastic limestones and dolomites were primarily deposited. In a large area, the well data does not appear to confirm the presence of the Douleb carbonates. As discussed for the Turonian play, the main seal for the Turonian-Coniancian play are the Turonian-Campanian Aleg shales and the two major source rocks are the Albian Lower Fahdene and the Cenomanian-Turonian Bahloul organic rich shales.

5.4.3 Campanian to Maastrichtian Play

The Campanian–Early Maastrichtian unit transgressively overlies the shale-prone Upper Aleg Formation, as well as occasionally, progressively older units down to the Aptian, with several local discontinuities, partly angular unconformities, and onlaps onto inversion and/or salt growth

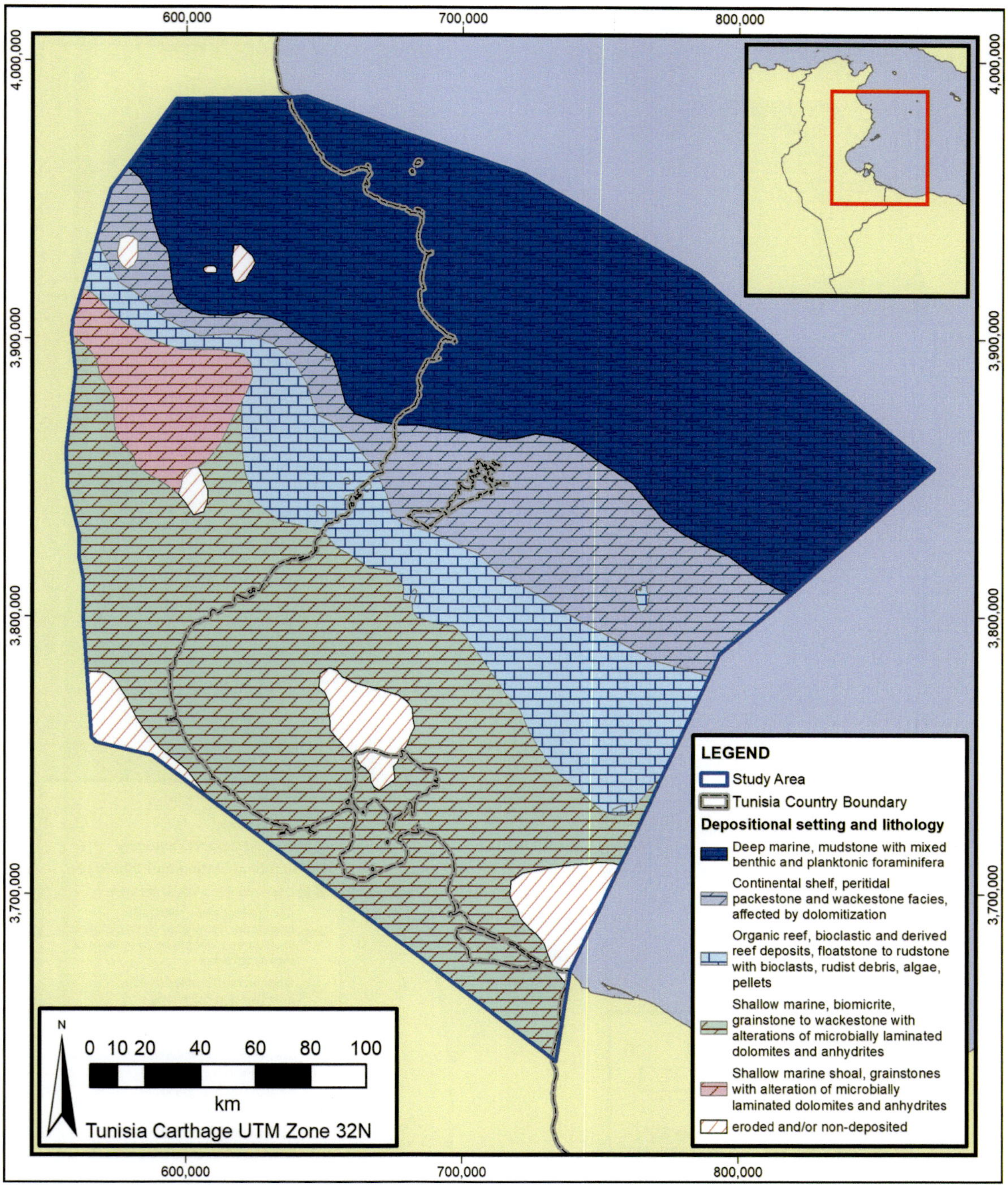

Fig. 5.28 Early to Middle Turonian Bireno reservoir depositional environment map

structures. The gradual transition into Late Maastrichtian El Haria shale is defined by a diminishing number of chalky limestone interbeds in only a few localities. Large areas are affected by significant erosion and subsequent transgressive onlap of Palaeocene often Thanetian age El Haria shale. For many well locations, a shaly Middle Abiod Member has been interpreted as Maastrichtian El Haria shale, resulting in the reduction of actual total thickness of Abiod chalky limestone whilst adding to the thickness of the Maastrichtian–Thanetian El Haria shale.

The Abiod predominantly consists of recrystallized chalky limestone with intercalations of partly calcareous shale. Ideally, complete Abiod sections comprise an upper and lower limestone member separated by an intermediate interbed of argillaceous limestones and calcareous shales. Locally, secondary dolomitization has affected the Abiod, either in association with intercalations of mainly basaltic volcanics or by exposure of meteoric waters around local islands due to salt growth. These hemipelagic chalks are present in the north eastern part of the study area (Fig. 5.30)

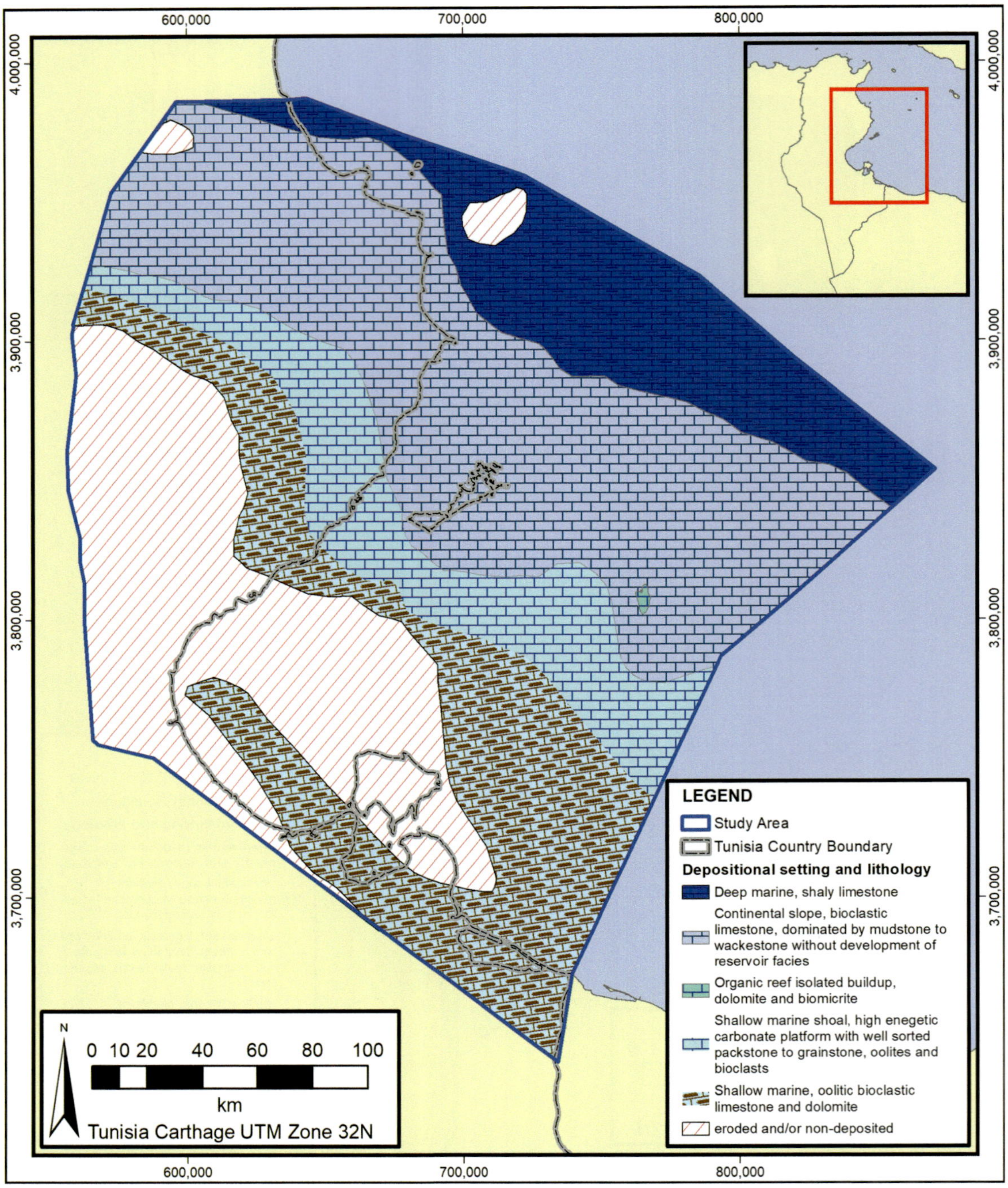

Fig. 5.29 Late Turonian to Santonian Douleb reservoir depositional environment map

which are considered the favourable reservoir facies. Abiod chalky limestones highly depend on the enhancement of reservoir quality by local fracturing, karsting and/or dolomitization (Troudi et al. 2000; Fourati et al. 2002; Deng et al. 2012; Abidi et al. 2015). The chalky limestones transform into a narrow NW-SE trending continental slope environment which shows higher dolomitization and sandy bioclasts in the northwestern part of the setting. In the southwestern part, the Campanian to Maastrichtian strata is not present due to non-deposition or erosion. Late Maastrichtian–Thanetian El Haria Formation shales are the main seal for the Campanian to Maastrichtian play. The unconformity discovered within the El Haria Formation is Mid Thanetian in age based on well data. Therefore, a 'Mid-Thanetian Unconformity' was regionally mapped. To date, a break had been recognised, but only locally (Haller 1983; Zargouni 1985; Zourai 1995; Mejri et al. 2006; Boussiga 2008; Taktak et al. 2010). Whilst there are areas where deposition has been almost continuous, seismic evidence clearly shows that the relative sea level change was

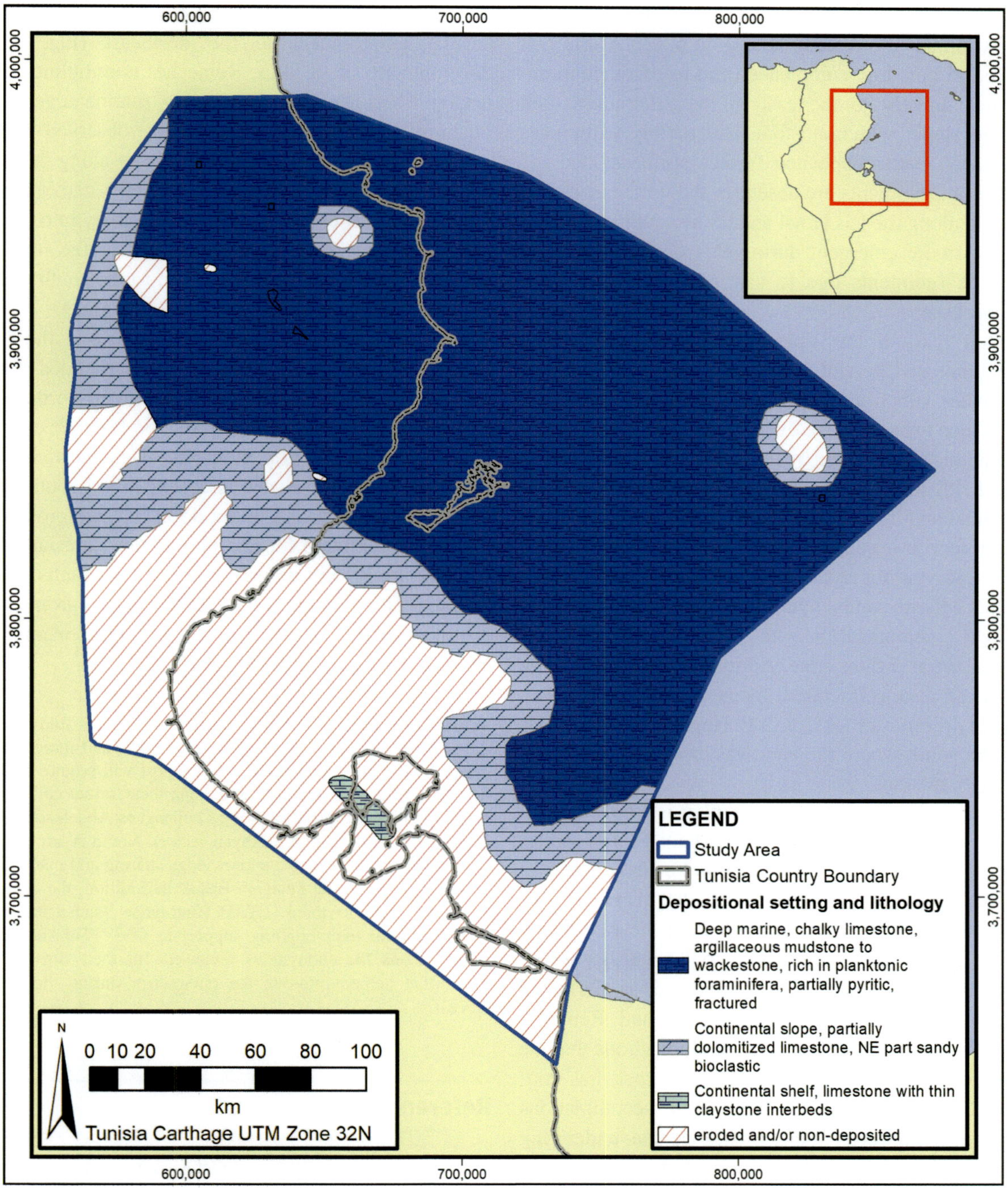

Fig. 5.30 Campanian to Maastrichtian Abiod Formation depositional environment map

also linked to tectonic uplift and was general in its extent (Burollet 1956; Bajanik et al. 1978; Fournié 1978; Ferjani et al. 1990; Bédir et al. 1992; Taktak et al. 2010). The variable nature of the seal is likely due to the continuing effects of salt diapirism (El Rabia et al. 2018). As for the other reservoirs of the upper Cretaceous plays, the charge for the Campanian-Maastrichtian play is from the two major source rocks, the Albian Lower Fahdene and the Cenomanian-Turonian Bahloul organic rich shales.

5.4.4 Ypresian Play

During the Early Eocene (Ypresian), the depositional environment mirrored earlier times, with a back basin, lagoon and land to the southwest, which becomes increasingly shallow marine along much of the Pelagian Basin area and deeper marine to the northeast. The structural lineament is predominantly NW-SE and the depositional systems follow this trend. The development of the El Garia nummulitic carbonate reservoir lies along the shallow marine edge offshore and onshore in the study area (Fig. 5.22). A thin belt of reworked nummulitic carbonates (non-reservoir nummuloclasts) lies directly to the east of the nummulitic belt defining the slope, with mudstone and muddy carbonates rich in organic matter deposited further northeast into the basin. Ypresian sediments are absent in the northeast due to erosion north along the Isis Horst and to the southwest where the area was more emergent during this time. Absence of Early Eocene sediments due to salt piercements are also observed onshore.

The main seal for the Ypresian El Garia play is the overlying Lutetian Cherahil A interbedded carbonates and Souar A shales which are deposited further into the basin. This seal is seen to be effective over the area as the carbonates are tight and this is confirmed by the number of commercially producing fields identified in this area (e.g., El Hajeb, Ashtart, Hasdrubal and Zarrat) (Macaulay et al. 2001). The main producing reservoir is capped by the Compact Micrite which is neither a reservoir nor an effective seal. In several instances the Compact Micrite can be hydrocarbon bearing if fractured and in direct communication with the underlying reservoir where it can act as a thief zone. Additionally, the overlying Cherahil A can also, if fractured, focus migration into much poorer quality reservoir rocks, as in El Hajeb field. The seal is absent in the south where the area was emergent and in the north through erosion.

The main charge for the Ypresian play is laterally from age equivalent Bou Dabbous organic carbonates and shales. Mapping of the near base Bou Dabbous Formation and El Haria shales has shown and confirmed that there are two kitchens offshore, to the south and north of the Kerkennah High and a further one is identified to the north onshore (Fig. 5.9). The Bou Dabbous is considered as the main source rock in the area for the El Garia and Reineche reservoirs. However, there are also some indications that the Upper Cretaceous Turonian Bahloul source rock has contributed to the hydrocarbon filling of these accumulations through vertical faulting. The presence of the underlying seals of Cherahil and Souar Formations may provide additional constraints to hydrocarbon migration routes into Eocene reservoir rocks. However, sourcing of the Ypresian play is predominantly through lateral migration from the adjacent Bou Dabbous organic rich shales.

5.4.5 Lutetian to Bartonian Play

The basin configuration of the middle Eocene Lutetian-Bartonian was very similar to that observed during Ypresian times. However, the facies zones are slightly offset towards the basin centre but still with a back basin, lagoon and land to the southwest, a thin NW-SE shallow marine belt along the central part of the area and a progressively deeper marine basinal environment to the northeast (Fig. 5.23). The development of a thin Reineche nummulitic carbonate reservoir belt lies along the shallow marine edge in both the offshore and onshore area, but is observed to be less extensive in the latter. Towards the east along the slope, the nummulitic facies is replaced by tight carbonates which progressively become mudstone further into the basin. Middle Eocene sediments are absent in the north due to erosion along the El Jem Horst and to the south west where the area was more emergent during this time.

The main seal for the Reineche play is the overlying Bartonian–Priabonian Cherahil B interbedded carbonates and Souar B shales which are deposited further into the basin. These seals are considered to be effective over the area as both members are more argillaceous rich than the underlying seals. The seal is absent in the south where the area was emergent and in the north through erosion.

The main charge for this play is the Ypresian Bou Dabbous. Sourcing will be predominantly via faults. Additional potential charging from the Cenomanian–Turonian Bahloul is also possible. To date, the focus of migration appears to be centred around the Kerkennah High.

Acknowledgements The work carried out for this study was done in 2015. It is understood that as time progresses and more data become available, the maps presented in this report will require further modification.The authors would like to thank their former colleagues Habib Troudi and Jomâa Friha, as well as colleagues Abdelkader Omri (who assisted our work from the Libyan sector), Noura Ayari and Giuseppe Cantarella for their extensive knowledge sharing and many discussions on the geology of the Pelagian Basin. In addition, the authors would especially wish to thank ETAP (Entreprise Tunisienne d'Activites Petrolieres) for their ongoing support to OMV Tunisia. Finally, the authors thank the anonymous reviewers for their constructive comments that helped improve the manuscript during the peer review process.

References

Abdallah H (2003) Genesis and diagenesis of the Gattar carbonate platform, lower Turonian, Northern Southern Tunisia. In: Gili E, Negra MH, Skelton PW (eds) North African cretaceous carbonate platform systems. NATO Science Series IV. Earth and Environmental Sciences, 28, pp 31–51

Abdallah H, Sassi S, Meister Ch, Souissi R (2000) Stratigraphie séquentielle et paléogéographie à la limite Cénomanien – Turonien dans la région de Gafsa-Chotts (Tunisie centrale) Cretaceous Research 21, 35–106

Abidi O, Inoubli MH, Sebei K, Boussiga H, Amiri A, Hamdi I, Ben Salem AH (2015) Geodynamic framework and petroleum potential of the cap bon-gulf of Hammamet Province–Tunisia. In: International conference on engineering geophysics, Al Ain, United Arab Emirates, 15–18 November 2015, pp 264–266, Society of Exploration Geophysicists

Acheche MH, Ghénima R, Saïdi M, Kilani HB (2003) Mesozoic and cenozoic petroleum systems in Tunisia. In: 1st EAGE North

African/Mediterranean petroleum & geosciences conference & exhibition, pp cp-8. European Association of Geoscientists & Engineers

Adouani F, Aissaoui S, Saïdi M (2003) Structural evolution of the cap Bon-Grombalia area, Northeastern Tunisia–effects on the hydrocarbon generation & expulsion histories. In: 1st EAGE North African/Mediterranean petroleum & geosciences conference & exhibition, pp cp-8. European Association of Geoscientists & Engineers

Affouri H, Montacer M, Disnar JR (2013) Organic geochemistry of the Cenomanian-Turonian Bahloul Formation petroleum source rock Central and Northern Tunisia. Resour Geol 63(3):262–287

Affouri H, Montacer M (2014) Organic matter appraisal of the lower Eocene (Ypresian) Bou Dabbous Formation in Tunisia. Arab J Geosci 7(12):5509–5542

Anketell JM, Mriheel IY (2000) Depositional environment and diagenesis of the Eocene Jdeir Formation, Gabes-Tripoli basin, Western Offshore Libya. J Petrol Geol 23(4):425–447

Badalini G, Redfern J, Carr ID (2002) A synthesis of current understanding of the structural evolution of north Africa. J Pet Geol 25(3):249–258

Baird DW (1967) The Permo-Carboniferous of southern Tunisia. In: 9th Annual field conference, Guidebook to the geology and history of Tunisia. Petroleum exploration society, Libya, Holland-Breumelhof NV, Amsterdam

Bajanik S, Stranik Z, Salaj J (1978) Notice explicative de la Feuille d'Enfidha au 1/50.000, edit. Serv. Géol, Tunisie

Baklouti S, Ahmadi R, Bougi MSM, Rasheed MA, Rao PS, Hasan SZ, Ksibi M (2017) Surface geochemical prospection for hydrocarbons in the oriental platform; the case of Guebiba oilfield, Sfax region, Tunisia. J Petrol Sci Eng 159:830–840

Beavington-Penney SJ, Racey A (2004) Ecology of extant nummulitids and other larger benthic foraminifera: applications in palaeoenvironmental analysis. Earth Sci Rev 67(3–4):219–265

Beavington-Penney SJ, Wright PV, Racey A (2005) Sediment production and dispersal on foraminifera-dominated early Tertiary ramps: the Eocene El Garia Formation Tunisia. Sedimentology 52(3):537–569

Bédir M (1995) Mécanismes géodynamiques des bassins associés aux couloirs de coulissements de la marge atlasique de la Tunisie: Séismo-stratigraphie, séismo-tectonique et implications pétrolières (Doctoral dissertation)

Bédir M, Zargouni F, Tlig S, Bobier C (1992) Subsurface geodynamics and petroleum geology of transform margin basins in the Sahel of Mahdia and El Jem (Eastern Tunisia) (1) AAPG Bull 76(9):1417–1442

Bédir M, Tlig S, Bobier C, Aissaoui N (1996) Sequence stratigraphy, basin dynamics, and petroleum geology of the Miocene from eastern Tunisia. AAPG Bull 80(1):63–80

Bédir M, Boukadi N, Tlig S, Timzal FB, Zitouni L, Alouani R, Slimane F, Bobier C, Zargouni F (2001) Subsurface Mesozoic basins in the central Atlas of Tunisia: tectonics, sequence deposit distribution, and hydrocarbon potential. AAPG Bull 85(5):885–907

Bédir M, Gabtni H, Azaiez H, Arbi A (2018) Meso-Cenozoic strike-slip salt tectonics, platform to basin structures, subsidence migrations, platform to basin structuring and new petroleum implications by seismic reflection analysis in the Eastern Tunisian margin. In: 7th Maghrebian Geophysic Congres. 20–22 February 2018. Algiers. Algeria: 41

Ben Ayed N (1986) Evolution tectonique de l'avant-pays de la chaîne alpine de Tunisie du début du Mésozoïque à l'Actuel (Doctoral dissertation, Paris 11)

Berrocoso ÁJ, Bodin S, Wood J, Calvert SE, Mutterlose J, Petrizzo MR, Redfer J (2013) Dynamic sedimentary conditions during periods of enhanced sequestration of organic carbon in the central southern Tethys at the onset of the Cenozoic global cooling. Sed Geol 290:60–84

Bishop WF (1975) Geology of Tunisia and adjacent parts of Algeria and Libya. AAPG Bull 59(3):413–450

Bishop WF (1988) Petroleum geology of east-central Tunisia. Am Asso Petrol Geol Bull 72(9):1033–1058

Bishop WF, Debono G (1996) The hydrocarbon geology of southern offshore Malta and surrounding regions. J Pet Geol 19(2):129–160

Boussiga H (2008) Géophysique appliquée aux séries paléogènes du Sahel de Tunisie. Tectonique de socle, halocinèse et implications pétrolières. Thèse en Sciences Géologiques, Université Tunis El Manar. 159 pp

Brahim GB, Brahim N, Turki F (2012) Tectonic evolution and hydrocarbon potential of the Aptian series in the Mahdia offshore area, Tunisia. J Petrol Sci Eng 90:175–190

Brahim N, Brahim S, Zargouni F (2003) The hydrocarbon potential of the Golf of Hammamet Basin, Tunisia. In: 1st EAGE North African/Mediterranean petroleum & geosciences conference & exhibition, pp cp-8. European Association of Geoscientists & Engineers

Brown J (2003) Miskar field–the value of information. In: 1st EAGE North African/Mediterranean petroleum & geosciences conference & exhibition, pp cp-8. European Association of Geoscientists & Engineers

Burollet PF (1956) Contribution à l'étude stratigraphique de la Tunisie centrale. Ann. Min. et Géol, No 18,352 p 93 fig.22 pl

Burollet PF, Mugniot JM, Sweeney P (1978) The geology of the Pelagian block: the margins and basins off southern Tunisia and Tripolitania. In: The ocean basins and margins, pp 331–359. Springer, Boston, MA

Burollet PF (1991) Structures and tectonics of Tunisia. Tectonophysics 195(2–4):359–369

Camoin G (1989) Les plates-formes carbonatées du Turonien et du Sénonien de Méditerranée centrale (Tunisie, Algérie, Sicile) (Doctoral dissertation, Aix-Marseille 1)

Camoin GF (1993) Turonian and Coniacian Carbonate Platforms from the African Tethyan Margin, Algeria, Tunisia: Chapter 13

Castany G (1947) Le problème des chotts tunisiens. C. R. Som. S; GF, No 8:166–167

Castany G (1951) Etude géologique de l'Atlas tunisien oriental. Annales Mines et Géologie, Tunis, no 8, vol 4: Thèse Doctorat ès Sciences, Paris, 632 p

Chalwati I, Berra F, Boukadi N (2018) Geological evolution of the offshore Tunisia (Gabes Basin, Pelagian Domain) since the Cretaceous: constraints from subsidence curves from hydrocarbon wells data. Mar Pet Geol 97:94–104

Chihi L (1995) Les fossés néogènes à quaternaires de la Tunisie et de la mer Pélagienne: une étude structurale et une signification dans le cadre géodynamique de la Méditerranée centrale. Univ de Tunis II, Fac Sc Tunis, These de Doctorat es Sciences Géologiques, p 324

Chikhaoui M (1988) Succession distension-compression dans le sillon tunisien, secteur de Nebeur, El Kef. Tunisie centre nord. Thèse. Sci. Univ, Nice, p 143p

de Lamotte DF, Leturmy P, Missenard Y, Khomsi S, Ruiz G, Saddiqi O, Guillocheau F, Michard A (2009) Mesozoic and Cenozoic vertical movements in the Atlas system (Algeria, Morocco, Tunisia): an overview. Tectonophysics 475(1):9–28

Deng Y, Lue L, Zhang J, Yang R (2012) Controlling Factors of Fractured Limestone Reservoir-forming of Upper Cretaceous in North Kairouan Region, Pelagian Basin, North Africa. J Southwest Petrol Univ (sci Technol Edn) 34(2):24–28

Dercourt J, Ricou EL, Vrielynch B (1993) Atlas Tethys paleoenvironmental maps, 307 pp, 14 sheets, 1 plate. Gauthier-Villairs, Paris

Dey Z, Kilani HB (2003) Structural evolution of the gulf of Gabes-Eastern Offshore-Tunisia. In: 1st EAGE North

African/Mediterranean petroleum & geosciences conference & exhibition, pp cp-8. European Association of Geoscientists & Engineers

Dkhaili N, Bey S, El Abed M, Gasmi M, Inoubli MH (2015) Integrated geophysical and geological study and petroleum appraisal of Cretaceous plays in the Western Gulf of Gabes, Tunisia. Tectonophysics 658:1–13

Donze P, Jardine S, Legoux O, Masure E, Méon H (1985) Les évènements à la limite Crétacé-Tertiaire: au Kef (Tunisie septentrionale), l'analyse palynoplanctologique montre qu'un changement climatique est décelable à la base du Danien. In Congrès national des sciences de la terre. 1, pp 161–171

El Asmi AM (2015) Strontium isotopes correlation of the Tunisian Late Cretaceous Abiod Formation: Comparison to previous biostratigraphic assignments. J Afr Earth Sc 111:375–386

El Euchi H, Saïdi M, Fourati L, El Maherssi C (2004) Northern Tunisia thrust belt: deformation models and hydrocarbon systems. In: Swennen R, Roure F, Granath JW (eds) Deformation, fluid flow, and reservoir appraisal in foreland fold and thrust belts: AAPG Hedberg Series, no 1, pp 371– 390

El Rabia A, Inoubli MH, Ouaja M, Abidi O, Sebei K, Jlailia A (2018) Salt tectonics and its effect on the structural and sedimentary evolution of the Jeffara Basin, Southern Tunisia. Tectonophysics 744:350–372

Elfessi M (2017) New insights into the stratigraphic, palaeogeographic and tectonic evolution and petroleum potential of Kerkennah Islands, Eastern Tunisia. J Afr Earth Sc 125:88–102

Ferjani AB, Burollet PF, Mejri F (1990) Petroleum geology of Tunisia. Enterprise Tunisienne dÁctivités Pétrolières

Ferjani AB, Ouachi A, Ghariani H, El Euchi M (2003) Upper Cretaceous Fractured Reservoir–M'Sela Prospect. In: 1st EAGE North African/Mediterranean petroleum & geosciences conference & exhibition, pp cp-8. European Association of Geoscientists & Engineers

Fiduk JC (2009) Evaporites, petroleum exploration, and the Cenozoic evolution of the Libyan shelf margin, central North Africa. Mar Pet Geol 26(8):1513–1527

Finetti I, (1982) Structure, stratigraphy and evolution of central Mediterranean: Bollettino di Geofisica Teorica ed Applicata, vol 24, no 96, pp 247–312

Fourati L, Hamouda F (1996) Abiod limestone reservoir characterisation in Beja block (northern Tunisia). In: Proceedings of 5th Tunisian Petroleum Conference: ETAP Memoir 10, pp 363–380

Fourati L, Troudi H, Saïdi M, Ben Boubaker K (2002) Petroleum Systems of the central eastern onshore Tunisia—North Africa (poster): ETAP, Tunis

Fourati L, Boubaker KB, Saïdi M (2003) Hydrocarbon generation and expulsion histories in Central Eastern Onshore Tunisia. In: 1st EAGE North African/Mediterranean petroleum & geosciences conference & exhibition, pp cp-8. European Association of Geoscientists & Engineers

Fournié D (1978) Nomenclature lithostratigraphique des séries du Crétacé supérieur au Tertiaire de Tunisie. Bull Cent Rech Expl Prod Elf Aquitaine 2:79–148

Gaaya A, Kharbachi S (1997) The carbon dioxide occurrences in the exploration óf the Tunisian basins-how to explore safely? In: 59th EAGE conference & exhibition, pp cp-131. European Association of Geoscientists & Engineers

Gharbi M, Masrouhi A, Espurt N, Bellier O, Amari EA, Youssef MB, Ghanmi M (2013) New tectono-sedimentary evidences for Aptian to Santonian extension of the Cretaceous rifting in the northern Chotts range (southern Tunisia). J Afr Earth Sci 79:58–73

Gili E, Masse JP, Skeleton PW (1995) Rudists as gregarious sediment dwellers, not reef-builders, on Cretaceous carbonate platforms. Palaeogeogr Palaeoclimatol Palaeoecol 12:205–222

Grasso M, Torelli L, Mazzoldi G (1999) Cretaceous–Palaeogene sedimentation patterns and structural evolution of the Tunisian shelf, offshore the Pelagian islands (central Mediterranean). Tectonophysics 315(1–4):235–250

Guiraud R (1998) Mesozoic rifting and basin inversion along the northern African Tethyan margin: an overview. In: Macgregor DS, Moody RTJ, Clark-Lowes DD (eds) Petroleum geology of North Africa. Geological Society, London, Special Publication 132, pp 217–229

Guiraud R, Bosworth W, Thierry J, Delplanque A (2005) Phanerozoic geological evolution of Northern and Central Africa: an overview. J Afr Earth Sc 43(1–3):83–143

Gumati YD, Kanes WH, Schamel S (1996) An evaluation of the hydrocarbon potential of the sedimentary basins of Libya. J Pet Geol 19(1):95–112

Haddad S, Echihi O, Ben Jemia H (2013) Petroleum prospectivity of the Jurassic strata of the Jeffara Basin, Tunisia. In: North Africa Technical Conference and Exhibition. Society of Petroleum Engineers

Haller P (1983) Structure profonde du Sahel tunisien. Interprétation géodynamique, thèse de 3ème cycle, Université de Besançon, 163 pp

Hallett D (2002) Petroleum geology of Libya. Elsevier Science BV 503:17–43

Hauptman MK, Bartholdy J, Bischoff R (2000) Depositional Conditions of Mid-Eocene Nummulite Accumulations in the Gulf of Gabes, Tunisia. In: EAGE Conference on geology and petroleum geology of the Mediterranean and Circum-Mediterranean Basins, pp cp-109. European Association of Geoscientists & Engineers

Hughes WB, Reed JD (1995) Oil and source rock geochemistry and exploration implications in northern Tunisia. In: ETAP (eds) Processings of the seminar on source rocks and hydrocarbon habitat in Tunisia, vol 10, pp 49–67 Mem

Jorry S, Davaud E, Caline B (2003) Controls on the distribution of nummulite facies: a case study from the Late Ypresian El Garia Formation (Kesra Plateau, central Tunisia). J Petrol Geol 26 (3):283–306

Khalifa Z, Affouri H, Rigane A, Jacob J (2018) The Albian oceanic anoxic events record in central and northern Tunisia: Geochemical data and paleotectonic controls. Mar Pet Geol 93:145–165

Khomsi S (2004) Géodynamique des bassins paléogènes et des réservoirs associés du Sahel et de Kairouan (Tunisie orientale); structuration, sismotectonique et organisation séquentielle. Thèse de Doctorat Université Tunis El Manar, Implications pétrolières, p 365

Khomsi S, Ben Jemia MG, de Lamotte DF, Maherssi C, Echidhi O, Mezni R (2009) An overview of the Late Cretaceous-Eocene positive inversions and Oligo-Miocene subsidence events in the foreland of the Tunisian Atlas: structural style and implications for the tectonic agenda of the Maghrebian Atlas system. Tectonophysics 475:38–58

Khomsi S, Echihi O, Slimani N (2012) Structural control on the deep hydrogeological and geothermal aquifers related to the fractured Campanian-Miocene reservoirs of north-eastern Tunisia foreland constrained by subsurface data. CR Geosci 344(3–4):247–265. https://doi.org/10.1016/j.crte.2012.02.002

Khomsi S, de Lamotte DF, Bédir M, Echihi O (2016) The Late Eocene and Late Miocene fronts of the Atlas Belt in eastern Maghreb: integration in the geodynamic evolution of the Mediterranean Domain. Arab J Geosci 9(15):650

Khomsi S, Roure F, Brahim NB, Maherssi C, Arab M, Khelil M (2019a) Structural styles, petroleum habitat and traps in the Pelagian-Sirt basins, Northern Africa: an overview and future exploration developments. In: Rossetti F et al (eds) The structural geology contribution to the Africa-Eurasia geology: basement and reservoir structure, ore mineralisation and tectonic modelling.

CAJG 2018. Advances in science, technology & innovation (IEREK Interdisciplinary Series for Sustainable Development). Springer, Cham. https://doi.org/10.1007/978-3-030-01455-1_33

Khomsi S, Roure F, Khelil M, Mezni R, Echihi O (2019b) A review of the crustal architecture and related pre-salt oil/gas objectives of the eastern Maghreb Atlas and Tell: need for deep seismic reflection profiling. Tectonophysics 766:232–248

Klett TR (2001) Total petroleum systems of the Pelagian Province, Tunisia, Libya, Italy and Malta—The Bou Dabbous-Tertiary and Jurassic-Cretaceous Composite. USGS Geolgical Survey Bulletin 2201-D

Kort HM, Gasquet D, Ikenne M, Ouazaa NL (2009) Cretaceous crustal thinning in North Africa: implications for magmatic and thermal events in the Eastern Tunisian margin and the Pelagic Sea. J Afr Earth Sc 55(5):257–264

Knott I, Moody R, Sandman R (1995) Coniacian-Turonian Carbonates of the Miskar Field, offshore Tunisia. AAPG Bull 79 (CONF-950995-)

Ligtenberg JH, Wansink AG (2001) Neural network prediction of permeability in the El Garia Formation, Ashtart oilfield, offshore Tunisia. J Pet Geol 24(4):389–404

Loucks RG, Moody RTJ, Bellis JK, Brown AA (1998) Regional depositional setting and pore network systems of the El Garia Formation (Metlaoui Group, Lower Eocene); offshore Tunisia. In: Macgregor DS, Moody RTJ, Clark-Lowes DD (eds) Petroleum geology of North Africa. Geol Soc Spec Publ 132:355–374

Lučić D, Bosworth W (2019) Regional geology and petroleum systems of the main reservoirs and source rocks of North Africa and the Middle East. In: Bendaoud A et al (eds) The geology of the Arab world-an overview. Springer, Cham, pp 197–289

Lüning S, Kolonic S, Belhadj EM, Belhadj Z, Cota L, Baric G, Wagner T (2004) Integrated depositional model for the Cenomanian—Turonian organic-rich strata in North Africa. Earth-Sci Rev 64:51–117

Mabrouk A, Jarvis I, Moody RTJ, Belayouni H, Brahim AB, De Cabrera S (2003a) Chemostratigraphy and correlation of a chalk petroleum reservoir–case of the Campanian-basal Maastrichtian Abiod Formation in Tunisia. In: 1st EAGE North African/Mediterranean petroleum & geosciences conference & exhibition, pp cp-8. European Association of Geoscientists & Engineers

Mabrouk A, Jarvis I, Moody RTJ, Belayouni H, Brahim AB, De Cabrera S (2003b) Isotopic study of a chalk petroleum reservoir–the late cretaceous abiod Formation, Tunisia–inferred sea-level change and d13C correlation with the Tethyan, Boreal Realm, and NW Pacific Ocean. In: 1st EAGE North African/Mediterranean petroleum & geosciences conference & exhibition, pp. cp-8. European Association of Geoscientists & Engineers

Macaulay CI, Beckett D, Braithwaite K, Bliefnick D, Philps B (2001) Constraints on diagenesis and reservoir quality in the fractured Hasdrubal field, offshore Tunisia. J Pet Geol 24(1):55–78

MacGregor DS, Moody R, Clark-Lowes DD (eds) (1998) Petroleum geology of North Africa, vol 132, pp 201–216. Geological Society, London, United Kingdom

Mejri F, Burollet PF, Ferjani BA (2006) Petroleum geology of Tunisia. In: A renewed synthesis. ETAP Mémoir, vol 22, 233 pp

Melki F, Zouaghi T, Ben Chelbi M, Bédir M, Zargouni F (2012) Role of the NE-SW Hercynian master fault systems and associated lineaments on the structuring and evolution of the Mesozoic and Cenozoic basins of the Alpine Margin, northern Tunisia. Tectonics—Recent advances: Rijeka, Croatia, InTech, pp 131–168

Mezni R, Khomsi S, Bédir M (2019) Structural styles, tectonic events, and deformation features along a surface–subsurface structural transect from the South Atlas Front (N–S axis) to the Eastern Sahel foreland basin of Tunisia. Arab J Geosci 12(6):186. https://doi.org/10.1007/s12517-019-4366-4

Morgan MA, Grocott J, Moody RTJ (1998) The structural evolution of the Zaghouan-Ressas Structural Belt, northern Tunisia. In: Macgregor DS, Moody RTJ, Clark-Lowes DD (eds) Petroleum geology of North Africa: Geological Society, London, Special Publication 132, pp 231–263

M'Rabet A (1981) Differentiation of environments of dolomite formation Lower Cretaceous of Central Tunisia. Sedimentology 28(3):331–352

M'Rabet A (1987) Stratigraphie, sédimentation et diagenèse carbonatée des séries du Crétacé inférieur de Tunisie centrale. Annale des mines et de la géologie No 30, Serv Géol Tunisie, 412 p

Nacer Bey R, Bergheul M, Guellal S (1995) Etude et characterisation des roches meres Crétacées du Bassin SE Constantinois. In: Proceedings of seminar on source rocks and hydrocarbon habitat in Tunisia, Tunis, pp 15–18

Nadhem K, Zahra N, Ménendez B, Salwa J, Jamel T (2017) Spatiotemporal sedimentological and petrophysical characterization of El Gueria reservoir (Ypresian) in Sfax and Gulf of Gabes Basins (SE-Tunisia). J Afr Earth Sci 130:202–222

Njahi Z, Kassabi N, Touir J (2017) Porosity and reservoir potentiality of the Cherahil Formation limestone (middle-upper Eocene) in the Gulf of Gabes (Tunisia). J Afr Earth Sci 131:166–178

Negra MH, Purser BH, M'Rabet A (1994) Permeability and porosity evolution in dolomitized upper Cretaceous pelagic limestones of Central Tunisia. In: Dolomites: a volume in honour of Dolomieu, vol 21, pp 309–323. Blackwell Publishing Ltd, Oxford, UK

Rabhi M, Chekhma HMR, Ben Haj Ali M, Et Mahjoub K (2001) Nouvelles données sur la tectonique compressive au sommet du Crétacé et au début du Tertiaire dans l'Axe Nord-Sud (Tunisie centrale): Notes du Service Géologique de Tunisie, no 68, pp 113118

Racey A (2001) A review of Eocene nummulite accumulations: structure, formation and reservoir potential. J Pet Geol 24(1):79–100

Racey A, Bailey HW, Beckett D, Gallagher LT, Hampton MJ, McQuilken J (2001) The petroleum geology of the early Eocene El Garia Formation, Hasdrubal field, offshore Tunisia. J Pet Geol 24 (1):29–53

Robaszynski F, Caron M, Dupuis G, Amedro F, Caladra F, Deloffre R, Gartner S, Gonzalez Denoso JM, Haedenbol J, Linares D (1990a) Litho-biostratigraphie, microfaciès et séquences stratigraphiques dans le Turonien d'un secteur de la Tunisie Centrale. Bull Soc Géol France 8:VI:1011–1024

Robaszynski F, Caron M, Dupuis G, Amedro F, Caladra F, Deloffre R, Gartner S, Gonzalez Denoso JM, Haedenbol J, Linares D (1990b) A tentative integrated stratigraphy in the Turonian of Central Tunisia: formation zones and sequential stratigraphy in the Kalaat Senan area. Bull CR Expl Prod Elf Aquitaine Pau 14(1):213–384

Roure F, Casero P, Addoum B (2012) Alpine inversion of the North African margin and delamination of its continental lithosphere. Tectonics 31(3)

Saïdi M (1993) Etude géologique et géochimique des roches mères albo-vraconiennes dans le domaine de la Tunisie centro-septentrionale: Thèse troisieme cycle. Université de Tunis, Faculté des sciences de Tunis, Tunisia, p 140

Saïdi M, Belayouni H (1994) Etude géologique et géochimique des roche mères albovraconienne dans le domaine de la Tunisie centro-septentrional. In: Proceeding of the 4th petroleum exploration conference, vol 7, pp 91–116. ETAP

Saïdi M, Philip RP (1997) Geochemical characterization and relationship of oil-source rock from northern Tunisia (abs.): AAPG annual meeting program, vol 6, pp A-100

Saïd A, Baby P, Chardon D, Ouali J (2011) Structure, paleogeographic inheritance, and deformation history of the southern Atlas foreland fold and thrust belt of Tunisia. Tectonics 30(6)

Scotese CR, Schettino A (2017) Late Permian-Early Jurassic paleogeography of western Tethys and the world. In: Permo-Triassic salt provinces of Europe, North Africa and the Atlantic margins, pp 57–95. Elsevier

Stampfli GM (2000) Tethyan oceans. Geol Soc Lond Spec Publ 173 (1):1–23

Stampfli GM, Borel GD (2004) The TRANSMED transects in space and time: constraints on the paleotectonic evolution of the Mediterranean domain. In the TRANSMED Atlas. The Mediterranean region from crust to mantle, pp 53–80. Springer, Berlin, Heidelberg

Stampfli GM, Kozur HW (2006) Europe from the Variscan to the Alpine cycles. In: Gee DG, Stephenson RA (eds) European Lithosphere dynamics: Geological Society, London, Memoirs, 32, 57–82

Swei GH, Tucker ME (2012) Impact of diagenesis on reservoir quality in ramp carbonates: Gialo Formation (Middle Eocene), Sirt Basin Libya. J Petrol Geol 35(1):25–47

Taktak F, Kharbachi S, Bouaziz S, Tlig S (2010) Basin dynamics and petroleum potential of the Eocene series in the gulf of Gabes, Tunisia. J Petrol Sci Eng 75(1–2):114–128

Taktak F, Bouaziz S, Tlig S (2012) Depositional and tectonic constraints for hydrocarbon targets of the Lutetian-Langhian sequences from the Gulf of Gabes—Tunisia. J Petrol Sci Eng 82:50–65

Tlig S, Sahli S, Er-Raioui L, Alouani R, Mzoughi M (2010) Depositional environment controls on petroleum potential of the Eocene in the North of Tunisia. J Petrol Sci Eng 71(3–4):91–105

Touir J, Ben Haj Ali N, Donze P, Maamouri AL, Memmi L, M'Rabet A, Razgallah S, Zaghbib-Turki D (1989) Biostratigraphie et sédimentologie des séquences du Crétacé supérieur du Jebel M'rhila (Tunisie centrale). Géol Méditerranéenne 16:55–66

Touir J, Soussi M (2003) Growth and migration of two Turonian rudist bearing carbonate platforms in central Tunisia. Eustatic and tectonic controls. In: Gili E, Negra MH, Skeleton PW (eds) North African Cretaceous Carbonate Platforms Systems. NATO Science Series IV. Earth Environ Sci 28:53–81

Touir J, Soussi M, Troudi H (2009) Polyphased dolomitization of a shoal-rimmed carbonate platform: example from the middle Turonian Bireno dolomites of central Tunisia. Cretac Res 30(3):785–804

Troudi H (1998) Les réservoirs et faciès associés du Crétacé supérieur en Tunisie centrale: Sédimentologie, Stratigraphie séquentielle et Diagenèse, Thèse de Doctorat. Université De Tunis I I:287p

Troudi H, M'Rabet A (1998) Deposition, Diagenesis and porosity development of the Lower-Middle Turonian carbonate reservoir in central Tunisia. In: Proceeding of the sixth Tunisia petroleum exploration and production conference, pp 371–385

Troudi H, El Euchi H, Tremolieres P (2000) Fracture origin, morphology and reservoir performance improvement: example of the Maastrichtian Chalky limestones in Tunisia (poster): ETAP, Tunis, P626

Troudi H, Saïdi M, Acheche MH, Abassi K (2002) Mid-Cretaceous platform carbonate in Tunisia: attributes and evidences from producing fields. In: Proceedings of the eighth Tunisia Petroleum exploration and Production conference, pp 329–350

Troudi H, Tari G, Alouani W, Cantarella G (2017) Styles of salt tectonics in Central Tunisia: an overview. In: Permo-Triassic Salt Provinces of Europe, North Africa and the Atlantic Margins, pp 543–561. Elsevier

Vennin E, Van Buchem FSP, Joseph P, Gaumet F, Sonnenfeld M, Rebelle M, Fakhfakh-Ben Hemia H, Zijlstra H (2003) A 3D outcrop analogue model for Ypresian nummulitic carbonate reservoirs: Jebel Ousselat, northern Tunisia. Pet Geosci 9(2):145–161

Wilson M, Guiraud R (1998) Late Permian to recent magmatic activity on the African-Arabian margin of the Tethys. IN: MacGregor DS, Moddy RTJ, Clark-Lowes DD (eds) Petroleum geology of North Africa. Geological Society, London, Special Publication 132, pp 231–263

Zaïer A, Beji-Sassi A, Sassi S, Moody RTJ (1998) Basin evolution and deposition during the Early Paleogene in Tunisia. Geol Soc Lond Spec Publ 132(1):375–393

Zargouni F (1985) Tectonique de l'Atlas méridional de Tunisie. Thèse Es Sciences, Université Louis Pasteur Strasbourg, Evolution géodynamique et cinématique des structures en zone de cisaillement, p 296

Ziegler PA, Roure F (1999) Petroleum systems of Alpine-Mediterranean foldbelts and basins. Geol Soc Lond Spec Publ 156(1):517–540

Zourai H (1995) Evolution géodynamique de l'Atlas centro-méridional de la Tunisie: stratigraphie, analyse géométrique, cinématique et tectono-sédimentaire: thèse es-Sciences, Fac. Sciences Tunis, Univ, Tunis II, p 251

Mesozoic and Cenozoic Tectonosedimentary Evolution and Subsidence History of South-Eastern Tunisia: Jeffara Basin Petroleum Prospectivity and Hydrocarbon Provinces

6

Radhouane Khouni, Mohamed Sabri Arfaoui, Mohamed Ghanmi, and Fouad Zargouni

Abstract

The Jeffara basin located in SE of Tunisia is part of the Pelagian Sea, recording the consequences of the rapprochement of the African plate and the European plate. It represents a typical example of a passive margin basin where petroleum exploration has promised much and working petroleum systems have proven. However, returning to the history of petroleum exploration in the Jeffara basin, we found that it is generally based on the types of simple structural traps that have led to modest deposits with few exceptions. Understanding tectonosedimentary history can provide new insights into geology and plays which in turn can change the spirit of exploration. In this study, we started with a synthesis of the lithostratigraphic aspect of the study area, of which we focused on the major unconformities and subsequently the paleostress which reigned from the Mesozoic to the Present. Analysis of the interpretation of available 2D seismic reflection and petroleum well data allowed us to discuss the structural systems that characterize the Jeffara basin which revealed three types of system: The first system, the second system and the third system. Five structural styles are the result: ancient horst or uplift, multiple stage faults slope, traditional slope, faulted slope break and flexure slope break. The assembly of the method of "Backtreeping" applied on a SW-NE seismic composite line and the vertical burial history method using the "Novva 1-D modeling" software on a petroleum well in the basin allowed us to deduce three types of evolution: a first with a high burial rate, a second with a low burial rate and a third with a minimal burial rate. The combination of the tectonosedimentary evolution in the Jeffara basin and the burial history from the Mesozoic to the Present firstly has confirmed the existence of the functional petroleum system which includes all the essential interdependent elements and the processes forming the functional unit. Secondly, has highlighted that more than simple structural traps, stratigraphic traps are showing a high probability of the existence of potential hydrocarbons which can push to a new spirit of petroleum exploration in this sector and around.

Keywords

Tunisia • Jeffara basin • Seismic data • Structural system • Structural style • Burial rate • Stratigraphic traps • Petroleum exploration

6.1 Introduction

The Tunisian Jeffara is located on the African edge of the Tethys of reconquest, at the limit of two different domains, the stable Saharan platform in the South and the intracontinental chain of the Saharan Atlas in the North (Fig. 6.1A). It began with the first tectonic manifestations linked to the Tethyan opening, and it was discreetly recorded in the Mesozoic sediments the deformations and the distribution of the fragmentation of the African margin during the opening of the Atlantic (Burollet 1956; Aubouin et al. 1980; Gabtni et al. 2005; Frizon et al. 2006; Gabtni et al. 2013).

The Jeffara basin is considered as a part of Ashtar-Tripolitania petroleum province with more than 60 years of oil and gas exploration. Many discoveries of hydrocarbon fields are exploited in the Jeffara basin (Mejri et al. 2006). Most of the oil fields are offshore with accumulations in bioclastic nummulitic limestone of the Lower Eocene reservoirs such as Didon and Hasdrubal fields and

R. Khouni (✉) · M. S. Arfaoui · M. Ghanmi · F. Zargouni
Laboratoire Géosciences, Energétiques Et Environnement, Département de Géologie, FST, Université de Tunis El Manar, Ressources Minérales, Tunis, Tunisie
e-mail: radhouanekhouni@gmail.com

R. Khouni · M. S. Arfaoui
PRIMOIL SA, Av. de La Bourse, Immeuble EmeraudeBloc B, 3Ème Étage Les Berges du Lac 2, Tunis, Tunisia

S. Khomsi and F. Roure (eds.), *Geology of North Africa and the Mediterranean: Sedimentary Basins and Georesources*, Regional Geology Reviews, https://doi.org/10.1007/978-3-031-18747-6_6

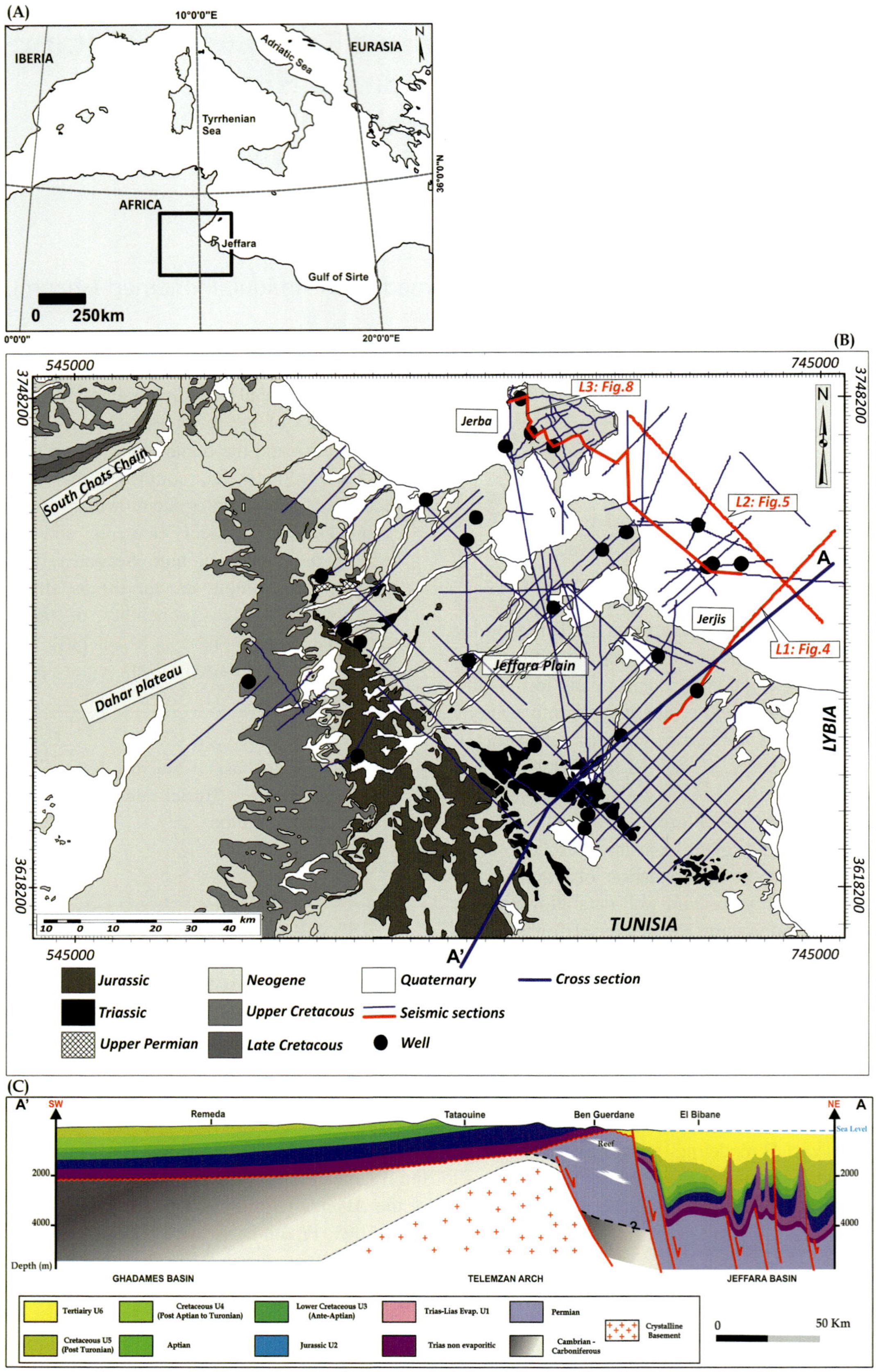

Fig. 6.1 **A** Study area location in southern Tunisia bordered by Algeria and Libya. **B** Simplified geologic map of southern Tunisia showing the outcrops of the principle stratigraphic series in Dahar and Jeffara costal, an overview of the 2D seismic data base position and petroleum wells used in seismic calibration and lithostratigraphic correlations. **C** Regional SW-NE cross section from seismic and well data in southern Tunisia showing the main structural and stratigraphic relations in the Ghadames and Jeffara basin separated by the Telemzane arch (Khouni et al. 2018)

Upper Cretaceous limestone: El Biban field or its equivalent at onshore the Ezzaouia field where Upper Jurassic sandstone reservoir is also distinguished. The identification of these potentialities is not possible without a good exploration by geophysical survey and wells drilling. It is considered also a widely studied domain with significant water resources in an arid Sahara area (Ben Alaya et al. 2016).

During the Mesozoic and Cenozoic rifting, the Pelagian Sea recorded the consequences of the rapprochement of the African and European plates. After the Hercynian event of the Permian-Carboniferous Age, a general extension took place, which gave rise to the Tethysian opening (Aubouin et al. 1980; Gabtni et al. 2005; Gabtni et al. 2013). This extension known by the abundance of normal syn-sedimentary faults of major NW–SE direction and where the Jeffara fault seems to be the major fault favored the individualization of the Jeffara basin in the South-East of Tunisia. This basin is characterized by a Horst and Graben structure with high subsidence going from the Jeffara fault toward the East. This variation in the thicknesses of the Lower Cretaceous series is the origin of the flowing of the salt material under the influence of the high lithostatic differential pressure leading to the formation of pillows, diapirs and salt walls which perse their sedimentary cover through normal faults, they have begun during Cretaceous as the Gulf of Sirte and the Neo-Tethys Ocean are underway opening (Khomsi et al. 2009; Khouni et al. 2018). Although the subsidence of the Jeffara basin played a major role in the activity of salt complexes along the Mesozoic and Cenozoic periods, these salt complexes, in turn, controlled the basin structuring and, sedimentation (Khouni et al. 2010, 2017, 2018) (Fig. 6.1C).

Although high, the subsidence of the sedimentary sequences underwent high lateral and vertical variation thus recording the involvement of syn-sedimentary tectonics. The interpretation of surface and subsurface data like the 2D seismic reflection and petroleum well data (Fig. 6.1B) shows new ideas on the tectonosedimentary evolution of the Jeffara basin during the Mesozoic and Cenozoic.

In this work, we will first study the impact of syn-sedimentary tectonics on subsidence, continuity and vertical and spatial distribution of the Mesozoic and Cenozoic sedimentary series within the Jeffara basin and secondly the relationship of the combination of all these elements with the hydrocarbon prospectivity of the study area.

6.2 Lithostratigraphy and Major Unconformities

Analysis of the oil drilling data available in the study area revealed that the sedimentation in the Jeffara basin recorded three major discontinuities (See Table 6.1).

Indeed, the series of the lower Cretaceous is based on those of the Jurassic by a major discontinuity named "**Unconformity 1 or U1**". Deposits of the Lower Cretaceous; Boudinar, Bouhedma or even the Continental Intercalaire, or Asfer qualified as "Wealdien", settled in a delta environment. The advanced age is upper Jurassic to Hauterivien but which can reach the Barrémien without foundation or tangible dating (Busson 1970; Peybernès et al. 1985; Haq et al. 1987; Ben Ismaïl et al. 1989; Ghanmi et al. 1993; Ben Youssef 1999; Touhami et al. 2008). On this series rests the Member Berrani capped by the medium sandstone term with fossiliferous marl intercalations Araguib and the upper sand-marly member Foum el Argoub (Fig. 6.2).

The Zebbag Formation, of upper Albian-Cenomanian age, is discordant on the Lower Cretaceous series, thus presenting the second Unconformity "**Unconformity 2 or U2**", it is subdivided into three members: The first consists of intercalations of marly limestone admitting at its base a level of limestone at Knémicéras. The second is enriched in gypsum in its middle compared to the first and the third is dolomitic (Busson 1967; Haq et al. 1987; Ghanmi et al. 1993; Ben Youssef 1999). The significant reduction in thickness of the Lower Cretaceous series, mainly in the South of the Gulf of Gabes and the fault which affect them and control their thickness and their distribution means that at this level the different terms of the Lower Cretaceous are undifferentiated and the bar of Barrani is confused with the slab of Radhouan.

Above the Zebbag Formation rests the Aleg Formation admitting at its base the term carbonate Biréno and the term Beida evaporitic. These two terms Biréno and Beida disappear at NNE. All of these deposits are absent, at the level of the southern part of the Gulf of Gabès, reflecting tectonic instability (Fig. 6.2).

On these series rests in Unconformity "**Unconformity 3 or U3**" the varicolored clay laterally evolving to marl with coquina limestone of Cherahil Formation of Eocene age, represented in onshore by conglomerates, red clay, lacustrine limestone and caliche attributed to Bou Loufa Fm (Bédir et al. 1996; Jamoussi et al. 2001; Frizon et al. 2009; Merzeraud et al. 2016; Khomsi et al. 2016). The lower Oligocene–Miocene interval is represented by the carbonate formations: Ketatna and Salammbô. The Ketatna Formation is carbonated with a dolomitic level at the base; the Salammbô Formation is essentially clay with a carbonated level with Nummulites at its base. The Middle Miocene-Pleistocene is marked by the deposition of a level of bioclastic Ain Grab limestones and by Mahmoud clays, all covered by the detrital and continental series, respectively, by the Oum Douil, Segui and Recent Quaternary groups (Fig. 6.2).

A regional correlation of the major discontinuities passing through several oil wells in a key NE-SW direction

Table 6.1 Illustrating the variation in thicknesses of the Lower Cretaceous, Upper Cretaceous and Cenozoic series from one well to another in the Jeffara basin and its surroundings

Wells	Thickness (m)		
	Lower cretaceous	Upper cretaceous	Cenozoïc
P-05	688	1750	675
P-06	561	255.5	647
P-07	1045	177	677
P-17	322	1388.5	669
P-19	300	650	3350
P-21	214	950	1370
P-24	342	1970	800
P-27	450	1100	2800
P-28	76	583	1960
P-29	0	94	1563
P-30	0	536	2376

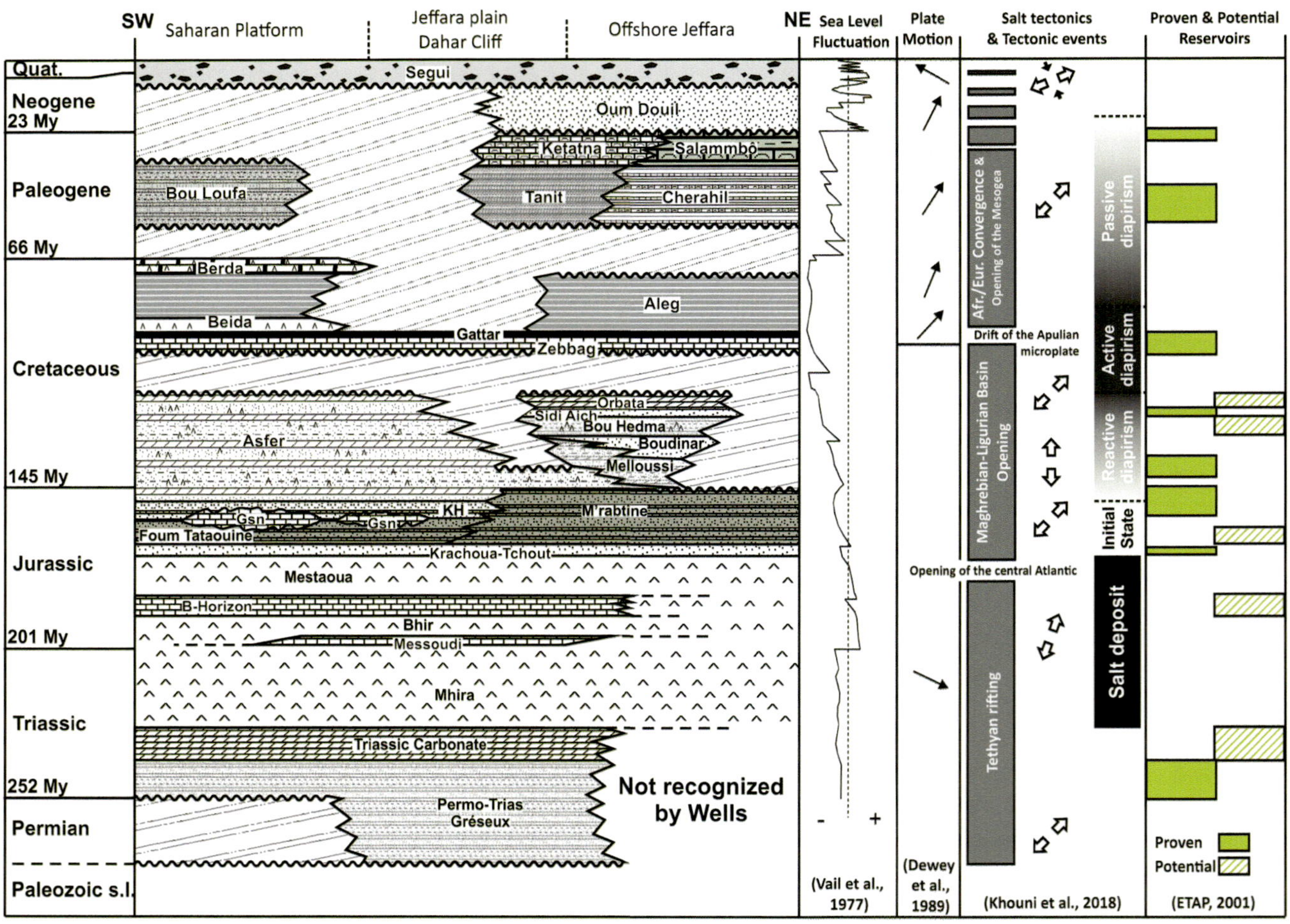

Fig. 6.2 Generalized lithostratigraphic evolution chart of Saharan basin of Mesozoic and Cenozoic formation compiled from field outcrops and sub-surface well data. Showing the variation of facies between Saharan platform and Jeffara basin from Mesozoic to Cenozoic, the major tectonic events, salt tectonic evolution and proven and potential hydrocarbon reservoirs in Jeffara basin (modified from Khouni et al. 2018)

(Figs. 6.4B, 6.5B) showed the distribution of the crossed series and the variation in their thickness summarized in Table I.

The first unconformity U1 separating the Lower Cretaceous series from the Jurassic series shows a great depth in the P-27 and P-30 wells, it is a high zone, marked by the absence of the Lower Cretaceous series (Figs. 6.2, 6.4) and where the Upper Cretaceous series rest in unconformity "U2" on the Jurassic series and subsident zones at the location of the P-06 and P-07 wells; delimitated to the East and SW, respectively, at the location of P-17 wells and the location of P-24 and P-27 wells (Fig. 6.1B).

In the zones marked by the maximum of lower Cretaceous deposits, in the central part of the sector at the site of P-07 well 1045 m, there is a little of upper Cretaceous deposits 177 m at the site of P-07 well 2946 m, and against a little of Lower Cretaceous deposits, 342 m at the location of P-24 well, we note 1970 m of Upper Cretaceous deposit. As for the tertiary and quaternary series, an unconformity on the Upper Cretaceous series "U3" (Figs. 6.2, 6.4B, 6.5B) fill the low areas of the Jeffara basin.

The Lower Cretaceous to Upper Cretaceous passage is marked by a reversal of subsidence: the depocentre of the Lower Cretaceous basin, located in the central part of the maritime Jeffara at the site of P-07 well, migrated during the Upper Cretaceous location of the P-24 well under the effect of a NE-SW rifting phase. This distension is responsible for the structuring in horst and graben thus creating high and low zones. To these extensional tectonics is added the rise of the Triassic bodies noted at the location of the wells P-28, P-29 and P-30 where these Triassic series are covered by tertiary series at the level of the first and by series of the Cretaceous above at the level of the second (Touhami et al. 2008).

The passage Lower Cretaceous-Upper Cretaceous is marked by an inversion of subsidence: the depocentre of the Lower Cretaceous basin, located in the central part of the maritime Jeffara at the site of well P-07, migrated during the Upper Cretaceous location of the well R under the effect of a NE-SW rifting phase, distension responsible for the horst and graben structuring thus creating high and low zones. In addition to these extensive tectonics, the rise of Triassic bodies noted SE of the basin played its role in the control of sedimentation during this interval (Khouni et al. 2018).

6.3 Structural Analyses

6.3.1 Available Data

This work is based mainly on the interpretation of several 2D reflection seismic lines, belonging to several seismic surveys and calibrated from the oil wells data available in the study area. To establish 2D geo-seismic sections showing the Triassic structures and the geodynamic evolution of the Jeffara basin, all available seismic sections were interpreted and three major composite lines are exhibited in this work (Fig. 6.1B).

The tectonosedimentary evolution is based on the interpretation of a regional composite line: The first regional composite line L1 with an NE SW direction and whose length is about 130 km is perpendicular to the Jeffara fault crossing the two areas, the Dahar plateau to the west and the Jeffara basin to the east where the salt complexes are individualized (Figs. 6.1B, 6.4). The second NW–SE trending composite line L2, whose length is about 75 km. This line spreads across three oil wells allowing precise calibration and is located offshore of the Jeffara basin and parallel to the coastline and the escarpment of the main NW–SE Jeffara fault (Figs. 6.1B, 6.5). The third regional NW–SE seismic composite line L3 with 90 km length has enabled to validate the geodynamic evolution detected from its neighbor and parallel and to construct a model of evolution based on the Back-stripping methodology (Figs. 6.1B, 6.8).

The well available in the study area allowing the calibrations of seismic lines provides also precious lithostratigraphic information for the identification of various Mesozoic and Cenozoic sedimentary series. However, it should be noted that the salt complexes in the Jeffara basin are traversed by very few wells, unlike those located in the Dahar plateau (Carpentier et al. 2016).

The following lithostratigraphic nomenclature and colors of the interpretation of the different stratigraphic units are inspired by the formations described in the central and meridional Atlas of Tunisia, the chotts fold belt, the Dahar outcrops and local lithological units identified in wells (Figs. 6.1, 6.2).

6.3.2 Syn-Sedimentary Deformations Analysis

The analysis of the interpretation of all available 2D seismic data in the study area coupled with the oil well data shows clearly that the Jeffara basin is structured in horst and graben, bounded by normal faults with major directions NW–SE, NNW-SSE and NW–SE and mixed flexures-faults, giving rise to a basin structured in tilted blocks (Khouni et al. 2010, 2017, 2018), associated with progressive unconformities (Figs. 6.3, 6.4A, 6.5A).

The structural analysis focuses on the characterization of the syn-sedimentary deformations which affected the Mesozoic and Cenozoic series. The analysis of the structural maps of the key levels from the seismic interpretation shows that this extensive deformation has undergone a lateral variation from one zone to another and vertical according to the period.

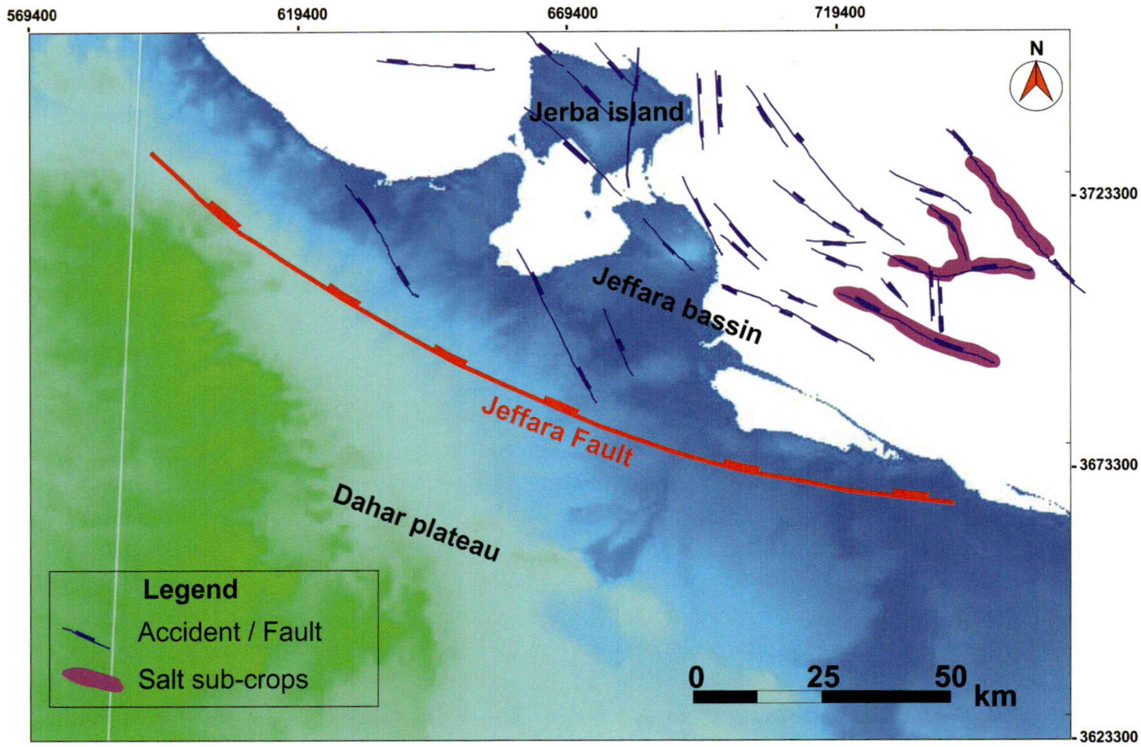

Fig. 6.3 Structural map based on the interpretation of the available seismic data on the Digital Elevation Model background, showing the right position of Jeffara fault et the distribution of the extensional network faults in Jeffara Basin

6.3.2.1 Syn-Sedimentary Deformations During the Jurassic

In the Jeffara Basin, Jurassic levels are affected by normal faults with variable discharges and the majority of which are toward the East. Locally, normal faults whose rejection is high appeared as faults affecting the Permo-Triassic basement and which will be reactivated several times during the Mesozoic, these faults most often have an arcuate appearance (Figs. 6.3, 6.4A, 6.5A). These normal faults have been highlighted in several sectors in the basin: The main fault is the Jeffara fault with NW–SE major direction, installed along the Jeffara mole of Medenine separating the Dahar plateau moderately stable in West from the Maritime Jeffara Basin in the East (Fig. 6.3). The other faults of the same family are scattered in several areas, like the eastern flank of the Jeffara fault, the Jerba area, the El Bibane area, the Ezzaouia area, the Rass Marmour area and the Rass Agil area (Khouni et al. 2018).

Along the Jeffara basin, the tectonic-sedimentation interaction made it possible to distinguish a syn-tectonic series (Figs. 6.4, 6.5). In this series, the syn-sedimentary character is attested by variations in the thickness of the Jurassic deposits, on both sides of the fault planes and by the continuity of the tectonic activity of the latter during the sedimentation of the sus-underlying levels. The variation in thickness of this series is generally small; there is a slight increase especially in the eastern center of the Jeffara basin.

A second minor family of normal faults with NE-SW major direction has been highlighted with a very low abundance localized especially in the Jerba and the Rass Agil zone, this network of faults is interpreted as synthetic NW–SE faults, thus announcing a normal dropout setup during this period.

6.3.2.2 Syn-Sedimentary Deformations During the Lower Cretaceous

During the Lower Cretaceous period, the tectonic aspect known during the Jurassic continues, recording an amplification of the activity of the NW–SE normal faults. The extensive Cretaceous regime produced a series of normal NW–SE major faults, the most important of which are located in the East of the Jeffara fault, others in the Jerba region in NE of the basin and the Rass Agil area in SE of the basin. These faults dominate the structural tissue observed today in this region and generally have normal listric fault type geometry and probably detach at a depth of 20 km. In plan view, their configuration varies according to the levels and relays (Figs. 6.4, 6.5).

To the NE of the Jeffara basin, at the Gerba region, the tectonic deformation is of fragile aspect, where the zone is

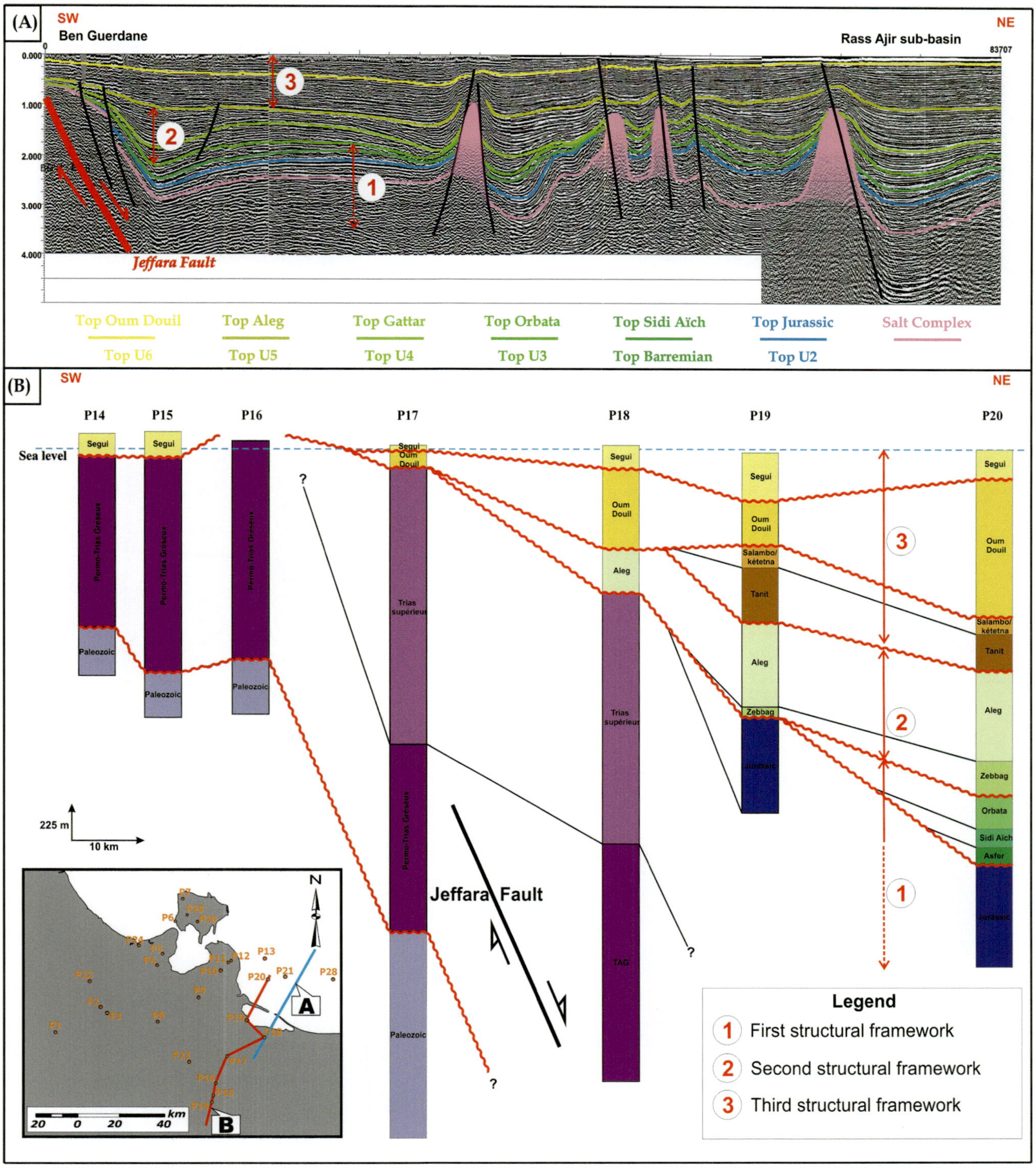

Fig. 6.4 A SW-NE interpreted 2D composite seismic line L1 crossing the Dahar plateau and Jeffara basin (location shown in Fig. 6.1), illustrating the geometry and structuration of the Mesozoic and Cenozoic strata, their relation with NW–SE trending fault and the arrangement of the resulting structural systems. **B** Lithostratigraphic petroleum wells correlation in the same direction showing the sedimentation evolution from the Dahar plateau to the Jeffara basin and the control of the NW–SE fault on the structuring of sedimentary substratum (modified from Khouni et al. 2018)

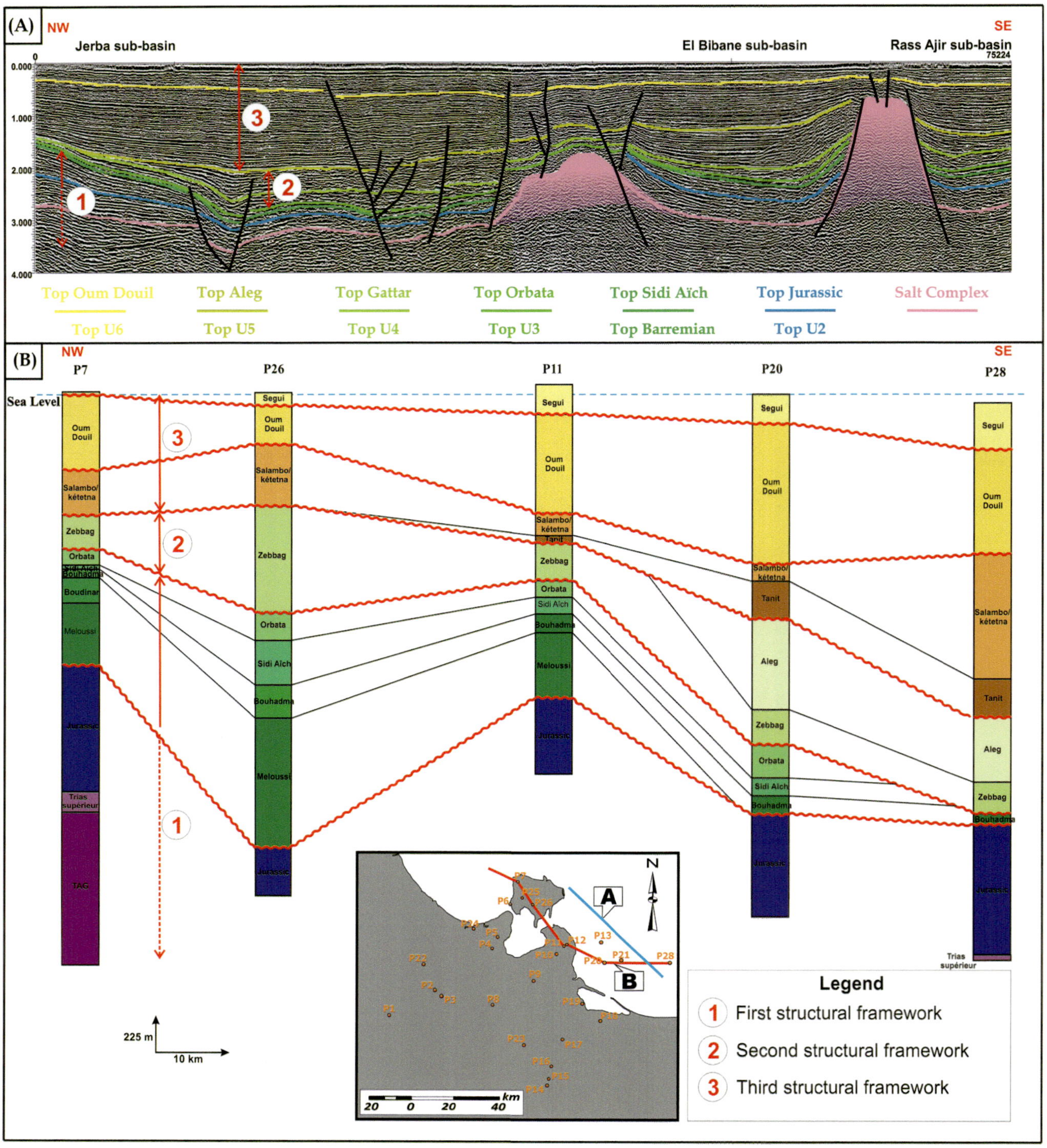

Fig. 6.5 **A** NW–SE interpreted 2D composite seismic line L2 across the Jeffara basin (location shown in Fig. 6.1), illustrating the geometry and structuration of the Mesozoic and Cenozoic strata, their relation with faults and associated salt complex bodies and the arrangement of the resulting structural systems. **B** Lithostratigraphic petroleum wells correlation in the same direction showing the evolution of subsidence in Jeffara basin and the control of the NE-SW fault on the structuring of sedimentary substratum (modified from Khouni et al. 2018)

dominated by structures in horsts and grabens limited by faults of the major direction parallel to the regional trend NW–SE and a minor direction NE-SW representing synthetic syn-sedimentary faults. It is also in this zone that the sequences of the lower Cretaceous which are the predominantly clastic Meloussi and Sidi Aich formations show a spectacular thickening compared to the zones located SE of the Jeffara Basin. It is believed that the thickening observed occurred in a half-graben inclined to the northeast "Djerba pit".

To the SE of the Jeffara basin, there is a halokinetic activity that began at the end of the Jurassic (Khouni et al. 2018), these diapiric structures were preferably aligned along normal faults in the NW–SE direction (Fig. 6.3). This halokinetic activity is the result of the sediment load built up since the Jurassic period favoring the appearance of a bald zone during the Lower Cretaceous observed since the absence of these series in the available oil well data in the area of interest (Figs. 6.4B; 6.5B), and it is translated by seismic toplap forms where a seismic reflection terminates against an overlying reflection without significant erosional truncation, reflecting the disappearance of an interval in upper part of lower Cretaceous due to sediment bypass or thinning of the bed to below seismic resolution. Erosional truncation is more abrupt, where a reflection exhibits an angular truncation against a younger surface, generally signifying an erosional contact (Fig. 6.4A).

6.3.2.3 Syn-Sedimentary Deformations During the Upper Cretaceous

During the Upper Cretaceous, the tectonic activity in the Jeffara basin is characterized by the abundance of three normal fault networks: The first fault network is in the NW–SE major direction, parallel to the regional trend. These faults are plunging toward the NE in the basin with a generally arched shape and high rejection reaching 1000 m especially in the South zone of the study area where they favored the reintroduction of salt intrusions.

A second network faults with NE-SW major direction, this network consists of NNE-SSW and NE-SW normal synthetic faults implanted on the first network and having an average rejection generally toward the SW, the main faults of this network concentrated to the NE of the Jeffara basin in Jerba zone and the SE of the basin in Rass Ajil zone.

The third network faults consists of E-W normal antithetic faults, they are very abundant in the central of the basin. These faults have an E-W-oriented dextral unhooking aspect resulting in the formation of NE-SW folds (Touati and Rodgers 1998), like the structure of Ezzaouia (Figs. 6.4A, 6.5A).

The tectonic activity in the Jeffara basin during the Upper Cretaceous recorded in turn its impact on the variation of the sedimentation rate, thus the Upper Cretaceous series increased in thickness away from the main Jeffara fault eastward across the basin. Examination of the Upper Cretaceous isopaque map shows that this variation reaches its maximum in the East center, it is linked to the NW–SE trending of the normal faults.

Analysis of the NW–SE direction seismic section and the isopaque map (Khouni et al. 2018) shows the influence of tectonic activity during the Upper Cretaceous on the distribution of sedimentation. These figures show that the variation in the thickness of the Upper Cretaceous series is controlled by the normal fault networks NNE-SSW and NE-SW. The distribution of subsidence during this period manifested itself in a reverse manner to that during the Lower Cretaceous, so in the SE part of the Jeffara basin in the Rass Ajil sub-basin, the Upper Cretaceous series are inconsistent on the Bouhedma series with the erosion of the Lower Cretaceous, this area is the site of significant sedimentation of the Upper Cretaceous marls with the individualization of an elevated area toward the North-West with reduced sedimentation up to in the Jerba sub-basin where we record a paleohigh limited to the South by a fault in the NE-SW direction (Khouni et al. 2018).

During the Triassic N-S distension, the different compartments limited by the E-W faults are tilted toward the south of the Jeffara basin. This structure creates a slope along which corridors can be established following the early dissolution of anhydrite. The dissolution allows the collapse of the Upper Cretaceous series in corridors armed by inherited faults and the slope allows the flow.

6.3.2.4 Syn-Sedimentary Deformations During the Cenozoic

After the Santonian period accompanied by phases of more or less significant erosion of the series of the upper Cretaceous, the Eocene–Oligocene period seems a relatively calm period corresponding to the post-Villafranchian where we notice an accentuation of the old folds. During the Miocene Pliocene period, we notice the reactivation of old faults signaled by the filling in the sagging areas (Figs. 6.4, 6.5).

6.4 Paleo-Stress Summary

6.4.1 Paleozoic-Triassic Extensive Episode

After the Hercynian orogeny, marine sediments from the Upper Paleozoic covered the Saharan platform in southwestern Tunisia. The Upper Permian deposits were concentrated in the Tebaga area of Medenine, which quickly subsided, where the subsidence of the Permian marine carbonates which were deposited reached 3500 m (Ben Ferjani et al. 1990; Touati and Rodgers 1998; Bouaziz et al. 2002) (Fig. 6.2).

During the end of Permian to lower Triassic, the study area was exposed to a NE-SW to NNE-SSW extension, which is responsible for the normal reactivation of the NW–SE fault systems and consequently led to the opening of the Jeffara subsidence basin to the SE of Tunisia and the individualization of a moderately resistant plateau in the West (Bishop 1975; Ben Ferjani et al. 1990; Bouaziz et al. 2002).

The study area reversed during the Upper Triassic, causing uplift and erosion up to 1700 m from the Upper Permian section at the end of the Middle Triassic. An evaporitic basin from the end of the Triassic formed the East Mount of Tebaga, in the Jeffara basin reaching 2000 m, so the sedimentary bevels during this interval are associated with normal synsedimentary play (Khouni et al. 2018) (Figs. 6.2, 6.4A, 6.5A).

6.4.2 Upper Jurassic-Lower Cretaceous Extensive Episode: Syn-Rift

The reconstruction of the paleo-stress shows that the extensive NE-SW tectonic regime still prevails in the study area (Bouaziz et al., 2002), highlighting an extensive deformation according to two normal fault networks.

The first network is defined by normal faults with NW–SE major direction, which plunges toward the NE of the basin. These faults follow the main Jeffara fault defined by the extensive NE-SW steering constraint. This continuous play of NW–SE faults leads to an extensive structuring in blocks tilted toward the NE part of the basin during the Jurassic and the Lower Cretaceous. The interpretation of geophysical and petroleum wells data shows that the maximum subsidence during the Lower Cretaceous period is toward the NE of the basin in the Jerba zone, this particular structure would result from the normal play of the NW–SE faults, which, from the made of the plasticity of the clay levels of the Lower Cretaceous, would be connected in depth with a level of detachment (Fig. 6.4A).

More to SE of the Jeffara basin, the Lower Cretaceous marks its absence in several petroleum wells, this phenomenon is identified at the level of the seismic lines by downlap shapes thus translating the resistivity of this at the end of the Lower Cretaceous which is probably due to halokinetic activity (Khouni et al. 2018).

The second network is defined by normal antithetical faults of NW–SE direction identified in some localities in the basin; these faults have low rejection toward the SW giving birth to structuring in horst and graben.

6.4.3 Aptian-Albian Period

Several studies like (Burollet and Desforges 1982; Ben 1986; Bedir 1995 and El Euchi et al. 1998) attributed the general intra-Albian unconformity to a regional transpressive event that is probably with NW–SE direction. The interpretation of geophysical data in the pelagic basin to the SE of Tunisia has shown the reactivation of NE-SW faults in reverse (Khomsi et al. 2004) (Fig. 6.2).

During this period, the end of the Lower Cretaceous strata and the Zebbag Fm are lifted, rotated, shouldered and eroded when the diapir crosses them by force through active diapirism. In this context, the upper Cretaceous is characterized by local depressions formed and developed around the salt walls (Khouni et al. 2018) (Fig. 6.4A).

6.4.4 Upper Cretaceous Episode

The distribution of transtensive stress tensors determines NE-SW trending sigma 1 trajectories (Bouaziz et al. 2002), oblique to the major NW–SE Jeffara fault. This trending of the trajectories of sigma 1 highlights the birth of a network of normal dextral faults to descending dextral E-W direction.

The NE-SW extension developed by this regime is mainly guided by the normal replay of the main and synthetic NW–SE direction faults of rejection toward the NE and the antithetical faults of rejection toward the SW (Fig. 6.3). This configuration mentioned a change in the distribution of the thicknesses of this series compared to that of the Lower Cretaceous where it shows an increase toward the SW of the basin by signaling uplifts in particular in the NE part of the study area until having sedimentation gaps in the Jerba area (Fig. 6.5A).

On the other hand, the normal faults of direction E-W were weakly activated in normal dexter in the new stress field established (figure). This last observation is in agreement with the work of (Touati and Rodgers 1998), emphasizing the fact that the southern Gulf of Gabes was under the influence of major lateral dexterity slip faults oriented E-W, which led to the formation of folds in the NE-SW direction and NW–SE oriented diapirs and salt walls (Khouni et al. 2018).

The NW–SE system of main and synthetic faults with rejection toward the NE and the antithetical faults with rejection toward the SW exerted a direct control on the sedimentation during this interval. The normal play toward the SW resulted in a reversal of subsidence from the Lower

Cretaceous to the upper Cretaceous, whose synclinal gutters tilted toward the SW control the sedimentation rates (Fig. 6.5A).

6.4.5 Paleocene-Eocene Extension

During the early Paleocene Eocene period, sedimentation is controlled by normal NE-SW and EW faults and consequently these synsedimentary faults developed under the influence of an NW–SE extension (Bouaziz 1995; Gabtni et al. 2006, 2012, 2013; Khomsi et al. 2016).

6.4.6 Eocene–Oligocene: Atlasic Event

It is the NW–SE Atlasic phase which was defined in the West of Algeria (Laffite 1939; Guiraud 1973; Khomsi et al. 2016), this inversion of the Mesozoic basins in the Maghreb is contemporary of the Pyrenean orogeny (Khomsi et al. 2016, 2019). A minor readjustment occurred over the Cretaceous paleo-highs with major pre-existing faults.

6.4.7 Oligocene–Miocene Extension

This NE-SW extension was produced during the Miocene-Pliocene periods, which reactivated the pre-existing NW–SE faults of the Jurassic, Lower Cretaceous and Upper Cretaceous. These normal faults played a very important role in the adjustment of the Tertiary especially toward the NE of the basin (Bouaziz et al. 2002; Khomsi et al. 2016).

6.4.8 The NNW-SSE Event (After Villafranchian Compression)

This NW–SE to NNW–SSE phase accentuated the old folds and NE-SW strike-slip faults of the Cretaceous and gave rise to the Villafranchian crusts tilting and the topographic surface anomalies (Zargouni 1985; Ben Ayed 1986, Khomsi et al. 2016) (Figs. 6.4A, 6.5A).

6.5 Synthesis and Discussion

According to the results given by the seismic analysis, the collapse of the Jeffara basin was controlled by a series of inherited normal faults controlling Mesozoic and Cenozoic sedimentation. In this region, after the Hercynian orogenesis, the subsidence began at the end of the Carboniferous (Memmi et al. 1986; Gabtni et al. 2009) with the installation, at the Permian, of a subsisting basin with marine sedimentation. Our results allow us to propose a new structural and kinematic model for the evolution of the Jeffara basin and to discuss its geodynamic significance in the evolution of subsidence.

6.5.1 Structural System of the Jeffara Basin

Our study is based on the combination of the results of the interpretation of 2D seismic data and the data of the regional paleostress prevailing in the South of Tunisia, deduced from previous work, to realize fault distribution models firstly and to constitute a distribution map of the structural system of the Jeffara from the Mesozoic to the Actual secondly. Overall, the distribution directions of the planes of most faults in the two structural systems are almost identical, and can be subdivided into three groups: the NW–SE group, the NE-SW group and the almost EW group. The fault system of the lower structural layer seemed to be distributed in the NW–SE direction (Figs. 6.4, 6.5, 6.6), these faults having the same direction as the major Jeffara fault, lose their frequency as they move east in the basin.

A second central zone shows a different and remarkable style, characterized by the abundance of NE-SW and E-W network faults distinctly. NW–SE faults appear to be frequent especially in the NE part of the basin, in the Gerba region giving rise to a paleohigh (Figs. 6.4, 6.5, 6.6).

The fault system of the upper structural layer also seemed to be distributed less frequently along the Jeffara Basin, which represented the NW–SE directions, which seems to be almost similar but simpler than that in the lower structural layer (Figs. 6.4, 6.5, 6.6).

The structural system of the Jeffara basin can be described in three aspects: (1) the direction of the fault system is mainly NW–SE, followed by the direction NE-SW, then from the direction almost E-W; (2) faults in the lower structural layer are very frequent throughout the pelvis, while those in the upper structural layer are less frequent and have little influence; (3) by applying the paleo-stresses summarized above, the upper structural system has undergone a certain rotation in a clockwise direction concerning the lower structural system (Fig. 6.6).

6.5.2 Structural Style of the Jeffara Basin

Based on the petroleum well data and 2D seismic reflection data available in the study area, we have summarized the structural styles that prevailed in the Jeffara Basin into five types below. Figure 6.7 presents descriptions of each structural style and provides interpretations of positions, backgrounds, and development formations.

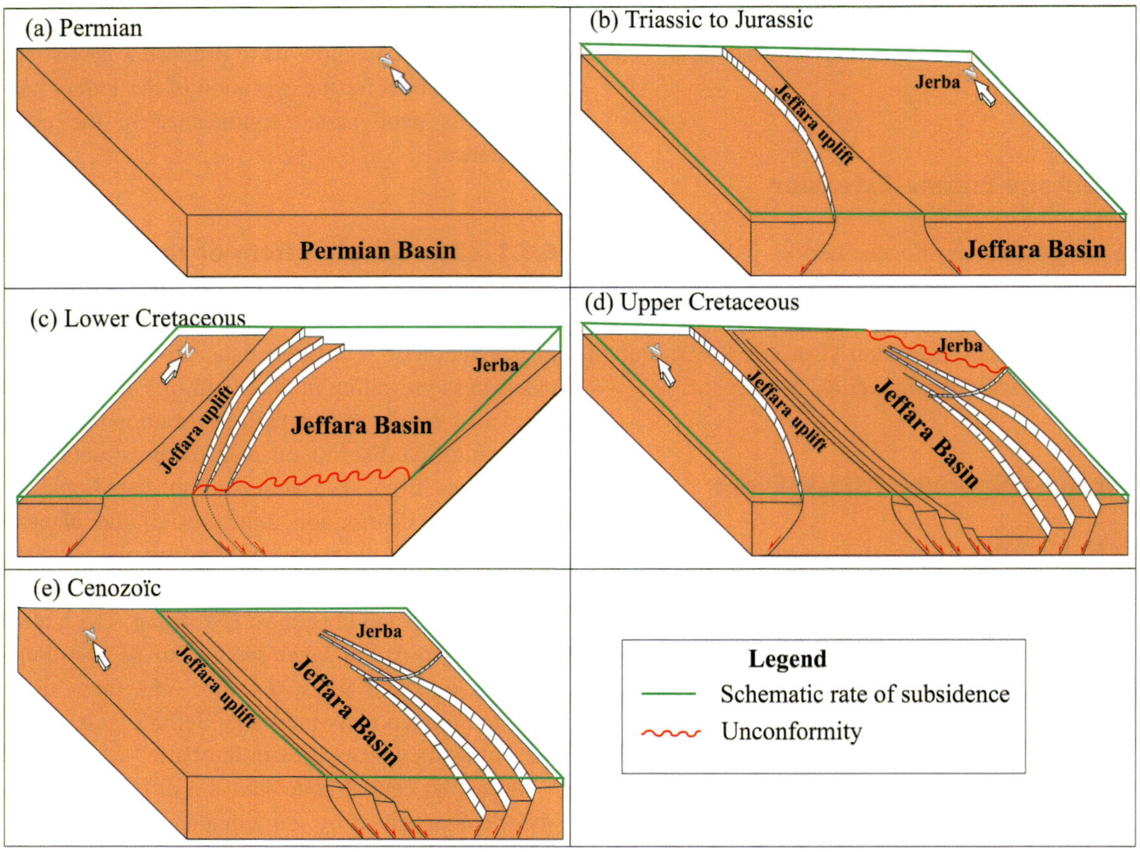

Fig. 6.6 Schematic three-dimensional model of geological history and associated paleostress results for inferred stress states in the Jeffara Basin from Permian to Cenozoic, illustrating the tectonic evolution and structural framework

Paleo-uplift: The Jeffara basin mainly developed east of the central transfer zone known by the Jeffara mole, considered as an ancient basin controlled by normal faults located mainly on the east side. The 2D seismic section (Fig. 6.4) shows that the ancient horst is a relatively low uplift in the Tébaga region of Medenine, plagued by exposure and erosion, providing deposits to surrounding areas as sub-provenance. Besides, it has deposit source supplements from the West and East slopes and continues to develop since the Triassic.

Multiple stage fault slope: During the Cretaceous period, rifting period, the Jeffara basin underwent strong subsidence controlled by the normal fault networks that were renounced in the region following an extensive NE-SW direction constraint. Analysis of the subsidence rates in the study area during this period revealed an inversion between the beginning of the Cretaceous and its end, the slope of the multistage faults is in the form of a staircase on a profile and a combination of a network of parallel faults on a plane.

During the Lower Cretaceous, the paleo-geomorphology of the multi-story fault belt is characterized by a fault terrace with several levels with a NW–SE major direction and looking toward the NE which is jointly controlled by several synthetic faults frequently active. Several main NW–SE faults have a ladder profile while faults develop at the base of each fault, where the sedimentation of the Lower Cretaceous series increases in subsidence toward the NE direction of tilting giving rise to shallow shallows "uplift" with low or no sedimentation toward the SW of the Jeffara basin.

Conversely to that of the Lower Cretaceous but with the same structural style, the Jeffara basin experienced a high increase in subsidence toward the SW in the Upper Cretaceous. This variation is controlled by the network of NW–SE normal antithetic faults looking toward the SW which are in the form of a rung on a plane tilting blocks of the basin toward the SW. This structure in turn generated paleohigh area with thin sedimentation and even without like Jerba region.

Traditional slope: After a period of intense rifting, the SE of Tunisia experienced a period of tectonic calm. During this period, the subsiding Jeffara basin is characterized by a type of gentle paleo-geomorphology and is generally associated with the belt of ancient faults with several floors, especially in the central-eastern part of the basin where a remarkable depression. Such a gentle slope is a prototype for different styles of slope paleogeomorphology, exhibiting the characteristics of wide distribution and also offering an ideal

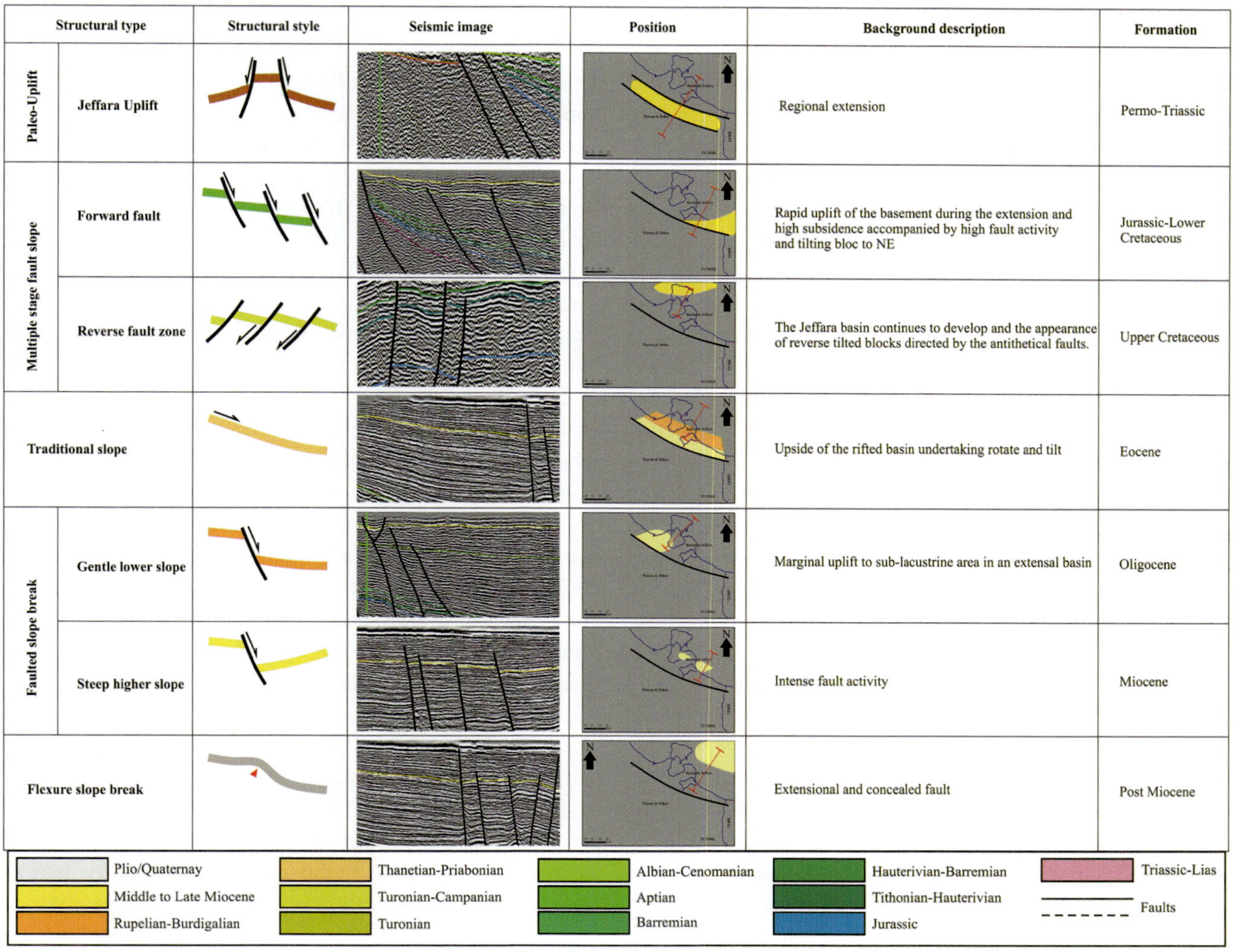

Fig. 6.7 Classification scheme for structural styles in the Jeffara basin, showing the seismic image, tectonic context and the concerned age

location for the braided deposition of the continental sedimentation has the lateral equivalent in offshore; the Tanit Fm made up of varicolored clay laterally evolving to marl with coquina limestone of Cherahil Fm.

Fault slope break: During the Oligo-Miocene period, the Jeffara basin was subjected to compressive stress from NW–SE to NNW-SSE direction generating a structural style generally formed due to the reformation of the traditional slope by reactivation of old fault networks that affected the basin. The difference in the position angle of the series occurred in parallel with the increase in the rate of displacement, which facilitates the formation of the Miocene. In the sub-depression in the eastern center of the Jeffara basin, we mainly identified two types of slope break: The first structural style is characterized by the gentle lower slope which developed mainly near the marginal basin faults and deposited a fan-shaped delta-shaped body of sand on the downstream side; The second structural style is characterized by the steep higher slope associated with a secondary fault and deposited near the shore in an underwater fan on the side of the NE Jeffara basin.

Flexure slope break: During the post-Miocene phase, the Jeffara basin is characterized by a structural style that follows architecture in inherited high and low zones. This style is the result of a break in flexion slope, generated by the draping on the paleo-crest, the paleo-faults and the buried paleo-hill, which shows an obvious thickening of the Plio-Quaternary series due to the filling by overlapping in sagging areas. Thinning and mismatch interfaces are observed on the upstream slope break strip (Figs. 6.4, 6.5). The flexural rupture belts mainly developed in the Rass Ajil zone SE of the basin, characterized by halokinetic activity and where the slope gradually increased toward the east center of the basin.

6.5.3 Discussion

The Jeffara basin presents a typical case of the passive margin showing a strong subsidence phenomenon, located at front of the Tunisian Atlas Fold and thrust belt domain and in the Eastern extension of the tectonically stable Saharan platform (Khouni et al. 2018) (Fig. 6.7).

To determine the tectono-sedimentary evolution of the Mesozoic and Cenozoic series in the Jeffara basin and the history of subsidence, we carried out the back-stripping method applied to the geo-seismic section prepared from the composite seismic lines L3. This evolution will be carried out in parallel with a model of vertical subsidence by "Novva 1-D Modeling" software from selected petroleum wells in the basin (Fig. 6.8).

During the Jurassic, sedimentation is moderately uniform along the Jeffara basin with a slight thickening toward the SE; this variation is essentially controlled by the pre-existing normal NW–SE faults. During the Lower Cretaceous period, the rate of sedimentation accelerated and the rate of tectonic subsidence increased by going from the Jeffara fault to the NE zone of the Jeffara basin without forgetting the individualization of a paleohigh in the SE part of the basin (Khouni et al. 2018).

During the upper Albian-Turonian period, the overburden is thin, noting a well-defined discrepancy between the Upper Cretaceous series and the various units of the Lower Cretaceous and Jurassic (Khouni et al. 2018). Meanwhile, the drift of Africa compared to Europe is made toward the northeast when the beginning of the propagation of the oceanic crust in the Atlantic and the increase of the ocean floor in the Maghreb basin are finished (Dewey et al. 1989). The sub-meridian extension becomes NE-SW and causes an inversion in the subsidence zones, giving rise to a paleohigh to the SE of the study area. The rate of extension is relatively reduced and consequently it is associated with the rate of sedimentation of this period (Figs. 6.8, 6.9).

The generalized section of the Tunisian margin continues in the Upper Cretaceous, the NE-SW extension encourages the mobilization of the NW–SE faults toward the inclination toward the north (Khouni et al. 2018). This extension of the Upper Cretaceous corresponds to an intracontinental rift which is accompanied by an increase in heat flow (Khomsi et al. 2004) with basic magma (Bédir et al. 1992; Bedir et al. 1995; Bédir et al. 2001; Khomsi et al. 2019). The sedimentation rate develops more in the SE part of the basin. During the Cenozoic, the sedimentation rate became increasingly important along the basin, which increased the lithostatic differential pressure as a result, due to the convergence of the African and Eurasian plates (Khouni et al. 2018).

The tectonic regime in Tunisia and the Pelagian Sea was extended to transtensive during the Paleogene. North-dip subduction along the southern margin of the Corso-Sardinian block was taken in favor of thermal subsidence. At the end of the Miocene and the beginning of the Pliocene, a tectonic extension event gave rise to faults in the Siculo-Tunisian strait, NW–SE trend (Boccaletti et al. 1987; Catalano et al. 2008, 2014; Civile et al. 2008). At the start of the Quaternary, post-Villafranchian compression affected the entire Tunisian Atlas and the Pelagian Sea along with an N-S trend (Zargouni 1985). These manifestations result in anticline structures at the top of the salt complexes and continuous and more or less rectilinear reflectors during the Miocene period (Khouni et al. 2018).

The burial-history model along the petroleum well-chosen in the Jeffara basin reflects practically the same subsidence characteristics deduced from the "Backstreeping" method. This burial-history model is characterized, especially by three types of evolution, low rate of burial, high rate of burial and periods of minimal burial at null. During the Triassic and the Jurassic period, the burial curve shows a moderate progressive and stable rate. Between the end of Jurassic and the Aptian, the steep decline of the curve represents the rapid increase in subsidence rate and consequently sedimentation rate. This period is known as the tectonic subsidence period, specifically describes that the rate of subsidence is mainly controlled by tectonic activities.

This phenomenon is practically observed on the model during the Upper Cretaceous and Cenozoic periods with more or less identical burial rates interrupted by periods where the rate is very low and even null: the Aptian-Albian interval, the Upper Cretaceous passage Tertiary and the Eocene–Oligocene passage representing the major uplifts and erosional events.

6.6 Petroleum Interest

6.6.1 Jeffara Basin: Structural Architecture and Stratigraphic Evolution

The Jeffara Basin was formed during the Triassic NNE-SSW rifting which affected a large part of southern Tunisia. This rifting was accompanied by volcanic activity especially toward the West, in the Chotts Basin. This period was characterized by pronounced subsidence. The lower Jurassic-Middle Jurassic interval corresponds to a slow and progressive period of subsidence with the establishment of normal faults oriented NW–SE and in line with the regional trend.

The Upper Jurassic period—Lower Cretaceous is marked by a significant extension phase; this phase was expressed by parallel NW–SE faults and the individualization of the Djerba furrow: a half-graben collapsed toward the SW and

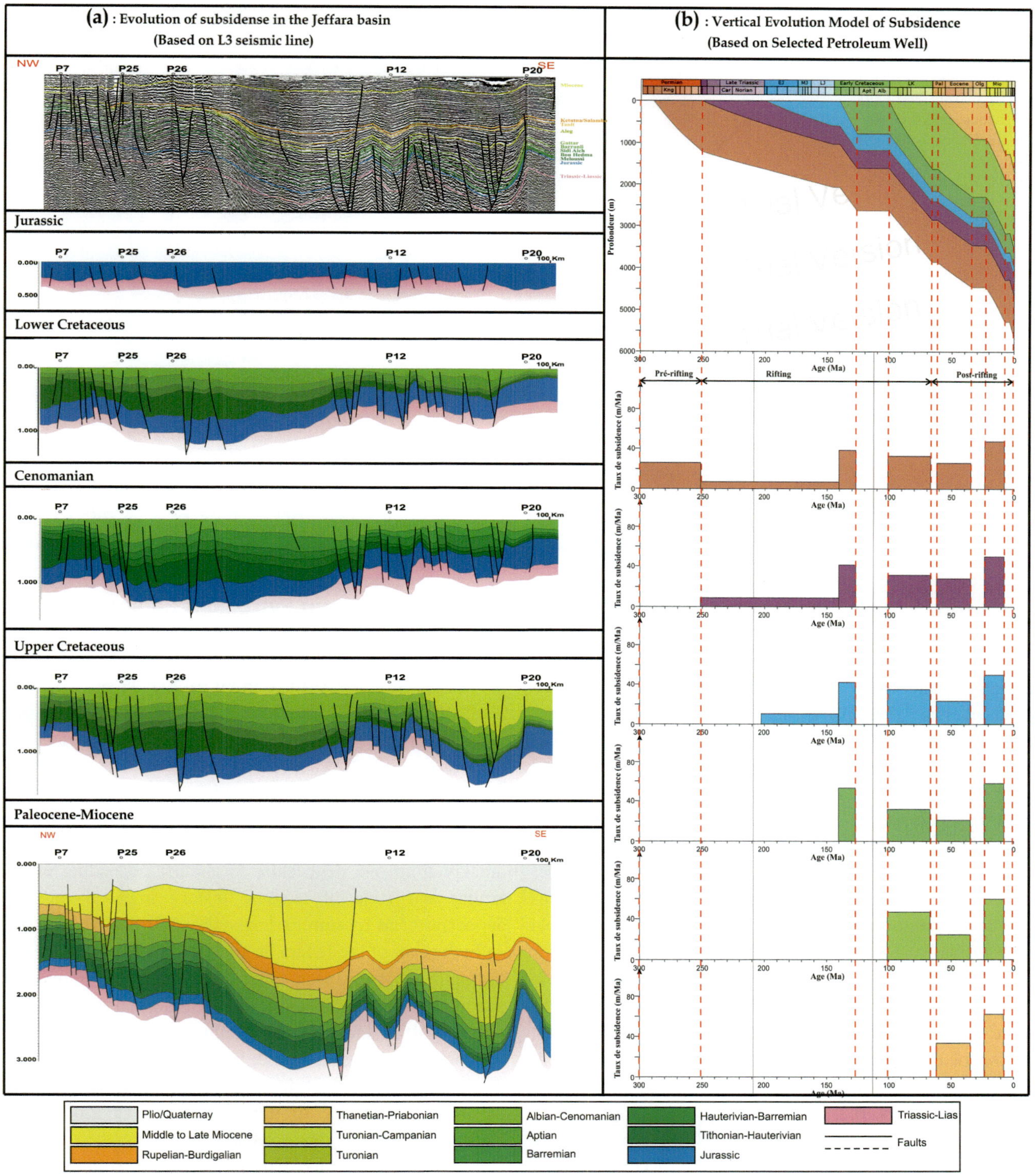

Fig. 6.8 **a** Model showing the evolution of the subsidence in the Jeffara basin from the Jurassic to the Tertiaire based on the back-stripping method applied to the L3 NW–SE composite seismic line (Khouni et al. 2018). **b** model of Vertical burial evolution since the Permian to the Actual by "Novva 1-D Modeling" software applied to a selected petroleum well in the basin

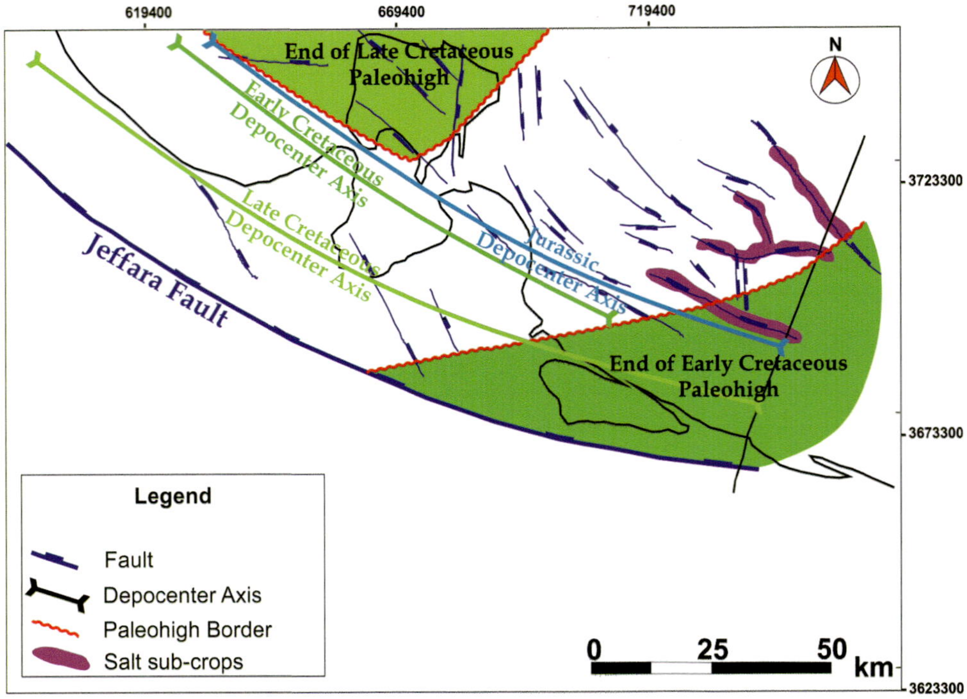

Fig. 6.9 Map showing the location of the paleohighs of two Cretaceous times and their borders, the depocenter axes of Jurassic, Lower and Upper Cretaceous, and the salt complexes positions in Jeffara basin

characterized by a high subsidence rate and predominantly silico-clastic sedimentation.

During the Aptian period, local emergences to the south of the study area are reported, which are the first indications of an inversion that will become more specific in the Santonian period (Bédir et al. 2001; Khouni et al. 2018). This extension phase was followed by the change from a detrital sedimentation regime to a predominantly carbonate regime (Figs. 6.2, 6.7, 6.8).

The Santonian saw the development of a compressive phase which was manifested by the establishment of several anticlines oriented NE-SW, with remobilization of the old faults. This compression introduced a component "strike slip" into the old faults (Touati and Rodger 1998; Khouni et al. 2018).

At the end of the Upper Cretaceous, the Jeffara basin recorded an uplift of anticlines formed during the Santonian phase accompanied by phases of more or less significant erosion of the Upper Cretaceous series marking an accentuation toward the northern part of the study area in the Gerba region (Figs. 6.2, 6.7, 6.8).

During the Miocene-Pliocene period, there was first an extension with the reactivation of old faults, accompanied by remarkable subsidence throughout the study area.

Besides, another tectonic regime is observed: halokinesis with the formation of salt walls, salt movements through the NW–SE faults in the SE part of the study area. This regime originated during the Jurassic (Khouni et al. 2018).

The last tectonic event is compressive and of post-Villafranchian age. It accentuated the Cretaceous folds and is responsible for the current geomorphology of the region.

6.6.2 Jeffara Basin Petroleum Geology

The tectonic and stratigraphic evolution described in the previous section has resulted in oil geology characterized by the development of numerous elements of hydrocarbon play in a variety of depocentres along the Jeffra basin. The combination of a functional source kitchen, tank-cover pairs and valid trapping geometries are key elements for a successful petroleum system. In a basin with a passive margin, the presence and effectiveness of these elements can vary quickly from one depocentre to another depending on the timing and style of tectonic history imposed on each depocentre and its surroundings (Fig. 6.9).

In the case of the Jeffara basin, three large depocentres are distinguished from each other following the evolution of tectonic activity dominated by NW–SE normal faults that the Jeffara fault ensures the major influence.

During the Jurassic Period, the Jeffara basin recorded the development of the first depocentre following the activity of the Jeffara fault due to the significant extension phase. During the lower Cretaceous and following the extension phase which persists, the depocentre changed slightly toward the West due to the Aptian inversion. During the Upper Cretaceous, there is a remarkable migration from the depocentre to the West, probably characteristic of the compressive phase of the Santonian.

This enlarged stratigraphy provided enough sediment overloads to ripen a hydrocarbon source cuisine, especially with the remarkable subsidence of the Miocene-Pliocene period.

The inventory of the various proven and potential reservoir rocks, the cover rocks and the possible source rock that supplied them and in the light of the various discoveries that have been made in the SE of Tunisia enabled us to identify the degree of prospectivity of exploration in the study area and consequently in the adjacent areas of the entire pelagic basin. This prospectivity is a function of the different paleogeographic conditions that influenced the distribution of the facies (Fig. 6.2).

Petroleum generation is from several source rocks, certainly, the existence of good quality source rocks is obvious and proven by the various deposits and discoveries made in the study area. The major Mesozoic source rocks in the study area are the clays of the Mestaoua formation of lower Jurassic age, the Callovian clays, and the most cited clays and carbonates from the base of the Aptian Age Orbata Formation (Mejri et al. 2006).

The known major Mesozoic reservoir rocks in the southern Tunisia Petroleum System begin from fluvial sandstone of the Triassic namely Ouled Chebbi Formation and Ras Hamia Formation in western part. On the other hand in the eastern part, the Jurassic can present two types of proven reservoir facies relating to the carbonate units M'Rabtine and Smida, containing massive and lenticular sandy-quartzitic intervals and fractured carbonate facies linked to the Krachoua and Tlalelett units. The potential Jurassic reservoirs in the study area are essentially summarized by the dolomitic member Tlalett, the sandstone levels of the Smida Member, and the carbonates of the Upper Nara Formation (Mejri et al. 2006).

The proven reservoir levels and main objectives of the Cretaceous are essentially summarized by the carbonates of the Zebbag formation, the Wealdian facies of the Meloussi formation and the sandstones of the Sidi Aich formation. The potential Cretaceous reservoirs are essentially summarized by the sandstone and carbonate levels of the Orbata formation on the one hand, and the sandstone and dolomite sequences of the Bouhedma formation on the other (Mejri et al. 2006).

6.6.3 Exploration History

Hydrocarbon exploration in the Jeffara basin and its surroundings experienced several phases of activity; it was particularly slow before the 1970s with some scattered wells and concentrated mainly on the Eocene limestones and the Upper Cretaceous where the structural trapping geometries could be seen from 2D seismic profiles. This exploration phase resulted in several small oil and gas discoveries, which are still in production until today.

Subsequently, these exploration activities mainly focused on the Lower Cretaceous objectives in addition to those confirmed, to test the potential indices encountered previously. Since the 1980s, exploration activity has undergone a remarkable change, targeting the Triassic sandstones as an objective.

At the end of the 1990s, exploration was slowed down due to the limited size of the traps, the lack of predictability of the quality of the reservoir and the probable absence of a good quality source rock capable of supplying the structures released by the seismic data available at that time. However, the exploration activity encountered several failures, which led to several petrophysical and geochemical studies which subsequently suggested firstly the presence of several potential reservoirs mentioned above and secondly the presence of source rock which may be kitchens in the Jeffara basin (Fig. 6.2).

Exploration of the number of Jurassic, Lower Cretaceous, Upper Cretaceous and even Eocene depocentres along and around the Jeffara Basin has been limited until recent years due to the absence of simple structural closures characteristic of success.

Despite the application of geophysical risk reduction techniques generally considered more robust in offshore, exploration in the Jeffara basin was generally limited to simple structural closures, thus neglecting other types of trapping that could be potential such as stratigraphic traps and rym-synclines on both sides of the salt walls. These exploration techniques are still considered to be higher risk traps, especially with the burial of objective reservoirs.

6.6.4 Evidence of Stratigraphic Traps and Potentiality

Bounded from the West by the great Jeffara fault, the Jeffara basin is made up of a sequence of blocks tilted toward the NE, delimited by faults with distensive play with an NW–SE regional trend. During the Lower Cretaceous period, the tectonic aspect known during the Jurassic continues, recording an amplification of the activity of NW–SE normal faults. The main Jaffera fault supports the Jeffara plain to the west while it plunges the sequences of the Lower Cretaceous toward the NE in deepwater of the basin where they are covered by sequences of the Upper Cretaceous, of which several sequences drape on this downlap slope and are then covered by a sequence of Maastrichtian sealing schists (Fig. 6.4). Downlap is commonly seen at the base of prograding clinoforms, and usually represents the progradation of a basin-margin slope system into deep water. Therefore, it represents a change from marine or lacustrine slope deposition to marine or lacustrine condensation or non-deposition (Roberts and Bally 2012).

The vertical evolution of subsidence reveals that the Jurassic source rocks underwent significant subsidence since the Upper Jurassic and especially during the Lower Cretaceous. Landfill is interrupted in the Upper Cretaceous by the late Cretaceous tectonic event which caused the erosion of part of the Upper Cretaceous series (Fig. 6.9). Locally in the Jerba region, the entire Upper Cretaceous is eroded. The thermal evolution resumes with the Oligocene or Miocene deposits. Likewise, the generation which probably started from the Santonian (85 MA) underwent a slowdown during the rest of the Upper Cretaceous before resuming with the Tertiary subsidence (Figs. 6.8, 6.10).

The embedding toward the NE of the Jurassic, Lower Cretaceous and upper series led to depocentres with NW–SE major directions, as shown by the isochron maps of the interpreted horizons. These depocentres can necessarily be storage centers and consequently they can be considered as centers of regional migration for the hydrocarbons generated; however, no previous exploration had tested this hypothesis in the Jeffara basin (Fig. 6.9).

Taking into consideration that the most common hydrocarbon traps are structural (anticlines, faults, salt dome) and stratigraphic traps (pinch-out, lens, and unconformity traps, the assembly of elements of the petroleum system can be targeted as a place likely to have a high probability of producing a trapping geometry in the Jeffara basin. This trapping geometry is spread out all along the slope of the Jeffara fault (Fig. 6.10A–E).

To the SE of the Jeffara basin, the interpretation of the seismic lines available showed the individualization of three salt walls in a major direction parallel to the regional trend. The analysis of the three salt complexes shows the unconformities and the rym-syncline forms along with the contacts of these salt bodies with the overlying series, which indicates a recording of the salt activity at the Mesozoic and the Cenozoic sedimentation (Khouni et al. 2018). After a phase of reactive diapirism, the rise of the diapir becomes more vigorous, the Lower Cretaceous strata and the Zebbag Fm are lifted, rotated, shouldered and eroded when the diapir crosses them by force through active diapirism. In this context, the upper Cretaceous is characterized by local depressions formed and developed around the salt walls so-called rim-synclines (Khouni et al. 2018).

These rim-syncline structures spread over twenty kilometers for each salt complex, can be geometries for hydrocarbon traps, and reinforce the potential of the Jeffara basin by three new zones (Fig. 6.10F).

The demonstration of the existence of the ingredients of petroleum system associated with structural and stratigraphic traps combined in the Mesozoic sequences of a passive margin such as the Jeffara basin shows the high probability of the existence of a considerable hydrocarbons potential in this basin and pushes for a new spirit of exploration likely to lead to additional discoveries in the future.

6.7 Conclusion

In this study, using available 2D seismic reflection and petroleum wells data, we analyzed the structural systems and styles that characterize the Jeffara passive margin basin, its tectonosedimentary evolution model and their roles in petroleum exploration. The result of this research shows:

Analysis of 2D seismic and well data allowed us to build the structural system that characterizes the study area. This structural system can be summarized in three distinct types: A first system reigned from the Triassic rifting until the end of the Lower Cretaceous, characterized by the abundance of NW–SE normal faults and favoring an average rate of subsidence which is accelerated toward the end. A second system reigned during the Upper Cretaceous where sedimentation is controlled mainly by NW–SE antithetical faults looking toward the SW. The third structural system reigned during the Cenozoic period where tectonics appears to have little influence on sedimentation.

During the Mesozoic and Cenozoic, the Jeffara basin was exposed to a polyphasic evolution which gave rise to a structuring essentially in horsts with stratigraphic gaps and grabens, the analysis of this evolution based on the seismic profiles, allowed us to identify five types of structures: ancient horst, traditional slope, rupture of slope in bending, rupture of slope in fault and slope of fault in several stages.

The analysis of the combination between the evolution of subsidence from the "Backstreeping" method and the historical model of vertical burial allowed us to deduce three types of evolution, the first with a rate of progressive and stable burial characteristic of the Triassic and Jurassic period, a second with an accelerated burial rate characteristic of the Lower Cretaceous, Upper Cretaceous and Cenozoic periods, variable from one period to another and a third with a minimal burial rate and even null during the Aptian-Albian interval, the Upper Cretaceous-Tertiary passage and the Eocene–Oligocene passage representing the main paleohigh and erosion events.

To guide petroleum research and exploration in the Jeffara basin, the analysis of the tectonosedimentary evolution and the burial history of the Mesozoic and Cenozoic series enabled us: Firstly to confirm the existence of the functional petroleum system which includes all the essential interdependent elements and the processes forming the functional

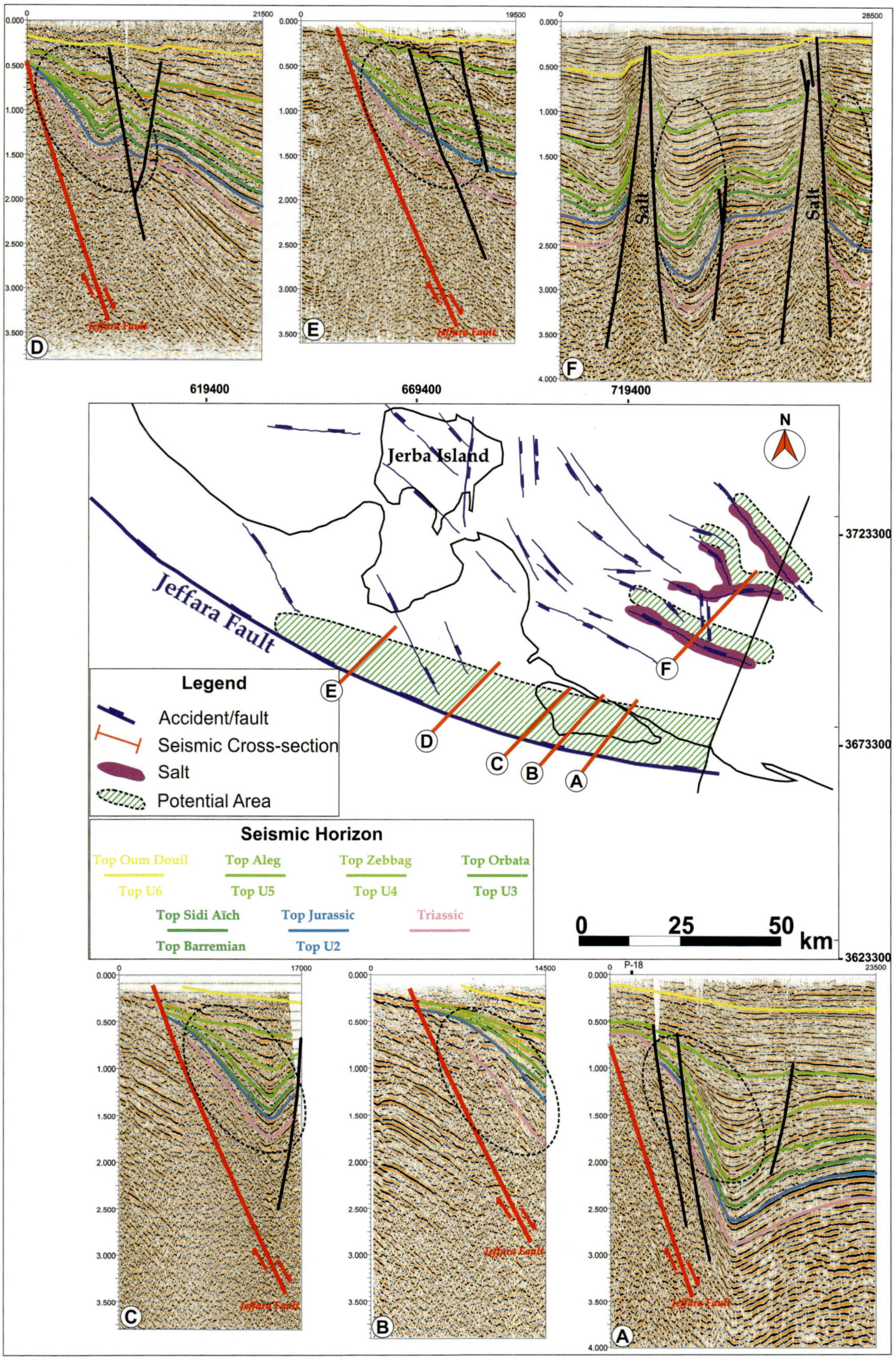

Fig. 6.10 Map illustrating the locations of potential stratigraphic traps for hydrocarbons and the position of seismic line extracts; **A–E** Seismic line extracts showing strata terminations and their geometries related to the Jeffara fault activity. **F** Seismic line extract showing strata terminations and their geometries related to salt activities

unit, secondly, the evidence of the existence of the ingredients of petroleum system associated to stratigraphic traps, showing the high probability of the existence of considerable hydrocarbon potential and pushing for a new spirit of exploration likely to lead to discoveries in the future.

Acknowledgements We thank the Tunisian Company of Petroleum Activities (ETAP) and particularly H. BEN KILANI and F. MEJRI, for providing and allowing the publication of the seismic and well data used in this study, obtained within the framework of the cooperation between the Faculty of Sciences of Tunis (FST) and The ETAP during the project of my masters. We are sincerely grateful to Dr. S. KHOMSI for his editorial assistance and the anonymous reviewer for providing constructive reviews that greatly improved this paper.

References

Aubouin J, Debelmas J (1980) L'Europe alpine : les chaines périméditerranéennes. Introd Mem B.R.G.M Paris, No 115:62–66

Bédir M, Zargouni F, Tlig S, Bobier B (1992) Subsurface geodynamics and petroleum geology of transform margin basins in the Sahel of Mahdia and El Jem (Eastern Tunisia). AAPG Bull 76(9):1417–1442. https://doi.org/10.1306/BDFF8A10-1718-11D7-8645000102C1865D

Bedir M (1995) Mécanismes géodynamiques des bassins associés aux couloirs de coulissement de la marge atlasique de la Tunisie. Seismo-stratigraphie, seismo-tectonique et implications pétrolières. Thése Doc. Etat, Tunis II, Tunis, Tunisia, 416 pp. https://hal.archives-ouvertes.fr/tel-01862051

Bédir M, Tlig S, Bobier B, Aissaoui N (1996) Sequence stratigraphy, basin dynamics, and petroleum geology of the Miocene from Eastern Tunisia. AAPG Bull 80(1):63–80. https://doi.org/10.1306/64ED8746-1724-11D7-8645000102C1865D

Bédir M, Boukadi N, Tlig S, Ben TF, Zitouni L, Alouani R, Slimane F, Bobier B, Zargouni F (2001) Subsurface Mesozoic basins in the central Atlas of Tunisia: tectonics, sequence deposit distribution, and hydrocarbon potential. AAPG Bull 85(5):885–907. https://doi.org/10.1306/8626CA2D-173B-11D7-8645000102C1865D

Ben Alaya M, Khouni R, Zemni T, Khemiri S, Zargouni F (2016) A multicriteria approach to study genesis of salted groundwater. Northern Jeffara (South Eastern Tunisia). Larhyss J 103–119. ISSN 1112-3680

Ben AN (1986) Evolution tectonique de l'avant pays de la Chaîne alpine de Tunisie du début du Mésozoïque à l'Actuel. Thèse Doc. Etat, Univ. Paris Sud, France, 328 p

Ben Ferjani A, Burollet PF, Mejri F (1990) Petroleum geology of Tunisia. Mem ETAP, vol 1. Tunis, ETAP, 194 p

Ben Ismail MH (1989) Les bassins mésozoïques (Trias à Aptien) du Sud de la Tunisie : Stratigraphie, caractéristiques géophysiques et évolution géodynamique.- Thèse de Doct. es Sci. Géologiques, Univ. Tunis II

Ben youssef M (1999) Stratigraphie génétique du Crétacé de Tunisie Micropaléontologie, stratigraphie séquentielle et géodynamique des bassins de la marge Sud et péri-téthysienne. Thèse d'état Univ. Tunis II. 402p

Bishop WF (1975) Geology of Tunisia and adjacent parts of Algeria and Libya. Assoc Pet Geol Bull 59(3):413–450

Boccaletti M, Cello G, Tortorici L (1987) Transtensional tectonics in the Sicily Channel. J Struct Geol 9:869–876. https://doi.org/10.1016/0191-8141(87)90087-3

Bouaziz S (1995) Etude de la tectonique cassante dans la plateforme et l'Atlas sahariens (Tunisie méridionale): evolution des paléochamps de contraintes et implications géodynamiques. Thèse de Doctorat.d' Etat, Univ. Tunis II, 486 p

Bouaziz S, Barrier E, Turki MM, Tricart M, Angelier J (2002) Tectonic evolution of the Southern Tethyan margin in Southern Tunisia. Tectonophysics 357:227–253

Burollet PF (1956) Contribution à l'étude stratigraphique de la Tunisie centrale. Ann Min et Géol 18:352

Burollet PF, Desforges G (1982) Dynamique des bassins néoCrétacés en Tunisie Livre Jub. Gabriel Lucas. Mem Geol Univ Dijon 381–389

Busson G (1967) Le Mésozoique saharien. 1ère partie: l'Extrême -Sud Tunisien, Publ. C.R.Z.A. Géologie 8, (CNRS), Paris, 194 p

Busson G (1970) Le Mésozoique saharien: 2ème partie. Essai de synthèse des données des sondages algéro-tunisiens. Publ. C.R.Z.A., Géologie 11, Paris (CNRS), 811 p

Carpentier C, Hadouth S, Bouaziz S, Lathuilière B, Rubino J-L (2016) Basin geodynamics and sequence stratigraphy of Upper Triassic to Lower Jurassic deposits of central and southern Tunisia. J Afr Earth Sci. https://doi.org/10.1016/j.jafrearsci.2016.01.029

Catalano S, De Guidi G, Romagnoli G, Torrisi S, Tortorici G, Tortorici L (2008) The migration of plate boundaries in SE Sicily: influence on the large-scale kinematic model of the African promontory in southern Italy. Tectonophysics 449:41–62. https://doi.org/10.1016/j.tecto.2007.12.003

Catalano R, D'Argenio B, Valenti V, Basilone L, Gasparo Morticelli M, Albanese C (2014) Neo-Tethys (or Palaeotethys arm?) Permian-Mesozoic carbonates in the Pelagian continental margin (Central Mediterranean), AAPG Europe Region, Mesozoic and Cenozoic carbonates of the Neo-Tethys: old and new concepts for petroleum exploration, pp 25–26

Civile D, Lodolo E, Tortorici L, Lanzafame G, Brancolini G (2008) Relationships between magmatism and tectonics in a continental rift: the Pantelleria Island region (Sicily Channel, Italy). Mar Geol 251:32–46. https://doi.org/10.1016/j.margeo.2008.01.009

Dewey JF, Helman ML, Turco E, Hutton DHW (1989) Kinematics of the western Mediterranean. In: Coward MP, Dr Etrich D, Park RG (eds) Alpine tectonics, geological society, London, Special Publication, vol 45, pp 265–283. https://doi.org/10.1144/GSL.SP.1989.045.01.15

El Euchi H, Fourati L, Hamouda F, Saidi M (1998) Structural style and hydrocarbon habitat in Northern Tunisia. Field Trip Guide Book. ETAP Mem., vol 13. Tunis, Tunisia, 72 pp

Frizon de Lamotte D, Michard A, Saddiqi S (2006) Quelques développements récents sur la géodynamique du Maghreb. C R Geosci 338:1–10

Frizon de Lamotte D, Leturmy P, Missenard Y, Khomsi S, Ruiz G, Saddiqi O, Guillocheau F, Michard A (2009) Mesozoic and Cenozoic vertical movements in the Atlas system (Algeria, Morocco, Tunisia): an overview. Tectonphysics 475:9–28. https://doi.org/10.1016/j.tecto.2008.10.024

Gabtni H, Jallouli C, Mickus KL, Zouari H, Turki MM (2005) Geophysical constraints on the location and nature of the North Saharan flexure in southern Tunisia. Pure Appl Geophys 162:2051–2069. https://doi.org/10.1007/s00024-005-2704-9

Gabtni H, Jallouli C, Mickus KL, Zouari H, Turki MM (2006) The location and nature of the Telemzan High-Ghadames basin boundary in southern Tunisia based on gravity and magnetic anomalies. J Afr Earth Sc 44(3):303–313. https://doi.org/10.1016/j.jafrearsci.2005.11.021

Gabtni H, Jallouli C, Mickus KL, Zouari H, Turki MM (2009) Deep structure and crustal configuration of the Jeffara basin (Southern Tunisia) based on regional gravity, seismic reflection and borehole data: how to explain a gravity maximum within a large sedimentary basin? J Geodyn 142–152. https://doi.org/10.1016/j.jog.2008.07.004

Gabtni H, Jallouli C, Mickus KL, Dhaoui M, Turki MM, Jaffal M, Keating P (2012) Basement structure of southern Tunisia as determined from the analysis of gravity data: implications for petroleum exploration. Pet Geosci 18(2):143–152. https://doi.org/10.1144/1354-079311-050

Gabtni H, Jallouli C, Mickus KL, Turki MM (2013) Geodynamics of the Southern Tethyan Margin in Tunisia and Maghrebian domain: new constraints from integrated geophysical study. Arab J Geosci 6:271–286. https://doi.org/10.1007/s12517-011-0362-z

Ghanmi M, Potfaj M, Ben Youssef M, Zargouni F, Louhichi L, Mamou A (1993) Notice de la carte géologique au 1/100.000°de Oglet Merteba. ONM, Tunis, feuille n°82, 44p.

Guiraud R (1973) Evolution post-triasique de l'avant-pays de la chaîne alpine en Algeérie d'aprés l'étude du bassin du Hodna et des régions voisines. Thése Doc. Etat, Univ. Nice, France, 265 pp

Haq BU, Hardenbol J, Vail PR (1987) Chronology of fluctuation sea levels since the Triassic. Science, Washinghton, vol 235, pp1156–1166. https://doi.org/10.1126/science.235.4793.1156

Khomsi S, Bédir M, Ben Jemi MG, Zouari H (2004) Discovery of a new thrust front in the oriental Atlas of Tunisia using seismic reflection data. Structural context and role of salt intrusions. C R Geosci 336(15):1401–1408. https://doi.org/10.1016/j.crte.2004.06.007

Khomsi S, Ben Jemia MG, De Lamotte FD, Maherssi C, Echihi O, Mezni R (2009) An overview of the Late Cretaceous–Eocene positive inversions and Oligo-Miocene subsidence events in the foreland of the Tunisian Atlas: Structural style and implications for the tectonic agenda of the Maghrebian Atlas system, vol 475, Issue 1, pp 38–58

Khomsi S, de Lamotte DF, Bédir M et al (2016) The Late Eocene and Late Miocene fronts of the Atlas Belt in eastern Maghreb: integration in the geodynamic evolution of the Mediterranean Domain. Arab J Geosci 9:650. https://doi.org/10.1007/s12517-016-2609-1

Khomsi S, Roure F, Khelil M, Mezni R, Echihi O (2019) A review of the crustal architecture and related pre-salt oil/gas objectives of the eastern Maghreb Atlas and Tell: Need for deep seismic reflection profiling. Tectonophysics 766(5):232–248. https://doi.org/10.1016/j.tecto.2019.06.009

Khouni R, Arfaoui MS, Ben Kilani H, Zargouni F (2010) Mise en place et evolution polyphasique des corps halocinetiques dans la Jeffara au Sud de latunisie du Jurassique jusqu'a l'Actuel. In: Conference paper, the organizing committee of the fifth international conference on the geology of the Tethys Realm At: South Valley University, Quena - Luxor, Egypt

Khouni R, Arfaoui MS, Zargouni F (2017) Évolution polyphasique de la Jeffara de la Tunisie, influence de l'halocinèse sur la structuration d'une marge passive au Mésozoïque. In: Conference paper, the 1st Atlas georesources international congress, Hammamet - Tunisia, vol 1

Khouni R, Arfaoui MS, Dridi S, Zargouni F (2018) Polyphasic evolution of the Jeffara basin in southern Tunisia, influence of halokinesis on the passive margin structuration in the Mesozoic and the Cenozoic. Arab J Geosci 11(4). https://doi.org/10.1007/s12517-017-3363-8

Laffite R (1939) Les plissements postnummilitiques dans l'Atlas saharien. Bull Soc Geol Fr 5 (IX):135–159

Mejri F, Burollet PF, Ben Ferjani A (2006) Petroleum geology of Tunisia: a renewed synthesis, memoir, vol 22. Enterprise Tunisienne d'Activités Pétrol., Tunis

Memmi L, Burollet PF, Viterbo I (1986) Lexique stratigraphique de la Tunisie, première partie: précambrien et paléozoïque. Notes du Service Géologique (Tunisie), No. 53, 64 p.

Peybernès B, Almeras Y, Ben YM, Mello J, Rey J, et Zargouni F (1985) Nouveau éléments de datation dans le Jurassique du Sud tunisien (Plateforme saharienne). C. R. Acad. C R Acad Sci Paris 300:113–118

Roberts DG, Bally AW (2012) Regional geology and tectonics: principles of geologic analysis. Elsevier, 900 p.

Touati MA, Rodgers MR (1998) Tectono-stratigraphic history of the Southern Gulf of Gabes and the hydrocarbon habitats. In: Proceedings of the 6 Tunisian petroleum exploration production conference, Tunis, Mai, 1998, pp 343–370

Touhami F, El Chahtour K, Khouni R, Ghanmi M, Fakhfekh H, Zargouni F (2008) Apport de la diagraphie a l'étude séquentielle des séries du Crétacé inférieur en Jeffara maritime et implication structurale. In: Conference paper, 22 colloquium of African geology At: Hammamet, Tunisia

Zargouni F (1985) Tectonique de l'Atlas méridional de Tunisie, évolution géométrique et cinématique des structures en zones de cisaillement.- Thèse d'Etat, Univ. L. Pasteur, Strasbourg

7 Ordovician–Upper Silurian–Triassic Petroleum System Assessment in the Chotts Area

S. Kraouia, A. Ben Salem, M. Saidi, K. El Asmi, and A. Mabrouk El Asmi

Abstract

The Upper Silurian Fegaguira Formation is thought to be an active source rock in the Chotts Basin of southern Tunisia and is thought to have probably supplied Ordovician and Triassic clastic reservoirs in the area. However, various debates are still held regarding either the source rock distribution and thermal maturity but also the reservoirs viability and extension. Within this scope, this study focussed on the evaluation and the characterization of Ordovician and Triassic reservoirs through the integration of well logging data of twenty wells drilled in the southern Chotts Basin aiming to better delineate prolific levels. Furthermore, a 1D BasinMod modelling was achieved to reconstruct the Fegauira source rock burial and thermal histories and also to estimate its hydrocarbon generation and expulsion potential. The El Atchane and Hamra Ordovician reservoirs generally bear low to fair petrophysical characteristics and mostly fall within nearly tight reservoirs. Paleozoic orogenic phases, especially the Hercynian phase had a major impact on their lateral distribution. The Triassic TAGI (Trias argilo-grèseux inférieur) reservoir, with good porosity, changes to volcanic material to the West which is believed to have been settled through faulting during the Tethyan rifting. Strikingly, this volcanic material bears also good porosities which could have been enhanced through fracturing and diagenetic processes. The 1D Basin modelling shows that the Fegaguira source rock is mature in real wells and has begun hydrocarbon generation during Early Cretaceous. Hydrocarbon expulsion at a SATEX of 10% took place since Paleogene, the earliest. This initial evaluation of Ordovician and Triassic (TAGI) reservoirs combined to Fegaguira burial and thermal history modelling point to a functioning petroleum system in the Chotts area with more targets to be discovered further West. This study is an anticipation for possible more plays to be discovered in an area which needs to be further thoroughly explored for petroleum accumulations.

Keywords

Chotts Basin • Petrophysical evaluation • Ordovician reservoirs • Triassic reservoir • Upper Silurian Fegaguira source rock • 1D modelling

7.1 Introduction

Proven hydrocarbon reserves in Tunisia are estimated of around 425 million barrels, half of which are located in the South (ETAP 2018) within oil fields such as El Borma, Bir Ben Tartar and Nawara. For that, southern Tunisia is seeing increased exploration interest in the hope of finding more promising targets. This is more encouraging considering the relatively tectonic stable character of the area especially further South where the Saharan platform occurs (Fig. 7.1).

Southern Tunisia comprises three main basins (Fig. 7.1): the Ghadames Basin, the Jeffara Basin and the southern Chotts Basin (e.g., Rezouga et al. 2012; Aissaoui et al. 2016; El Rabia et al. 2018; Khomsi et al. 2019).

The Jeffara Basin is geologically separated from the Ghadames Basin by the Jeffara flexure zone (Burollet and Desforges 1982; Ben Ayed 1986; Bouaziz et al. 2002;

S. Kraouia (✉) · A. Mabrouk El Asmi
Laboratory of Sedimentary Basins and Petroleum Geology (BSGP), Faculty of Sciences of Tunis, El Manar University, 2092 Tunis, Tunisia
e-mail: safa.kraouia@fst.utm.tn

A. Mabrouk El Asmi
e-mail: amina.mabrouk@fst.utm.tn

A. Ben Salem · M. Saidi
Entreprise Tunisienne d'activités Pétrolière, 2035 Charguia II, Tunisia

K. El Asmi
Faculty of Sciences of Bizerte, Carthage University, Zarzouna, Tunisia

S. Khomsi and F. Roure (eds.), *Geology of North Africa and the Mediterranean: Sedimentary Basins and Georesources*, Regional Geology Reviews, https://doi.org/10.1007/978-3-031-18747-6_7

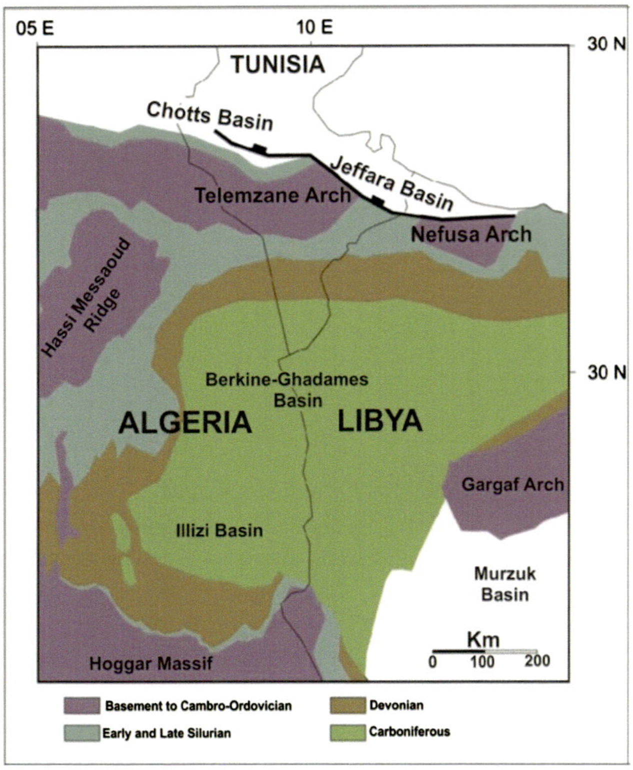

Fig. 7.1 Localization of the three basins in southern Tunisia: Chotts Basin, Jeffara Basin, and Ghadames Basin (Galeazzi et al. 2010)

Gabtni et al. 2009; Soua 2012). The Chotts Basin is separated from the Ghadames Basin by a lower Paleozoic uplift termed the Telemzane Arch (Cunningham 1989; Echikh 1998; Gabtni et al. 2006; Soua 2012). This Arch was affected by mild and periodic uplift, first during the Middle-Late Ordovician and then during the Hercynian orogeny (Ghénima 1993).

The Chotts Basin constitutes a prolific hydrocarbon province for oil and gas production (Belhaj Mohamed et al. 2015) and it is currently considered a very important petroleum objective for the Tunisian economy. Several wells have already been proven productive in multiple fields of the region (e.g. Sabria, Franig and Baguel) from two promising reservoir levels which are the Ordovician Hamra reservoir (Fig. 7.2) and the Triassic TAGI reservoir (Trias Argilo-Grèseux Inférieur). Still, the Ordovician El Atchane reservoir is also considered as an objective, even though it hasn't been proven oil bearing yet.

In the Chotts area, the facies association of El Atchane Formation (Figs. 7.2b and 7.3a) reflects the development of several prograding parasequences that characterize the intertidal to subtidal zone (internal platform). It is composed of two different facies (Fig. 7.3a): (i) The first facies is formed by mixed sands that are characterized by a low sand/clay ratio with frequent sandstone layers, showing mud drapes and small-scale cross-laminate structures, due to current ripples migrations; (ii) the second facies is a silty interval that is moderately bioturbated (Troudi 2006). The Hamra reservoir (Figs. 7.2b and 7.3a) is separated from the sandstone of the El Atchane reservoir by a regional erosive surface. It is formed by homogeneous facies dominated by clean quartz sands which are generally cemented by quartzites (Troudi 2006).

The Triassic reservoir (TAGI: Trias Argilo-Grèseux Inférieur) represents a good reservoir rock in southern Tunisia, especially in the Baguel region. All productive and potential wells in the region produce from this interval. Usually, overlying the Hercynian unconformity, the TAGI deposits were strongly influenced by eustatic movements (Echikh 1998). Massive bodies of sandstone (Fig. 7.3b), with their tabular geometry, were deposited during periods of low marine sea level. During high sea levels, clays, shale and vases have settled down in lagoons and in flood plains (Echikh 1998).

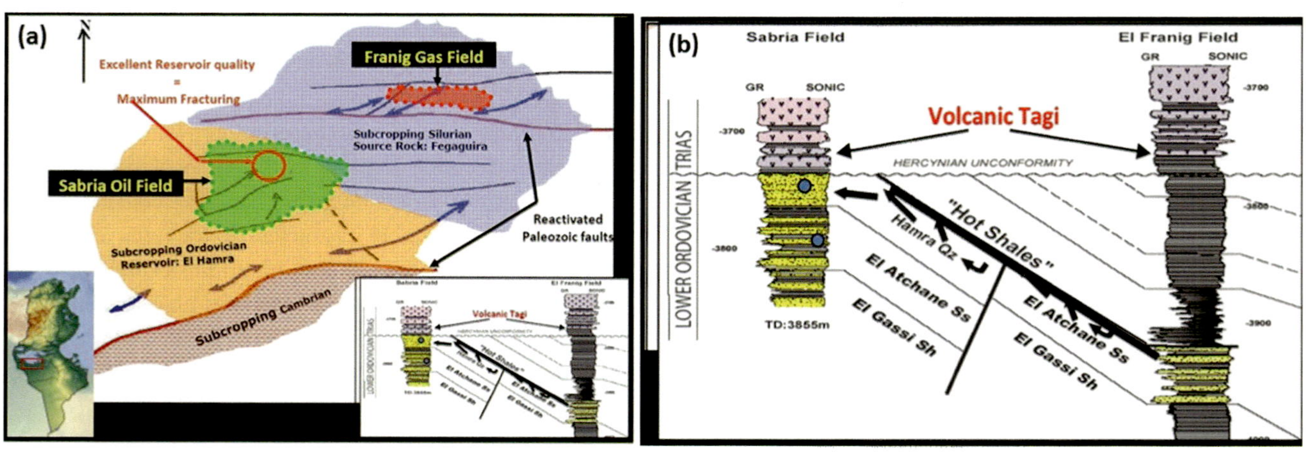

Fig. 7.2 **a** Ordovician play as proposed by Troudi (2006); **b** lateral variation of El Atchane and Hamra reservoirs between Sabria and El Franig Fields (Troudi 2006)

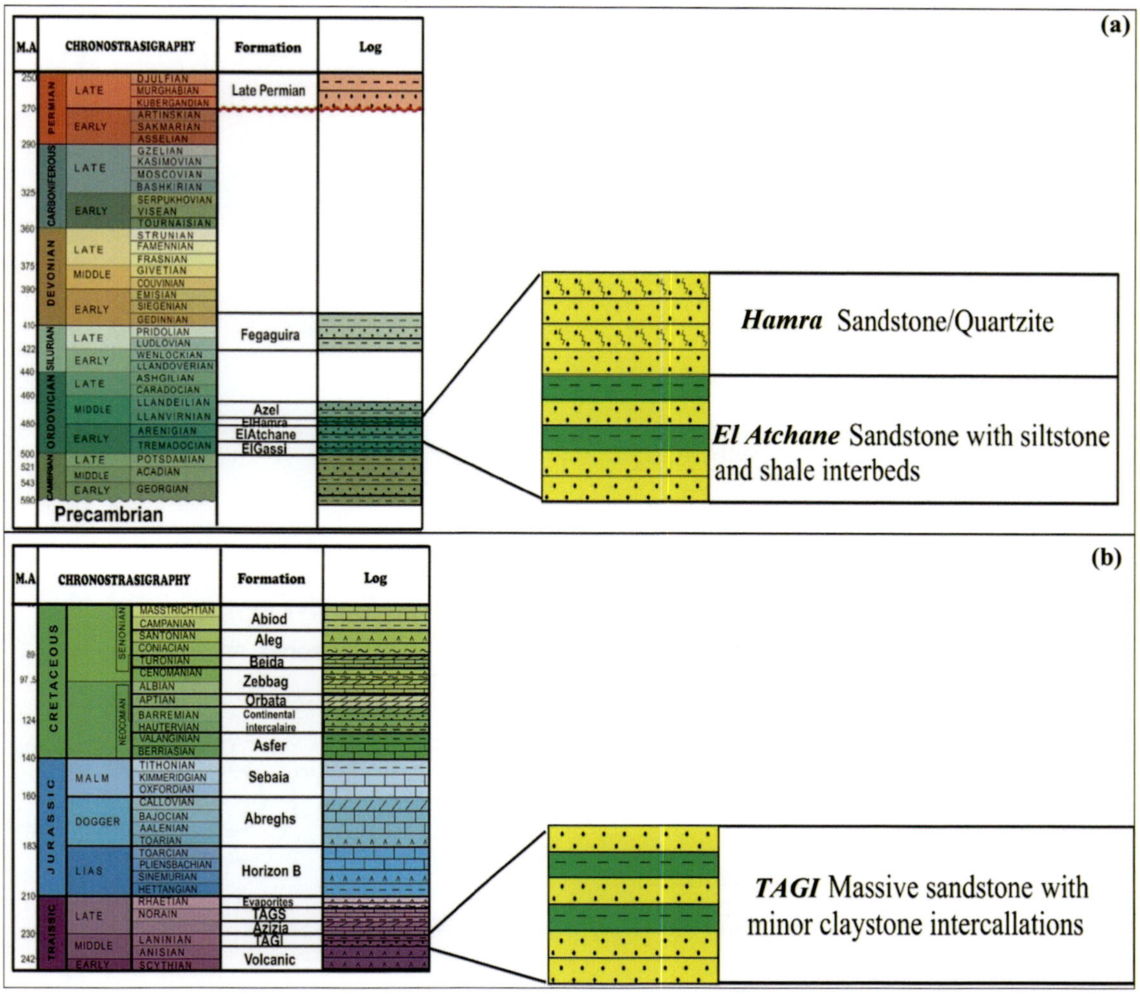

Fig. 7.3 **a** Lithofacies of El Atchane and Hamra reservoirs in the Chotts area; **b** lithofacies of the TAGI reservoir in the Chotts area

Excluding proven producing wells, oil indices have also been encountered in these reservoirs, which pledges in favour of functioning petroleum systems in the area (Bruna et al. 2019). Furthermore, in the Chotts region, the Upper Silurian Fegaguira is relatively rich in organic matter (Saïdi et al. 1998; Soua 2014; Belhaj Mohamed et al. 2015) and is thought to be the main source rock of this zone. It mainly consists of black shale with silt and carbonate interbeds. The basal member of this Formation is called "hot shale" and is formed by radioactive clays rich in organic matter (Fig. 7.4). The hot shale member of the Fegaguira Formation appears to be a good to excellent source rock (Saïdi et al. 1998; Mejri et al. 2006; Belhaj Mohamed et al. 2015) with Total Organic Carbon (TOC) values ranging from 1 to 20%. The Petroleum Potential (PP) and the Hydrogen Index (HI) values average 8 kg HC/t rock and 225 mg/g of TOC, respectively, implying a good oil and gas generating potential (Belhaj Mohamed et al. 2015).

Taking all this into consideration, this study tries to assess some elements of this petroleum system. Indeed, this work will evaluate the petrophysical characteristics of the Ordovician reservoirs (El Atchane and Hamra) and the Triassic reservoir (TAGI) in South of the Chotts Basin, aiming to delineate the most interesting reservoirs levels.

The study will also evaluate the Fegaguira source rock, its burial and thermal maturity history through 1D modelling using BasinMod software. Such assessment will allow to determine the generation, timing and possible hydrocarbon expulsion of the source rock towards the surrounding reservoirs.

7.2 Geographical, Geological and Structural Settings

The Chotts Basin is an intra-cratonic basin identical in appearance to the Ghadames Basin. It is located in the south-west of Tunisia, near the Tunisian-Algerian border. It is located between longitudes 7° 30 E and 9° 30 E and latitudes 34° N and 33° N (Fig. 7.5).

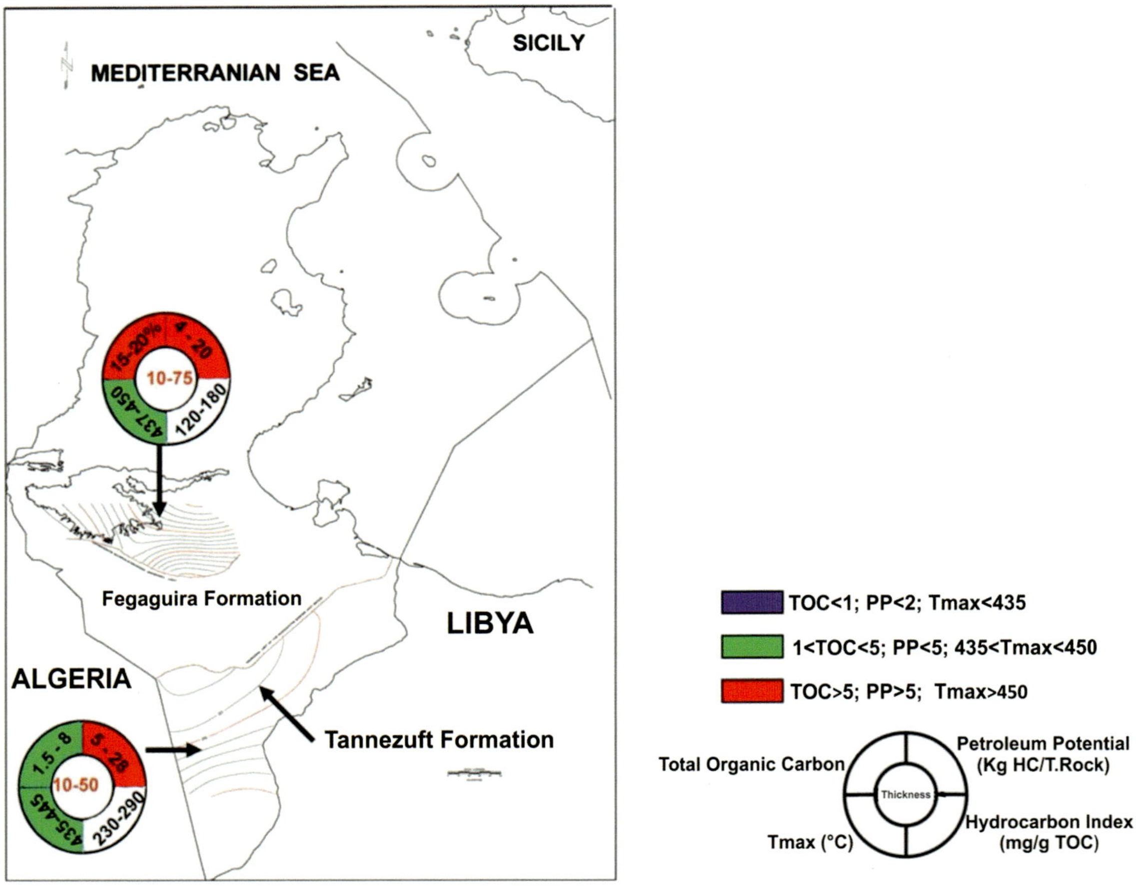

Fig. 7.4 Distribution and geochemical characteristics of the Fegaguira source rock as proposed by Saïdi et al. 1998

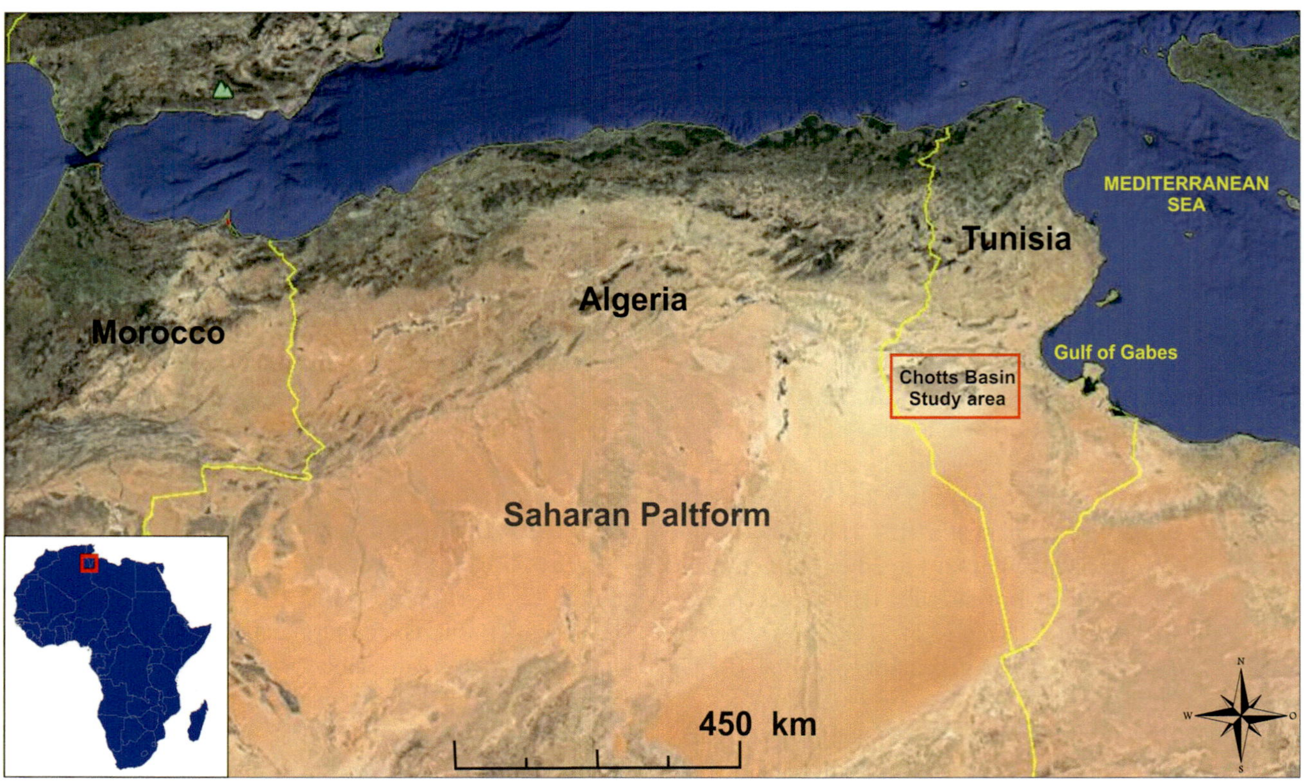

Fig. 7.5 Geographical location of the Chotts Basin (google earth)

It is believed that it was formed during the opening of the Tethyan at the end of the Carboniferous (Ben Ferjani et al. 1990). A maximum subsidence occurred during the Middle Triassic and Jurassic when Europe separated from Africa (Van de Weerd and Ware 1994; Echikh 1998). It is geologically bounded to the North by the North–South chains of the Chotts, to the East by the Dahar plateau, to the South by the Telemzane Arch and to the West by the "Grand Erg Oriental" (Fig. 7.6).

The tectonic history of the area is related to the regional structural context of southern Tunisia (Figs. 7.7 and 7.10). The Pan-African orogeny was the first orogeny phase which took place during late Pre-Cambrian as a result of the collision between the West African Craton and the East African Craton since 1800 Ma (Pogacsas et al. 1996, 1998; Hallett and Clark-Lowes 2016). The boundary between these two cratons forms a N–S axis that has been reactivated through time (Pogacsas et al. 1996, 1998). Following this pan-African orogeny, an extensive phase occurred during the Cambrian-Ordovician period and was marked by a facies and thicknesses lateral variation following a N–S direction (Zulauf et al. 2007). The Early Ordovician is characterized by a discordance at its base represented by the Azel shaly Formation (El Euchi et al. 2003). The Late Ordovician is marked by a second compressive phase corresponding to the Taconic phase occurring after the Pan-African movement and was materialized by the uplift and the erosion of Cambrian sediments (Pogacsas et al. 1998). This was followed by the creation of series overlapping the erosional troughs, which were later filled by periglacial deposits materialized by Late Ordovician micro-conglomeratic shales (Le Heron et al. 2012). The Silurian time was characterized by a major transgression caused by the melting of icecap and sea level increase (Fello et al. 2006). This transgression favoured the deposition of organic matter rich facies with the Fegaguira Formation to the North of the area and the Tannezuft Formation to the South (Cunningham 1989; Lüning et al. 2000; Belhaj Mohamed 2010; Rezouga et al. 2012). The Caledonian compression phase was initiated during the Pridoli–Gedinnian (Devonian) as a result of the collision between West Africa and North America (Balintoni et al. 2011). The Hercynian compression event was the most important compression phase during late Carboniferous, resulting in the uplift and the erosion of Telemzane Arch and placing Silurian and Devonian source rocks in direct contact with Triassic (Echikh 1998). The Hercynian orogeny mainly follows a N–S trend illustrating the collision between Laurasia and Gondwana and subsequent regional uplifts, folding, and intense erosion of Paleozoic rocks (Aliev et al. 1971; Boote et al. 1998). The Triassic-Jurassic extensional phase is the result of the general rifting related to the breakup of Pangea and the opening of the Tethys and Atlantic oceans (Aliev et al. 1971; Boudjema 1987; Khomsi et al. 2009). The E–W Austrian compression phase related to the early stages

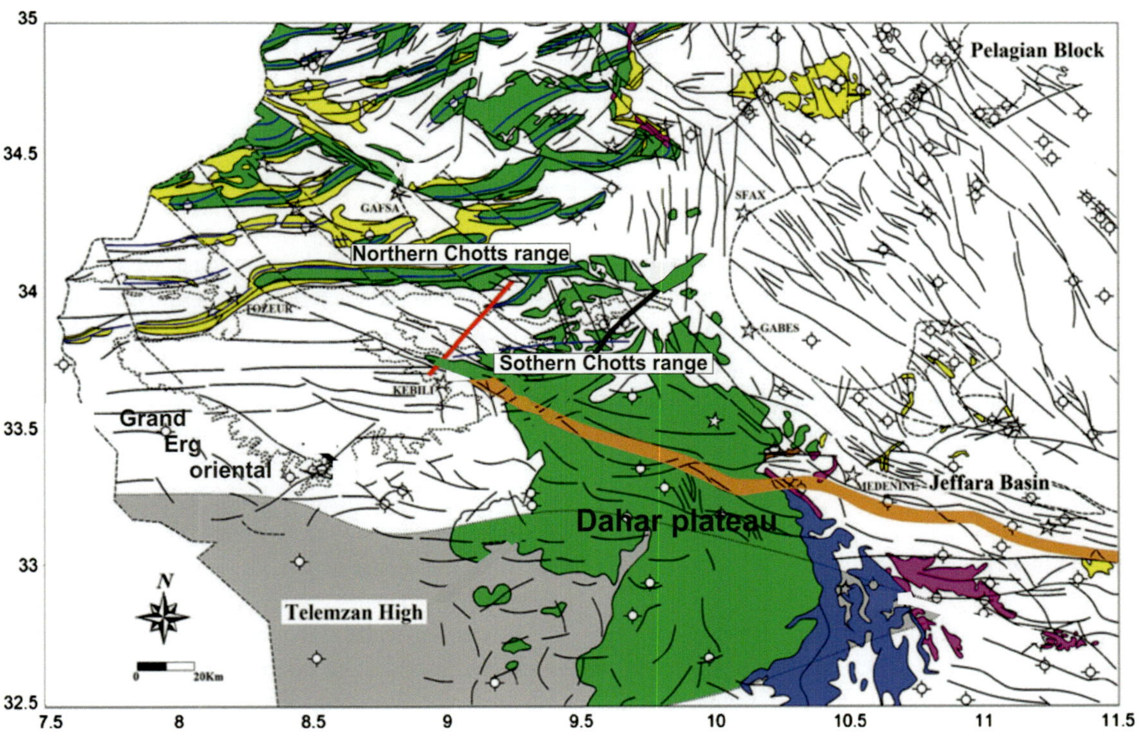

Fig. 7.6 Geological and structural map of Central and southern Tunisia (Zargouni 1985; Ben Ferjani et al. 1990; Boukadi 1994; Bédir 1995; Zouari 1995; Hlaiem 1998; Hlaiem 1999; Gabtni et al. 2006)

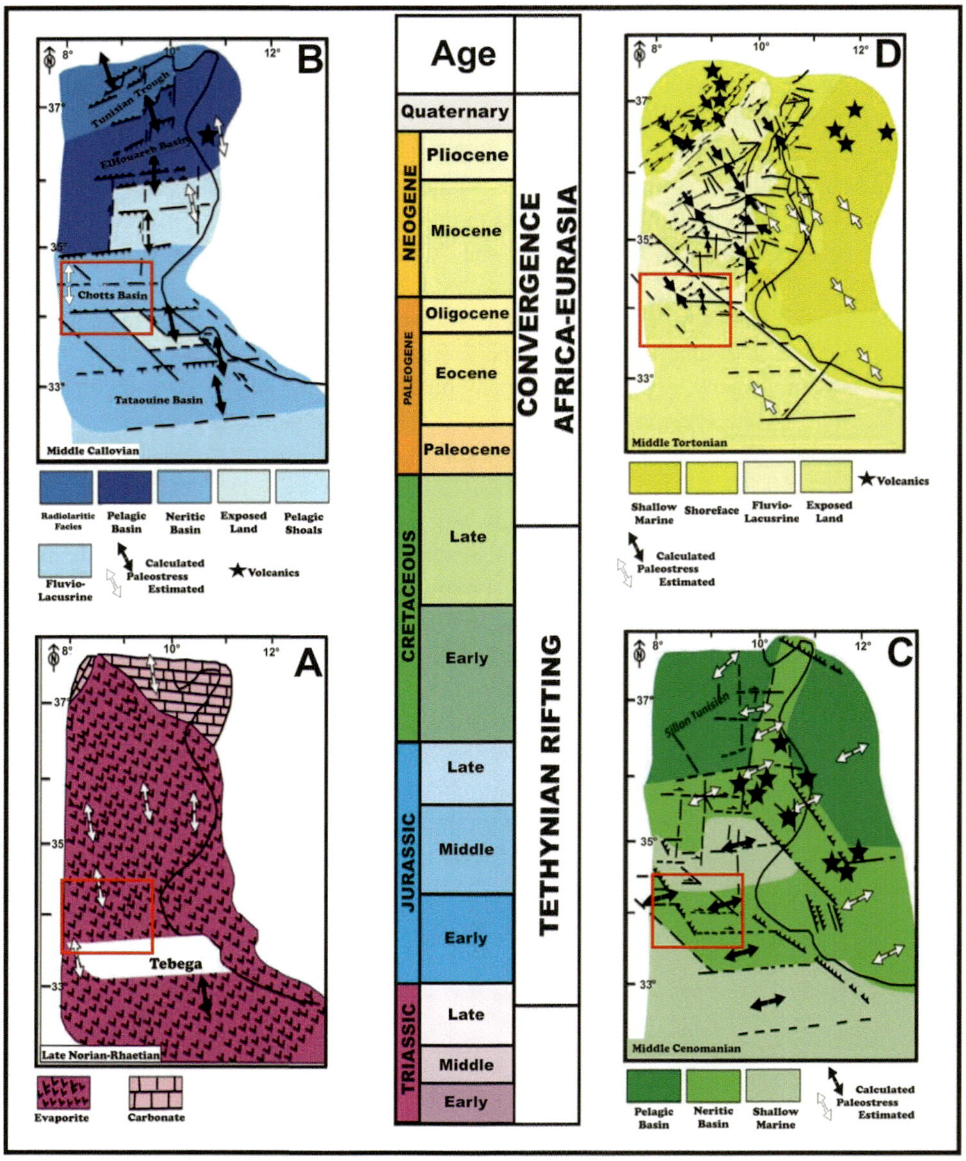

Fig. 7.7 Geodynamic evolution of the Chotts Basin as proposed by Bouaziz et al. (2002); Sarsar Naouali et al. (2016); modified

of the opening of Central Atlantic is well marked by the Aptian unconformity (Lazzez et al. 2008). The NW–SE Pyrenean/Alpine compressive phase is related to the opening of the North Atlantic and to the convergence/collision between the African and European plates, illustrated through compression, inversion, uplifting then erosion (Zargouni 1985; Khomsi et al. 2007, 2009).

7.3 Materials and Methods

Twenty wells drilled South of the Chotts Basin were evaluated in this study (Fig. 7.8). They are precisely located in the delegations of Nafta (governorate of Tozeur), El Faour (governorate of Kébilie) and Beni Khedeche (governorate of Medenine). Wells are termed from W1 to W20 (Fig. 7.8).

In all wells, well logging data were used in order to evaluate and characterize the Ordovician reservoirs (El Atchane and Hamra) and the Triassic reservoir (TAGI). Various petrophysical parameters such as porosity, water saturation and volume shale using logging tools (GR, Neutron curves, density and sonic) and available records were estimated. This method is based on the determination of petrophysical parameters as well as thicknesses and top and bottom depths of each reservoir, which subsequently allowed the construction of iso-thickness, iso-depth and iso-porosity maps.

In chosen wells, 1D modelling of the Fegaguira source rock via BasinMod software was performed. This modelling allowed reconstruction of the Fegaguira burial and thermal history and the prediction of hydrocarbon generation and expulsion as well as the quantification of expelled oil amounts.

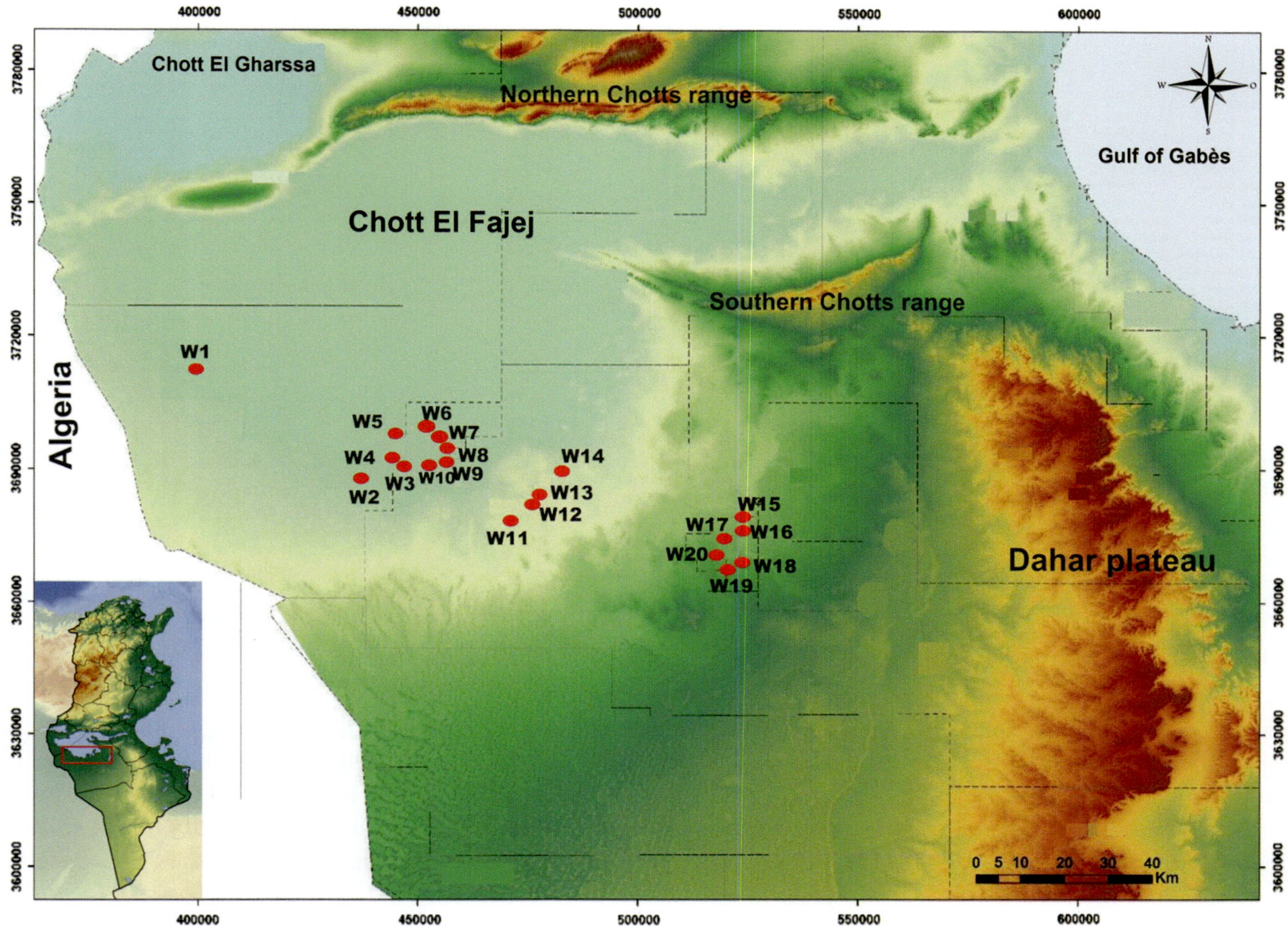

Fig. 7.8 Location map of studied wells, South of the Chotts area

7.4 Lithostratigraphic Correlation

In order to comprehend the basin architecture and the lateral extension of reservoir rocks (El Atchane, Hamra and TAGI) and their supplying Fegaguira source rock in the study area, a WNW-ESE lithostratigraphic correlation, covering 8 wells (W1, W2, W3, W5, W8, W13, W14, and W17) was established (Fig. 7.9). Through this correlation, we note:

- EL Atchane reservoir is only present in three wells with thicknesses varying from 36 m (W5), 38 m (W8) to 66 m (W3);
- Similarly, Hamra reservoir is only encountered in two wells (W3–W8) with a thicknesses range between 15 and 46 m;
- The Fegaguira Formation appears from W5 and then disappears into W17, it is thicker at W13 with a maximum thickness of 283 m.
- The TAGI reservoir is crossed by four wells (W1, W13, W14, and W17) showing a thickness variation from 14.5 to 23.4 m.

Ordovician reservoirs and Fegaguira source rock thickness variation and their absence from one well to another could be explained by normal Paleozoic faults (Lüning 2005). The first phase took place during the Cambrian-Ordovician period, which is marked by a lateral thicknesses and facies variation (Mejri et al. 2006; Belhaj Mohamed et al. 2014, 2015). The second phase occurred during Early Silurian and was associated to volcanism as proven in some wells drilled in southern Tunisia (Belhaj Mohamed et al. 2014, 2015). These faults were reactivated during Mesozoic with the opening of Tethys characterizing the North African margin since Triassic. It shows NW–SE to E–W Horst and graben structures following syn-sedimentary normal faults (Zargouni 1985; El Euchi et al. 2003; Zouaghi et al. 2011; Amri et al. 2017; Khomsi et al. 2019). On the

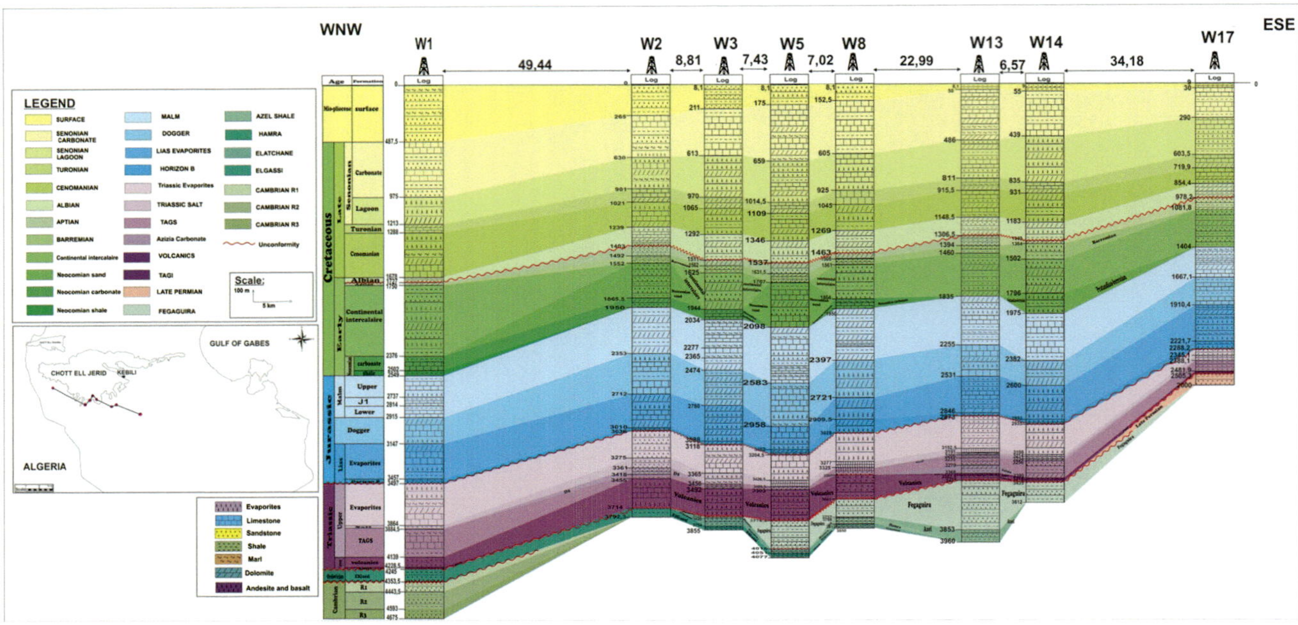

Fig. 7.9 WNW-ESE lithostratigraphic correlation between chosen wells in the South of Chotts area

other hand, Paleozoic series thickness variation could also be explained by erosion during Taconic, Caledonian and the Hercynian phases (Busson 1967).

The absence of TAGI reservoir in other wells is explained by its replacement by volcanic rocks resulting from magmatic material uprising through normal faults during the tethysian rifting (Kebaier et al. 2003; Laaridhi-Ouazaa et al. 2005). This volcanism is detected in most wells with an important thickness at W5 (271.5 m), decreasing towards the East until its total absence at W17 (Fig. 7.10).

7.5 Petrophysical Characterization of Reservoirs in the Study Area

– Ordovician reservoirs

When available, volume shale, water saturation and porosity were calculated using the Neutron-Density method. We present here, as an example, the Ordovician reservoir petrophysical results for well W8 (Figs. 7.11 and 7.12, Tables 7.1 and 7.2).

Figure 7.12 shows the cut-off plot in well W8 where Net reservoir is shown in green colour and Net/Pay in red.

The Net/Pay of the El Atchane reservoir is 30.10 m, with a porosity of 11.6% and a high-water saturation value reaching 97.6%, hence the El Atchane reservoir is generally water saturated. On the other hand, the Net reservoir of the Hamra represents 86.4% of the entire reservoir. It has a Net/Pay of the order of 9.85 m, an average porosity of 10.1% and a water saturation of 40.9%.

The iso-thickness map of the El Atchane reservoir shows a NW–SE thickness variation (Fig. 7.13b). Its maximum thickness in the NW part is around 66 m (W3) and passes to 36 m to the SE except in the W16 well where it reaches 65 m thick. The reservoir deepens to the NW, reaching 4109 m depth at W6 (Fig. 7.13c). The iso-porosity map shows values ranging between 5 and 12%. The highest porosity (12%) is recorded at W4 and decreases towards the SE (Fig. 7.13d).

The iso-thickness map of the Hamra reservoir shows a W–E thickness variation, with a maximum value of 65 m at W6 (Fig. 7.14b) and where it is encountered at 4044 m depth (Fig. 7.14c). The iso-porosity map shows a slight variation between 7.1 and 10.1% with maximum values occurring towards the East at W7 and W8 (Fig. 7.14d).

The Ordovician reservoirs (El Atchane and Hamra) generally show the same petrophysical characteristics. The porosity of the two reservoirs varies between 5 and 12% with a decrease in thickness from West to East until the total disappearance of the Hamra reservoir. This variation can be related either to the bevelling of the Ordovician on the Telemzane Arch or to the erosive effects of the orogenic phases during Paleozoic (Cunningham 1989; Busson 1967).

– Triassic reservoir (TAGI)

The petrophysical characterization of TAGI reservoir in well W13 is given as an example (Figs. 7.15 and 7.16, Tables 7.3 and 7.4).

Results show a low Net/Pay of 2.13 m. This level shows a maximum porosity of 21.5% and a water saturation of

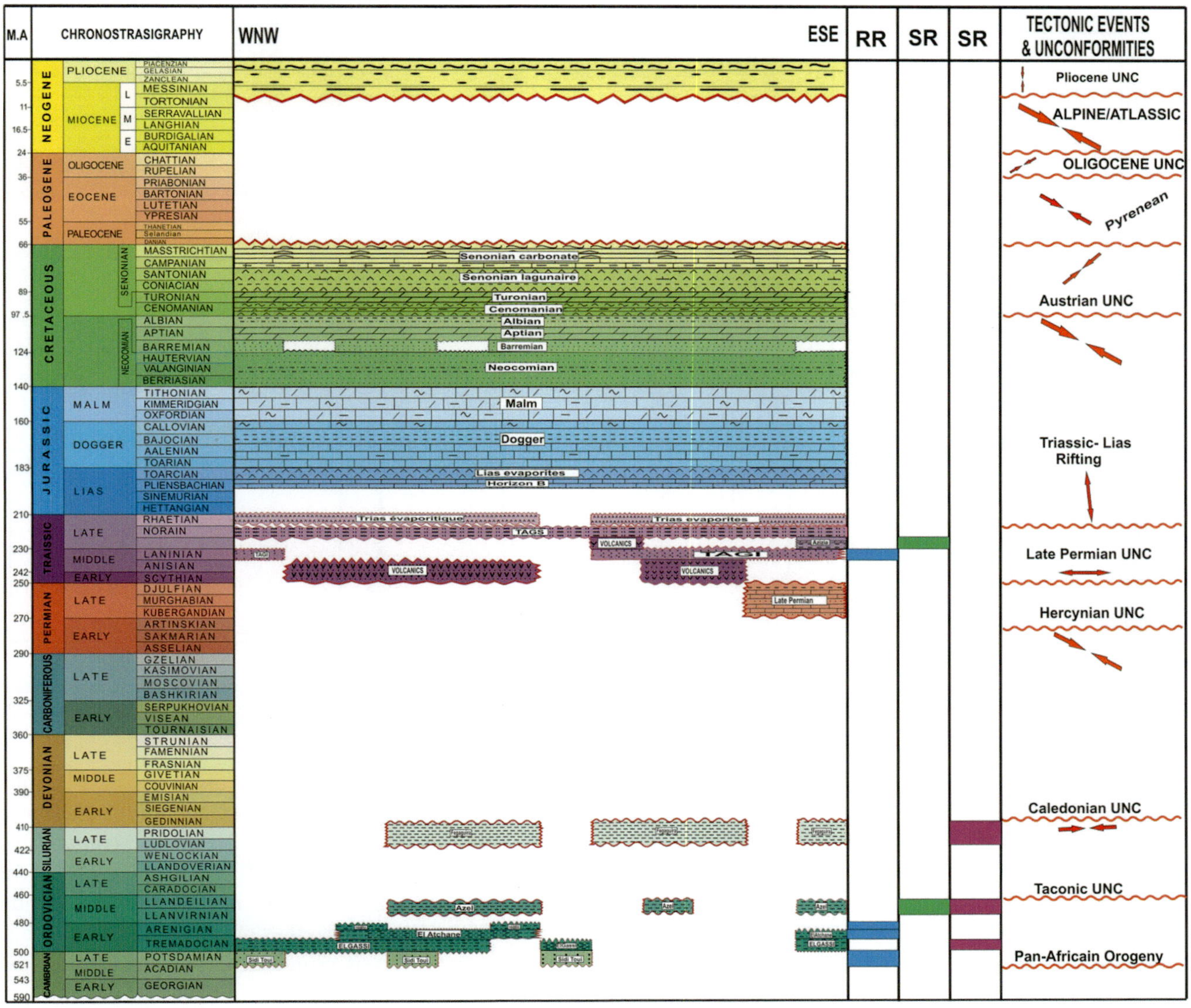

Fig. 7.10 Proposed lithostratigraphic chart of the study area

around 23.8%. In addition, petrophysical results (Figs. 7.15 and 7.16) indicate hydrocarbon indices at this level.

The iso-thickness map of the TAGI reservoir (Fig. 7.17b) shows a thickness variation from West to East. A maximum thickness of 50 m is recorded at W18 and a minimum thickness of 14.5 m is recorded at W13 with the TAGI deepening to the West (Fig. 7.17c). The iso-porosity map (Fig. 7.17d) shows a minimum of 9.5% at W12 and a maximum of 23.2%, recorded at W13.

The TAGI reservoir thickness variation and its absence in the western part of the study area can be explained by its replacement by volcanic materials which arose through faulting and took place during the tethysian rifting (Laridhi-Ouazaa 1994; Laridhi-Ouazaa et al. 2005). The TAGI is sometimes found below the volcanic body, sometimes above and is sometimes completely absent. In the Chotts region, well reports suggest that the reservoir is being sandwiched by volcanic material, occurring above and below. This probably argues in favour of two volcanic phases during Triassic which are succeeded in time.

This petrophysical characterization indicates that the Ordovician series and the Triassic series are overall good reservoirs in the southern zone of the Chotts but show the best petrophysical characteristics towards the West of this area.

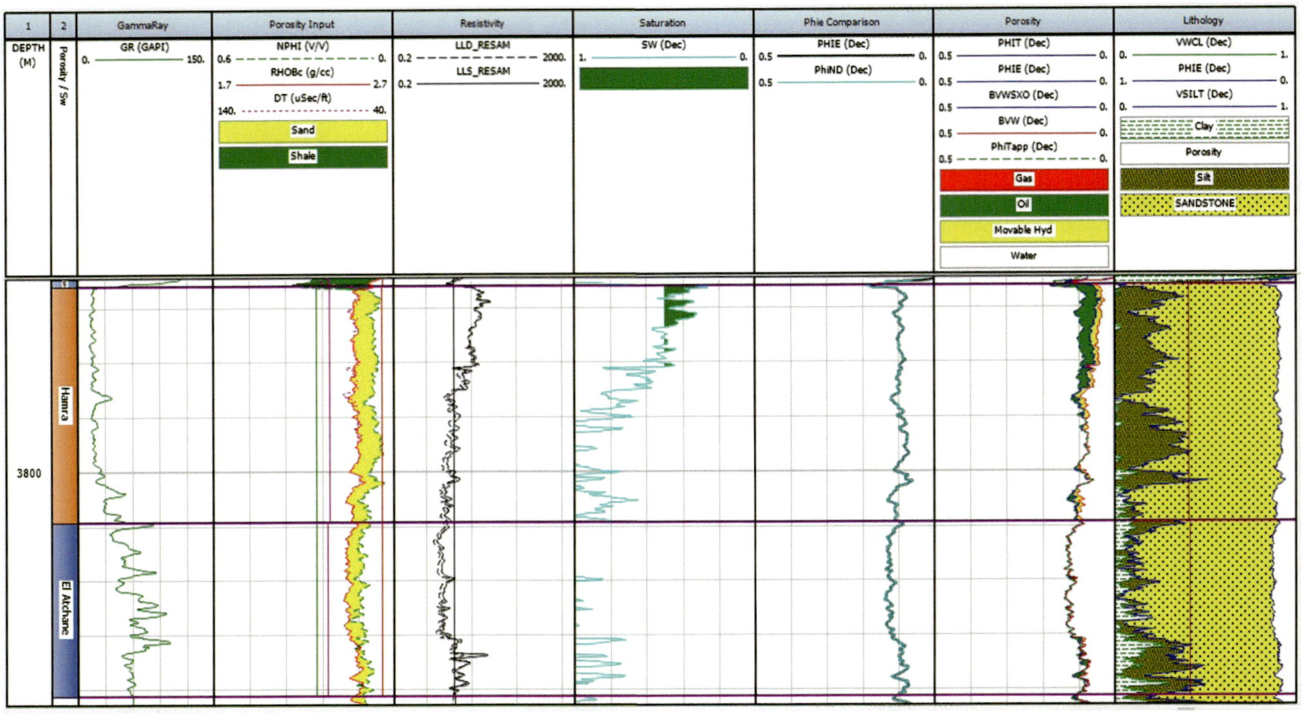

Fig. 7.11 Petrophysical interpretation of well 8 showing the different responses of GR, neutron, density, sonic, resistivities, water saturation (SW) and effective porosity through the Ordovician reservoirs

7.6 1D Modelling of the Fegaguira Formation in the Southern Region of the Chotts Basin

As mentioned above, the Fegaguira source rock is believed to have contributed to reservoir oil impregnation in the area. It is considered a good to excellent source rock in North Africa (Saïdi et al. 1998; Mejri et al. 2006; Belhaj Mohamed et al. 2015) and it is interpreted to be the main petroleum source rock in the Chotts Basin and possible reservoir for shale gas and shale oil accumulations (Soua 2014).

For that, the second aspect of this study is to perform 1D modelling via BasinMod software of the Upper Silurian Fegaguira Formation. The modelling allows burial and thermal history reconstruction of the basin and predicts the timing of hydrocarbon generation and expulsion. It also allows calculation of generated and expelled hydrocarbon amounts.

Burial history reconstruction

Inputs of well lithological data, measured depths of each formation in a given well were initially set. The depositional, non-depositional and erosional events were estimated from stratigraphic data and well reports.

Based on the model burial history charts results (Fig. 7.18), the study area shows two phases of subsidence which were interspersed by the Hercynian compressive phase. The latter, dated 300 Ma (Underdown and Redfern 2007), has played a major role in sediment distribution (Echikh 1998). The erosion effect of this phase on the area is very important and was estimated in this study of around 1000 m.

The first phase of subsidence occurred during the Paleozoic and the second phase during the Mesozoic age (Triassic-Cretaceous). The latter resulted in the deepest burial of sediments following the influence of the Tethyan rifting (Pogacsas et al. 1998).

Thermal history reconstruction

Model calibration for thermal history reconstruction was achieved based on various corrected BHT temperatures (Bottom Hole Temperature), Tmax temperatures issued from Rock Eval Pyrolysis, vitrinite reflectance and volcanism estimated temperatures.

Curves above parameters and those measured are almost superimposed (Fig. 7.19). The best calibration overlay was obtained for a heat flow of 40 mW/m^2.

Based on thermal history charts, the Fegaguira source rock was subjected to temperatures allowing it to reach

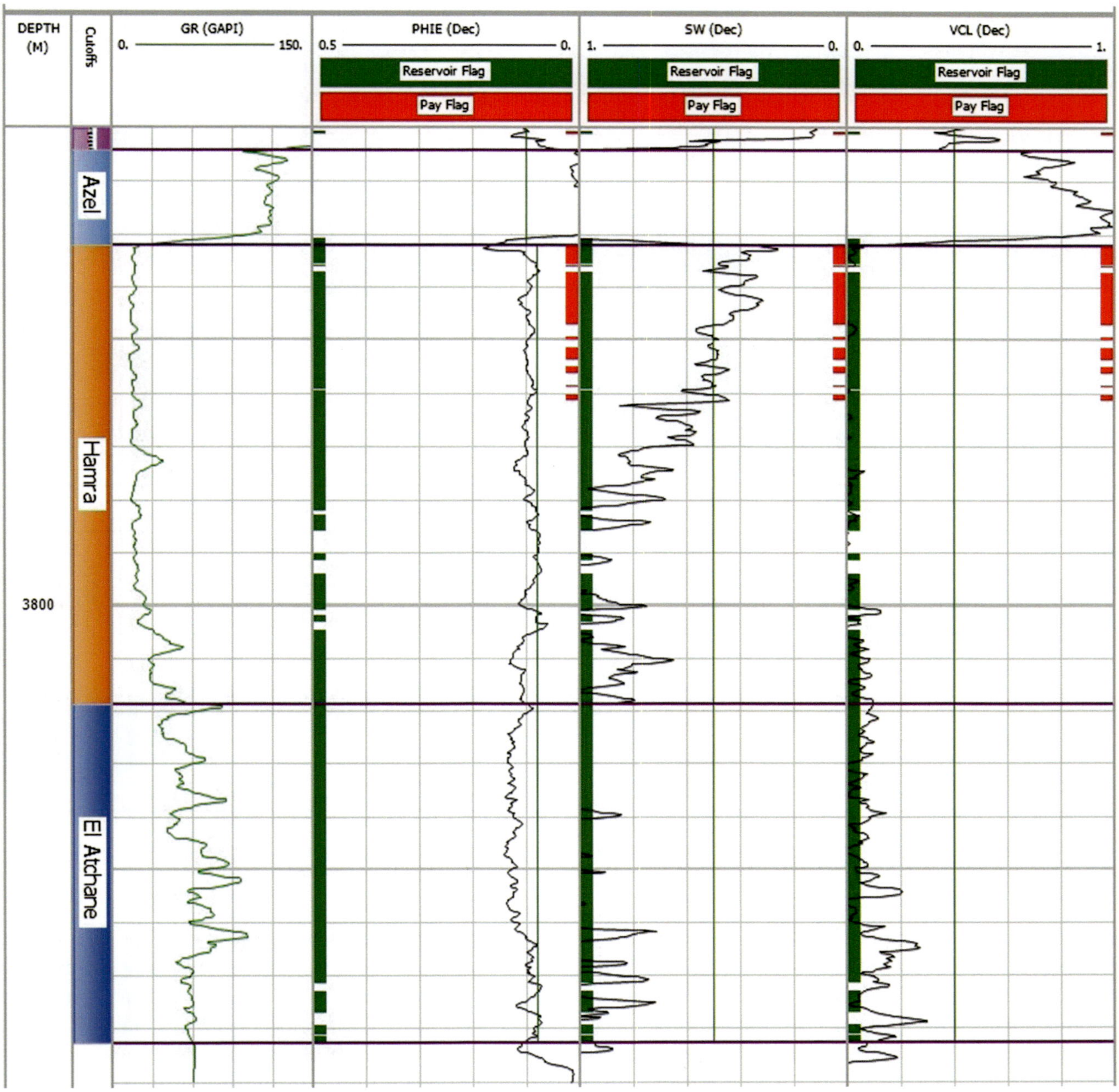

Fig. 7.12 Cut off plot showing the Net/Pay in W8

Table 7.1 Petrophysical characteristics of the Hamra and El Atchane reservoirs in well 8

Zone	Top	Bottom	Gross	Net	N/G	Avr Phi moy (%)	Rw Ohmm	Avr Sw (%)	Avr Vcl (%)
Hamra	3766	3809.30	43.30	37.40	0.86	10.1	0.0966	68.2	1.5
ElAtchane	3809.3	3841.40	32.10	30.10	0.94	11.6		97.6	7.4

Table 7.2 Petrophysical characteristics of the Net Pay part of the Ordovician reservoirs in well 8

Zone	Top	Bottom	Gross	Net	N/G	Phi moy (%)	Sw moy (%)	Vcl moy (%)
Hamra	3766	3809.30	43.30	9.85	0.227	10.1	40.9	0.6
ElAtchane	3809.30	3841.40	32.10	0.00	0.00	–	–	–

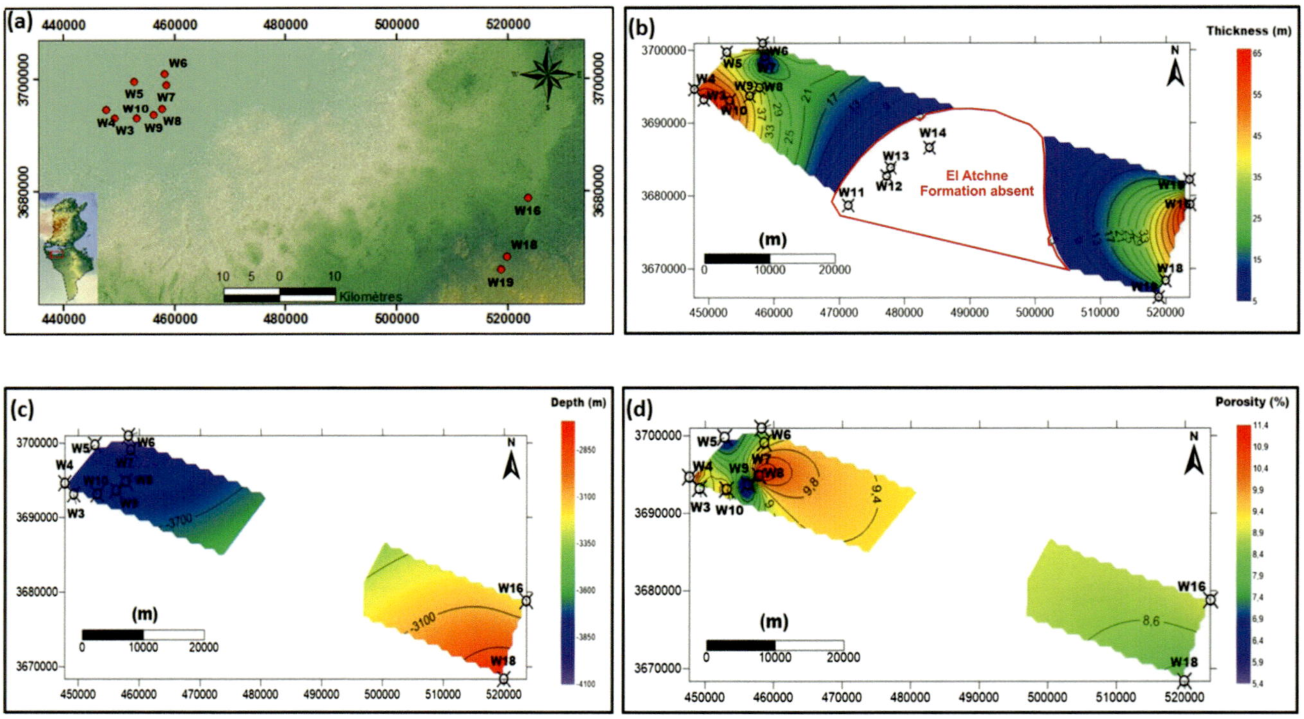

Fig. 7.13 **a** Map showing only wells recording El Atchane reservoir; **b** isopach map of the El Atchane reservoir; **c** isobath map of the El Atchane reservoir; **d** iso-porosity map of the El Atchane reservoir

generation and also expulsion (Fig. 7.20). Indeed, the maturity history modelling of simulated wells in the South of the Chotts Basin shows that the Fegaguira source rock is mature and has began generating hydrocarbons during the Early Cretaceous. Generated amounts range between 11.4 and 223.35 bbls/acre * ft varying with wells. These amounts could be higher in more subsident areas (Figs. 7.21 and 7.22).

Overall evaluation of the study area

The petrophysical study based on well logging data shows that the Ordovician and Triassic series are good reservoirs in the South of the Chotts Basin and have the best characteristics especially to the West of this zone:

Indeed, porosity of Ordovician reservoirs varies between 5 and 12%. This variation is related to diagenesis, compaction and cementation. Porosity is improved by fractures due to the play of faults during extensive phases (Rigo 1996). However, towards the West of the area, maximum water saturations are recorded (water saturation reaching 97% for the El Atchane reservoir and 68.2% for the Hamra reservoir). This is due to the relative burial of these series compared to the East of the study area. In fact, Series (Ordovician reseroirs) deepen to the West and reach depths exceeding 4000 m. However, the Ordovician reservoirs thin from West to East until the even total disappearance of the Hamra reservoir. This is related in one hand to the bevelling of the Ordovician on the Telemzane Arch and on the other hand to the erosive effects of the orogenic phases during the Paleozoic Taconic, Caledonian and Hercynian phases (Busson 1967, 1970; Lüning 2005).

The TAGI reservoir thickness varies between 14.5 and 50 m, with a maximum of 50 m at W18 to the East of the

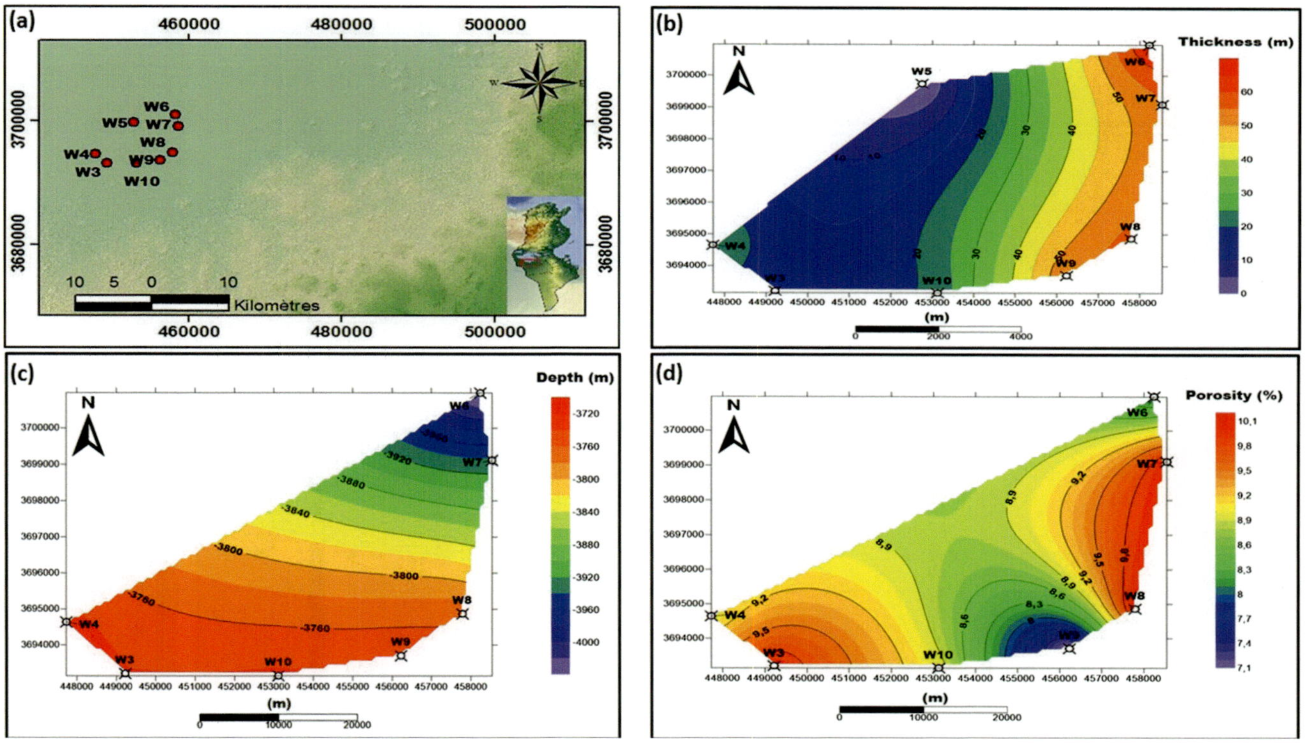

Fig. 7.14 **a** Map showing only wells encountering the Hamra reservoir; **b** isopach map of the Hamra reservoir; **c** isobath map of the Hamra reservoir; **d** iso-porosity map of the Hamra reservoir

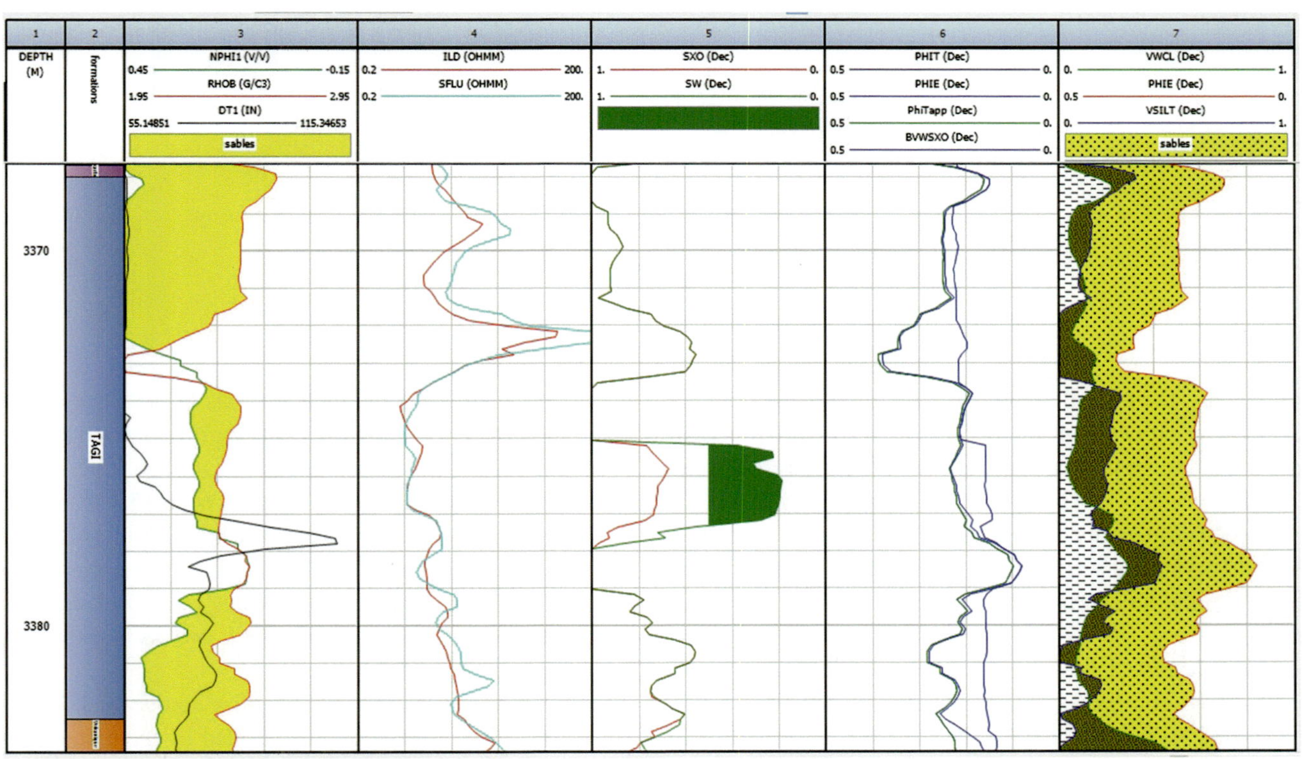

Fig. 7.15 Petrophysical interpretation of well 13

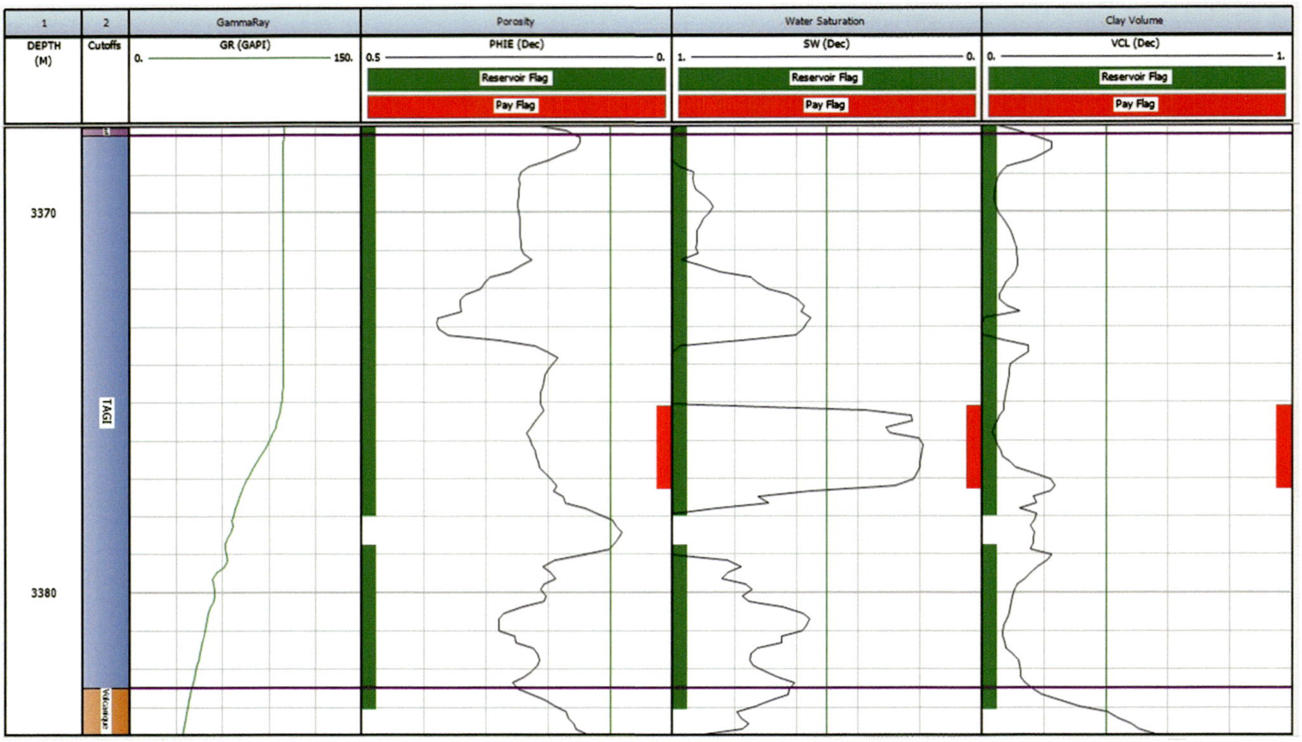

Fig. 7.16 Cut off plot showing the Net/Pay in well 13

Table 7.3 Petrophysical characteristics of the TAGI reservoir in well 13

Zone	Top	Bottom	Gross	Net	N/G	Phi moy (%)	Rw Ohmm	Sw moy (%)	Vcl moy (%)
TAGI	3368	3382.5	14.5	13.74	0.947	23.2	0.0966	71.3	10.1

Table 7.4 Petrophysical characteristics of the Net/Pay part of the TAGI reservoir in well 13

Zone	Top	Bottom	Gross	Net	N/G	Phi moy (%)	Sw moy (%)	Vcl moy (%)
TAGI	3368	3382.5	14.5	2.13	0.147	21.5	23.8	0.91

study area. However, the TAGI is absent to the West and is replaced by volcanic materials as they are believed to have migrated through normal faulting that has played during the Tethyan rifting.

The TAGI porosity varies between 9.5 and 23.2% with a maximum value at well W13 (23.2%). This variation is due to facies distribution during the Mesozoic period (Amri et al. 2017).

Volcanic materials in the Chotts Basin are found sometimes above, sometimes below, sometimes sandwiched with the TAGI reservoir. The TAGI reservoir is occasionally absent and is replaced by volcanic rocks in the western part of the basin following normal faults during Tethyan rifting.

In well W13, the TAGI and the volcanic material are characterized by good porosity ranging from 15.4 to 23.2%. This can be explained by two hypotheses: (i) Volcanism did not affect the porosity of TAGI but porosity was improved by fracturing and diagenetic processes; (ii) Or volcanism affected the porosity of the TAGI but later was improved by fracturing and diagenetic processes.

The burial history of the Fegaguira source rock in the Chotts Basin has been largely controlled by tectonic events

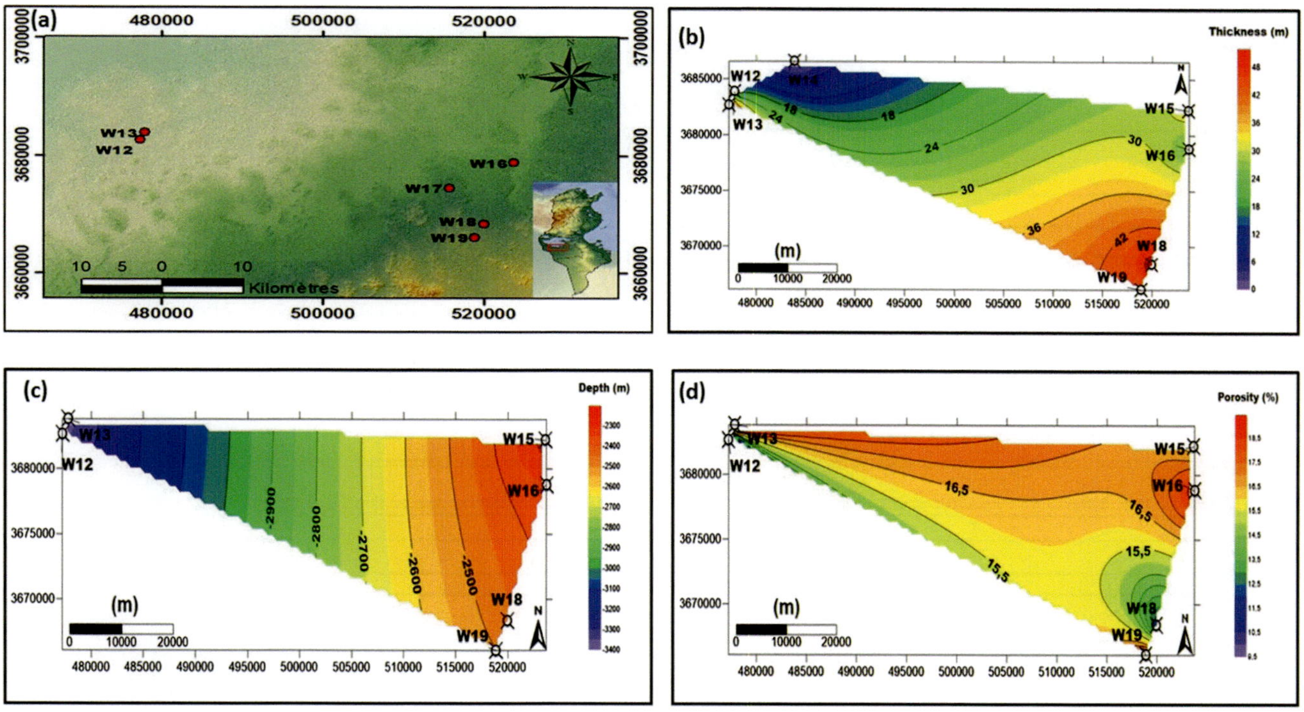

Fig. 7.17 a Map showing only wells recording the TAGI reservoir; b isopach map of the TAGI reservoir; c isobath map of the TAGI reservoir; d iso-porosity map of the TAGI reservoir

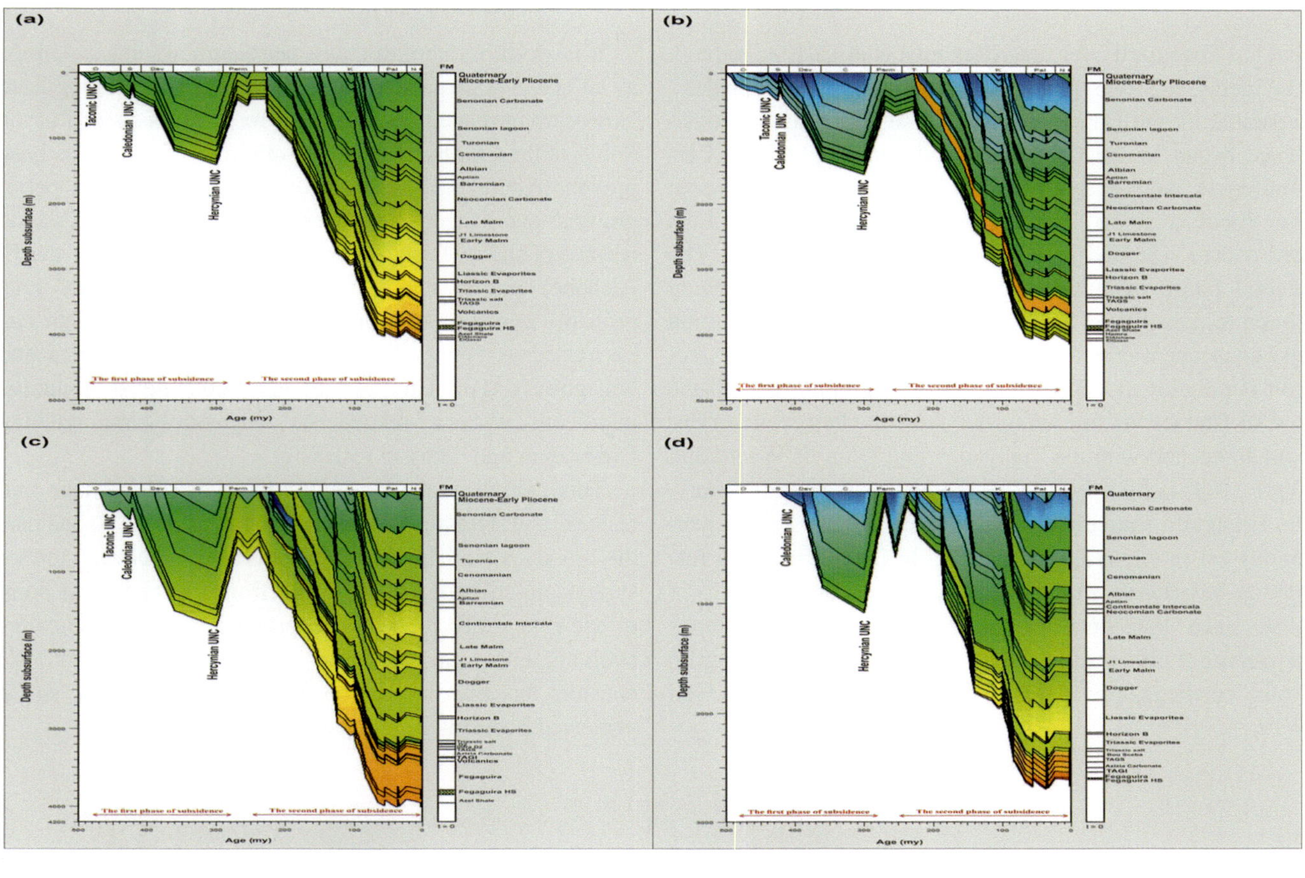

Fig. 7.18 Burial history of the Fegaguira source rock in wells 5 (a), 7 (b), 13 (c), and 19 (d)

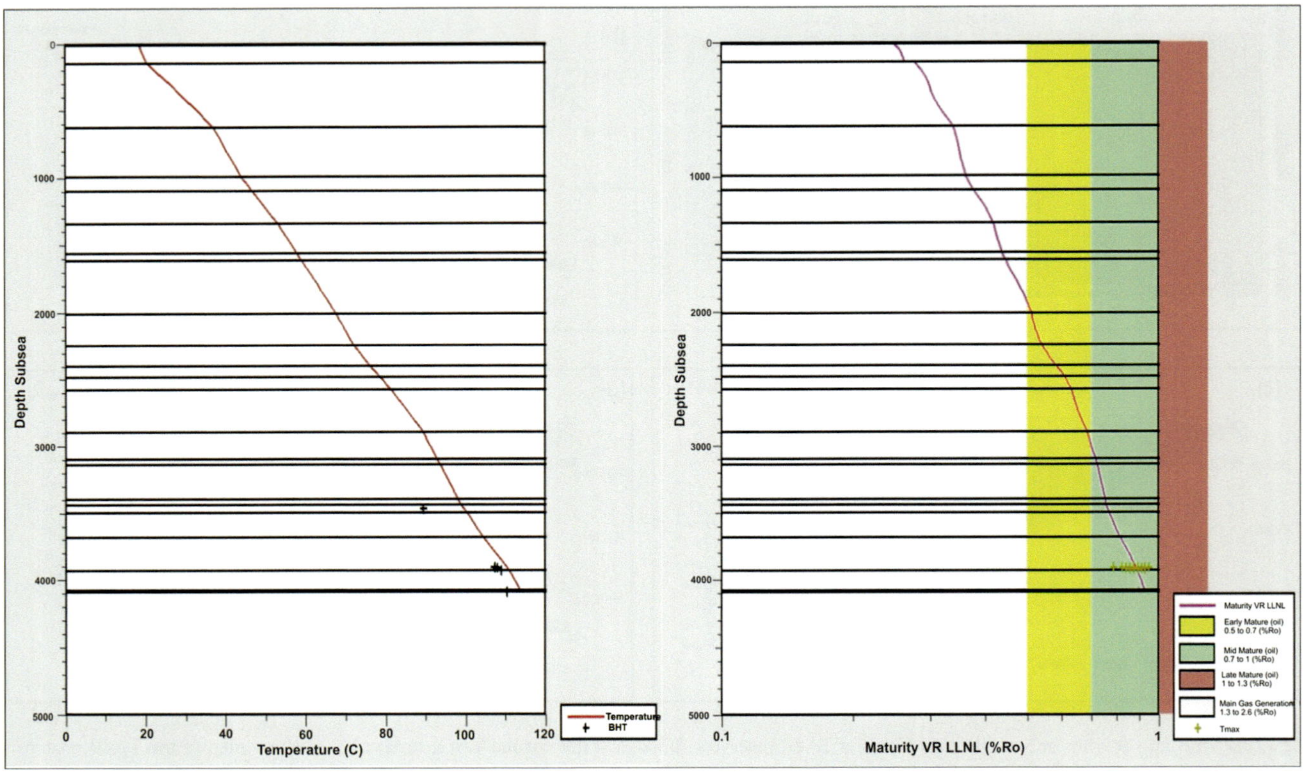

Fig. 7.19 Example of calibration of the thermal history based on corrected BHT, Tmax and R_0 in the well 7

that have occurred over time (Rezouga et al. 2012). Through 1D modelling, two phases of subsidence were depicted, separated by the Hercynian phase. The first phase of subsidence began in the Ordovician and continued to the Carboniferous and was interrupted by two compression phases: the Taconic phase and the Caledonian phase of Late Ordovician age and Late Silurian age successively. The second phase of subsidence is limited by the Hercynian phase, which played a major role in sediment distribution and consequent erosion of large parts of the Paleozoic section (Underdown and Redfern 2008). It continued until the end of Cretaceous following the Triassic-Cretaceous extension associated with the Tethyan rifting (Van de Weerd and Ware 1994; Echikh 1998). Consequent to this burial history, the Fegaguira source rock is simulated to have started generating oil in all wells since the Early Cretaceous. The formation has started expelling hydrocarbons at a SATEX of 10% since Paleogene, the earliest. Expelled oil varies between 6.9 bbls/acre ft and 147.8 bbls/acre, depending on wells. The Fegaguira source rock has not expelled oil in well W19.

It is worth reminding that a petroleum system is defined as an active source rock and all generated and impregnated hydrocarbon related to that source rock. In this study, in the South of the Chotts area, the Fegaguira source rock has been proven mature and has already expelled oil. The petrophysical assessment undertaken on Ordovician and Triassic reservoirs shows fairly good reservoirs properties of the El Atchane, Hamra and TAGI reservoirs. The study area probably encompasses three petroleum systems which can be named as Fegaguira-El Atchane, Fegaguira-Hamra and Fegaguira-TAGI. This can be also considered a unique petroleum system related to Fegaguira impregnating Ordovician and Triassic reservoirs.

This overall evaluation of Ordovician (El Atchane and Hamra) and Triassic (TAGI) reservoirs is given to indicate the good potential of these intervals. However, further studies are still needed especially from a geophysical point of view in order to delineate promising closed structures evolving either in one or in all of these reservoirs. We consider that this publication is to draw attention to so far neglected horizons in this particular area.

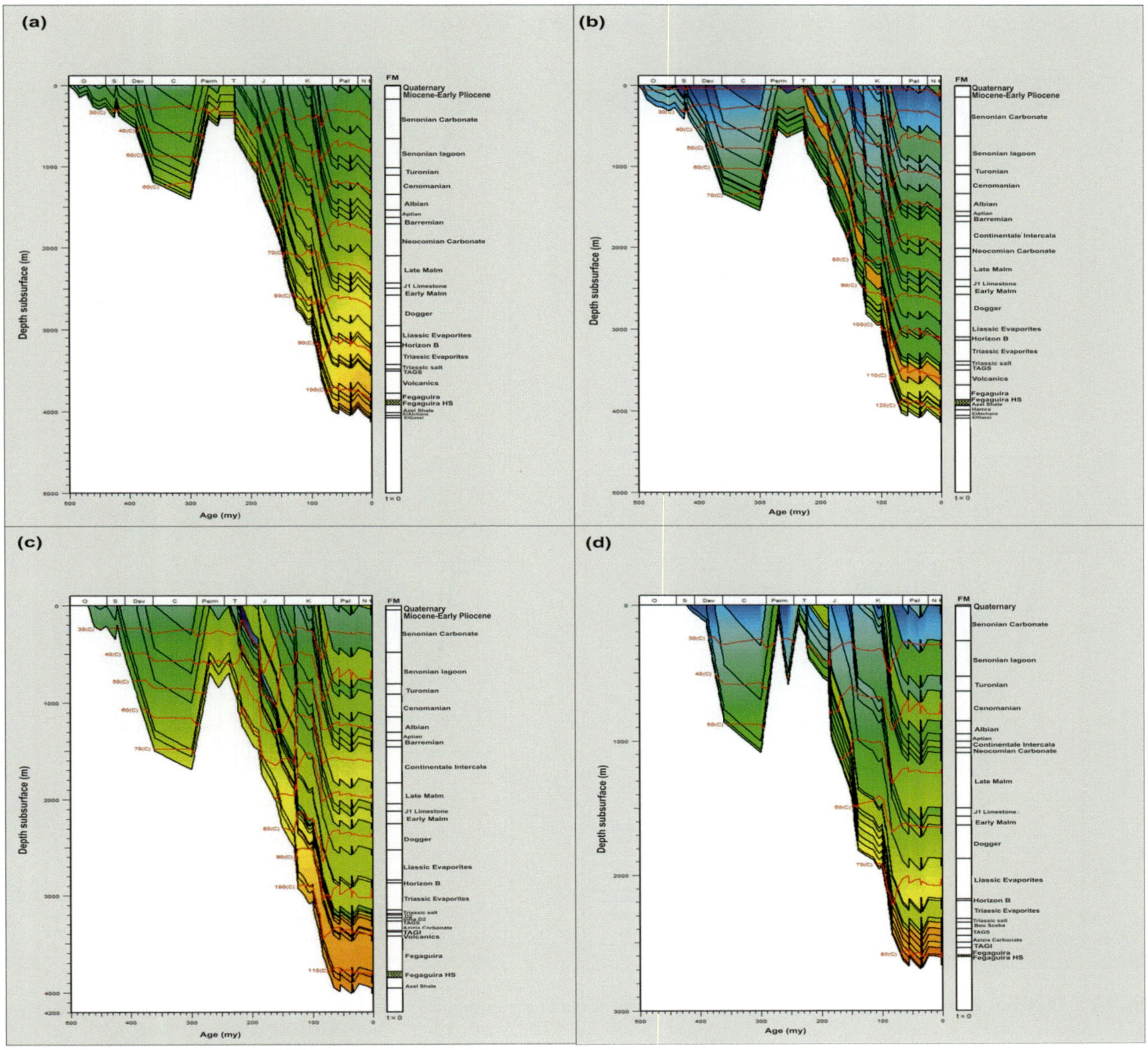

Fig. 7.20 Maximum temperatures of the Fegaguira source rock at wells 5 (**a**), 7 (**b**), 13 (**c**), and 19 (**d**)

7.7 Conclusions

The Chotts Basin is considered a prolific hydrocarbon zone. We believe that the combination of petrophysical characterization of the Ordovician and Triassic reservoirs and 1D modelling of their supplying source rock (Fegaguira Formation) can solve certain exploration problems and improve the hydrocarbon perspectivity of the study area.

Results show that the Ordovician reservoirs (El Atchane and Hamra) have similar and fair petrophysical characteristics towards the West of the study area. The TAGI reservoir is replaced by volcanic material to the West of the study area which took place through faulting during the Tethyan rifting. Its porosity varies between 9.5 and 23.2%.

The study area burial history modelling shows two phases of subsidence which are interspersed by a compressive phase corresponding to the Hercynian phase and which played a major role in sediment distribution.

The maturity history modelling of real wells in the South of the Chotts region shows that the Fegaguira source rock is mature and began generating hydrocarbons during the Early Cretaceous and expelled its hydrocarbons since the Paleogene, the earliest.

A petroleum system related to the Fegaguira Formation impregnating the El Atchane, Hamra and TAGI reservoirs is

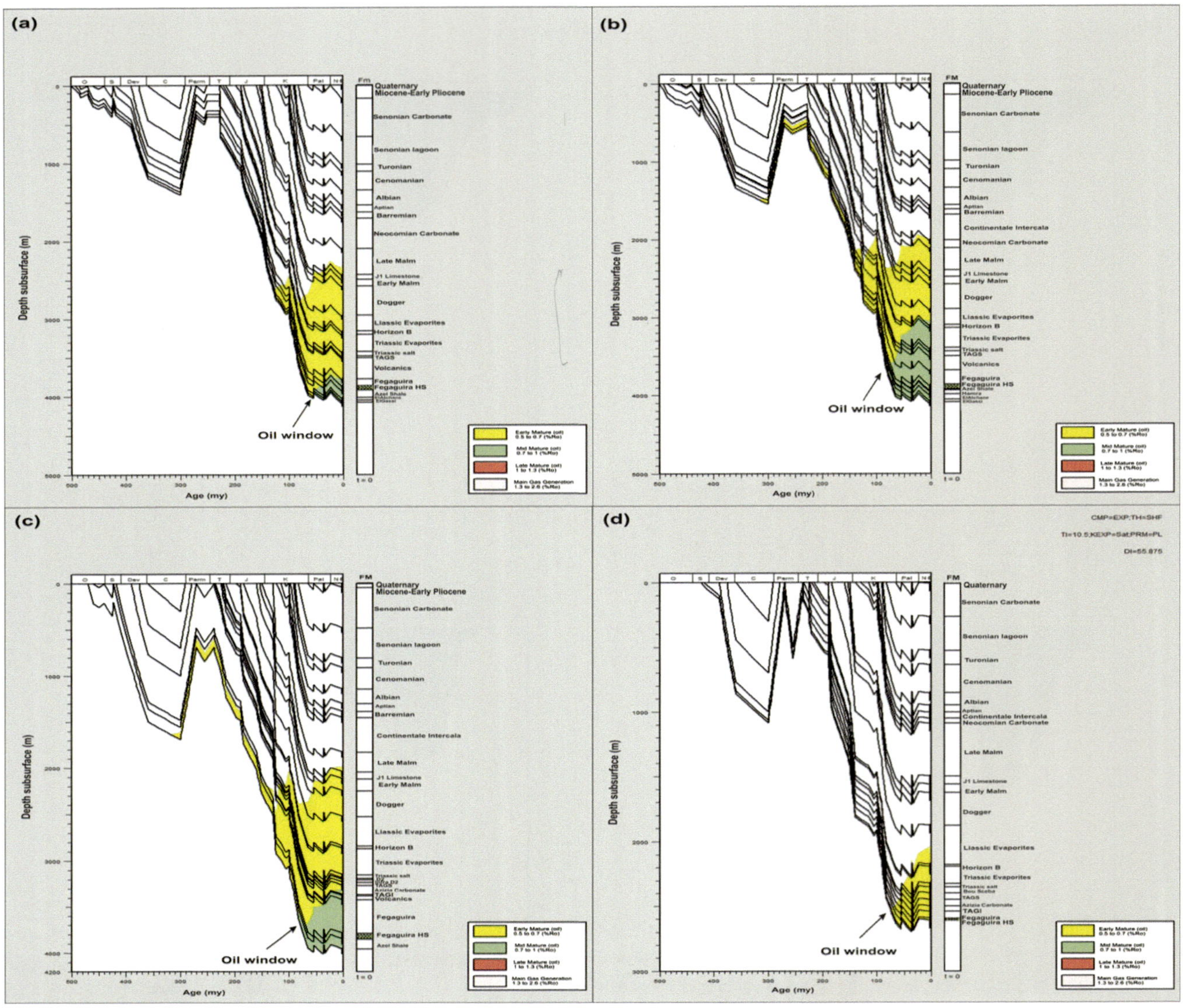

Fig. 7.21 Diagrams showing burial history and maturity windows of the Fegaguira source rock at wells 5 (**a**), 7 (**b**), 13 (**c**), and 19 (**d**)

very sustained. The seal rock is provided by clays, and Triassic evaporitic rocks series. All the ingredients of functioning petroleum systems are gathered with a prominent mature and expulsing Silurian source rock surrounded by good and trap-shaped Ordovician and Triassic siliciclastic reservoirs. These traps have to be further sustained by robust geophysical studies.

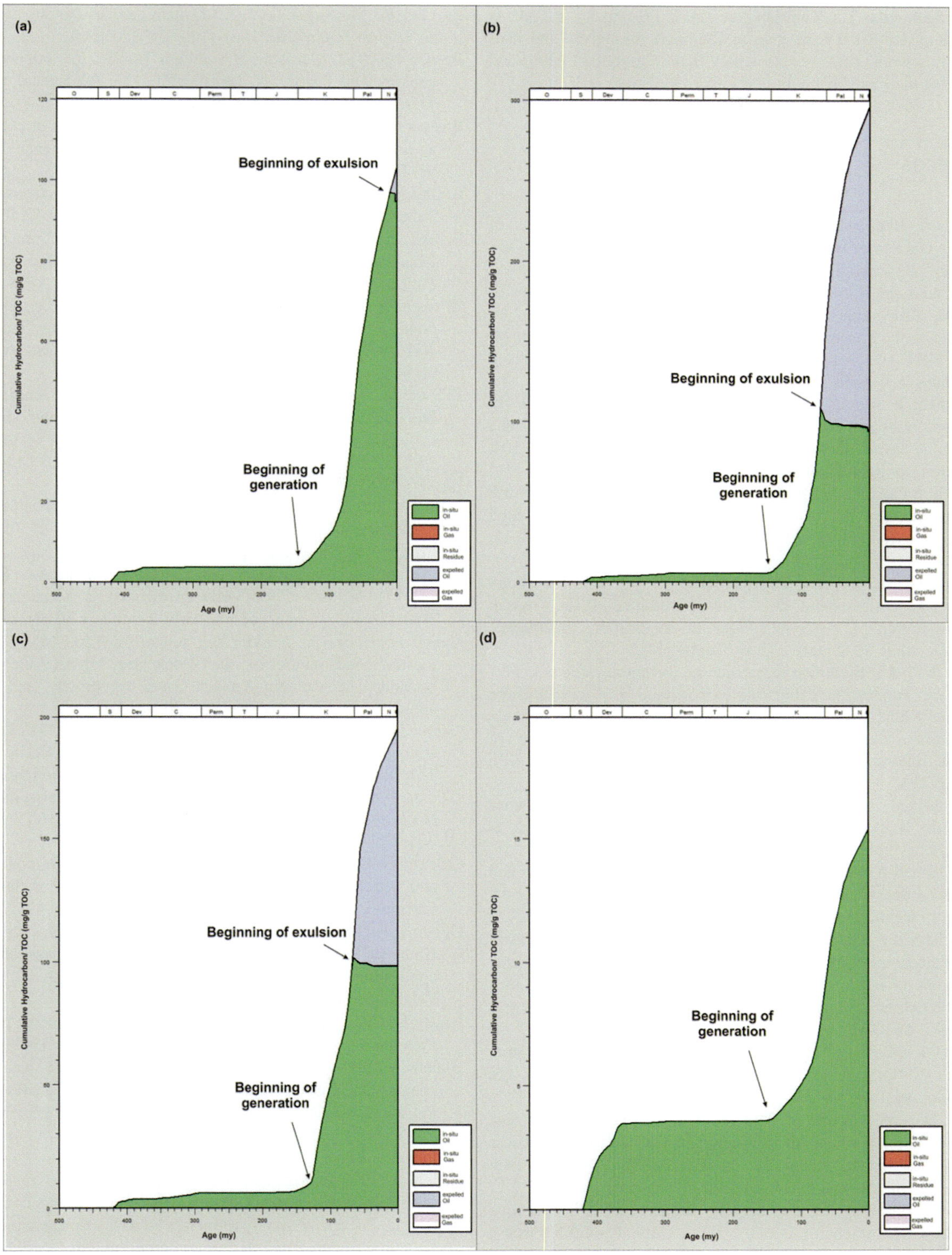

Fig. 7.22 History of generation and expulsion of hydrocarbons by the Fegaguira source rock in wells 5 (**a**), 7 (**b**), 13 (**c**), and 19 (**d**)

Acknowledgements The Tunisian National Oil Company (**ETAP**) is greatly acknowledged for the help and support throughout the realization of this work. Authors also kindly thank the journal editor and reviewers for their valuable comments and ms editing.

References

Aissaoui NM, Bedir M, Gabtni H (2016) Petroleum assessment of Berkine-Ghadames Basin, Southern Tunisia. AAPG Bull 100:445–476. https://doi.org/10.1306/01141612083

Aliev M, Laoussine N, Avrov V, Aleksine G, Barouline G, Lakovlev B, Korj M, Kouvykine J, Makarov V, Mazanov V, Medvedev E, Mkrtchiane O, Moustafinov R, Oriev L, Oroudjeva D, Oulmi M, Saïd A (1971) Geological structures and estimation of oil and gas in the Sahara in Algeria. Altamira-Rotopress, S.A, Spain, p 265

Amri D, Dhahri F, Soussi M, Gabtni H, Bedir M (2017) The role of E–W basement faults in the Mesozoic geodynamic evolution of the Gafsa and Chotts basins, south-central Tunisia. J Earth Syst Sci 126:104. https://link.springer.com/article/10.1007/s12040-017-0876-1

Balintoni I, Balicia C, Hann HP (2011) Peri-Gondwanan – North African enlarged acceptance of the Caledonian Orogeny. Stud Univ J Geosci Front 56:29–32. https://doi.org/10.1016/j.gsf.2013.09.002

Bédir M (1995) Mécanismes géodynamiques des bassins associés aux couloirs de coulissement de la marge atlasique de la Tunisie: sismo-stratigraphie, sismo-tectonique et implications pétrolières. Thèse Doctorat Es Sciences, Univ. Tunis, p 412

Ben Ayed N (1986) Evolution tectonique de l'avant-pays de la chaîne alpine de Tunisie du début, du Mésozoïque à l'Actuel. Thèse Doctorat d'Etat. Univ. Paris Sud, Centre d'Orsay, p 286

Ben Ferjani A, Burollet P, Mejri F (1990) Petroleum geology of Tunisia. Entreprise Tunisienne des Activités Pétrolières (ETAP), Tunis, p 194

Belhaj Mohamed A (2010) Caractérisation géochimique des roches mères de la région des Chotts et corrélation huiles/huiles et huiles/roches mères. ETAP, p 81

Belhaj Mohamed A, Saidi M, Bouazizi I (2014) Oil-oil correlation and potential source rocks in the Chotts Basin, Tunisia. In: International petroleum technology conference, Doha, Qatar, 20–22 Jan 2014

Belhaj Mohamed A, Saidi M, Soussi M (2015) Organic geochemistry of the Paleozoic source rocks in the Chotts Basin, Southern Tunisia. In: Society of Petroleum Engineers North Africa technical conference and exhibition, Cairo, Egypt, 14–16 Sept 2015. https://doi.org/10.2118/175830-MS

Boote DRD, Clark-Lowes DD, Traut MW (1998) Palaeozoic petroleum systems of North Africa. Geol Soc, Lond, Spec Publ 132:7–68. https://doi.org/10.1144/GSL.SP.1998.132.01.02

Bouaziz S, Barrier E, Soussi M, Turki MM, Zouari H (2002) Tectonic evolution of the northern African margin in Tunisia from paleostress data and sedimentary record. Tectonophysics 357:227–253. https://doi.org/10.1016/S0040-1951(02)00370-0

Boudjema A (1987) Evolution structurale du bassin pétrolier triasique du Sahara nord-oriental (Algérie). Thése Univ. Paris XI (Orsay), p 290

Boukadi N (1994) Evolution de l'Atlas de Tunisie: signification géométrique et cinématique des noeuds et des zones d'interférences structurales au contact de grands couloirs tectoniques. Thèse Doctorat Es Sciences, Univ. Tunis, p 250

Bruna PO, Bertotti G, Ben Amor S, Nasri A, Ouahchi S (2019) Vertical movements and petroleum system modelling in 1 the Southern Chotts Basin, Central Tunisia. In: Conference of the Arabian Journal of Geosciences, 25–28 Nov 2019

Burollet PF, Desforges G (1982) Dynamique des bassins neo crétacés en Tunisie. Mém Géol Univ Dijon 7:381–389

Busson G (1967) Mesozoic of Southern Tunisia. In: Guidebook to the geology and history of Tunisia, P.E.S.L., 9th annual field conf. Tripoli, pp 131–152

Busson G (1970) Rapports entre terrains mésozoïques et paléozoïques au Sahara Algéro-tunisien: La discordance hercynienne. C R Acad Sci Paris 269:685–688

Cunningham SM (1989) Gothlandian source rocks discovered north of Talemzane arch, Tunisia. Bull Am Assoc Pet Geol 72:505–540

Echikh K (1998) Geology and hydrocarbon occurrences in the Ghadames Basin, Algeria, Tunisia, Libya. Geol Soc, Lond, Spec Publ 132:109–129. https://doi.org/10.1144/GSL.SP.1998.132.01.06

El Euchi H, Kebaier D, Ferjaoui M (2003) Discontinuités et déformations tectoniques dans le Paléozoique de Sud Tunisien. In: ATEIG-Chronologie des évènements tectoniques en Tunisie, pp 9–13

El Rabia A, Inoubli MH, Ouajaa M, Abidi M, Sebei K, Jlaillia A (2018) Salt tectonics and its effect on the structural and sedimentary evolution of the Jeffara Basin, Southern Tunisia. Tectonophysics 744:350–372. https://doi.org/10.1016/j.tecto.2018.07.015

ETAP (2018) Annual report

Fello N, Lüning S, Štorch P, Redfern J (2006) Identification of early Llandovery (Silurian) anoxic palaeo-depressions at the western margin of the Murzuq Basin (southwest Libya), based on gamma-ray spectrometry in surface exposures. GeoArabia 11:101–118. https://pubs.geoscienceworld.org/geoarabia/article/11/3/101/566911/identification-of-early-llandovery-silurian-anoxic

Gabtni H, Jallouli C, Mickus KL, Zouari H, Turki MM (2006) The location and nature of the Telemzane High-Ghadames basin boundary in southern Tunisia based on gravity and magnetic anomalies. J Afr Earth Sci 44(3):303–313

Gabtni H, Jallouli C, Mickus KL, Zouari H, Turki MM (2009) Deep structure and crustal configuration of the Jeffara Basin (Southern Tunisia) based on regional gravity, seismic reflection and borehole data: how to explain a gravity maximum within a large sedimentary basin. J Geodyn 47:142–152. https://doi.org/10.1016/j.jog.2008.07.004

Galeazzi S, Haddadi N, Mather J, Druesne D (2010) Regional geology and petroleum systems of the Illizi-Berkine area of the Algerian Saharan Platform: an overview. Mar Pet Geol 27:143–178

Ghenima R (1993) Étude des roches mères paléozoïques du Bassin de Ghdames. Modélisation de la migration des hydrocarbures et application à l'étude du gisement d'El Borma. Thèse de Doctorat de l'Université d'Orléans, p 276

Hallett D, Clark-Lowes D (2016) Chapter 2 - Plate tectonic history. In: Petroleum geology of Libya, 2nd ed. Elsevier, pp 35–53

Hlaiem A (1998) Etude géophysique et géologique des bassins et des chaînes de Tunisie centrale et méridionale durant le Mésozoïque et le Cénozoïque: évolution structurale, modélisation géothermique et implications pétrolières. Thèse Doctorat de troisième cycle, Univ. Pierre & Marie Curie, Paris VI, p 315

Hlaiem A (1999) Halokinesis and structural evolution of major features in eastern and southern Tunisian Atlas. Tectonophysics 306:79–95

Kebaier D, Laridhi Wazaa N, Hamouda F (2003) Volcanic intrusions in Southern Tunisia–characterisation and development mechanism. In: 1st EAGE North African/Mediterranean petroleum & geosciences conference & exhibition, 06 Oct 2003. EEGS, EAGE

Khomsi S, Bédir M, Soussi M, Ben Jemia MG, Ben Ismail-Lattrache K (2007) Reply to comment on the paper Mise en évidence en subsurface d'événements compressifs Eocène moyen-supérieur en Tunisie orientale (Sahel): Généralité de la phase atlasique en Afrique du Nord. C R Géosci (2006) 338(1–2):41–49, C R Géosci 339:173–177

Khomsi S, Ben Jemia MG, Frizon de Lamotte D, Maherssi C, Echihi O, Mezni R (2009) An overview of the late cretaceous–Eocene positive inversions and oligo-Miocene subsidence events in the foreland of the Tunisian atlas: structural style and implications for the tectonic agenda of the Maghrebian atlas system. Tectonophysics 475:38–582. https://doi.org/10.1016/j.tecto.2009.02.027

Khomsi S, Roure F, Khelil M, Mezni R, Echihi O (2019) A review of the crustal architecture and related pre-salt oil/gas objectives of the eastern Maghreb Atlas and Tell: need for deep seismic reflection profiling. Tectonophysics 766:232–248. https://doi.org/10.1016/j.tecto.2019.06.009

Laridhi-Ouazaa N (1994) Etude minéralogique et géochimique des épisodes magmatiques mésozoïques et miocènes de la Tunisie. Unpublished thesis, Doctorat d'Etat, Université Tunis II, p 466

Laridhi-Ouazaa N, Mattoussi Kort H, Kassaa S, Bougadir B, Saidi A (2005) Activités magmatiques en Tunisie associées à l'ouverture de la Téthys. In: First international conference on the geology of the Tethys, Cairo University, Egypt, vol 2

Lazzez M, Zouaghi T, Ben Youssef M (2008) Austrian phase on the northern African margin inferred from sequence stratigraphy and sedimentary records in southern Tunisia (Chotts and Djeffara areas). C R Geosci 340:543–552. https://doi.org/10.1016/j.crte.2008.05.005

Le Heron DP, Craig J, Sutcliffe OE, Whittington R (2012) Late Ordovician glaciogenic reservoir heterogeneity: an example from the Murzuq Basin, Libya. In: Regional geology and tectonics: principles of geologic analysis, pp 452–489

Lüning S, Craig J, Loydell DK, Storch P, Fitches B (2000) Lower Silurian 'hot shales' in North Africa and Arabia: regional distribution and depositional model. Earth Sci Rev 49:121–200. https://doi.org/10.1016/S0012-8252(99)00060-4

Lüning S (2005) AFRICA | North African Phanerozoic. In: Selley RC, Cocks LRM, Plimer IR (eds) Encyclopedia of geology. Elsevier, Oxford, pp 12–25

Mejri F, Burollet PF, Ben Ferjani A (2006) Petroleum geology of Tunisia. A renewed synthesis. Memoir ETAP no 22, p 233

Pogacsas GY, Rumpler J, Koncyz I, Hassi J, Samu L (1996) Tectonostratigraphic evolution and related hydrocarbon habitat of the Kibili area, Central Tunisia. ETAP Memoire no 10, pp 209–223

Pogacsas GY, M'rabet A, Hassi J, Samu L, Vakarcs G (1998) Paleozoic-Mesozoic facies evolution and related hydrocarbon system of the Kebili area (Central Tunisia). ETAP Memoire no 12, pp 285–307

Rezouga N, Belhaj MA, Saidi M, Bouazizi I (2012) Assessment of unconventional shale reservoir the Fegaguira Fm Chotts Basin Tunisia. In: North Africa technical conference and exhibition, 20–22 Feb 2012. Society of Petroleum Engineers, Cairo, Egypt

Rigo F (1996) Quartzarenite reservoir quality related to structural deformation. In: Proceedings of the 5th Tunisian petroleum exploration and production conference. ETAP Memoir no 10, pp 225–232

Saïdi M, Inoubli M, Ghenima R, Ben Nasr D (1998) Caractéristiques géochimiques des roches mères en Tunisie. Rapport interne ETAP

Sarsar Naouali B, Guellala R, Bey S, Inoubli MH (2016) Gravity data contribution for petroleum exploration domain: Mateur case study (Saliferous Province, Northern Tunisia). Arab J Sci Eng 42:339–352

Soua M (2012) Orbital forcing and Milankovitch cyclicity constraint during Permian deposition on the Northern margin of Africa, Southern Tunisia. North Afr Tech Conf Exhib, Soc Pet Eng J 2012:15

Soua M (2014) Palaeozoic oil/gas shale reservoirs in southern Tunisia: an overview. J Afr Earth Sci 100:450–492. https://doi.org/10.1016/j.jafrearsci.2014.07.009

Troudi H (2006) Integrated study of the Norian Tags and Paleozoic (Upper Silurian F6 and Ordovician) reservoirs in Hamra Perimeter/Illizi Basin -Algeria. Internal reports: vol no 1, vol no 2, vol no 3

Underdown R, Redfern J (2007) The importance of constraining regional exhumation in basin modelling: a hydrocarbon maturation history of the Ghadames Basin, North Africa. Pet Geosci 13:253–270. https://doi.org/10.1144/1354-079306-714

Underdown R, Redfern J (2008) Petroleum generation and migration in the Ghadames Basin, north Africa: a two-dimensional basin-modeling study. AAPG Bull 92:53–76. https://doi.org/10.1306/08130706032

Van de Weerd AA, Ware PLG (1994) A review of the East Algerian Sahara oil and gas province (Triassic, Ghadames and Illizi basins). First Break 12:363–373. http://www.earthdoc.org/publication/publicationdetails/?publication=28350

Zargouni F (1985) Tectonique de l'Atlas méridional de Tunisie, évolution géométrique et cinématique des structures en zone de cisaillement. Thèse d'état, Université Louis Pasteur. Strasbourg. Edit Mém INRST 3(1986):302

Zouaghi T, Guellala R, Lazzez M, Bédir M, Ben Youssef M, Inoubli MH, Zargouni F (2011) The Chotts fold belt of Southern Tunisia, North African margin: structural pattern, evolution, and regional geodynamic implications. IntechOpen 55

Zouari H (1995) Evolution géodynamique de l'Atlas centro-méridional de la Tunisie. Thèse d'Etat. Univ. Tunis II, F.S.T, Tunis

Zulauf G, Romano SS, Doerr V, Fiala J (2007) Crete and Minoan terranes: age constraints from U-Pb dating of detrital zircons. Geol Soc Am Spec Pap 423:401–411

An Overview of the Eastern Atlas Fold and Thrust Belt and Its Foreland Basin Along the North–South Axis and Chorbane–Ktitir Platform: Surface/Subsurface Major Structures and Tectonic Evolution (North Africa)

8

Riadh Mezni and Sami Khomsi

Abstract

We present a structural overview of the major structures of the Eastern foreland basins of the Eastern Atlas and the N–S Axis, the Northern branch of the South Atlasic Front (SAF) and its relations with the Ktitir platform and Chorbane major structural anomaly. Different subsurface interpretations allow deciphering the major tectonic pulses in the Eastern Atlas and its foreland basin, from Jurassic–Early Cretaceous rifting to Upper Cretaceous–Late Eocene Atlas folding events well expressed in the surface and in the subsurface. Seismic interpretations allow visualizing the main control on the Triassic salt Diapirism related to deep-seated faulting affecting the Pre-Triasic and Meso-Cenozoic sedimentary sequences. Surface and subsurface interpretations show that the continuous reactivation of the deep-seated faults has trigged the lateral flow of the Triassic evaporites toward the salt diapirs. Furthermore, the salt material has played the role of a general décollement level of the post-Triassic cover during the shortening phases: Late Eocene Atlas phase and the Late Miocene–Quaternary Alpine phase. The Kairouan flexural basin, an intriguing subsiding zone at the front of the Atlas, coincides with a roughly E–W basin cutting orthogonally the N–S Axis and recording more than 8500 m of Meso-Cenozoic sedimentary sequence and recording the main Atlas tectonic pulses.

R. Mezni (✉) · S. Khomsi
Laboratoire Géoressources, Centre des recherches et des technologies des Eaux (CERTE), Pôle technologique de Borj Cédria, Université de Carthage, 8020 Soliman, Tunisia
e-mail: mezniriyadh79@gmail.com

R. Mezni
Faculté des Sciences de Tunis, Université de Tunis El Manar II, Tunis, Tunisia

S. Khomsi
Faculty of Earth Sciences, Geo-exploration Department, University King Abdulaziz (KAU), Jeddah, Saudi Arabia

Keywords

Eastern Atlas • N–S axis • Thrusting • Deep-seated faults • Triassic salt • Tectonic inversion • Major structures

8.1 Geological Framework of the Eastern Maghreb and Tunisian Atlas

The Atlas fold and thrust belt constitutes a major FABT system at the scale of Northern Africa (Fig. 8.1) with dominant morphology and thrusting structures striking roughly NE–SW. The Atlas corresponds also to an inversion intra-continental fold and thrust belt system issued from the positive uplift of Jurassic–Cretaceous sedimentary Tethyan basins of the Maghrebian margin (Bracène and Frizon de Lamotte 2002; Frizon de Lamotte et al. 2009; Khomsi et al. 2009b, 2019; Roure et al. 2012; Mezni et al. 2019). This paleo-rift system was mainly controlled by a roughly E–W and locally N–S bounding faults system (Khomsi et al. 2019). These extensional structures controlled the sedimentary thicknesses and facies of the sedimentary packages.

Regionally, the Maghrebian Atlas is running from Tunisia in the east to Morocco in the west (Fig. 8.1). In eastern Maghreb (Fig. 8.2), it is limited in its eastern and southern sides by the eastern Tunisia foreland, called Sahel, and by the Sahara Platform, respectively (Busson 1971; Frizon de Lamotte et al. 1998, 2000; Khomsi et al. 2016; Mezni et al. 2019).

The structural boundary separating the Atlas from its foreland (Fig. 8.2) basins in the east and south, corresponds to the South Atlas Front (Frizon de Lamotte et al. 1998, 2000; Khomsi et al. 2016; Mezni et al. 2019) which is an important, thrust fault system and buttressing domain along which the Atlas belt is overhanging tectonically its adjacent foreland basins (Frizon de Lamotte et al. 1998; Khomsi et al.

S. Khomsi and F. Roure (eds.), *Geology of North Africa and the Mediterranean: Sedimentary Basins and Georesources*, Regional Geology Reviews, https://doi.org/10.1007/978-3-031-18747-6_8

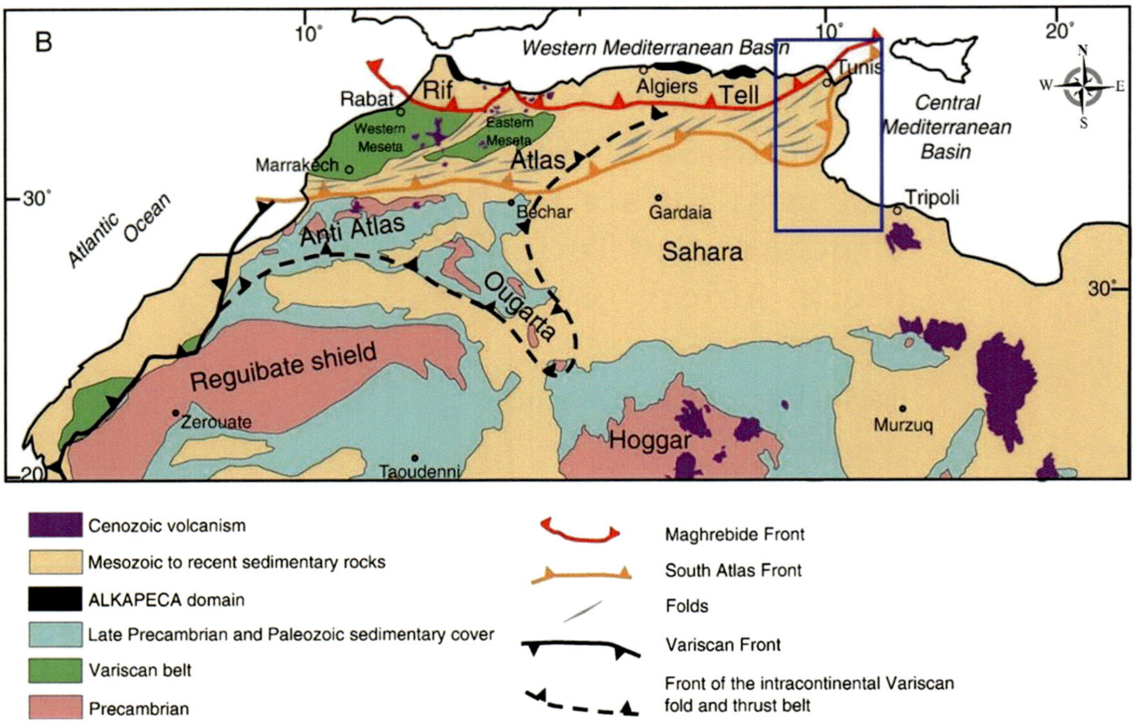

Fig. 8.1 Geological map of the Maghreb showing the main structural domains (Frizon de Lamotte et al. 2009)

2004a, b, 2016, 2019) and admitting the Triassic salt as major décollement level.

Globally, Tunisia can be divided into five structural domains that are from north to south (Fig. 8.2):

- The Tell domain interpreted as an "Alpine" belt and constituted by several thrust nappes displaced from northwest to southeast (Rouvier 1977; Zargouni and Abbès 1987; Khelil et al. 2021).
- The Mejerda valley–Kechabta Basin corresponding to a complicated structural domain with Triassic salt plugs displaced horizontally (Rouvier 1977; Jauzein and Perthuisot 1981; Rigo et al. 1996; El Euchi et al. 2002, 2004; Khomsi et al. 2012; Khelil et al. 2020).
- The Atlas belt corresponds to a fold and thrust belt generally trending NE–SW (Turki 1985; Haller 1983; Zargouni and Abbès 1987; Bédir 1995; Hlaiem 1999; Mezni et al. 2019; Khomsi et al. 2004a, b; 2019; Fig. 8.2).
- The Eastern Tunisia Foreland is composed by the Gulf of Hammamet (Messaoudi and Hammouda 1994; Bédir 1995; Patriat et al. 2003; Gharsalli et al. 2013; Ben Brahim et al. 2014; Bédir et al. 2016), the Sahel plain and the Gulf of Gabes (Blanpied and Bellaiche 1983; Patriat et al. 2003; Taktak et al. 2012; Khomsi et al. 2019) to the south (Fig. 8.2). This foreland domain represents an important deformation zone with thrusting and folding (Haller 1983; Turki et al. 1988; Hlaiem 1999; Khomsi et al. 2004b).
- The Sahara platform corresponds to Paleozoic–Mezosoic basins (Busson 1971; Zargouni and Abbès 1987; Burollet et al. 2000; Bouaziz et al. 2002; Raulin et al. 2011; Rigane and Gourmelen 2011).

The Atlas fold and thrust belt of the Maghreb (Figs. 8.2 and 8.3) represents an active orogenic domain since the Upper Cretaceous (Frizon de Lamotte et al. 2009; Khomsi et al. 2009a, 2022; Roure et al. 2012).

8.2 Stratigraphy Overview

The Pre–Late Cretaceous deposits correspond to the Triassic, Jurassic, and the Early Cretaceous series. These Pre–Late Cretaceous sequences are made up of sands, evaporites, and dolomite rocks for the Triassic (Figs. 8.4, 8.5, and 8.6). The Jurassic is represented by an important carbonate platform deepening to the north and controlled by grabens, hemi-grabens, and horsts (see Frizon de Lamotte et al. 2009), allowing transitional facies between deep marine environments to the north and an internal platform to the south. The major faults controlling the Jurassic platform correspond to basement faults (Bédir et al. 2001; Khomsi et al. 2004b) as the Kairouan–Sousse fault (FKS, see Khomsi et al. 2004b) a major E–W fault cutting the Sahel foreland and From the N–S axis (Fig. 8.2).

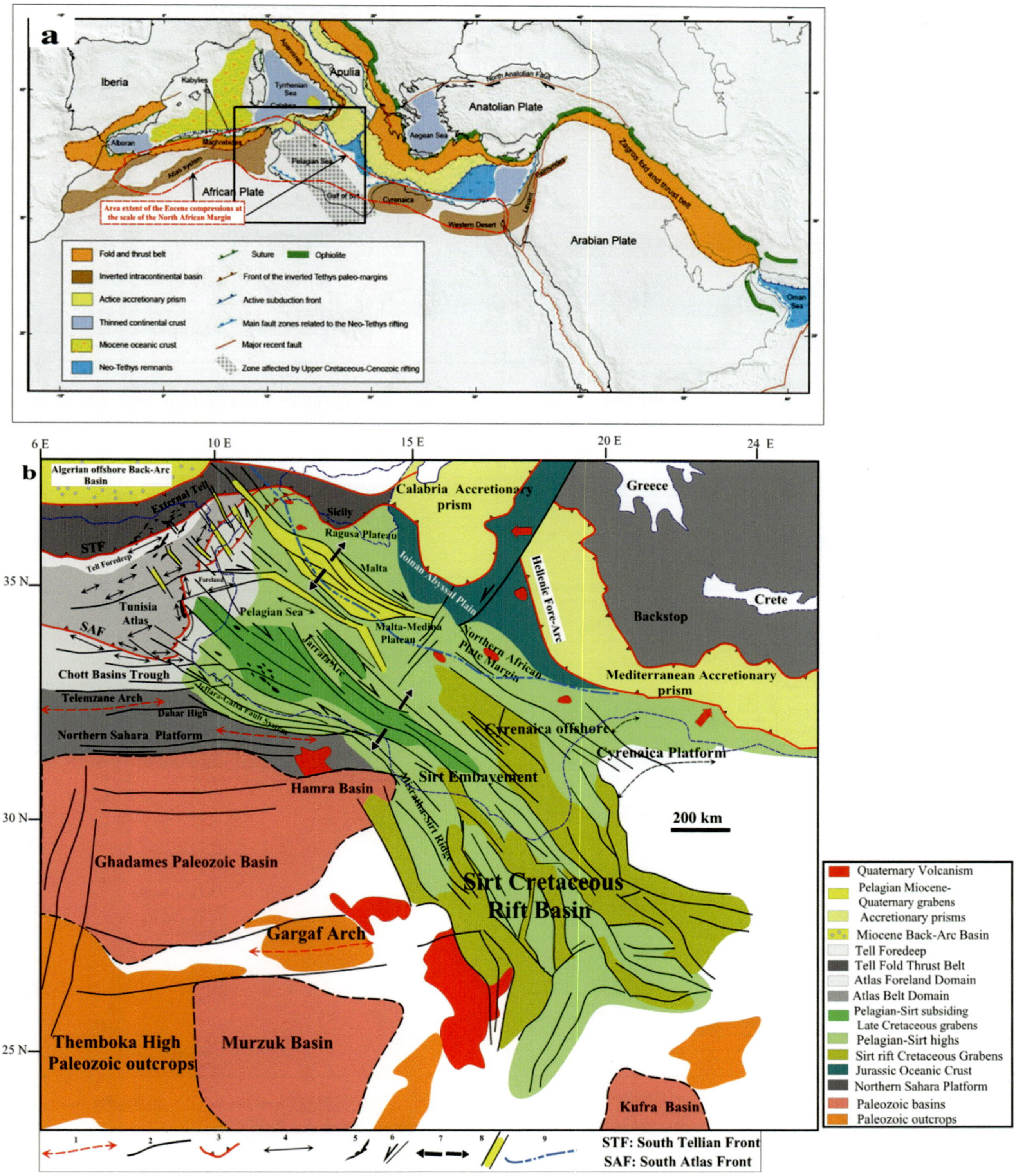

Fig. 8.2 Geotectonic maps of the Circum Mediterranean domains and Central Mediterranean. **a** Structural domains of the Mediterranean Area and North Africa (modified from Frizon de Lamotte et al. 2011). **b** Detailed structural features of the Eastern Atlas and its foreland basins with the major structures in the Pelagian-Sirt domain. Captions. 1: major arches, 2: major fault, 3: major thrust, 4: major anticline, 5: salt bodies, 6: major strike-slip, 7: strike of opening of grabens, 8: graben (Khomsi et al. 2016). Notice that the Sirt graben is orthogonal to the Atlas trending NE structure in the Eastern Maghreb foreland. Notice the SAF

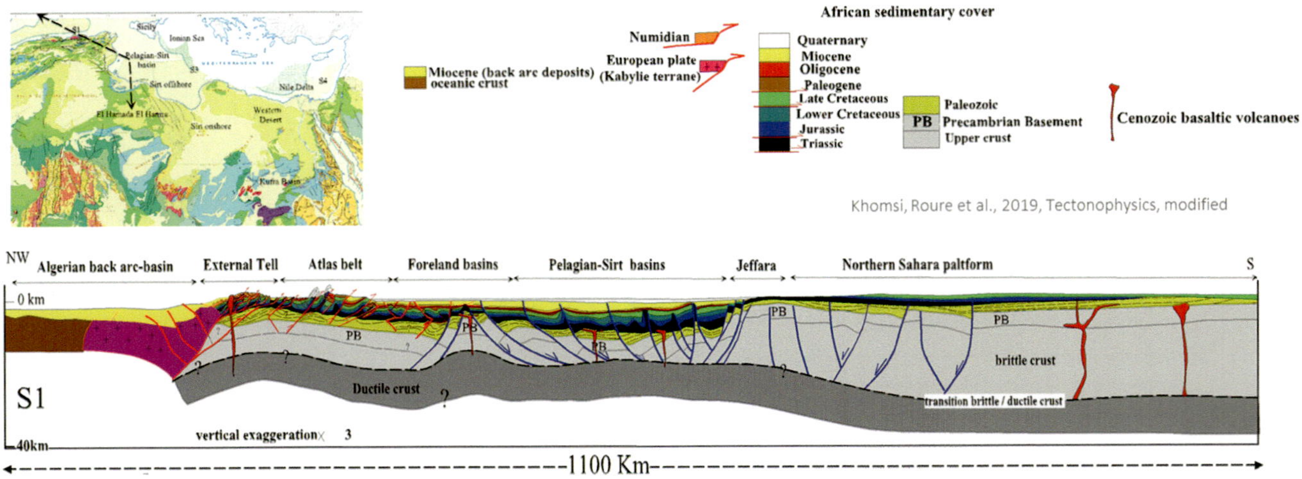

Fig. 8.3 Regional structural cross-section along the Eastern Maghreb Domain, throughout the major structures from the Tell to the Sahara Reprinted from Khomsi et al. (2019). Copyright (2019), with permission from Elsevier

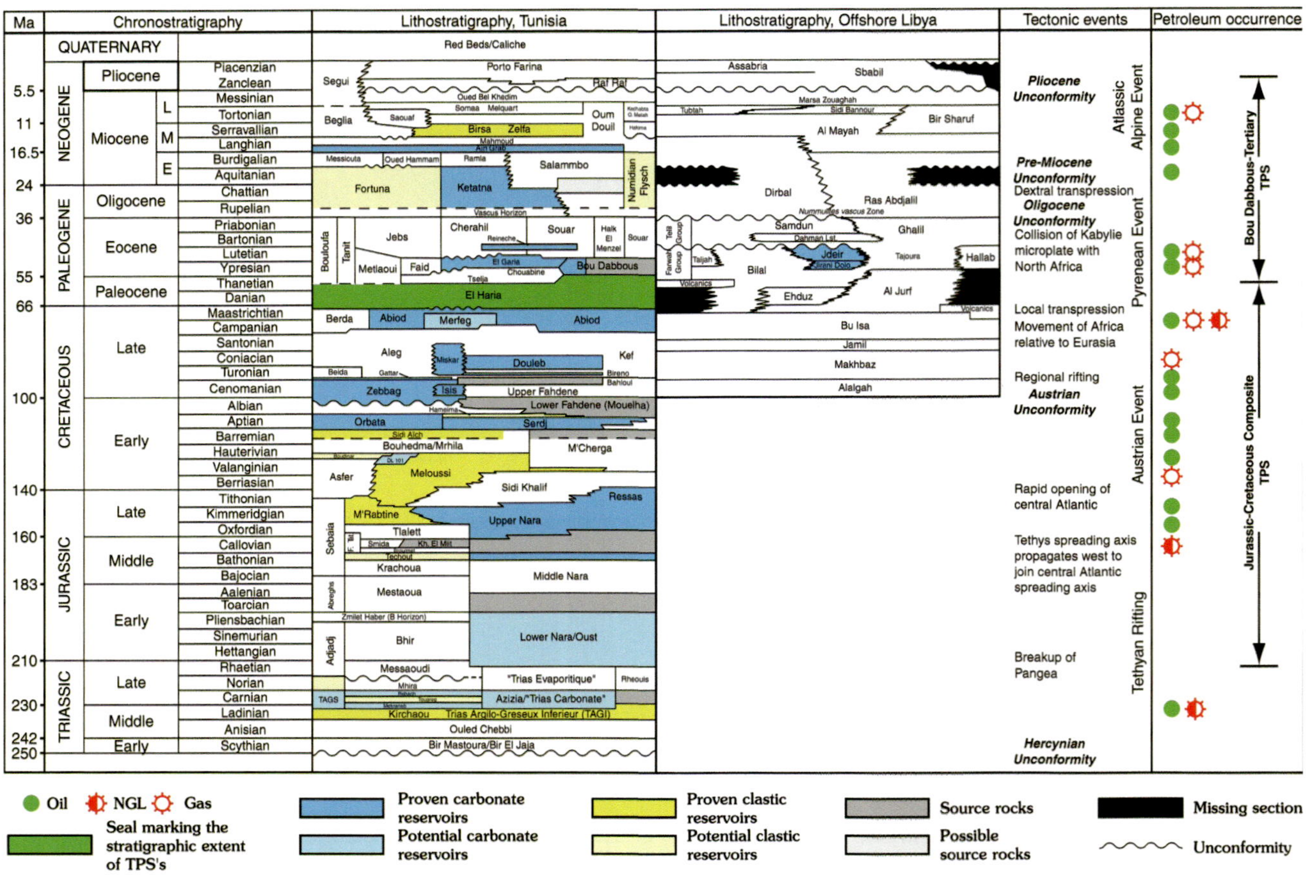

Fig. 8.4 Lithostratigraphy Nomenclature, petroleum systems, reservoirs, and plays of the Triassic–Neogene of Tunisia and Pelagian Sirt of Lybia (Etap 1998)

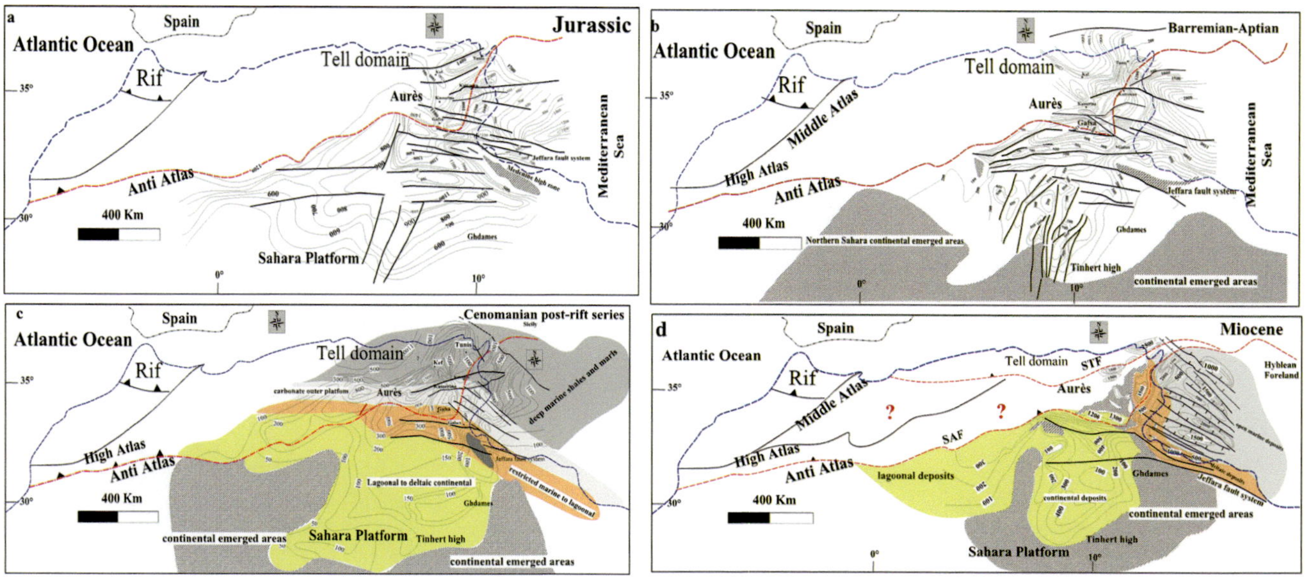

Fig. 8.5 Iso-thickness maps at the scale of Eastern Maghreb for the Jurassic series (**a**), the Barremian–Aptian (**b**), the Cenomanian (**c**), and the Miocene (**d**), respectively. Maps combined from Busson (1971) and Ben Ferjani et al. (1990). Notice that the Jurassic and Barremian–Aptian periods correspond to the syn-rift phases of the Tethyan realm. The maps show that the predominant majors' faults controlling thickness variations are trending to the east and to the south. East-trending faults are predominant in the Atlas belt and at the northern edge of the Saharan platform. The iso-thickness map of Miocene series shows that the Miocene deposits extend to the south over the Saharan platform and correspond to continental sediments essentially derived from the erosion of the uplifted areas. Notice also the great sedimentary thickness recorded south of the SAF. The map shows also the Pelagian-Sirt fault system

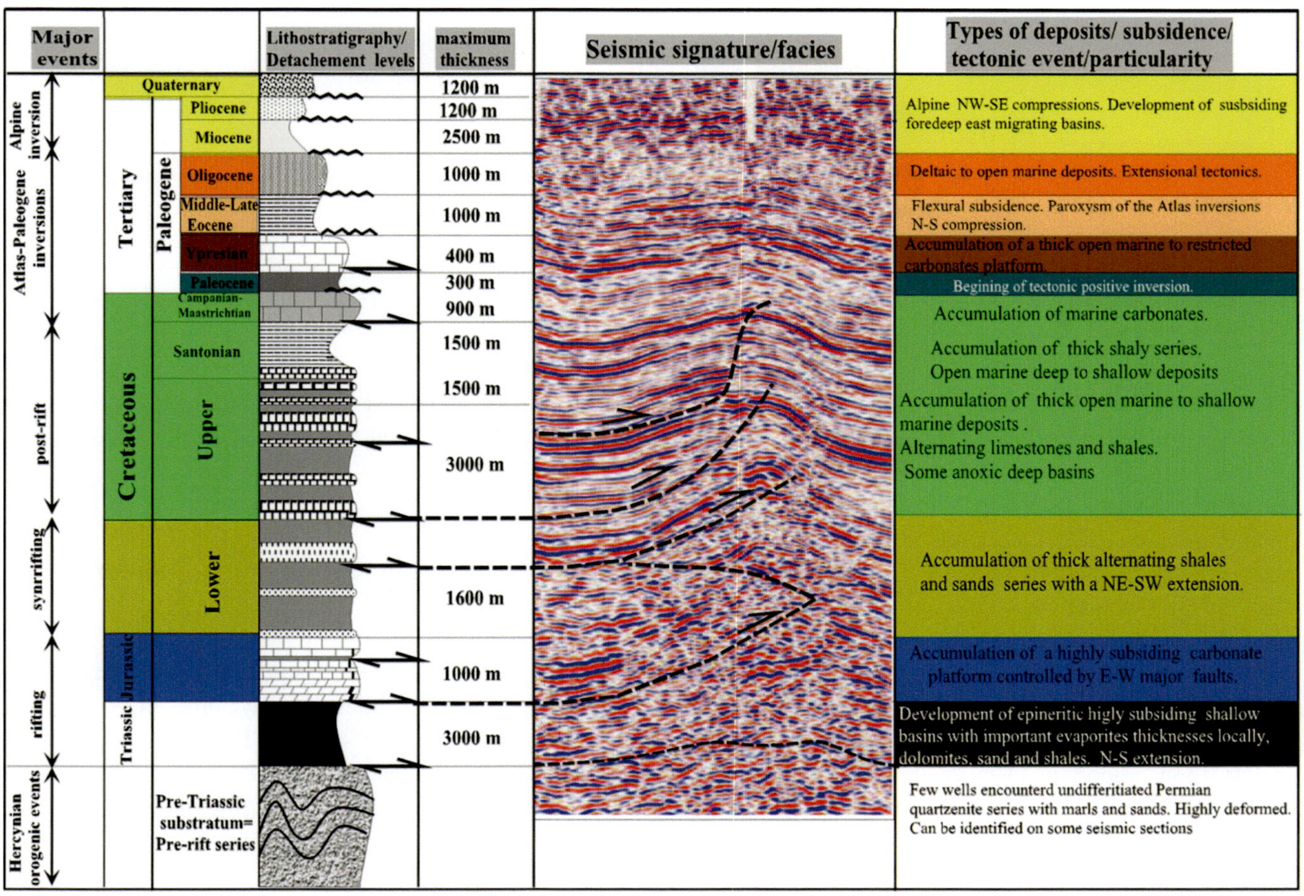

Fig. 8.6 Major décollement levels and types of deposits for the Triassic–Quaternary series in Eastern Atlas foreland. Reprinted by permission from (Springer-Nature) [Arabian Journal of Geosciences] Khomsi et al. (2016). Copyright (2019)

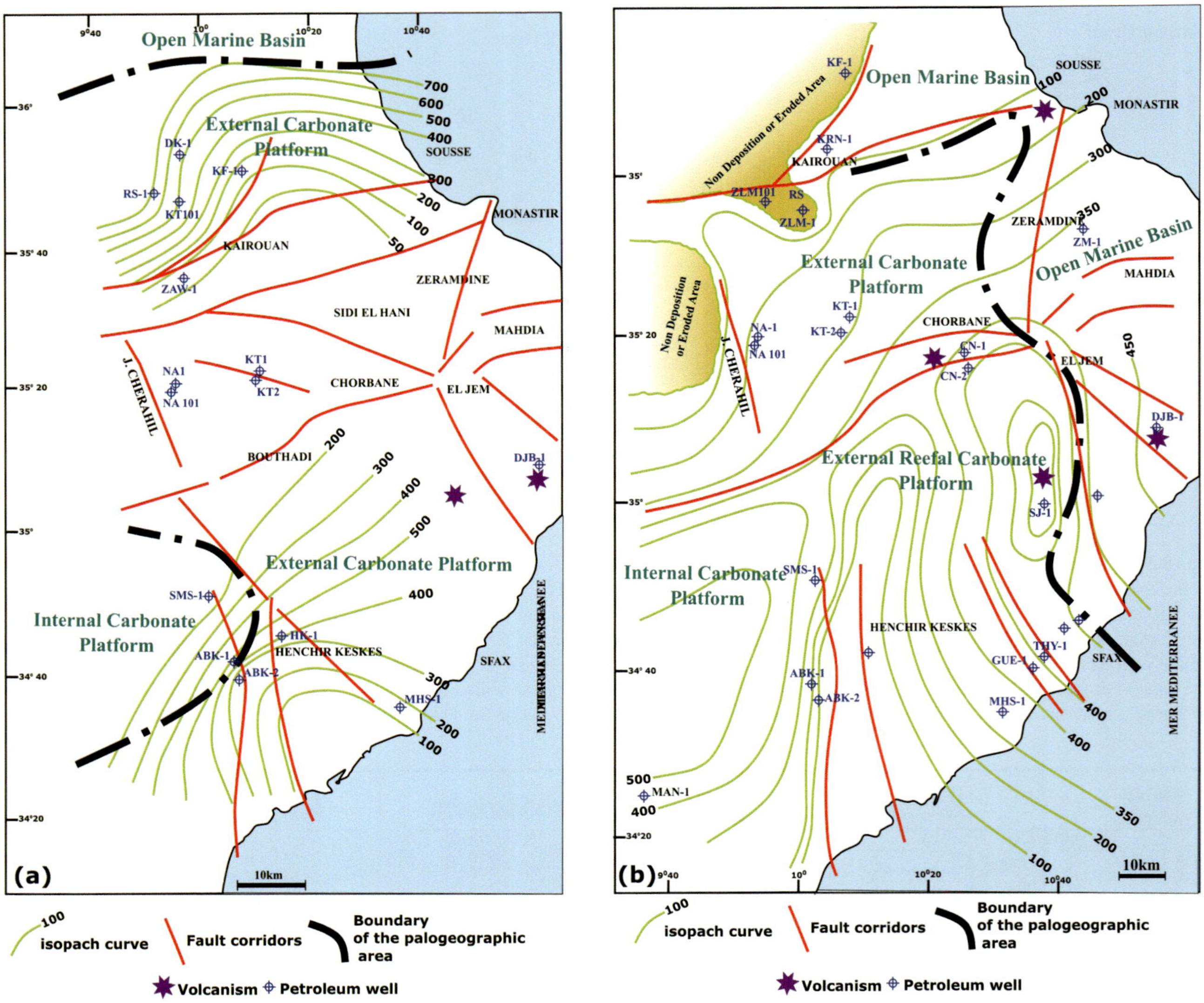

Fig. 8.7 Iso-thickness and paleo-sedimentary environments in Eastern Tunisia, respectively, for the Aptian (left) and Cenomanian (right). From Bédir (1995)

The Late Cretaceous interval (Figs. 8.7, 8.8, and 8.9) includes the Fahdene Formation (Clansayesian to Cenomanian, including the terminal horizon Bahloul (Late Cenomanian to Earliest Turonian)), the Aleg Formation, and Abiod Formation. The Late Cretaceous sequences in the Sahel (the central part of Oriental Tunisia) include thick units (locally reaching more than 1800 m) of carbonates, marls, and argillaceous limestones that were deposited in shallow marine to deeper environments ranging from an inner platform (Bireno member) to a pelagic and hemipelagic environment (Aleg and Abiod Formation). These deposits show important changes in lithologic facies and thickness, especially for the Abiod Formation (Negra 1994; Negra et al. 1996) related to the sysnsedimentary control by the major E–W and N–S inherited faults (Bédir 1995; Khomsi et al. 2004a, b).

8.2.1 Clansayesian–Cenomanian: Fahdene Formation

This sequence is made up of alternating black clays, limestones beds, and fossiliferous clay limestones with the presence of levels of black marls. Then at the base, some sandstone beds intercalation (Fig. 8.9). At the top of the Fahden Formation, a bed of black bituminous limestones

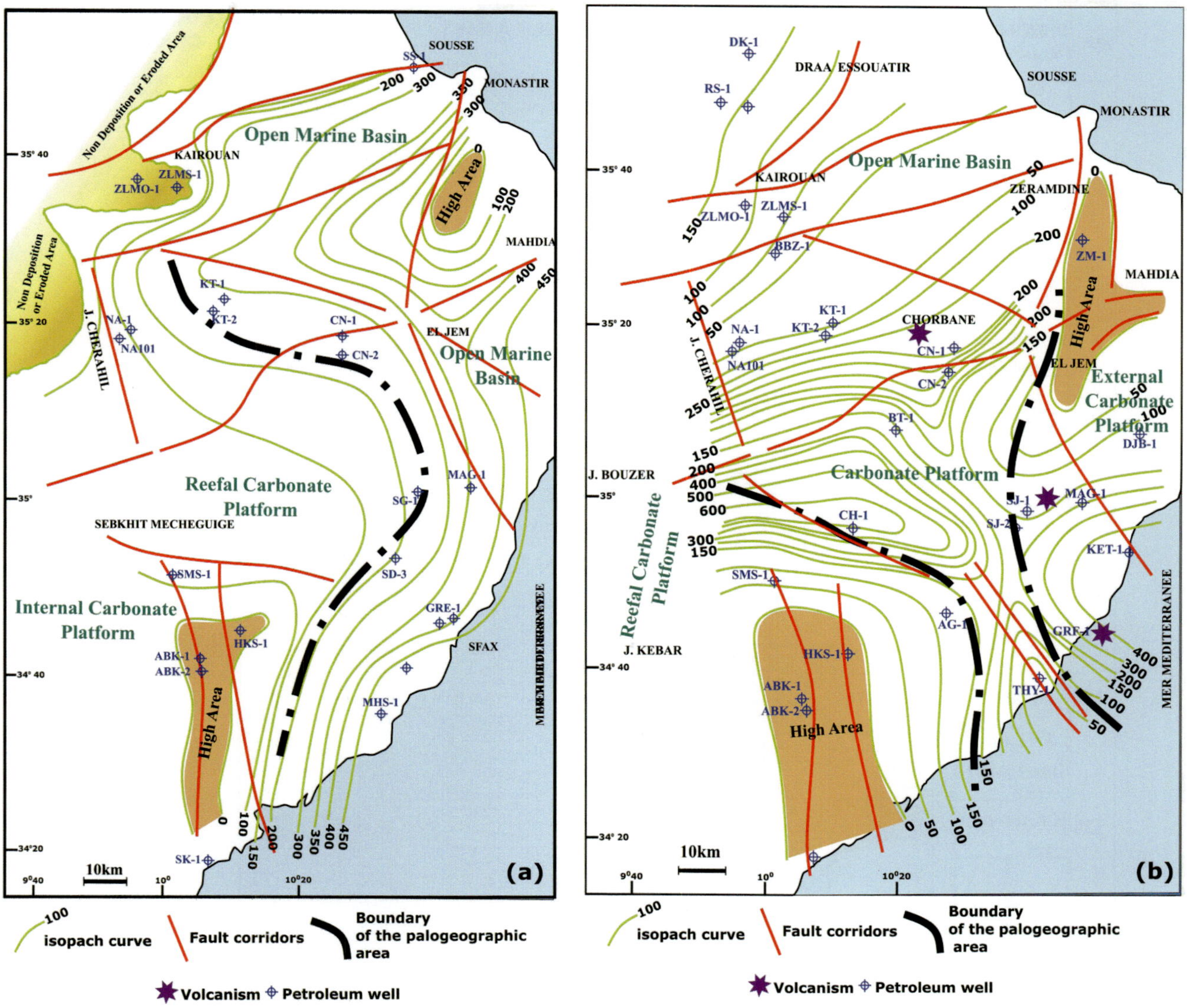

Fig. 8.8 Iso-thickness and paleo-sedimentary domains environments in Eastern Tunisia, respectively, for the Turonian (left) and Campanian–Maastrichtian (right). From Bédir (1995) and Bédir et al. (2018). The Bireno limestones, well-characterized to the northwest of the Kasserine Island and in Eastern Tunisia, is generally a coarse calcarenite; in some places, there are reefs of Rudistids

rich in fossils corresponds to the Bahloul member and represents an excellent source rock at the scale of Northern Africa basins.

8.2.2 Turonian–Santonian: Aleg Formation

The Aleg Formation is a thick series of gray marl and shale, interbedded between the Fahdene Formations and the Abiod Formation (Figs. 8.8 and 8.9). Aleg should begin at Turonian base, including Coniacian and Santonian series, and finish in the Lower Campanian. The Aleg Formation includes two sequences in the Lower part: the first sequence includes Annaba and Bireno Members, more or less Lower and Middle Turonian; the second sequence is the Douleb sequence, including Upper Turonian and Lower Coniacian series.

Near Sfax is the reservoir of three oil fields, with packstone, wackestone, or oolitic facies. Bishop (1988) gave a good evaluation of the Annaba–Bireno interval in Tunisia, especially around Sfax, both on land and offshore. The Upper parts of the Aleg Formation may also form good reservoirs. A good example is the Coniacian Rudistid reef and the Oyster coquinas developed at Jebel Bou Zer, south of Faid in the North–South axis (Bismuth et al. 1985).

Unconformities are known in the Aleg Formation and at the base of Abiod limestones; the main unconformities at the base of Santonian shales: in the North–South Axis and

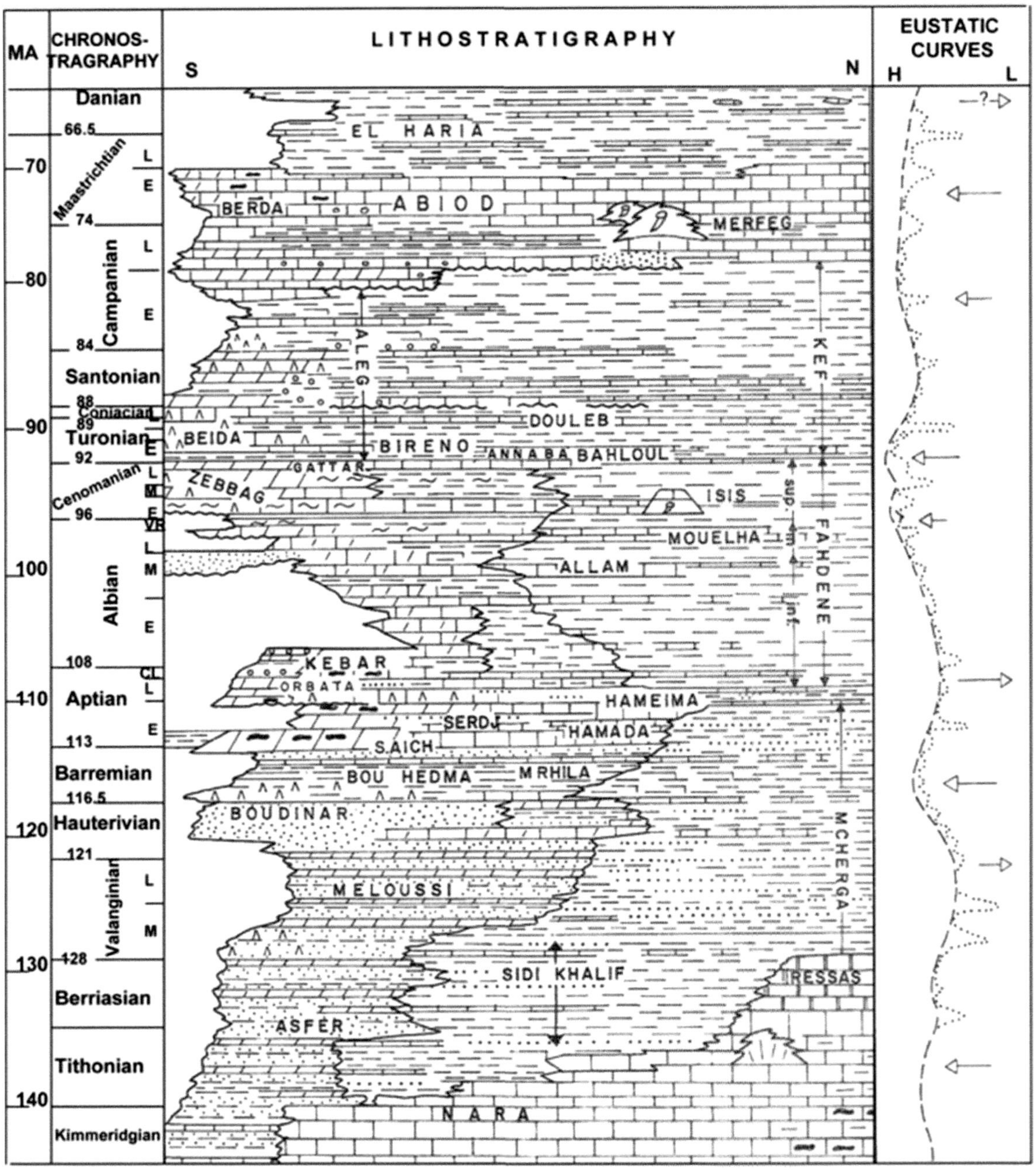

Fig. 8.9 Lithostratigraphic chart of the Upper Jurassic–Cretaceous facies distribution and formations (from Mejri et al. 2006)

subsurface basins of Eastern Tunisia (Bramaud et al. 1976; Khessibi 1978; Fournié 1978). Intense submarine erosion cut the previous series, often tilted by the salt Diapirs.

8.2.3 Campanian–Early Maastrichtian: Abiod Formation

The Abiod Formation, Campanian–Maastrichtian in age, is essentially made up of carbonates (Fig. 8.9), generally chalky limestones, and it is subdivided into three members:

- Lower carbonates member: it is made up of chalky mudstone, rich in foraminifera (Globigerinidae, Globotruncanidae, and Orbitoides).
- Middle member: it is generally made up of marl and clay. The green clay contains interbedded argillaceous limestones in the lower part, rich in benthic foraminifera.
- Upper carbonates member: this member shows three sub-members from the base to the top:
  - Chalky limestones with argillaceous levels; begins with a calcareous glauconitic sandstone.
  - Interbedded chalky limestones, white, and light gray marl.

Fig. 8.10 Field photos of the Campanian–Maastrichtian Abiod Formation outcropping in the Cherahil fold with **a** panoramic view of the east limb of the Cherahil north showing the stratigraphic unites: Abiod Formation, El Haria shale Formation, Metlaoui Formation which overlies by the shale with interbedded limestones beds of the Cherahil (A) Formation. **b** The Abiod limestones sequence with interbedded calcarenite beds. **c** Limestones bed surface with presence of trace of Inocerames. **d** Photo showing a slump indicating instability of the sedimentary floor

– Massive chalky limestones with rare green shale horizons, irregularly stratified with slumps, Inocerames prints, and frequent Echinids in the upper part. In the middle, it shows locally turbidites made of calcarenite and nodular fragments of coral (Figs. 8.10 and 8.11).

The Abiod Formation extends over Northern and Central Tunisia and is known in the boreholes in Eastern Tunisia and the Pelagian Sea.

In fact, along the N–S Axis, the Abiod Formation is sandy, at least partially, due to the reworking of Lower Cretaceous sandstones. Then, at Jebel Merfeg, on the southwest flank of Jebel Kebar, a reefal facies was described by Khessibi in 1978. There are organic reefs, with Rudistid and Coral calcareous mud mounds, talus, grading to breccias, and chalky mudstone, rich in Calcisphaeridae: it is the Merfeg Formation (Negra 1994).

In numerous places, the sedimentation of the Abiod Formation overlies various older series. These unconformities are especially frequent along the NOSA and on apical parts of Diapirs. In addition, Turbidites, mudflows, and slumps are frequent in the Abiod Formation, which indicate instability of the deposit floor (Fig. 8.10).

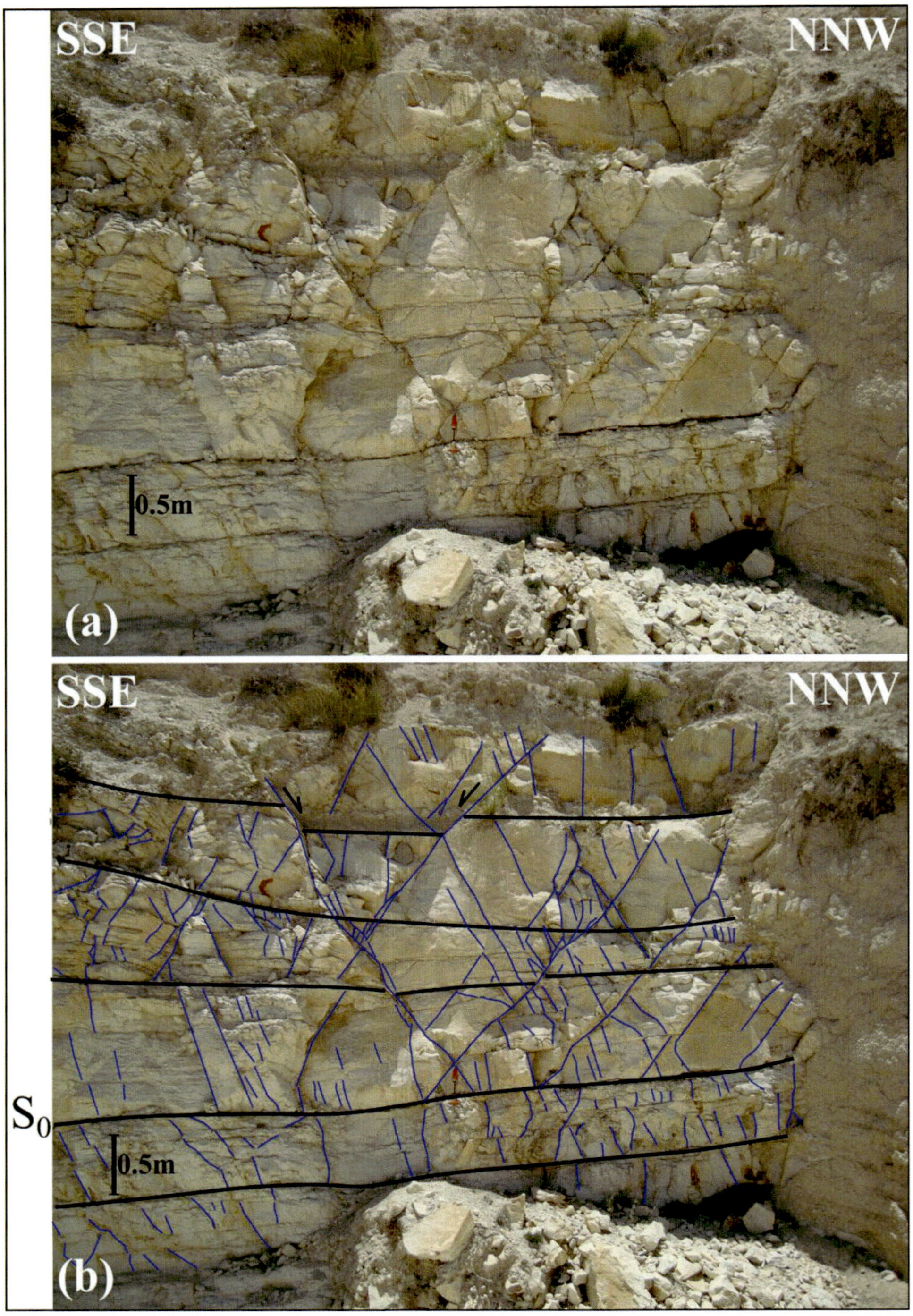

Fig. 8.11 Field photo of the southern part of the Cherahil structure showing limestones beds of the Abiod Formation cross cut by a set of systematic conjugate extension fractures

8.2.4 Late Maastrichtian–Paleocene: El Haria Formation

Throughout Tunisia, the Cretaceous–Paleogene transition is materialized by the El Haria Formation (Fig. 8.12). Burollet defined this latter in 1956, which is Upper Maastrichtian–Paleocene in age (Burollet 1956). The Haria Formation is formed by black to gray clays with intercalation of a few limestones and clay-limestones beds.

These infra-neritic marine facies are only complete in northern Tunisia, with a thickness that varies between 400 and 1000 m (Burollet 1956; Khomsi et al. 2021). Meanwhile, the El Haria Formation is thinning in Central and Central-Eastern Tunisia. Moreover, it is represented locally by a very thin

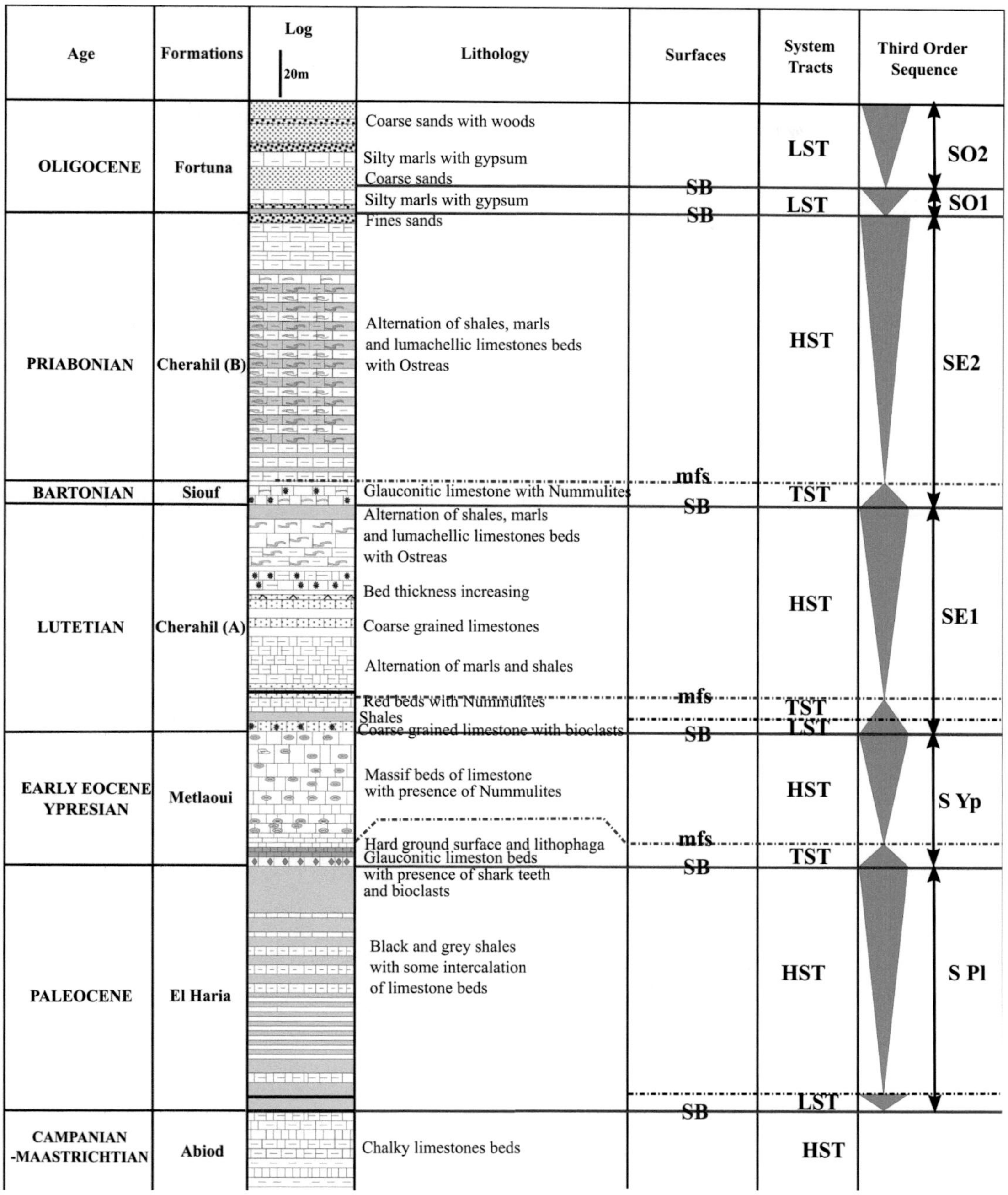

Fig. 8.12 Sequence stratigraphy of the Late Cretaceous–Oligocene deposits in the Jebel Cherahil, N–S axis with the formations and lithologies (Khomsi 2005)

level of lagoonal-littoral clays and phosphate with benthic foraminifera (Comte and Dufaure 1973; Yaich 1984).

The lateral facies and thickness variations of the El Haria Formation are linked to the paleogeography inherited from the Upper Cretaceous tectonic regime, (Turki 1985) with the starting of the Upper-most Cretaceous compressions and vertical inversions (Khomsi et al. 2019).

8.2.5 Ypresian (Early Eocene): Metlaoui Formation and Its Equivalents

The Metlaoui Formation, Late Paleocene–Early Eocene in age, is formed by limestones to dolomitic limestones beds with intercalations of clay and conglomerate lenses of Debris flow deposits at the top (Abbés 2004).

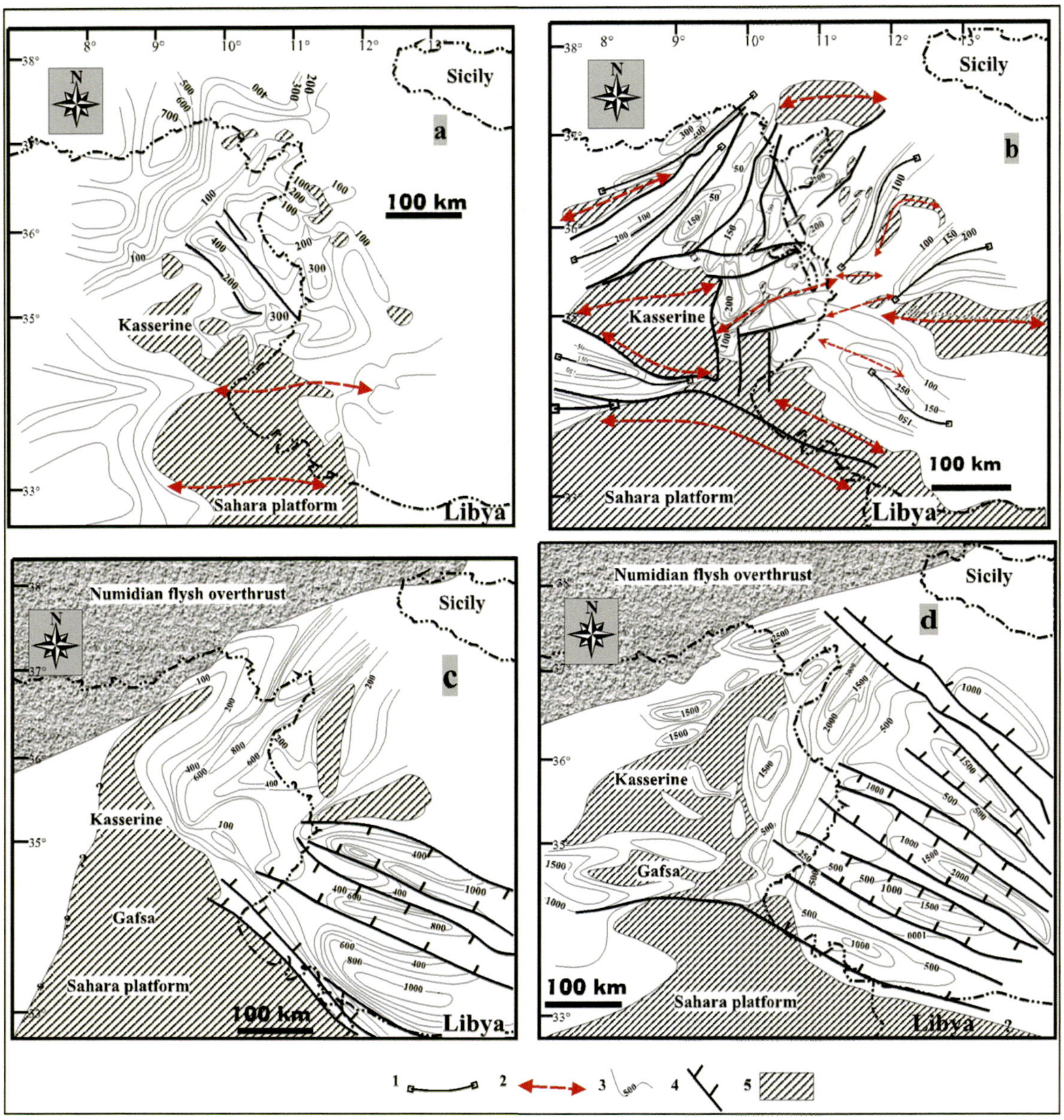

Fig. 8.13 Paleogeographic summary maps with iso-thickness curves of Ypresian to Miocene series in Tunisia and bordering areas, modified from Bishop (1988); Ben Ferjani et al. (1990); and Zaier et al. (1998). **a** Lower Eocene, **b** Middle–Late Eocene, **c** Oligocene, **d** Miocene. 1: Subsiding synforms, 2: active folds

According to Ben Ismail-Lattrache (1994), the Paleocene–Eocene transition is generally marked by a micro-conglomerate beds and phosphate nodules.

In the northern part of the North–South axis range belt and Jebel Cherahil, the Metlaoui Formation changes facies where it is represented by Nummulite limestones known as the El Gueria Formation (Fournie 1978) (Figs. 8.13, 8.14, 8.15, 8.16, and 8.17). While toward the north of Tunisia, the El Gueria facies becomes formed by a series of clay limestones with Globigerines. The latter characterizes the Bou Dabbous Formation (Fournie 1978). The transition from Nummulitic facies to globigerina facies occurs via a bioclastic transition facies known as the Nummulitoclasts (Ben Jemia Fakhfakh 1991; Rigane 1991; Abbés 2004). Toward the southern part of the N–S axis range belt, the Métlaoui Formation becomes formed essentially by evaporites and gypsum, which corresponds to the Faidh Formation (Burollet 1956): continental to lagoonal deposits. This facies changes indicate a school continental zone around Central-Eastern Tunisia (Figs. 8.13, 8.14, 8.15, 8.16, and 8.17). Finally, Burollet described at Jebel Reinèche a predominantly phosphate Formation of late Paleocene–Lower

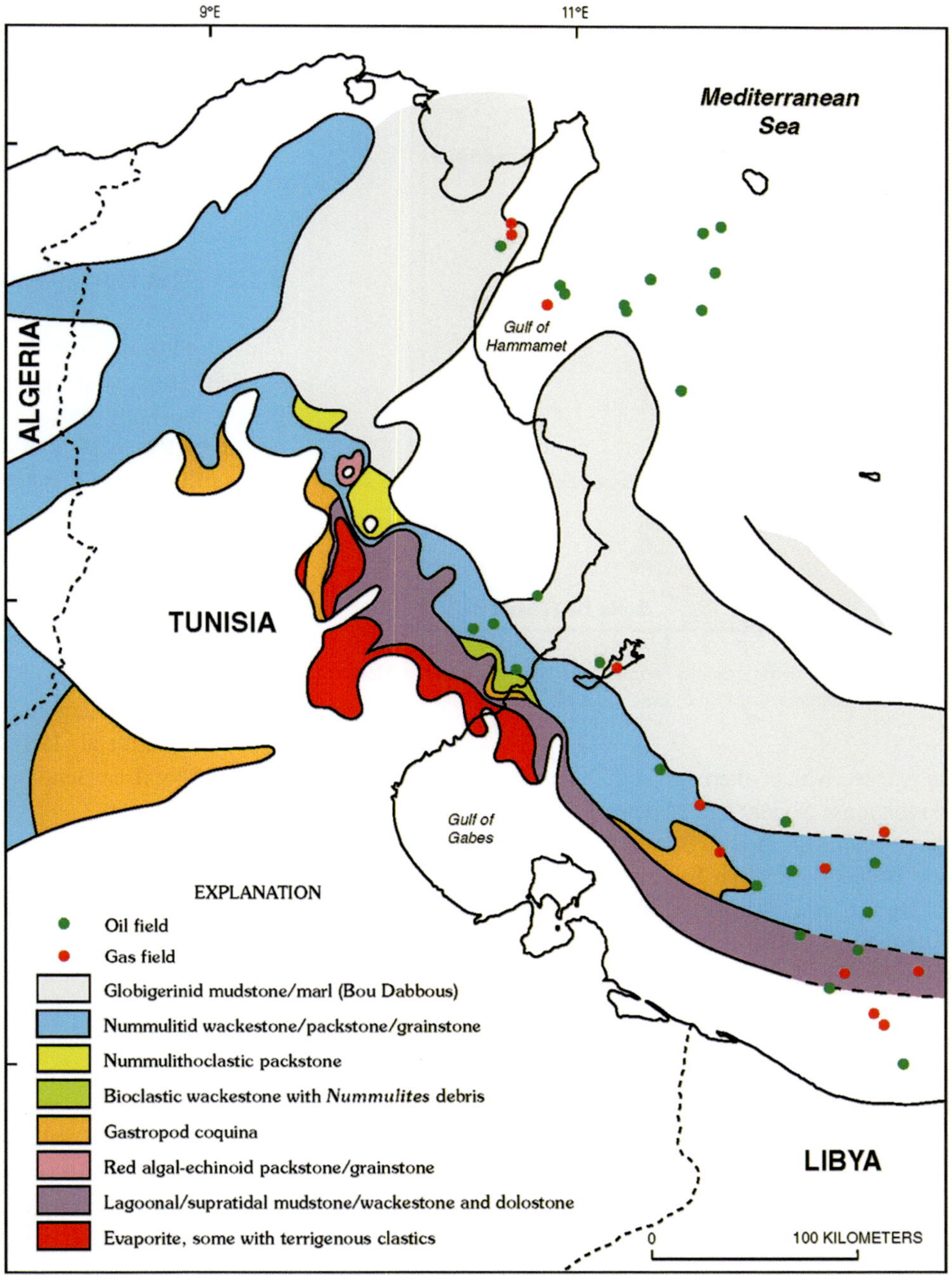

Fig. 8.14 The distribution of the Metlaoui Ypresian reservoirs and source rocks, as well as locations of the main oil/gas fields situated within the Bou Dabbous reservoirs. Notice oil occurrence in the Boudabbouss reservoirs in the Gulf of Hammamet and Gulf of Gabès and in the onshore Sahel (map from Klett 2001)

Lutetian in age. It corresponds to the Chouabine Formation, a lateral equivalent of the Metlaoui Formation in the central and northern part of the North–South Axis (Burollet 1956).

8.2.6 Middle–Late Eocene: Souar Formation and Its Equivalents

The Souar Formation, deposited in a deep environment, includes a thick clay series rich in planktonic foraminifera (Burollet 1956; Ben Ismail-Latrache 1981); within the middle a limestones bank with shells known as the Reinéche member (Burollet 1956). The Cherahil Formation (Figs. 8.18 and 8.19), the lateral equivalent of the Souar Formation, is essentially made up of alternating marls and limestones with oysters and lamellibranches (Burollet 1956), and in the middle of this Formation, we distinguish the limestones bar of Siouf (Burollet 1956), which together with the Reinèche member represent a good c geologic mapping reference (Ben Ismail-Lattrache and Bobier 1996; Ben Ismail-Lattrache 2000; Abbés 2004).

Northwest of Mezzouna, Burollet (1956) defines the Djebs Formation, a lateral equivalent of the Cherahil Formation, essentially made up of levels of gypsum and

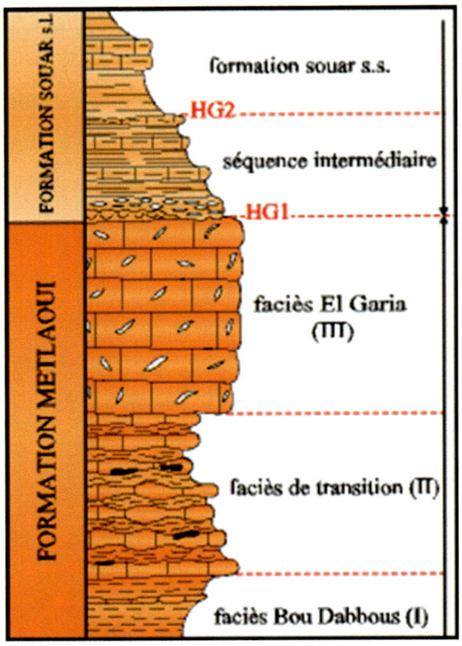

Fig. 8.15 Lithostratigraphic section of the Ypresian: Metlaoui Formation (without scale) (Gourmelen et al. 2000)

anhydrite, which alternate with blackish clays. The three Formations: Souar, Chérahil, and Djebs are Middle Lutetian–Priabonian in age (Burollet 1956; Ben Ismail-Lattrache and Bobier 1996; Ben Ismail-Lattrache 2000).

On the other hand, the lithostratigraphic column of the Atlas series shows a set of detachment levels in the sedimentary pile (Fig. 8.6). The Triassic evaporites are the most important due to their wide-spreading, large extent, and important thickness (Fig. 8.19). A second important detachment level is located in the thick Lower Cretaceous shales and marls. Secondary detachment levels are located in the Upper Cretaceous shales, Paleocene, and Middle–Late Eocene shales, respectively (Fig. 8.6). These detachment levels (Anderson 1996; Morgan et al. 1998; Khomsi et al. 2004a, 2006, 2007) played a dominant role in the control of the structural style during the Paleocene–Eocene positive inversions related to what is known as the Atlas compressional phases and also during the so-called Alpine compressional phases extending during the Late Miocene to the Recent. Therefore, the stratigraphy column of the Atlas series presents multi-detachment layers allowing vertical decoupling of structures issued from successive reactivations.

8.3 Structural Framework of the Eastern Atlas and Its Foreland Basins

The study area belongs to the Central-Eastern area of Tunisia. It is represented by the Sahel plains that are limited to the west by the outcrops of the North–South Axis.

8.3.1 The North–South Axis (NOSA)

The notion of the N–S Axis was introduced first by Burollet in 1956. It is a complex structural trend representing a geographical and geological boundary between the NE–SW central-eastern Atlas fold and thrust belt to the west and the plains of the Sahel domain to the East (Figs. 8.20 and 8.21). It represents the Eastern Front of the Atlas thrust belt (Bédir 1995; Ouali et al. 1987; Abbés 2004; Ouali 2007; Khomsi et al. 2009a) (Fig. 8.21).

Striking North–South in its southern part, this N–S axis mountain belt has a NE–SW orientation toward the North (Fig. 8.21), where it extends into the massifs of the Tunisian ridge (Zargouni et al. 1979; Truillet and Turki 1980; Truillet et al. 1981; Abbés et al. 1981; Turki 1981; Ouali 1984; Ben Jemia 1986; Rabhi 1999; Abbés 2004; Ouali 2007).

This N–S megastructure is related to a deep-seated fault which manifests in the gravity maps by positive anomaly striking N–S (Gabtni 2005). This North–South configuration is related to the pre-Triassic substratum's faulting since the Jurassic which evolved into a large thrust fault system during the orogenic events (Ouali 2007).

The NOSA shows thin-skin tectonic style (Figs. 8.22, 8.23, 8.24, and 8.25) (Zargouni et al. 1979; Truillet and Turki 1980; Truillet et al. 1981; Abbès et al. 1981; Delteil and Truillet 1983; Ouali 1984; Ben Jemia 1986; Rabhi 1999; Ouali 2007; Khomsi et al. 2009a, b). It results from the tectonic thrusting/compression of the Atlas domain over the eastern domain. This thrusting is highlighted by fold-thrust, duplexes, and fold propagation fault (Creusot et al. 1993; Abbés 2004; Khomsi et al. 2004a, b; Ouali 2007; Mezni et al. 2019). The different works highlight that the Meso-Cenozoic series lie tectonically on the clay and carbonates series of the Upper Eocene (Cherahil Formation). The saliferous Triassic series plays the role of a detachment level.

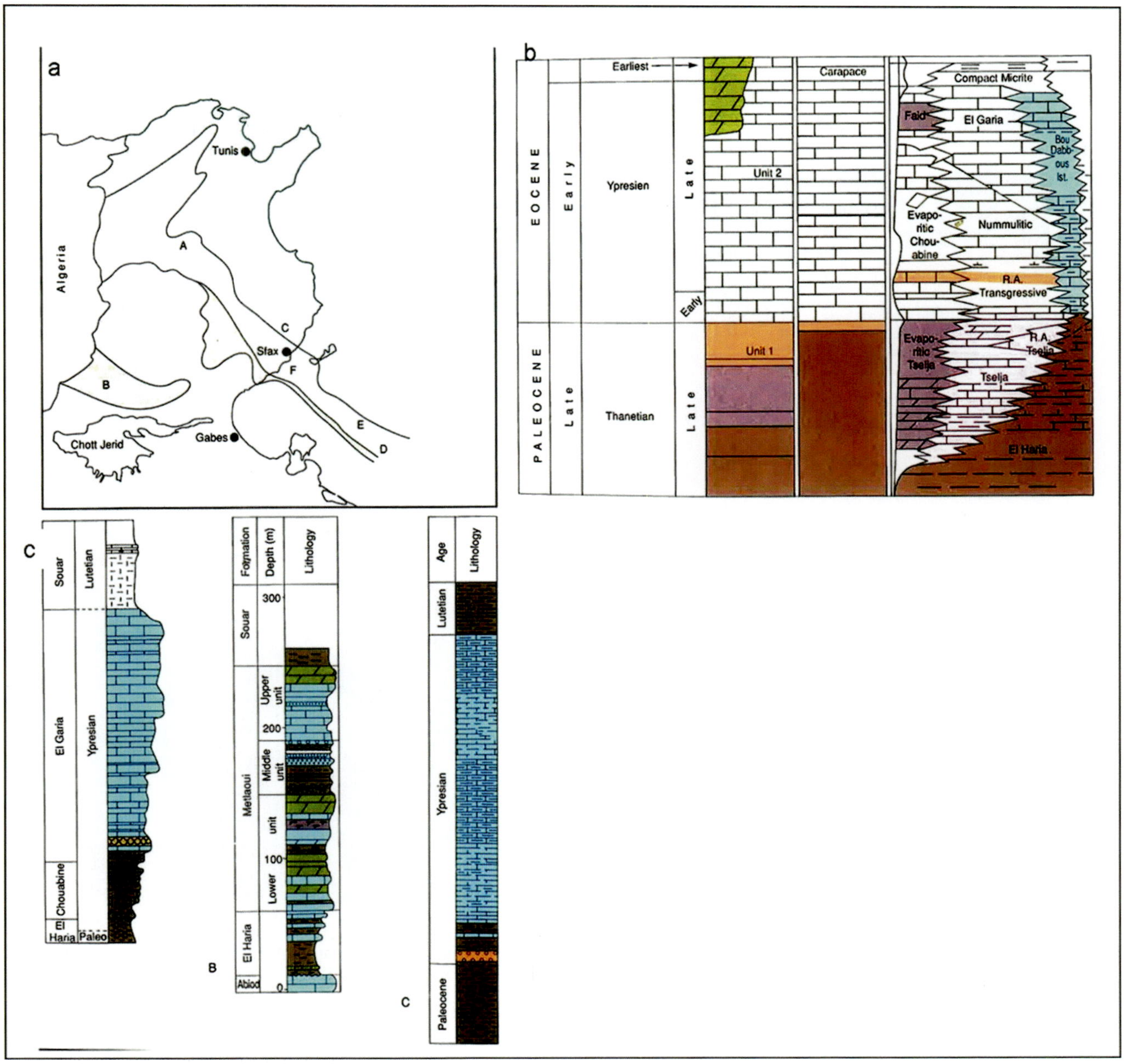

Fig. 8.16 Ypresian Reservoirs litho-facies (**a**), distribution map facies (**b**), representative cross-sections (**c**) in different domains. From Etap Book (1991)

On the other hand, different intraformational and angular unconformities were recognized (Figs. 8.17, 8.18, 8.19, 8.25, and 8.26) along the N–S Axis, especially that at the base of the Oligocene on top Eocene and that of Ypresian on Campanian–Maastrichtian.

8.3.2 The Sahel Foreland Domain

The different field investigations conducted in the area describes the Sahel domain as a stable and slightly deformed domain with a thick Mio-Plio-Quaternary cover (Castany 1951; Burollet 1956; Haller 1983; Ellouz 1984; Bédir 1988; Amari and Bédir 1989). However, beneath this Neogene sedimentary pile, this domain shows a structural complexity revealed by the various subsurface investigations (Khomsi et al. 2004a).

The Sahel domain can be subdivided into three sub-domains separated by two major subparallel faults, at the junction between the Atlas belt and its foreland basins which are, respectively, from north to south: The Kairouan–Sousse fault (Khomsi et al. 2004a, b) (Fig. 8.27) and the Bouthadi–Chorbane fault (Haller 1983; Mezni et al. 2019; Khomsi et al. 2019) (Fig. 8.27). These major faults limit

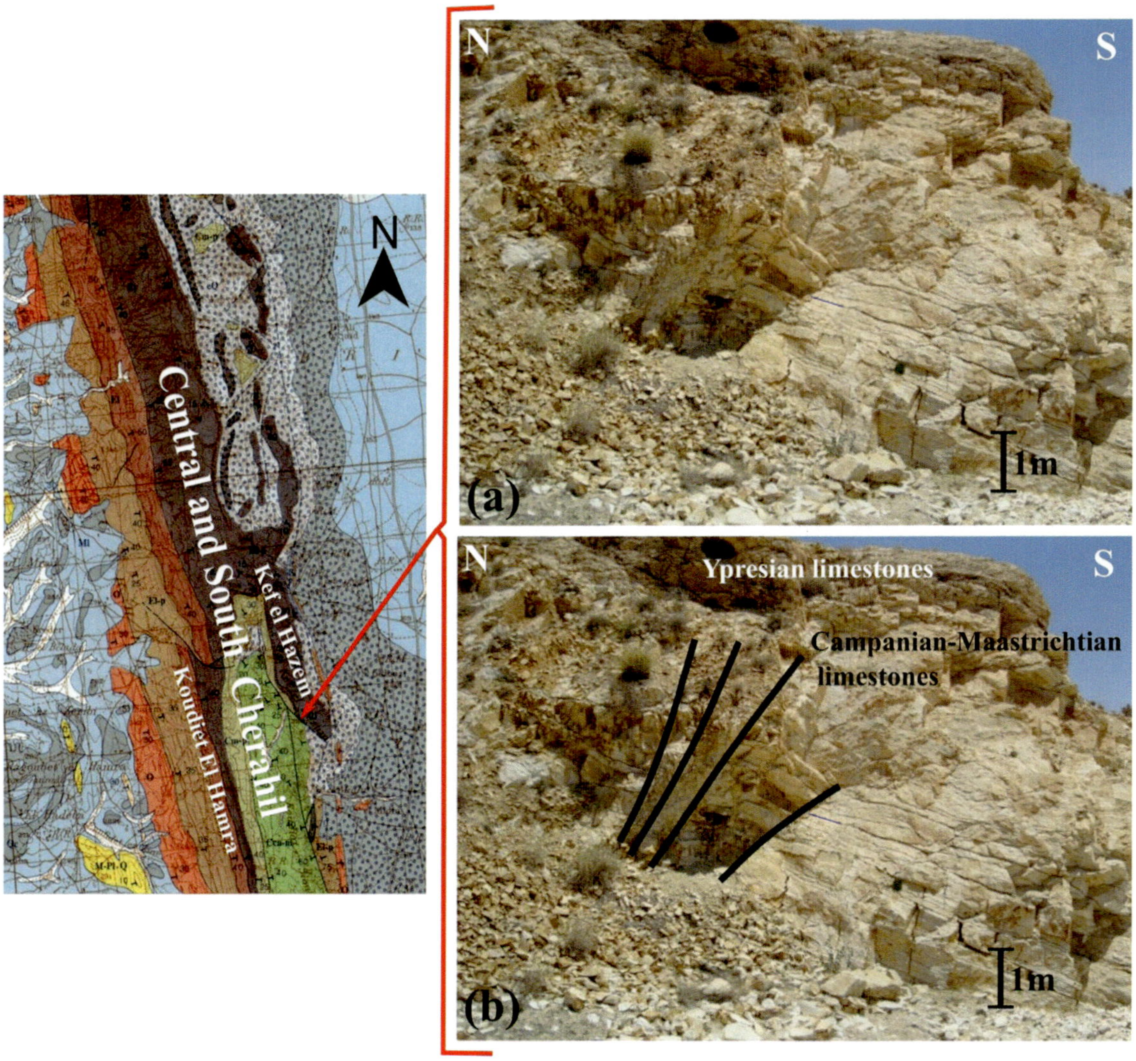

Fig. 8.17 Angular/erosional unconformity of the Ypresian limestones on the Campanian–Maastrichtian Abiod Formation with a gap of the Paleocene shales in Jebel Cherahil. N–S axis

various geological structures such as grabens, fishtail structures (Khomsi et al. 2009a, b; Khomsi et al. 2016; Mezni et al. 2019), and flower structures associated with strike-slip faults (Bédir 1995, Bédir et al. 2020), as well as synclines, duplexes, and detachment folds (Khomsi et al. 2009a, b, 2016; Mezni et al. 2019) buried underneath a thick Cenozoic series. In this tectonic context, the Triassic Salt plays the role of general detachment level of the Meso-Cenozoic series on their substratum (Burollet 1973; Haller 1983; Khomsi et al. 2004a, 2019; Khomsi 2005; Khomsi et al. 2009a, b; Mezni et al. 2019). In addition, tectonic knots, where the master faults intersect, correspond to tectonic weakness zones of the Meso-Cenozoic cover (Fig. 8.8), where the salt Triassic, recognized by its chaotic seismic facies (Haller 1983; Bédir 1995; Khomsi 2005; Khomsi et al. 2019) moves upward through this cover and forms diapiric like-structures buried beneath the Mio-Plio-Quaternary deposits (Fig. 8.28).

8.3.3 Pre-Cretaceous Period

The Pre–Late Cretaceous period corresponds to a period of rifting linked to the extensional movement between the Eurasiatic and African plates, allowing the opening of the

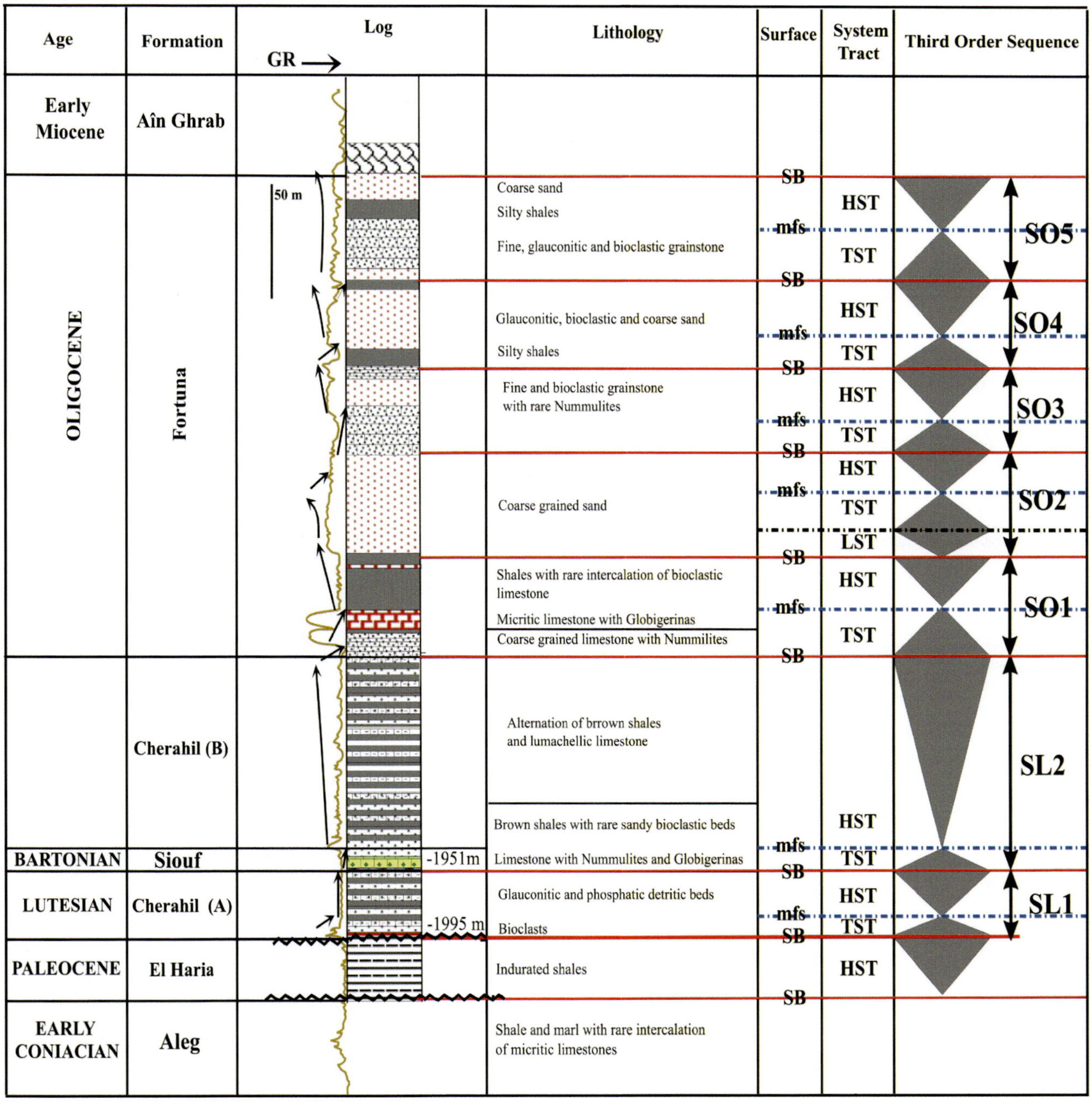

Fig. 8.18 Angular/erosional unconformity of the Ypresian limestones on the Campanian–Maastrichtian Abiod Formation with a gap of the Palocene shales in Jebel Cherahil. N–S axis

Tethys and the activation of the deep faults in the E–W and N–S directions along the Eastern Maghreb. These major faults, inherited from Paleozoic tectonics, control grabens, half-graben, and horsts. These basins are filled with a thick sedimentary pile of the Lower Triassic–Cretaceous in age. In addition, Triassic material has been mobilized from the Middle Jurassic (Bédir 1995), guided by the activity of the pre-Triassic faults.

8.3.4 The Late Cretaceous–Eocene: Tectonic Inversion and the First Atlas Phase

The Cretaceous period is characterized by an extensional regime related to the opening of the Pelagian and the Sirt domains (Frizon de Lamotte et al. 2009; Khomsi et al. 2009a, b, 2016) (Figs. 8.29 and 8.30) and which persisted until the Upper-most Cretaceous (Khomsi et al. 2016). This

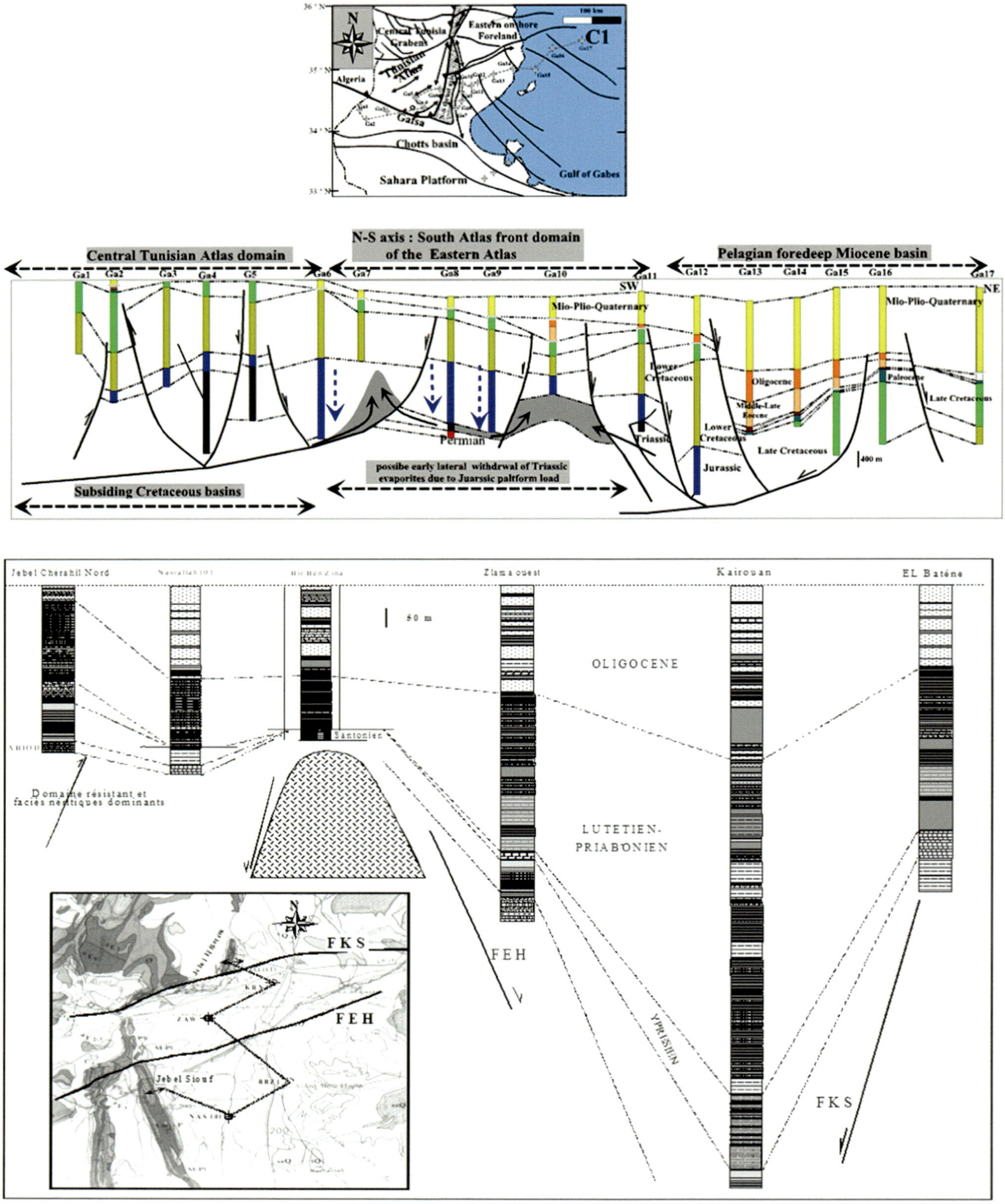

Fig. 8.19 Stratigraphic correlations in the Eastern Foreland basin of the Atlas from the N–S axis to the Pelagian Domain (above) from Khomsi et al. (2016) and stratigraphic correlations from the N–S axis through the Kairouan flexural basin showing the control of the subsidence (below). Notice the Diapirism. From Khomsi (2005). Notice that the Kairouan Basin crosscuts orthogonally the N–S axis. Reprinted by permission from (Springer-Nature) [Arabian Journal of Geosciences] Khomsi et al. (2016). Copyright (2019)

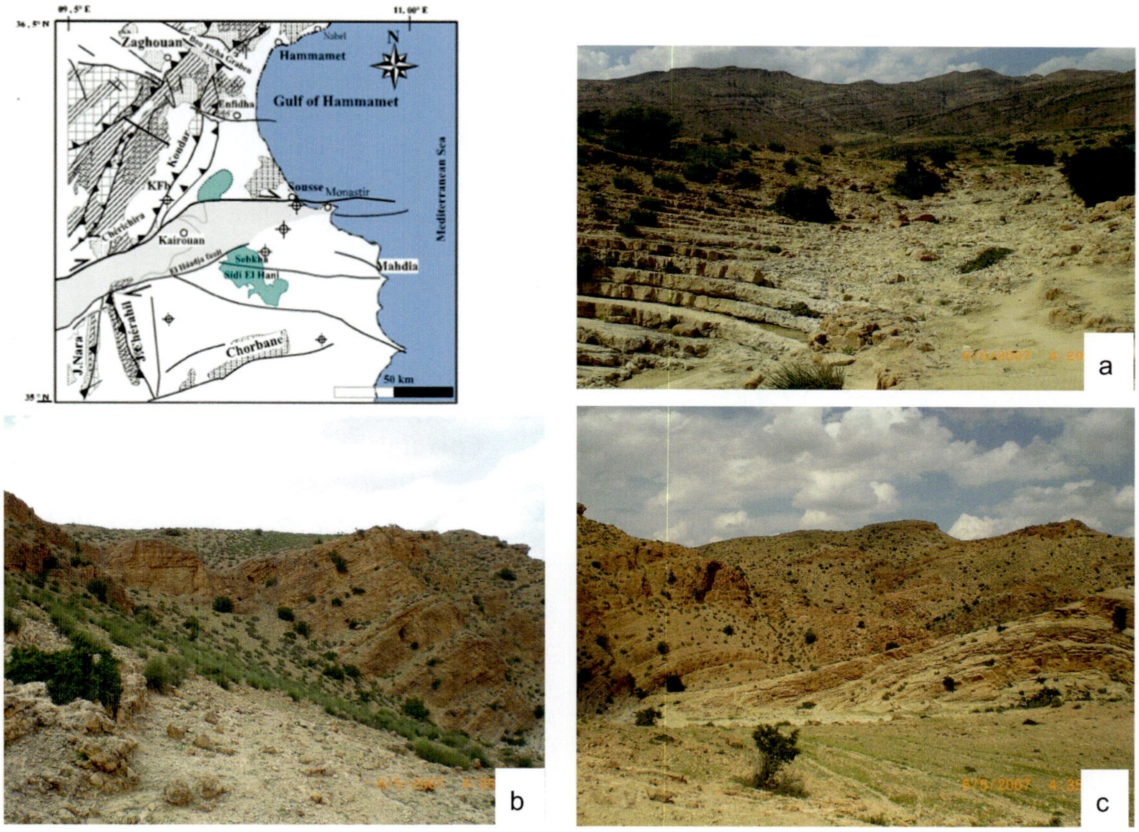

Fig. 8.20 Field sketches in the western N–S axis in Jebel Nara anticline axis (see map). **a** Panorama view looking NW showing the Jurassic limestones in the Wadi and the Cretaceous sequences above in the cliff. **b** Synsedimentary Jurassic detachment normal fault controlling thickness variations of the Jurassic sequence. **c** Thrusting affecting the Jurassic limestones in the anticline axis with local duplication

extensional phase is coupled with thick clay-limestones series deposited in a deep to shallow environment. It is accompanied by tilted blocs limited by NW–SE faults (Figs. 8.29, 8.30, and 8.31) and controlling in some areas, uprising Triassic salt movements (Khomsi et al. 2016, 2019).

In the Sahel area, a compressive event starts in the Maastrichtian–Campanian, amplifies in the Paleocene, and reaches its paroxysm in the Middle-Upper Eocene (Khomsi et al. 2006). This compressive phase is expressed in the whole Maghrebian belts by folds, inverted grabens, and thrust folds, admitting the Triassic salt as a décollement level as well as angular unconformities (Figs. 8.17, 8.18, 8.19, 8.25, and 8.32). This compressive phase represents the Atlas phase corresponding to a major orogenic event (Laffitte 1939) generalized throughout the Maghreb according to the work of Khomsi et al. (2006) and it is characterized by the mobilization of Triassic material as diapiric structures.

8.3.5 Post-inversion Phase: Oligocene—Early Miocene Period

During the Oligocene, the foreland of the Atlas belt is characterized by a period of tectonic quiescence characterized by a thick Oligocene–Lower Miocene series deposited in grabens and hemi-grabens (Khomsi et al. 2009a) (Figs. 8.32, 8.33, and 8.34), and locally the development of listric normal faults (Fig. 8.32) branched at depth on "décollement" levels (Khomsi et al. 2009a). This negative inversion continued during the Early Miocene (Khomsi et al. 2009a, b) with the occurrence of synsedimentary normal faults related to an increase of the subsidence rate. In the subsurface, the seismic investigations outline these extensional structures (Khomsi et al. 2009a, 2016).

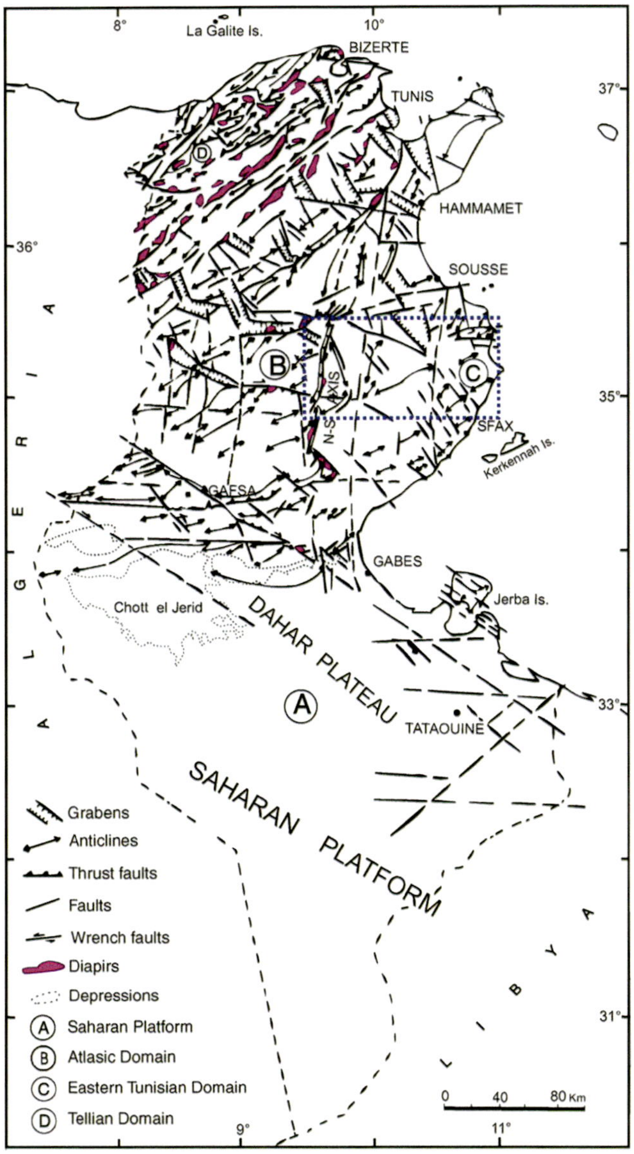

Fig. 8.21 Structural sketch of Tunisia showing the main lineaments and the different structural domains (Bouaziz et al. 2002)

8.3.6 Studied Area

The studied area, in part, corresponds to the major Bouthadi–Chorbane anticline running roughly ENE (Fig. 8.35). At the surface, it is a dissymmetrical anticline striking from E–W to NE–SW direction (Bramaud and Burollet 1981). Its axis is occupied by Mio-Pliocene continental molasses deposits and Upper Miocene–Quaternary deposits (Fig. 8.36) which are also folded (Fig. 8.35). According to Soyer and Tricart (1989), the Bouthadi–Chorbane structure belongs to the major structural trend intersecting orthogonally the North–South Axis, resulting from the reactivation of a major deep-seated infra-Triassic fault. In the subsurface, this zone is affected by the Bouthadi–Chorbane fault (BCF).

It represents a tectonic weakness area which is associated with diapirs made up of Triassic salt and locally with some volcanic material (Haller 1983; Bédir 1995; Laridhi Ouazaa and Bédir 2004), which is underlined by gravimetric and geothermal anomalies (Gabtni 2005; Khomsi et al. 2012). Our structural interpretations by the means of seismic studies focus on the tectonic events (Fig. 8.36) as well as on the structural styles along the Bouthadi–Chorbane structure.

8.4 Wells Data and Lithostratigraphic Correlations

8.4.1 Lithostratigraphic Correlation (C1): Kairouan–El Hancha

Striking N–S (Fig. 8.35), this section is 46 km long and admits the sea level as datum plan (Fig. 8.37). It includes wells (Fig. 8.37) belonging to different structural sub-domains. In fact, the correlation goes from north to south through: the Kairoun–El Hedadja graben limited to the south by El Hdadja fault (Fig. 8.37) (Khomsi 2005). Further south, the Sidi El Hani basin (Fig. 8.37), then the anticlinal structure of Chorbane (Fig. 8.37), and finally, the subsiding structures of El Hancha–El Jem (Bédir 1995; Khomsi et al. 2016; Mezni et al. 2019).

The well S14 drilled in the Chorbane structure shows a duplication of the Cenomanian–Turonian series which is related to the N90 Chorbane thrust fault system (Khomsi et al. 2016; Mezni et al. 2019). The inversion movements of the Chorbane fault is associated with changes in thickness and stratigraphic unconformities indicated in borehole S15 by the deposits of the Oligocene series lying directly on reduced series of the Campanian–Maastrichtian (Fig. 8.37) with gap of the entire Paleocene–Eocene. On the other hand, the well S14 underlines the unconformity of the Upper Eocene over the carbonates deposits of the Abiod Formation (Campanian–Maastrichtian in age) (Fig. 8.37). This unconformity is also accompanied by the gap of the Paleocene–Ypresian series.

Thus, the gaps in the Paleocene and Eocene deposits detected in S14 and S15 indicate a high zone related to the positive inversion of the Chorbane fault (Figs. 8.37 and 8.38) which is accompanied by the non-deposition or erosion of the Paleocene–Eocene deposits. However, the lateral correlation of the Upper Cretaceous and Eocene series to the north (boreholes S13 and S11) and to the south (borehole S17; Fig. 8.37) of Chorbane indicates a subsiding domain characterized, respectively, by thick Cenomanian–Maastrichtian (Fig. 8.37) and Paleocene–Eocene (Fig. 8.37) deposits.

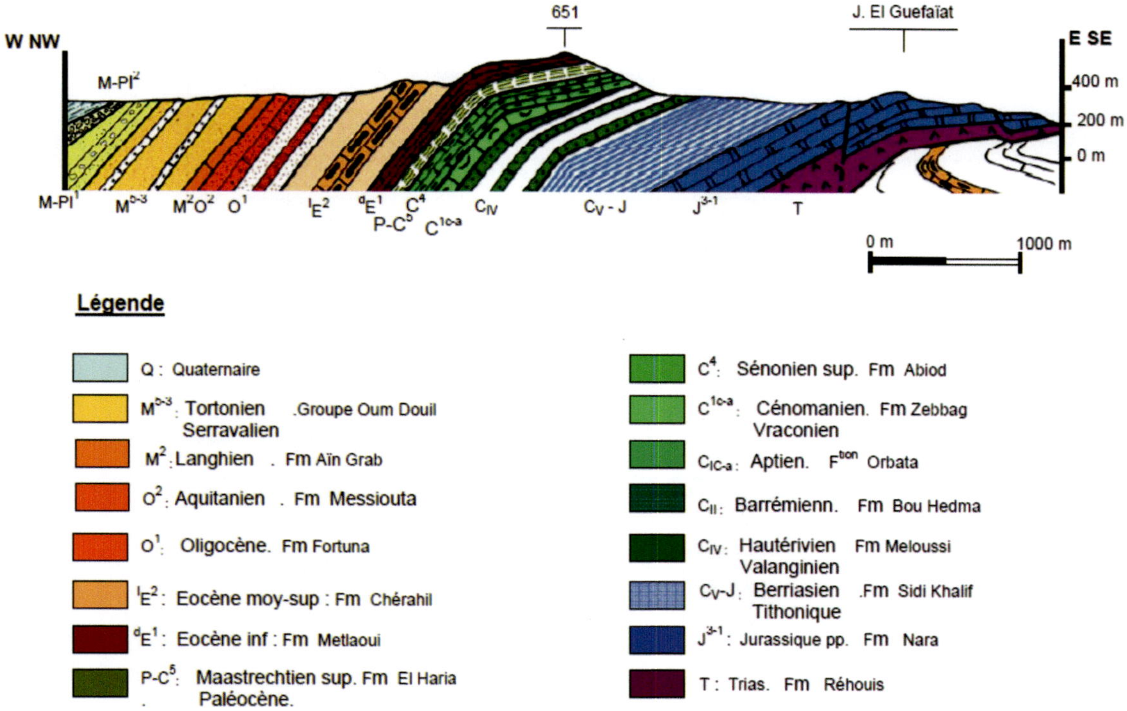

Fig. 8.22 Geological section of Jebel Bou Gabrine (the North–South axis) showing an overlap of the Jurassic on the Neogene series, and the décollement level is ensured by the Triassic salt (Ouali 2007)

8.4.2 Lithostratigraphic Correlation (C2): Nasrallah–El Hancha

It extends over 100 km (Fig. 8.35) and admits the sea level as a datum plane (Fig. 8.39). It includes wells over the Central-Eastern area of the Sahel domain (Fig. 8.39) from Nasrallah (borehole S4) to the west, cutting through the platform of Ktitir (Figs. 8.39, 8.40, and 8.41) characterized by the outcrops of Ktitir (boreholes S7 and S8), of Mechertate (borehole S9) and the Mio-Pliocene structure of Bouthadi–Chorbane (Fig. 8.39).

The Upper Cretaceous sequences (Cenomanian–Maastrichtian) are more developed at the east of Ktitir–Mechertate (Figs. 8.39 and 8.40), where the boreholes cut through a thicker series than at the west of Ktitir (Fig. 8.40). This suggests that the Ktitir–Mechertate lineament oriented NNE–SSE represents a paletectonic boundary between a high zone characterized by reduced sedimentation to the West (Figs. 8.39, 8.40, and 8.41), and to a subsiding area to the East (Khomsi 2005).

The well S7 cuts through the Cenomanian series, which rest directly on the Jurassic dolomites (Fig. 8.39), with a gap of the entire Lower Cretaceous. At the location of S16, the Turonian sequences rest directly on the Triassic salt (Fig. 8.39). Thus the upward movement of the salt material began since the Middle Jurassic (Bédir 1995; Hlaiem 1999; Khomsi 2005; Khomsi et al. 2016, 2019; Mezni et al. 2019). The deposition rate during the Jurassic period is controlled by the normal movements of the major deep-seated faults of Ktitir and the Bouthadi–Chorbane fault (Fig. 8.39) (Bédir 1995; Hlaiem 1999; Khomsi 2005; Khomsi et al. 2016, 2019; Mezni et al. 2019).

Toward the south, at the location of S15, 4 km from S15 (Fig. 8.39), we have a thinning of the lower Campanian–Maastrichtian (58 m) which is surmounted by a thick siliciclastic series of Oligo-Mio-Plio-Quaternary in age (Fig. 8.39).

Toward the West, the correlation shows the thickenings of the Campanian–Maastrichtian series. In fact, Abiod Formation measures 436 m in well S18 (Fig. 8.39). Then it becomes thinner as shown by borehole S16 and S4 where it measures, respectively, 138 m and 24 m (Fig. 8.39). However, toward the East, we have a thickening of the Campanian–Maastrichtian, which reaches 436 m at the level of S17 drilled in the depression of El Hancha (Fig. 8.39).

The Upper Maastrichtian–Paleocene series represented by the clays of the El Haria Formation (Fig. 8.39), are crossed by different wells either entirely (S4, S16, S18, and S17) (Fig. 8.39) or in part (S8, S9, and S10). In S15 and S14, the Upper Maastrichtian–Paleocene is missing indicating\that the Chorbane zone (S15 and S14) represents a paleotectonic high area during the Late Cretaceous–Paleocene time (Mezni et al. 2019).

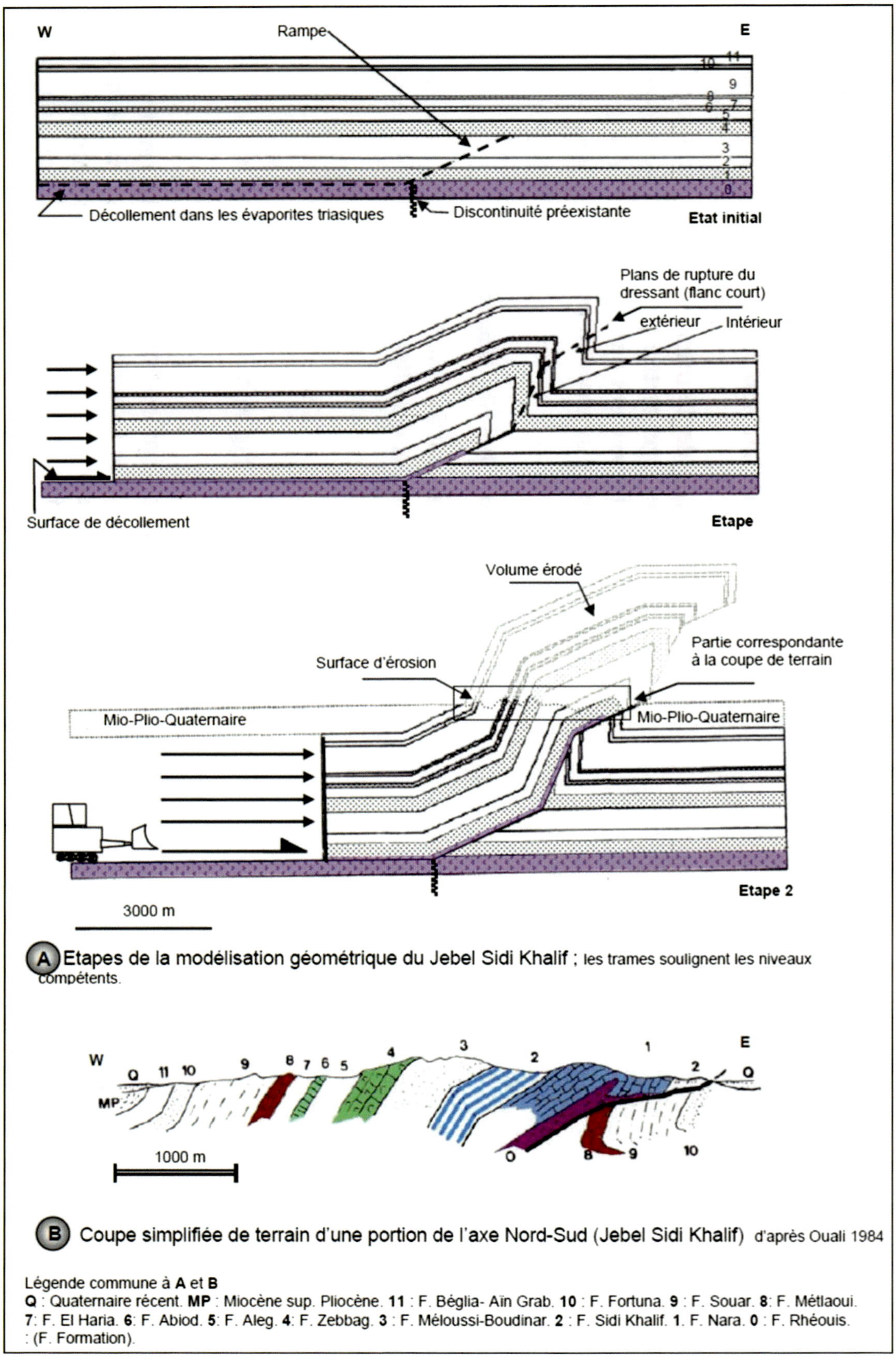

Fig. 8.23 Geometric modeling of a section of the North–South axis structure shows fault-propagation fold models for the N–S axis (Creusot et al. 1993, modified by Ouali 2007)

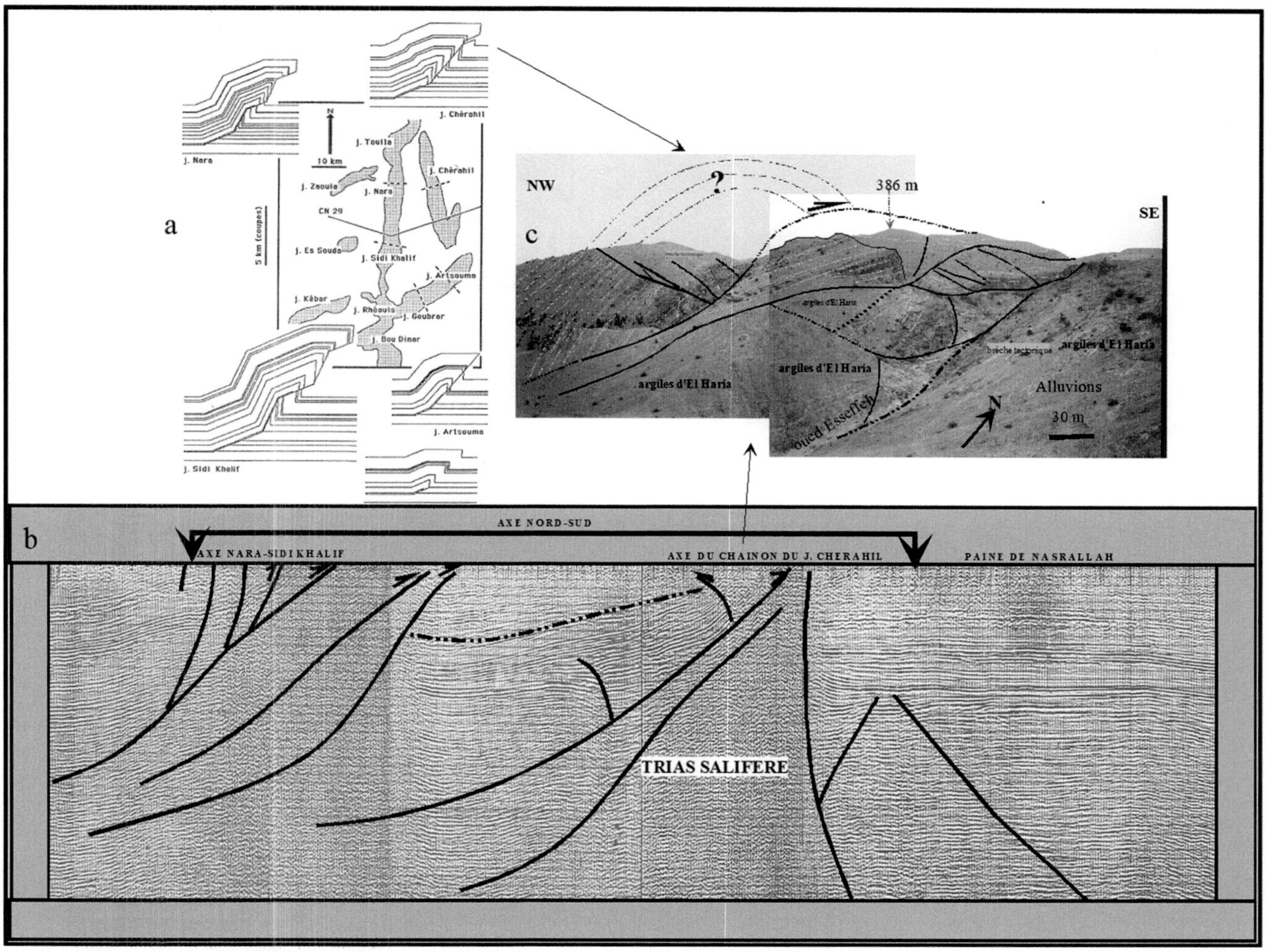

Fig. 8.24 Field observations in Jebel Cherahil, North N–S axis showing thrust tectonics affecting the Ypresian limestones duplicated and detached over the Paleocene shales. **a** Tectonic modeling from Creusot et al. (1993). In Khomsi (2005). **b** Thrust tectonics underlined by seismic interpretations along the N–S axis. **c** Field panorama looking N along the axial zone of the Cherahil anticline

The Ypresian series (Lower Eocene) are absent from well P18 to well P14 toward the east and within the well P16 to the west. This indicates that the Chorbane–Bouthadi zone was a high paleotectonic trend during the Ypresian time span too. It persists during the Middle-Upper Eocene with reduced thickness that does not exceed 40 m, as shown by well S14 (Fig. 8.39).

This structural architecture described above is controlled by the positive inversion of the Bouthadi–Chorbane thrust fault accompanied by emersions and erosions outlined by the non-deposition of a number of series ranging from the Maastrichtian to the Upper Eocene. This positive structure appears contemporary with the compression Atlas phase, which began in the Late Cretaceous and reached its paroxysm during the Upper-Middle Eocene (Khomsi et al. 2006, 2009a, b, 2016, 2019).

8.5 Presentation and Interpretations of Representative Seismic Sections

8.5.1 Section L1

The seismic line L1 crosses the N–S Axis structure through the Sidi Khalif anticline, Sidi Saad basin, Cherahil anticline, and continues eastward along the platform of Ktitir (the Sahel foreland domain) (Fig. 8.42). This Geoseismic section deciphers the structural styles of the North–South axis which belongs to the South Atlasic front (Khomsi et al. 2009a, b; Khomsi et al. 2019; Mezni et al. 2019). The Sidi Khalif corresponds to a thrust fold anticline developed over a thrust fault which corresponds to the south Atlasic thrust. The back-limb of this thrust fold is affected by a set of back-thrusts.

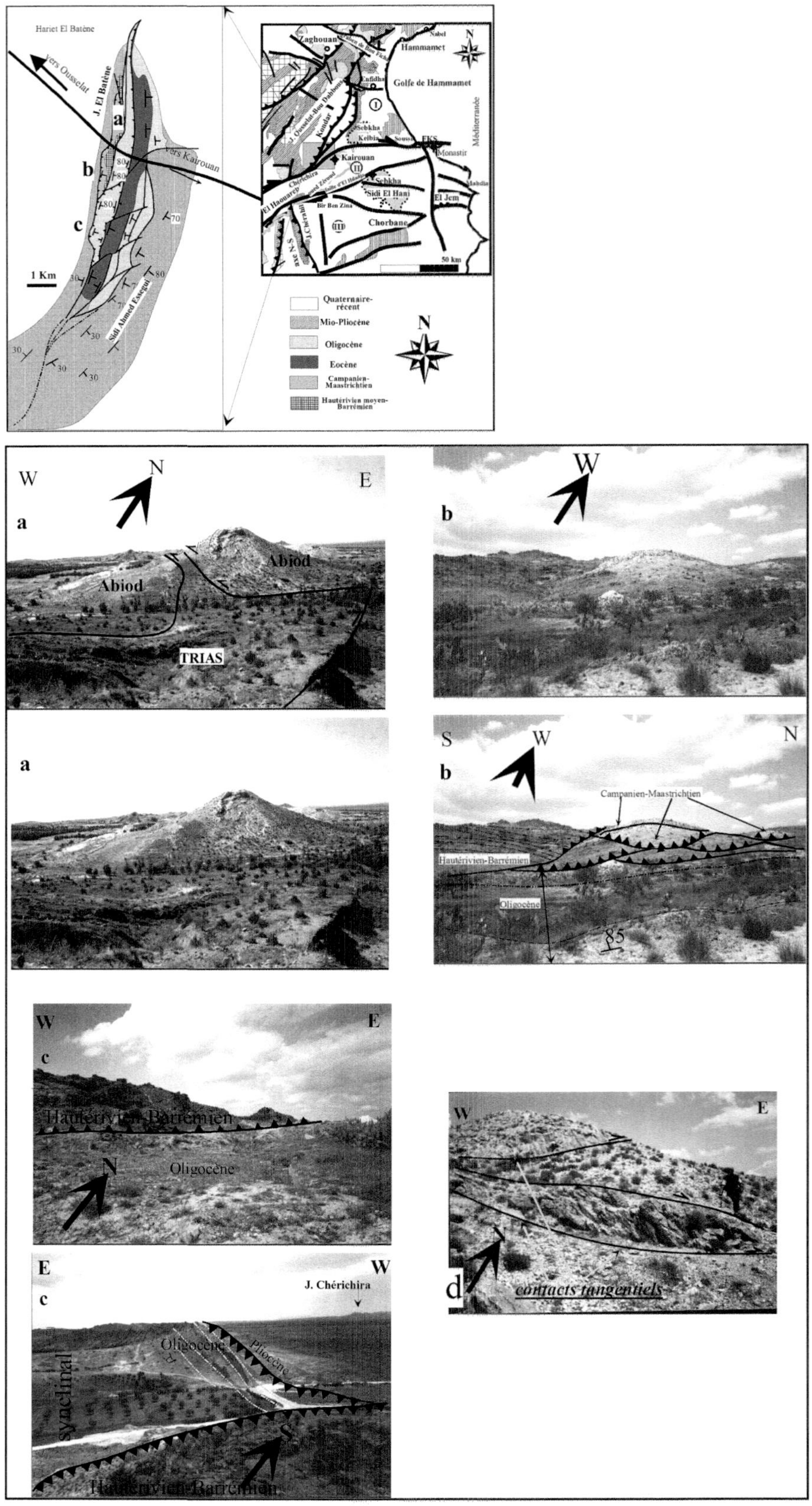

Fig. 8.25 Thrust tectonic affecting the Baten Mountain, Northernmost N–S axis. Above: position map and simplified geologic map. **a**, **b**, and **c** are field observations located on the map

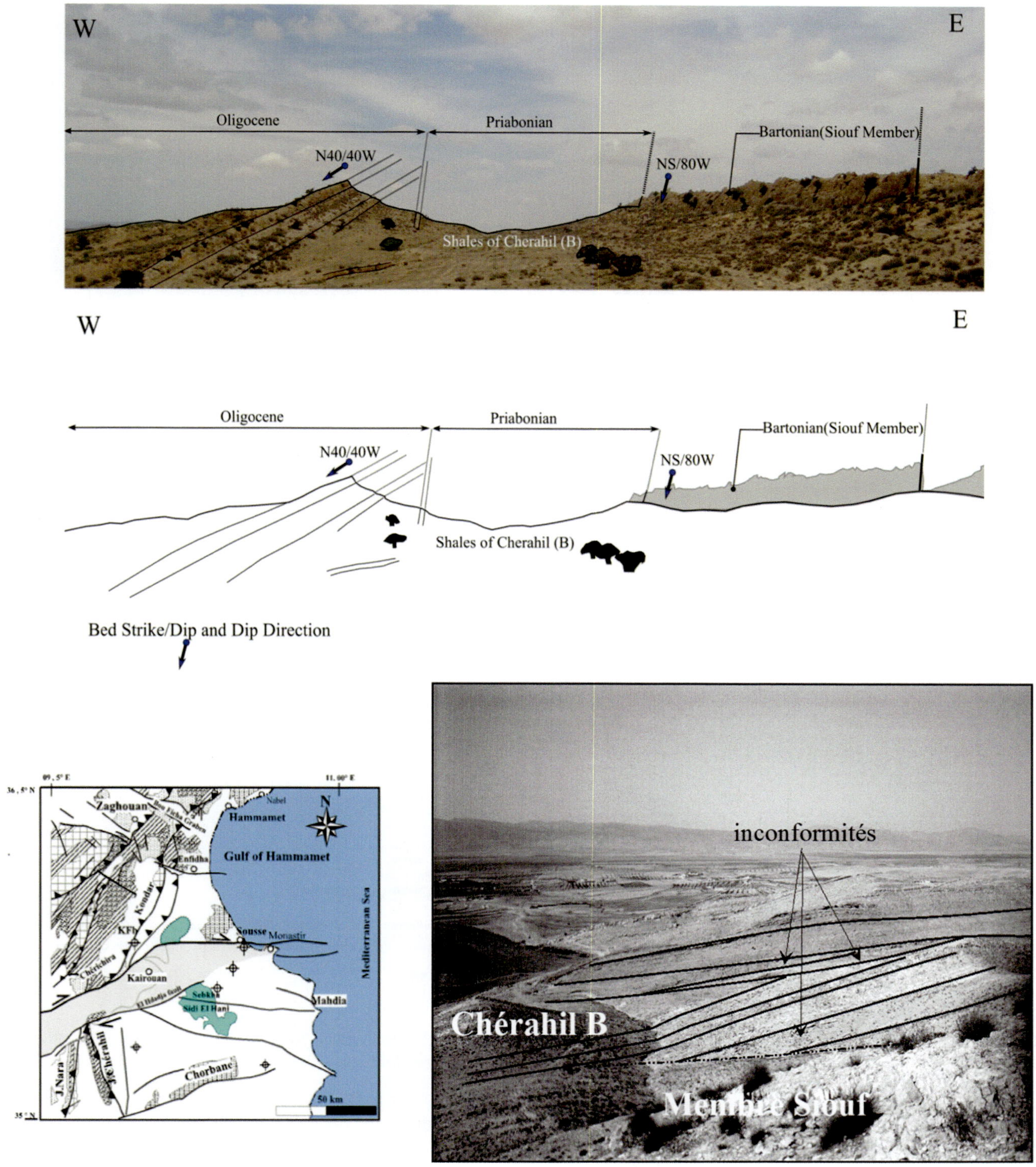

Fig. 8.26 Panorama view looking NNW along the back-limb of the Cherahil anticline showing intraformational unconformities of the Upper Eocene sequences corroborating tectonic pulses

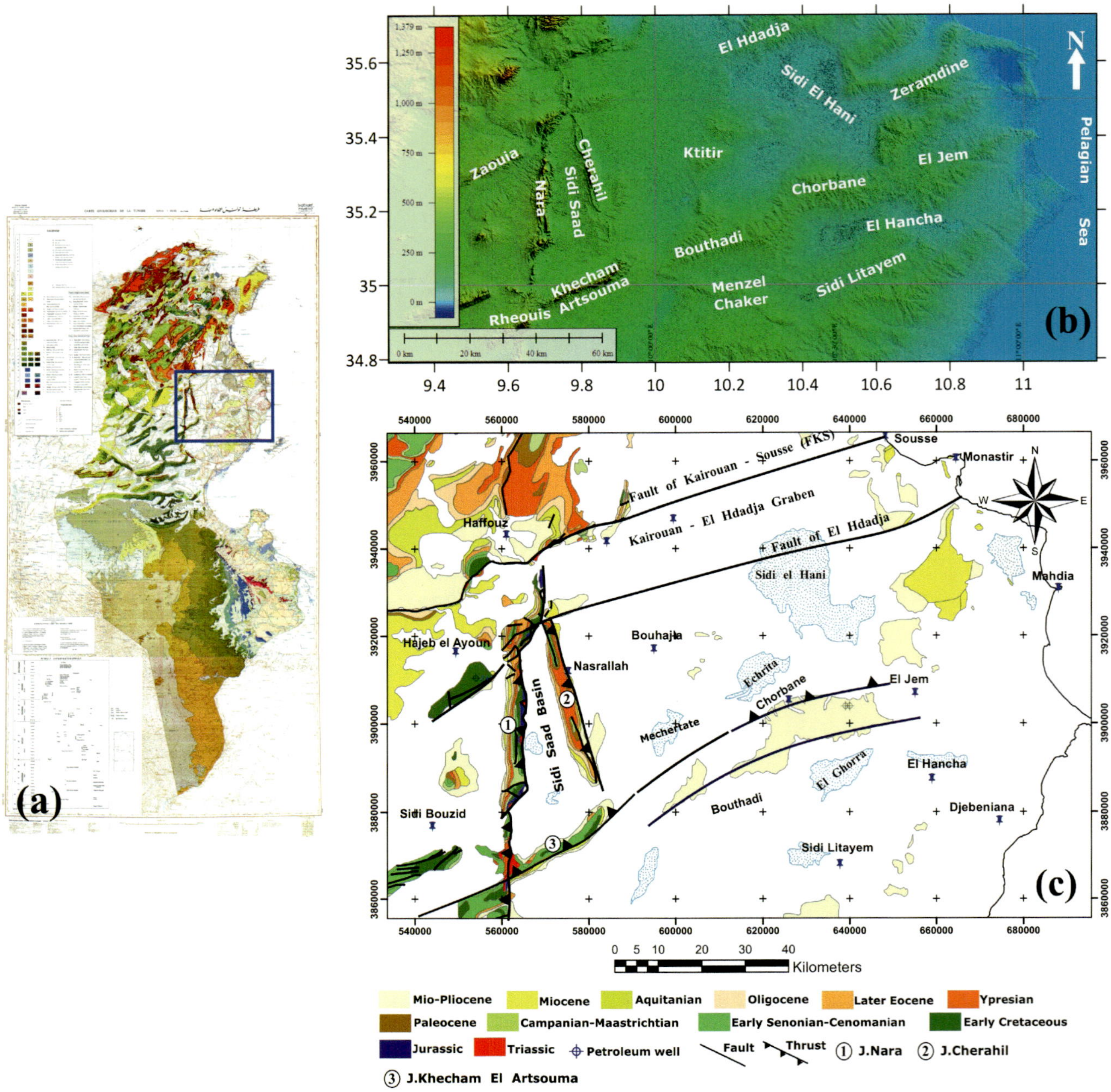

Fig. 8.27 Geological map of the Sahel domain showing a little deformed area characterized by NE–SW to E–W elongated Mio-Pliocene structures. (**a**) geological map of Tunisia showing the study area. (**b**) topographic image of the study area. (**c**) geological map of the Sahel domain

The horizons extending from the Triassic to Campanian–Maastrichtian highlight a subsidence inversion. The Early Cretaceous horizons show thinning to the east (Fig. 8.42). However, in the same direction, the Paleocene horizons show thickening. Moreover, the Ypresian series show from the East to the West a thinning till disappearing.

The Cherahil anticline structure shows, in the subsurface, a thrust fold developed over the Cherahil thrust (Fig. 8.42). The Middle–Late Eocene horizons are presumably thinner toward the top of the Cherahil fold which indicates that this compressional structure occurred during the Middle–Late Eocene.

Field investigations in the Cherahil belt corroborate such features with intraformational angular unconformities affecting the Cherahil layers (Fig. 8.42).

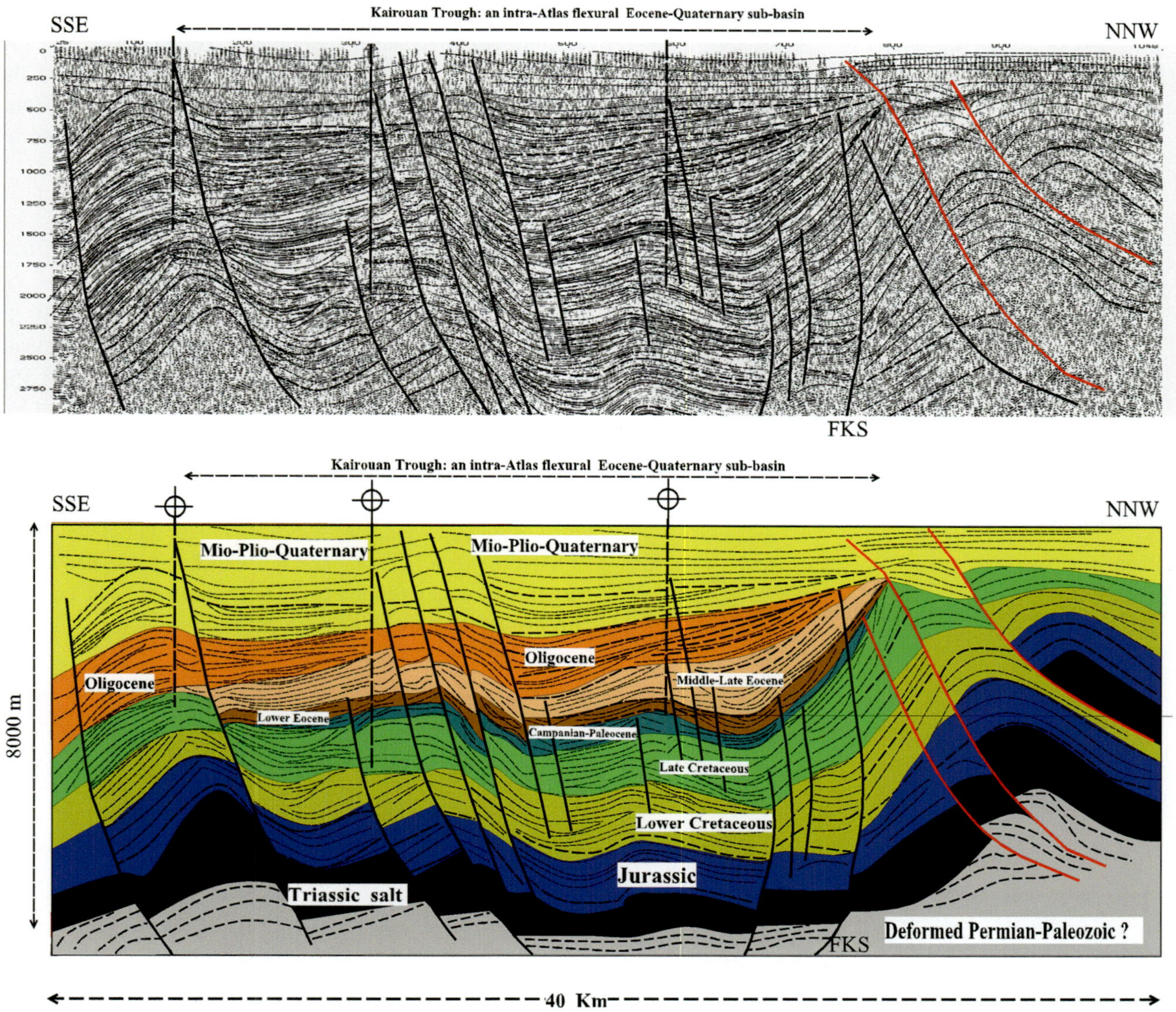

Fig. 8.28 Examples of the Sahel subsurface structures buried beneath a thick Molasses sequence. Here the example of the Flexural basin of Kairouan. Modified from Khomsi et al. (2004a, b, 2019)

8.5.2 Section L2

This section L2 passes between the N–S axis (Rheouis Mountain) and the south of Cherahil fold (Fig. 8.43) measures 29.5 km and is oriented NE–SW. The subsurface configuration is characterized by two salt diapirs (Fig. 8.43). The first is recognized in the southwest part of this cross-section and localized in the North–South Axis. This salt diapir is bounded by the N–S fault (Fig. 8.43). The second salt Diapir is located on the northeastern side of the cross-section and it is bounded by the reverse fault of Cherahil. The picking of the reflectors of the Jurassic to Mio-Plio-Quaternary shows a thinning toward the two salt Diapirs which indicates that the upward movement of salt material occurred since the Jurassic (Bédir 1995; Khomsi et al. 2009a, b) and has controlled the sedimentary packages (Fig. 8.43).

8.5.3 Section L3

This N–S cross-line L3 measures 19 km (Fig. 8.44). It is located in the southern part of the Sidi Saad Basin. It shows an amplification of the fracturing from North (west flank of Cherahil fold) to South in Khechem El Artsouma fold (Fig. 8.44). The faulting system is characterized by a set of branched sub-vertical faults corresponding to deep-seated faults (Fig. 8.33). The thickness variations in the Jurassic

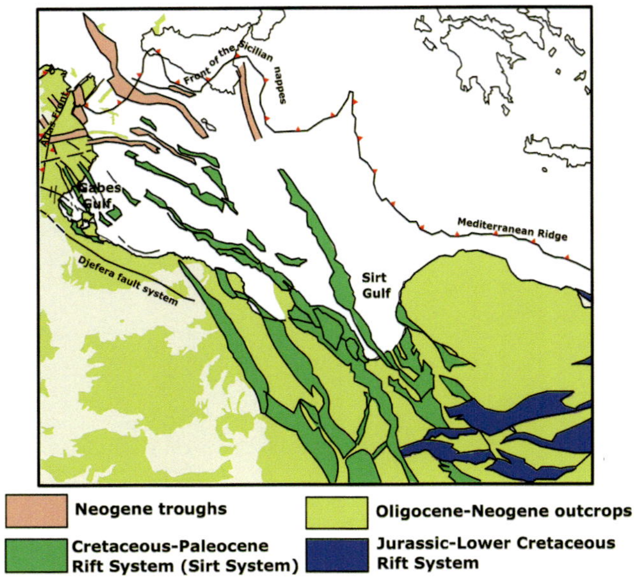

Fig. 8.29 Main structures linked to the Late Cretaceous Sirt rifting from Libya to Eastern Tunisia. Notice the creation of NWSE fault coeval to NWSE elongated basins from (Frizon de Lamotte et al. 2009)

and Early Cretaceous horizons indicate that the halokinetic activity occurred since the Jurassic (Bédir 1995; Khomsi et al. 2009a, b). The migration of salt material toward the Diapir is coupled with the development of a rollover structure over a normal detached-listric fault which is sealed by the Turonian series (Fig. 8.44).

8.5.4 Section L5

This NW-trending section is calibrated to the Well S2 (Figs. 8.25 and 8.45). It crosses three structural blocks, labeled A, B, and C from the NW to the SE, respectively (Fig. 8.35). These blocks are delimited by two major faults, F1 and F2, which constitute the Bouthadi–Chorbane fault system (Fig. 8.35). The seismic section tied to well shows that the Triassic salt intrudes Jurassic–Cretaceous series along fault F2. On the other hand, the structural interpretation underlines a horst between faults F1 and F2. This horst is sealed by the progressive unconformity of the Miocene over the Paleocene–Eocene reflectors with an angular unconformity.

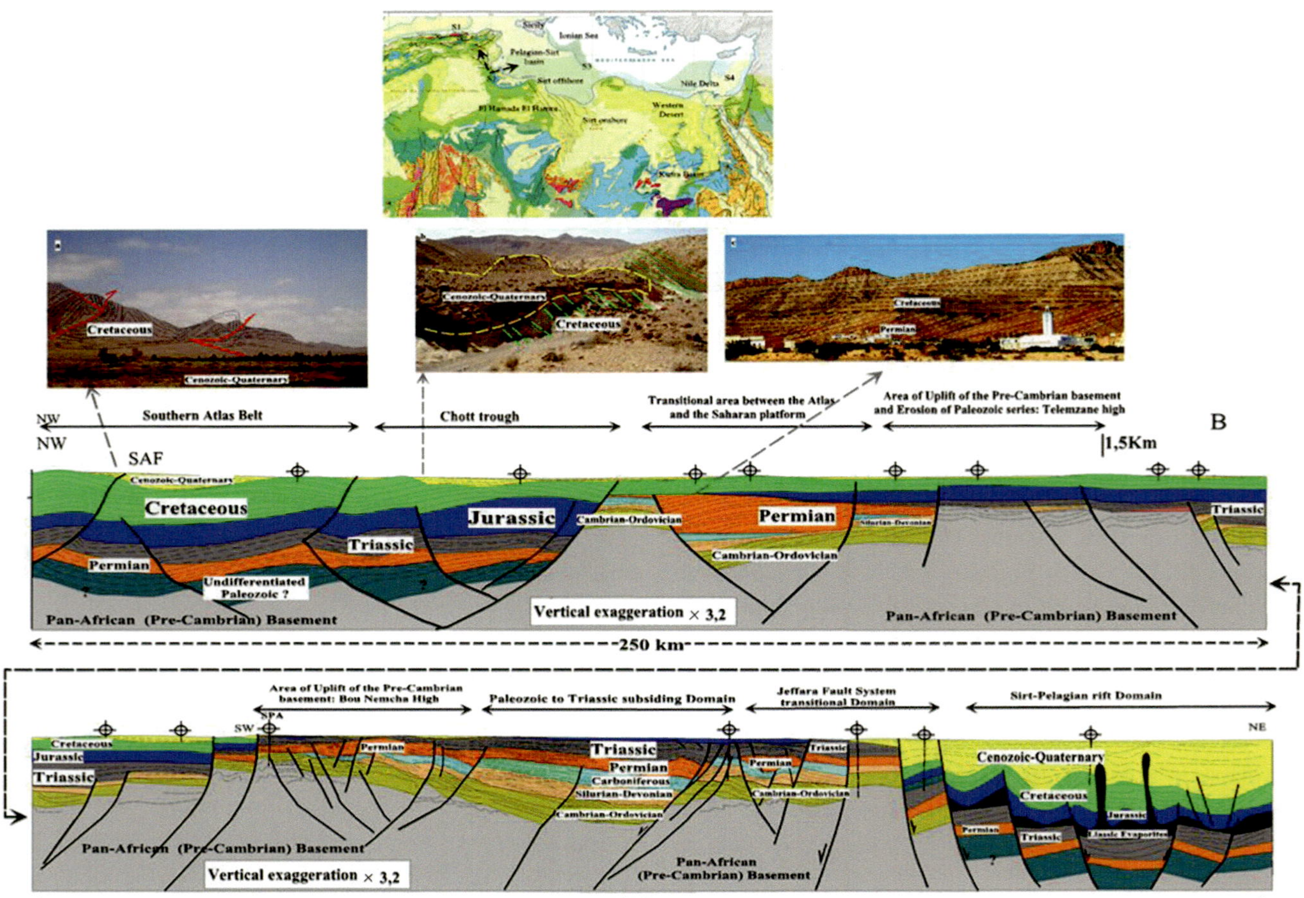

Fig. 8.30 Main structures linked to the Late Cretaceous Sirt rifting from Libya to Eastern Tunisia along a regional cross-section showing grabens and tilted blocks related to the rifting of the Sirt-Pelagian domain (modified from Khomsi et al. 2019), with Liassic–Triassic Diapirs in the Sirt-Pelagian Domain. Reprinted from Khomsi et al. (2019). Copyright (2019), with permission from Elsevier

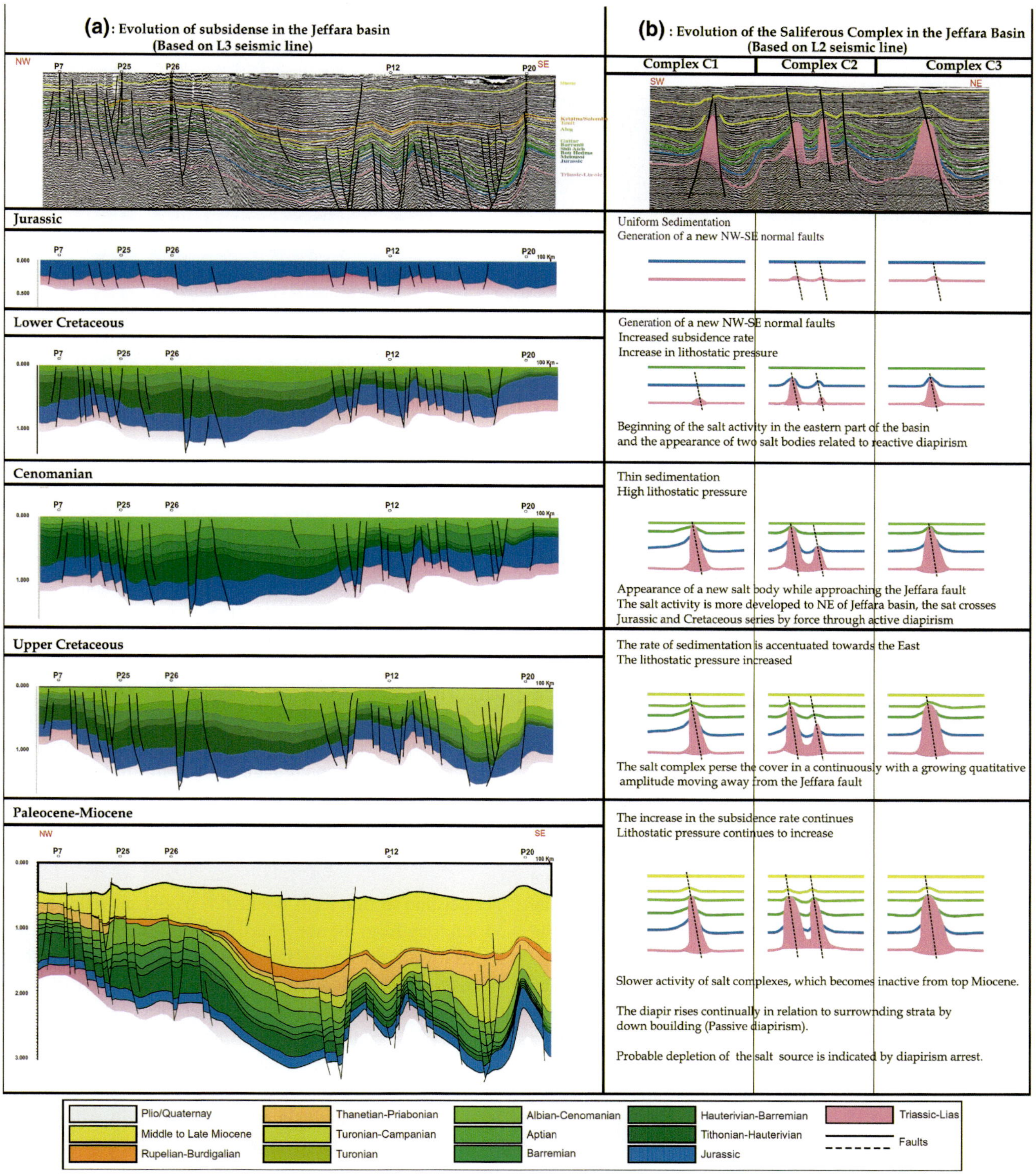

Fig. 8.31 Geoseismic interpretations and basin restorations from Jurassic to Miocene in the Gulf of Gabès, Pelagian Sea. The interpretations show early Diapirism during Upper Jurassic–Early Cretaceous. From Khouni et al. (2018)

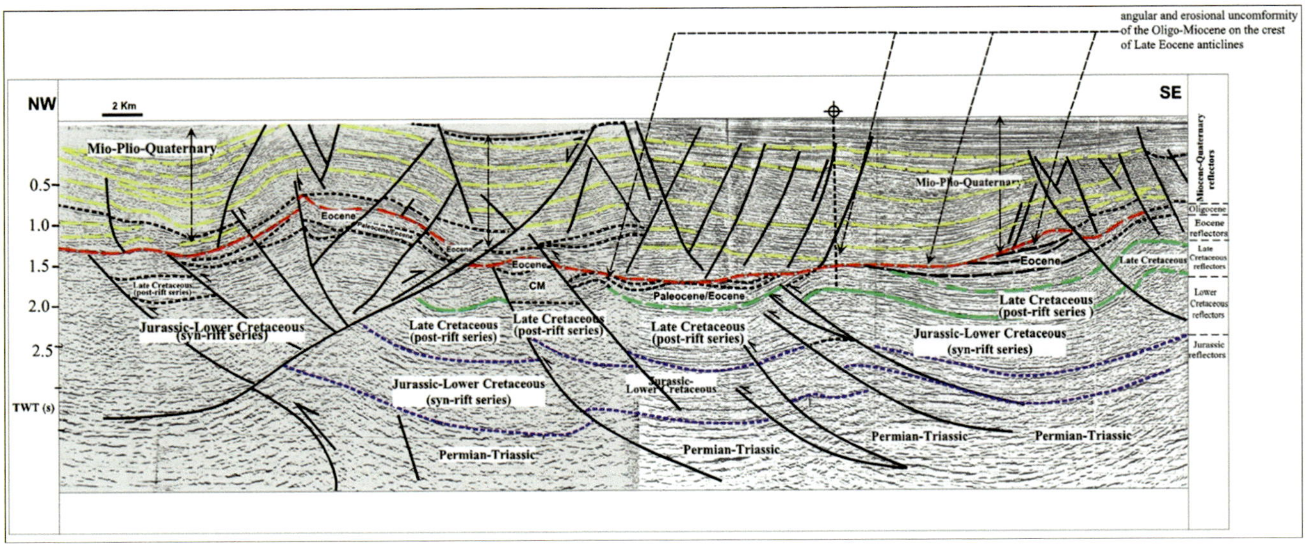

Fig. 8.32 Seismic cross-section in eastern Tunisia, which illustrates the Eocene compression event corroborated by the angular unconformities and sealing of the Aquitania sequence on the Eocene inverted structures. Reprinted by permission from (Springer-Nature) [Arabian Journal of Geosciences] Khomsi et al. (2016). Copyright (2016)

To the East, block C corresponds to a rollover anticline, with the Miocene lying directly on Paleocene strata with thin Eocene reflectors. This configuration indicates that during Upper Miocene–Quaternary, the major fault F2 was activated as normal fault, limiting the rollover anticline to East. In fact, compared to the other structures shown on the seismic, the bloc C recorded the most important thickness of Miocene–Quaternary underlining the synsedimentary activity of F2 during Miocene–Quaternary compressions.

The deformation of the Quaternary horizons and the intra-Quaternary unconformity recorded at the crest of the anticline structure (Fig. 8.35) result from the Late Miocene–Tortonian and Villafranchian (Quaternary) compressional events. These events are well-documented in the entire Sahel region, both in surface outcrops (Ouali et al. 1987; Boukadi and Bédir 1996; Bédir et al. 1992; Hlaiem 1999; Ouali 2007; Khomsi et al. 2016) and in the subsurface (Bédir and Bobier 1987; Bédir 1988). In addition, the comparison of the sedimentary infilling between blocks A and B indicates a reverse offset of these faults during Late Cretaceous–Paleocene and Mio-Pliocene times (Fig. 8.35).

8.5.5 Seismic Section L6

The seismic section L6, striking SW–NE, is calibrated by the Well S9 (Fig. 8.46). This section crosses the Mechertate anticline which corresponds to a pop-up anticline affected by normal faulting in its axial zone (Fig. 8.36). This anticline is limited to the south and to the north by two listric faults, F1 and F'1, respectively (Fig. 8.36). The Campanian–Maastrichtian limestones reflectors are folded and keep the same thickness along the section (Fig. 8.36). By contrast, the Paleogene series show a thinning toward the apical culmination of this fold with a gap of the Ypresian horizons and an unconformity between the Late Eocene and the Paleocene strata (Fig. 8.36). Then, the Mio-Pliocene seals this structure, with a lack of Oligocene (Fig. 8.36).

The pop-up anticline of Mechertate can be dated as Paleocene–Lower Eocene, because it is sealed by the Late Eocene reflectors with an angular unconformity at the top.

8.5.6 Section L9

The seismic section L9 is oriented NW–SE and extends over 34 km long from the south of Cherichira structure in the N–S axis toward the depression of Sidi El Hani. In the South of Cherichira, the fault system is characterized by thrust faults that bound a system of imbricate slices propagating eastward (Fig. 8.47). To the east, this thrust is connected to a deep-seated fault in a fishtail configuration (Khomsi et al. 2016, 2019; Mezni et al. 2019). On this oblique deep fault branches a set of antithetic and synthetic faults which affect the Early–Late Cretaceous and Eocene and sealed by the Middle–Late Eocene series (Fig. 8.48).

In the Eastern part of the section, the Sidi El Hani is characterized by strike-slip faults with a flower structure configuration (Fig. 8.48). This fault system is formed by a set of deep-seated faults and bounding a fold structure which is collapsed in the Miocene as corroborated by the thick Miocene series (Fig. 8.48). Thus, in the Bouhajla–Sidi El Hani domain, the fault system is characterized by a set of oblique faults affecting the Sidi El Hani anticline.

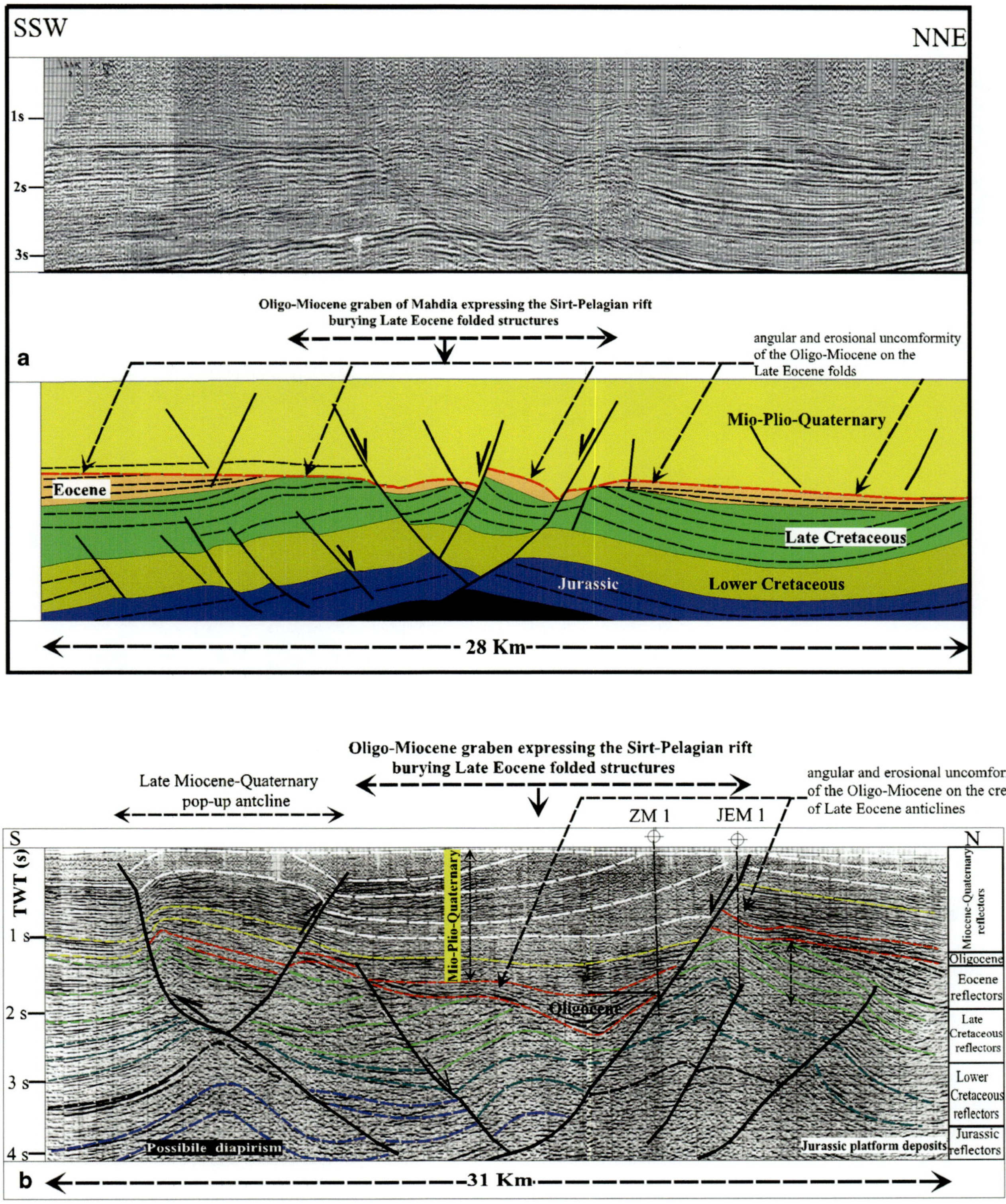

Fig. 8.33 Seismic cross-section in eastern Tunisia, which illustrates the Eocene compression event corroborated by the angular unconformities followed by a quiescence period outlined by collapse faults limiting Oligo–Miocene filled grabens (Khomsi et al. 2016). Notice the angular/erosional unconformity of the Oligo–Miocene on the Upper Cretaceous. Reprinted by permission from (Springer-Nature) [Arabian Journal of Geosciences] Khomsi et al. (2016). Copyright (2016)

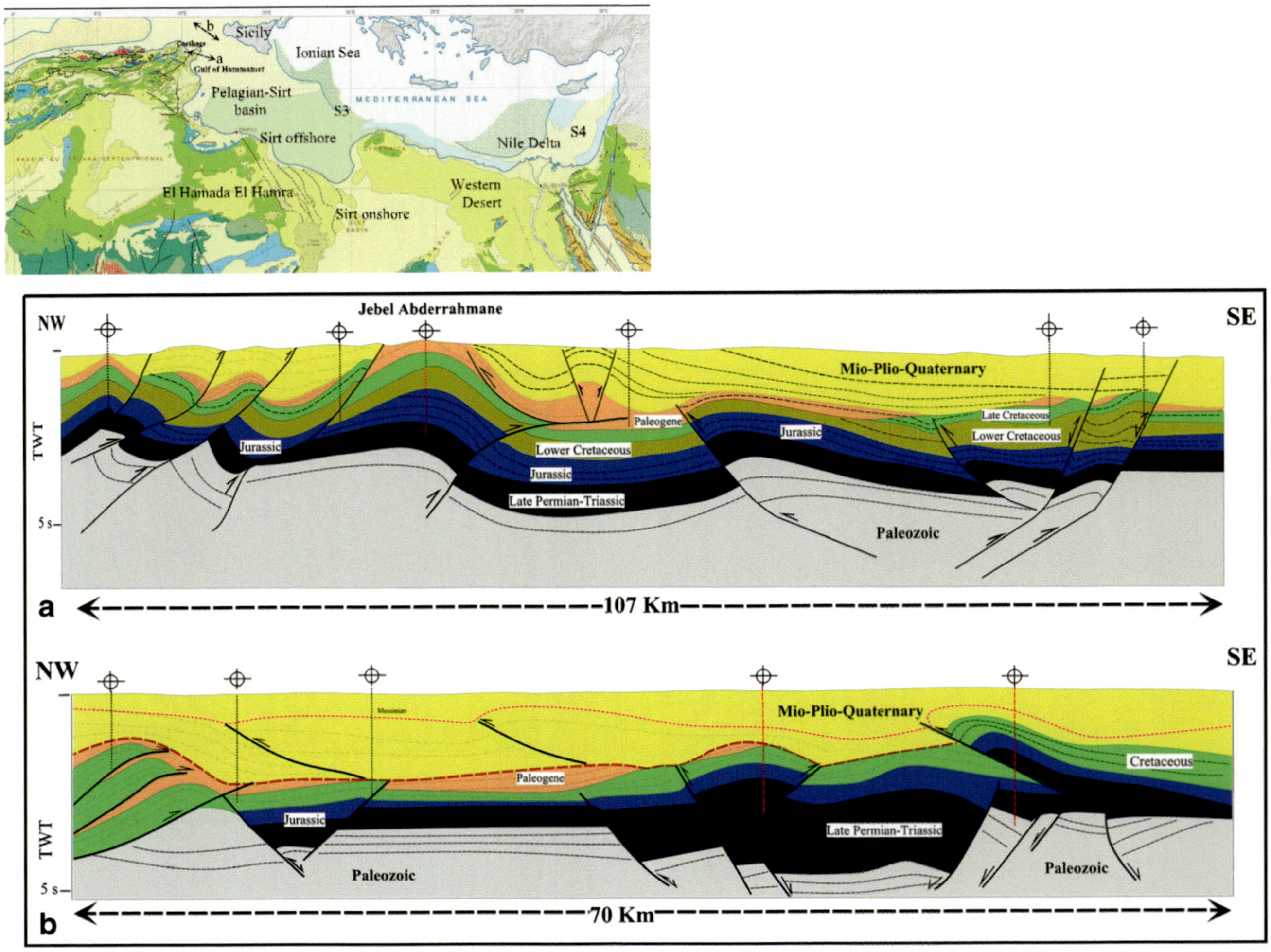

Fig. 8.34 Geoseismic sections in the Gulf of Hammamet and Sicily–Tunisia channel outlining Miocene flexural basins (from Khomsi et al. 2016, modified from Casero and Roure 1994). Reprinted by permission from (Springer-Nature) [Arabian Journal of Geosciences] Khomsi et al. (2016). Copyright (2016)

8.5.7 Section L10

This section is located in the Bouhajla area (for localization, see Fig. 8.35). It extends from South to North, respectively, from Southeast Ktitir to Northwest edge of Sidi El Hani depression (Fig. 8.49). The section, measuring 22 km long, shows two folds separated by a subsiding syncline (Fig. 8.45). In fact, anticline A is affected by an oblique fault that cuts the Cretaceous–Quaternary horizons. A single fault shows a few milliseconds offset between the same horizons terminations. In addition to this fault, we note, at 2.5 s (TWT) (Fig. 8.49), the existence of an oblique fault sealed by the Cretaceous series and continuing downward in the Triassic (Fig. 8.49) and probably through the pre-Triassic series (Permian series). This fault controls presumably the halokinetic activity of the upward movements of the Triassic salt deeply underneath Anticline A. The thinning of the Upper Eocene (Fig. 8.49) suggests that tan upward movement of the salt material occurred during the Late Eocene.

In the northern Part of the section we observe anticline (B) affected by multiple oblique faults (Fig. 8.49). This oblique faults system indicates a possible flower structure (Fig. 8.49) and it is characterized, by the development of a set of oblique faults that are sealed by Lower and Upper Cretaceous series to the Upper Eocene. The rejuvenation of this faults system has potentially trigged the upward movement of Triassic salt and set up this salt-cored fold (anticline B). This anticline occurred presumably in the Late Eocene because of the thinning of the Late Eocene series on the anticline crest. In addition to that, it is noticeable that the presence of compensations basins filled with thick Late Cretaceous, Paleocene, and Late Eocene series. These compensation basins are controlled by the upward growth of salt intumescences underneath anticlines A and B.

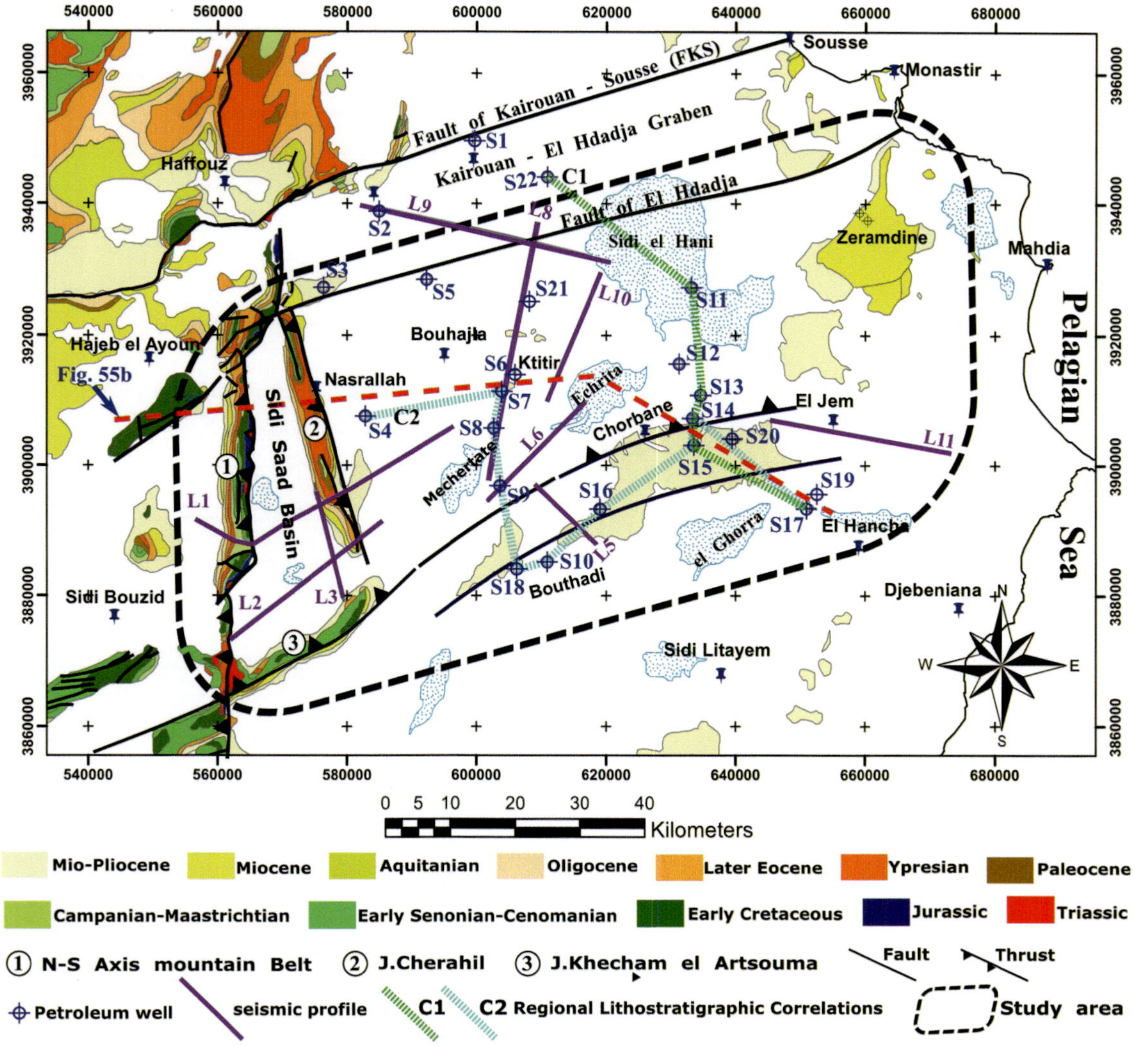

Fig. 8.35 Geological map of the N–S axis and Sahel domain with the location of petroleum wells, seismic profiles, and regional cross-sections (Mezni et al. 2019)

8.5.8 Section L11

This section is located in the easternmost part of the Sahel domain, in the El Jem area and East of Chorbane anticline. It highlights the structural style and fault system that affects the Meso-Cenozoic layers. This section measures 31.4 km long and oriented WNW–ESE (Fig. 8.50). At the western part of the section, the fault set limits the graben of El Jem (Fig. 8.50) with deep-seated faults. It is noticeable the Triassic doming and the Jurassic. This configuration indicates early Diapirism, deep below. The Diapiric Dome seems to be controlled by deep-seated faults presumably linked to paleo-faulting system of the Tethyan rift.

8.6 Subsurface Isochrone Maps of the Turonian–Coniacian, the Campanian–Maastrichtian, and the Early Eocene Formations

The Sahel domain is crosscut by a set of major faults/Thrust (Khomsi et al. 2009a, b, 2016, 2019) and transtension/transpressive strike-slip faults accounting for the Quaternary readjustments during the Alpine compressions (Khomsi et al. 2004a, b). This fault system limits different geological structures represented by inverted grabens, pop-up, thrust fold, salt-related structures, and a strike-slip-related fold with

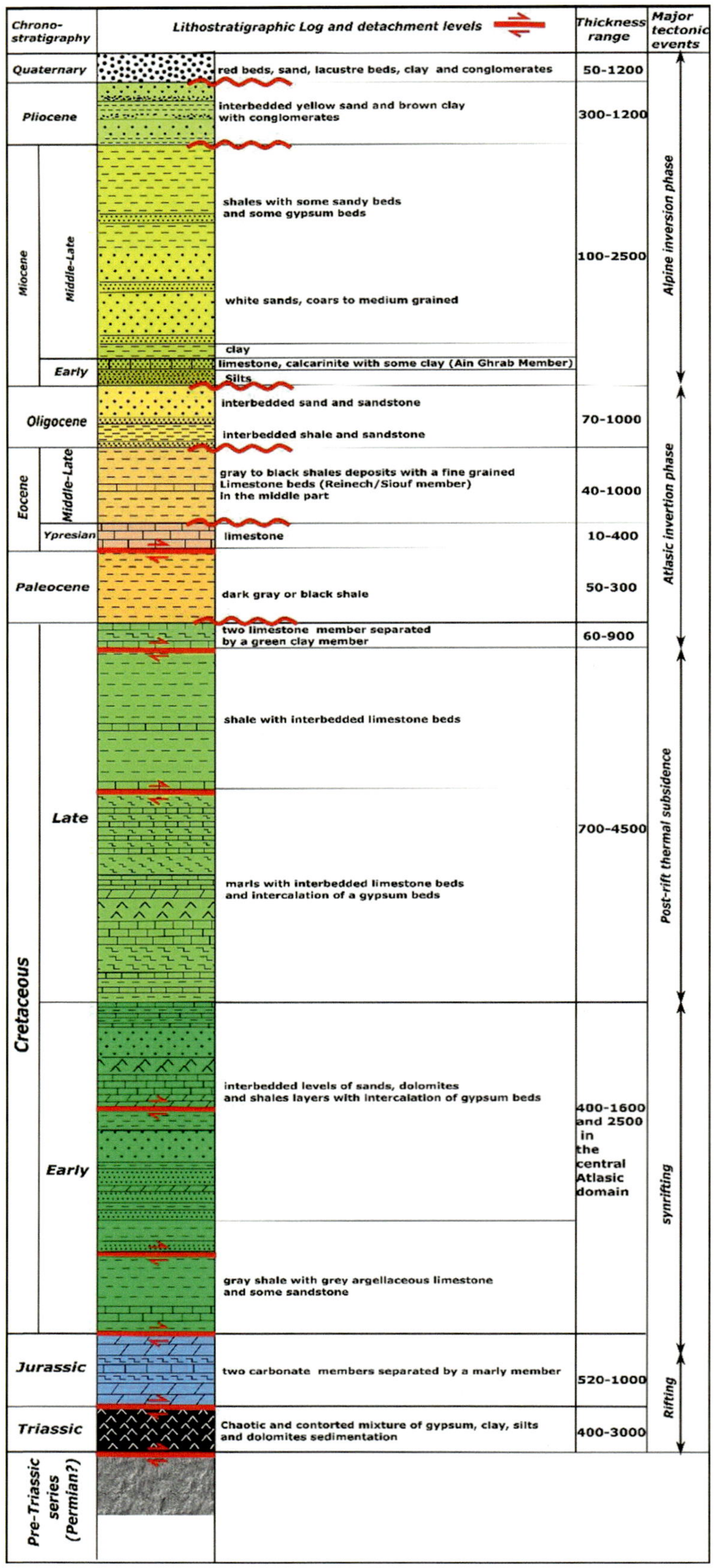

Fig. 8.36 Lithostratigraphic column of eastern Tunisia in Chorbane anticline showing also the major tectonic events and major detachment levels. Reprinted by permission from (Springer-Nature) [Arabian Journal of Geosciences] Mezni et al. (2019). Copyright (2019)

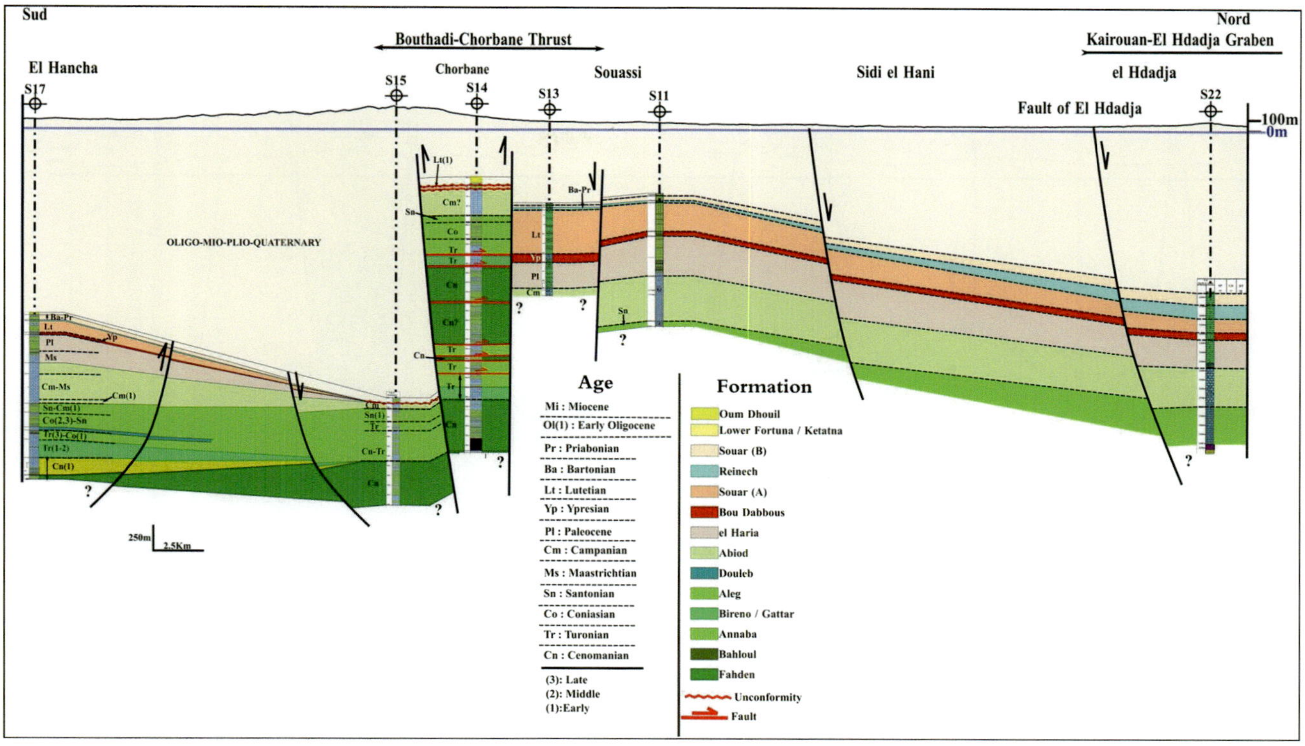

Fig. 8.37 North–South lithostratigraphic correlation between El Hdadja and El Hancha. See location in Fig. 8.35

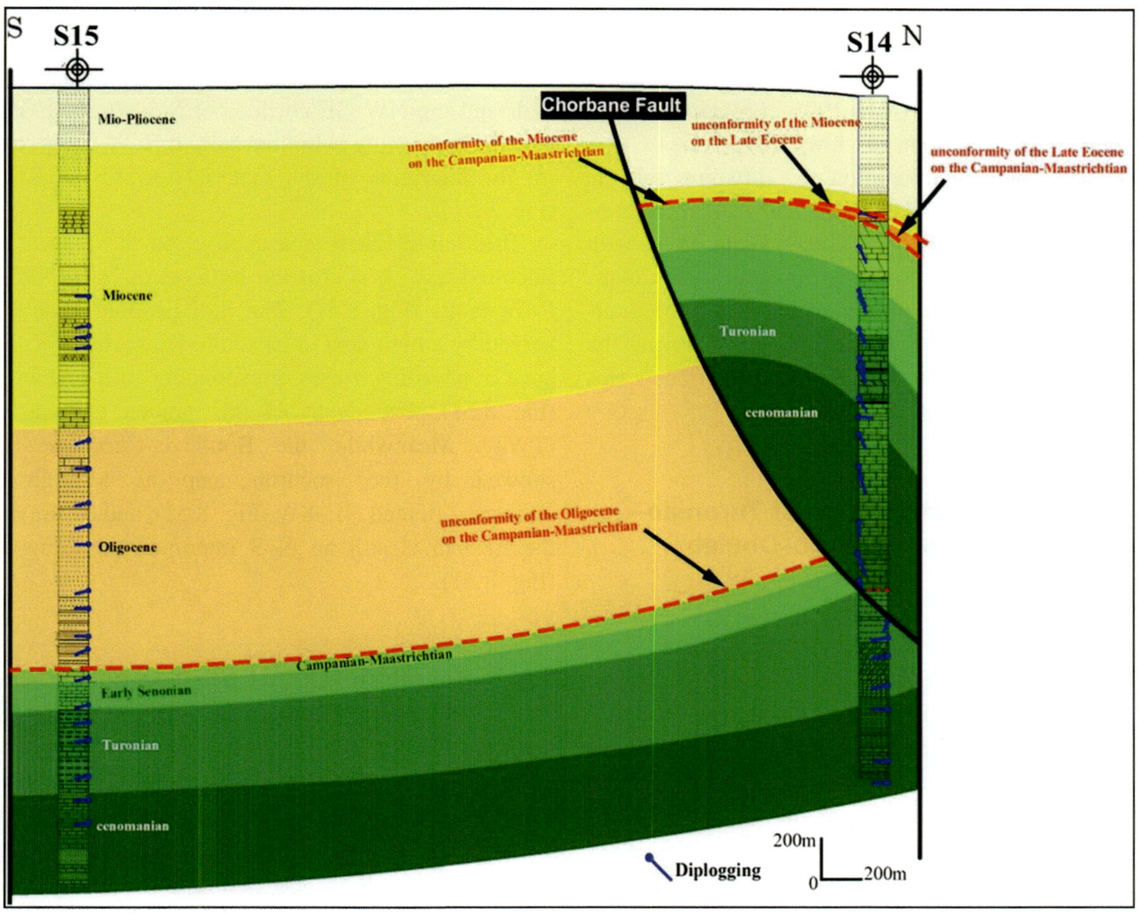

Fig. 8.38 Interpretative structural cross-section along the Bouthadi–Chorbane fault based on dipmeter logging, showing the Chorbane anticline which is interpreted as a fault-detachment fold (Mezni et al. 2019). See location in Fig. 8.35. Reprinted by permission from (Springer-Nature) [Arabian Journal of Geosciences] Mezni et al. (2019). Copyright (2019)

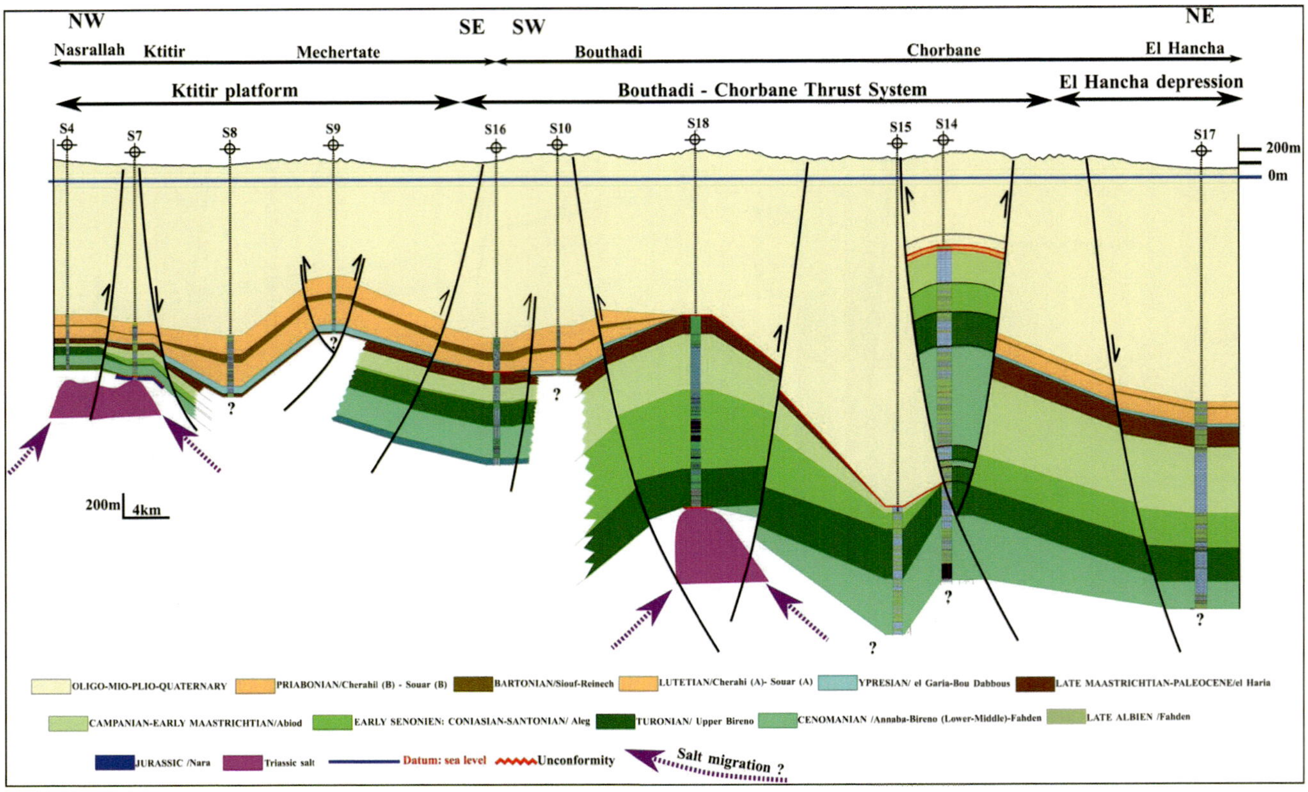

Fig. 8.39 N–S then SW–NE correlation between Nasrallah and El Hancha. See location in Fig. 8.35. Modified from Mezni et al. (2019)

positive and negative flower structures detected in the central and eastern part of the Sahel (Bédir 1995). From a structural reservoir characterization point of view, this part deals with seismic analysis and interpretation of different seismic structural maps of the Late Cretaceous and Eocene carbonates series, which extends in the Sahel domain along with the Chorbane–NS axis fault corridors domains. These maps are created after the calibration, the analysis, and the interpretation of a set of seismic reflection profiles covering the study area. The result comes in different subsurface maps enclosing the isochron maps.

8.6.1 Isochron Map of the Top of Turonian–Coniacian Horizon: Bireno–Douleb Formations

The map underlines that the Sidi Saad basin is a subsiding area-oriented N–S overhanged tectonically by the Nara thrust belt (Fig. 8.51). Towards the East, we have a high resistant area related to the Cherahil fold and a second high resistant area corresponding to the lineament of Rheouis–Khechem El Artsouma (Fig. 8.51). The Sidi Saad basin is affected by the NW–SE and N–S, and NE–SW faults (Fig. 8.51).

To the East of Cherahil, the area of Nasrallah is characterized by high structures-oriented E–W. To the North, we have a NE–SW high structure bounded by the El Hdadja fault and the NW–SE corridor of Sidi El Hani (Fig. 8.51). The Ktitir structure is limited by two E–W faults responsible for the migration of salt material toward the Ktitir diapir (Fig. 8.51).

The Sidi El Hani area appears on the map as a highly fractured area. It is crosscut by a set of N–S, NW–SE, and E–W faults (Fig. 8.51). The Sidi El Hani zone is characterized by a high area in its south. However, at its north, we have a subsiding zone corresponding to a graben structure (Fig. 8.51). The depth of this graben exceeds 3000 ms (TWT). Meanwhile, the Bouthadi–Chorbane alignment showed by the isochron map as a high resistant structure-oriented NE–SW (Fig. 8.51), and on the other side, the El Jem is still an N–S important subsiding structure (Fig. 8.51).

8.6.2 Isochron Map of the Top of the Abiod Horizon (Campanian–Maastrichtian)

The isochron map of the top of the Campanian–Maastrichtian Abiod horizon shows different features and variously directed faults (Fig. 8.52). In fact, on the western side of the map (Fig. 8.52), the Sidi Saad basin corresponds to a foredeep basin thrusted by the Nara fold-thrust belt and it is

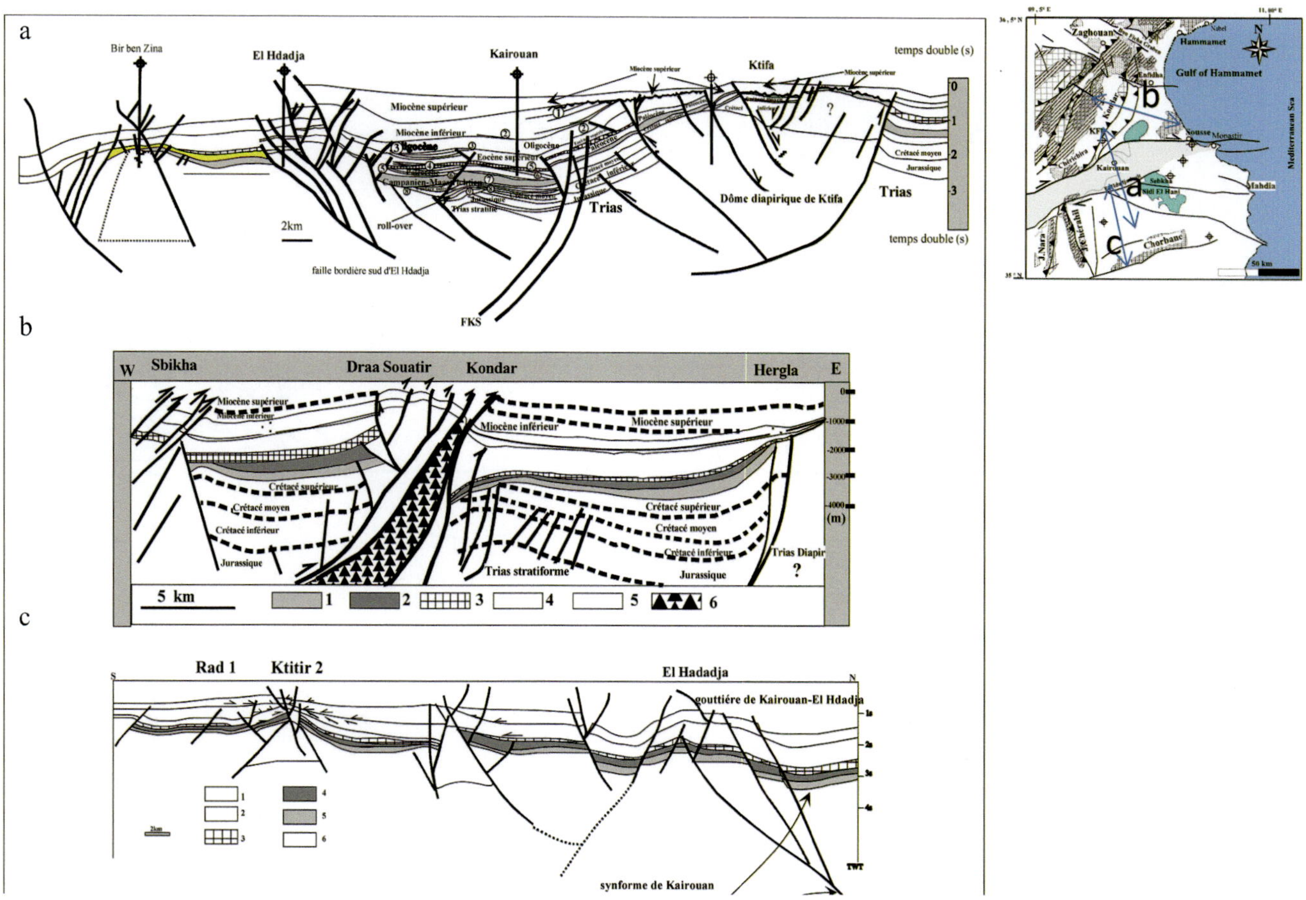

Fig. 8.40 Interpretative structural cross-section along the different domains of the Eastern foreland showing the importance of salt tectonics in the control of the structural styles. **a** Through Kairouan flexural basin, **b** along the Northern foreland basin, and **c** throughout Ktitir–El Hdadja platform (from Khomsi 2005)

deepening westward in the direction of the Nara belt (Fig. 8.52). Besides, the isochron map shows that the Sidi Saad basin is affected by multidirectional faults NW–SE, NE–SW, and N–S (Fig. 8.52).

The central part of the map corresponds to the Sahel domain and represents the foreland basin of the Atlasic thrust belt. In this foreland domain, the Abiod horizon is highly deformed and fractured (Fig. 8.52). This carbonates platform is crosscut by N–S, E–W, NE–SW, and NW–SE faults (Fig. 8.52). In the Nasrallah area, we notice the existence of a high zone-oriented E–W limited by two faults (Fig. 8.52). Then to the North, there is a NE–SW high structure (Fig. 8.52). On the other side, the zone of Ktitir corresponds to an N–S high structure related to the Ktitir corridor (Fig. 8.52), then, we recognize a subsidente basin affected by NNW–SSE to N–S faults. The N–S elongated basins are limited by NW–SE fault as the grabens of El Jem and Sidi El Hani (Fig. 8.52). To the south of the study area, the lineament of Bouthadi–Chorbane represents a structural limit between the domain of the Sahel of Kairouan and the domain of Sidi Litayem in South (Fig. 8.52).

8.6.3 Isochron Map of the Top of Ypresian: Bou Dabbous/El Gueria Horizon

The isochron map of the roof of the Bou Dabbous/El Gueria horizon shows a structural change in comparison with the structural repartition of the different geological features described in the isochron map of the top of Abiod horizon.

The Sidi Saad basin shows a westward deepening where the top of the horizon goes downward for more than 2400 ms (Fig. 8.53). This tectonic subsidence is related to the uplifting of the Nara thrust belt during the orogenic events. Then to the South part of this foredeep basin (Sidi Saad basin), this area's color is red on the isochron map, indicating that it is a high zone. Thus, the uplifting of the south area of Sidi Saad is caused by the reactivation of the N–S axis thrust and Khechem–Artsouma thrust which forms a tectonic weakness zone and trigs the rise of the salt diapir of Rheouis (Fig. 8.53). In addition, we notice that the southernmost part of Cherahil becomes a high structure (Fig. 8.53) caused by the reverse movement of the Cherahil fault.

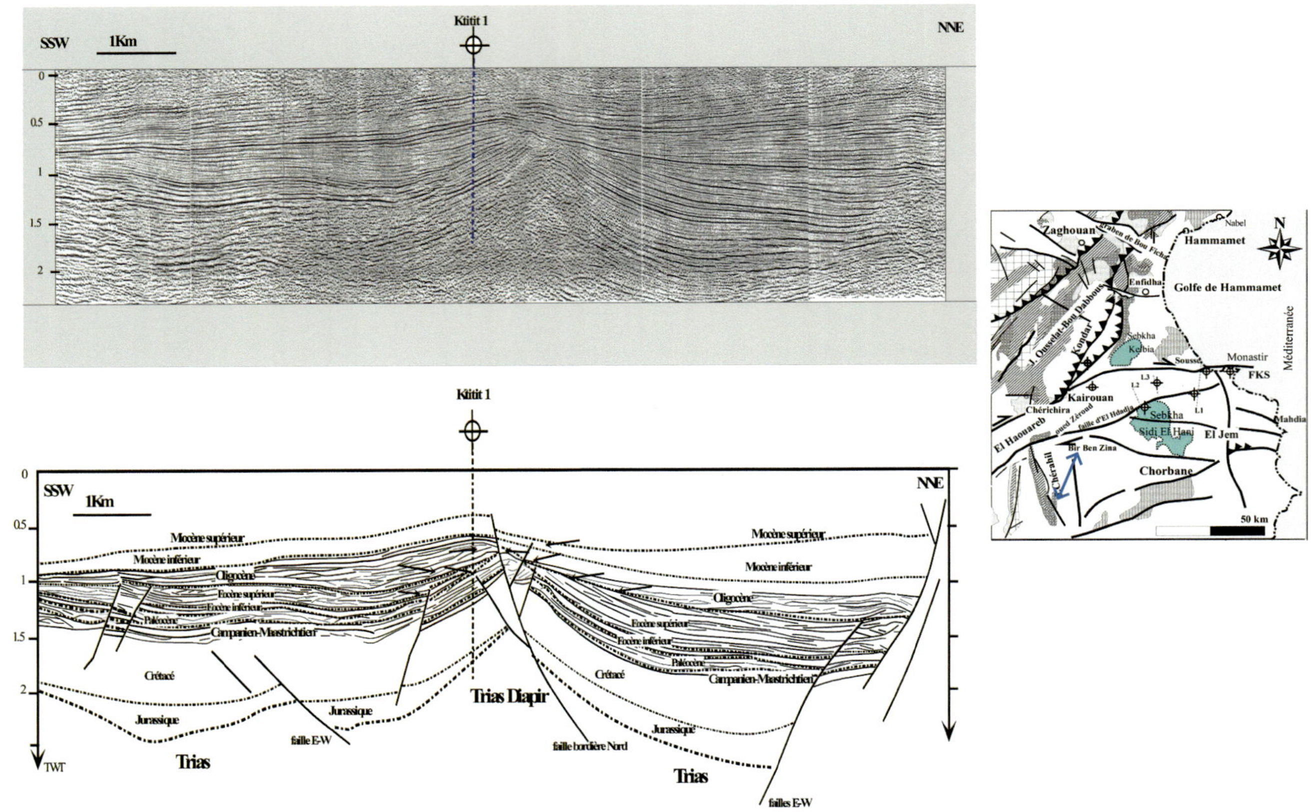

Fig. 8.41 Interpretative structural cross-section along the Ktitir Diapric structure (Khomsi 2005). Notice the sysnsedimentary faulting activities controlling thicknesses variations of Jurassic–Cretaceous. Notice that salt Diapirism is mainly controlled by deep-seated faulting controlling thicknesses variations of the syn-rift series (Jurassic–Cretaceous). Notice also onlapping on the apical part of the Diapir

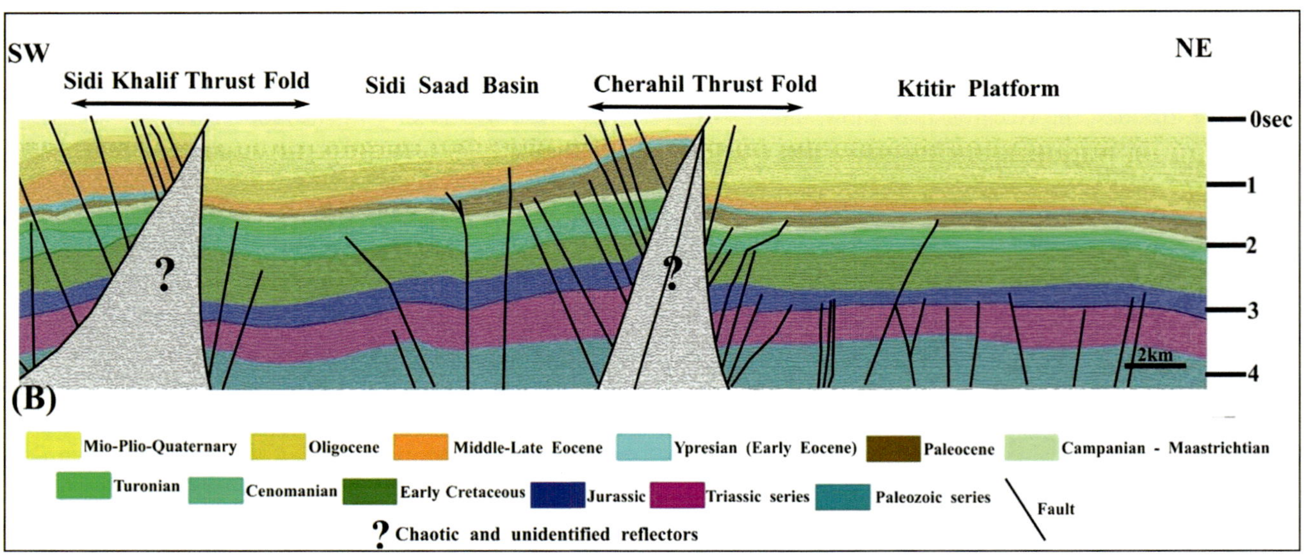

Fig. 8.42 Seismic section L1 oriented SW–NE running from Sid Khalif thrust to the Ktitir platform. The Meso-Cenozoic series are little deformed which indicates the dominance of the sliding mechanism during the deformation regime. The fault system is characterized by vertical deep-seated faults. See location in Fig. 8.35

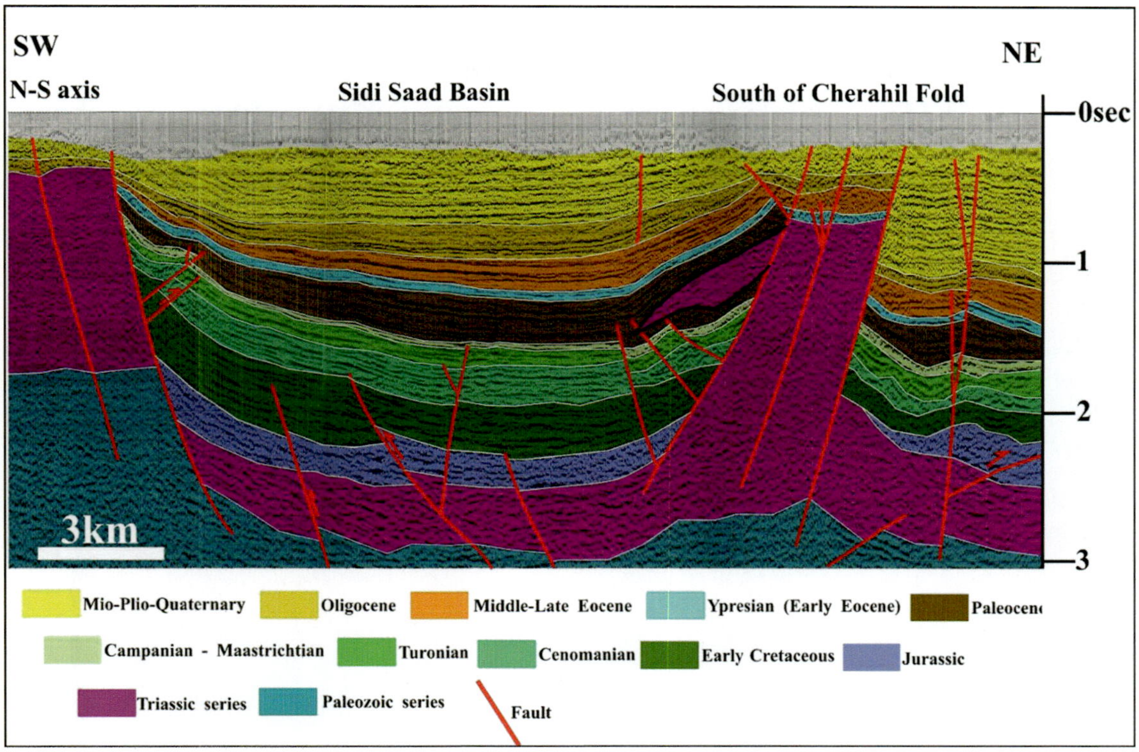

Fig. 8.43 Seismic section L2 oriented SW–NE running from the NS axis to the South of Cherahil fold showing two diapiric structures related to deep-seated faults (modified from Zouaghi 2008). See location in Fig. 8.35

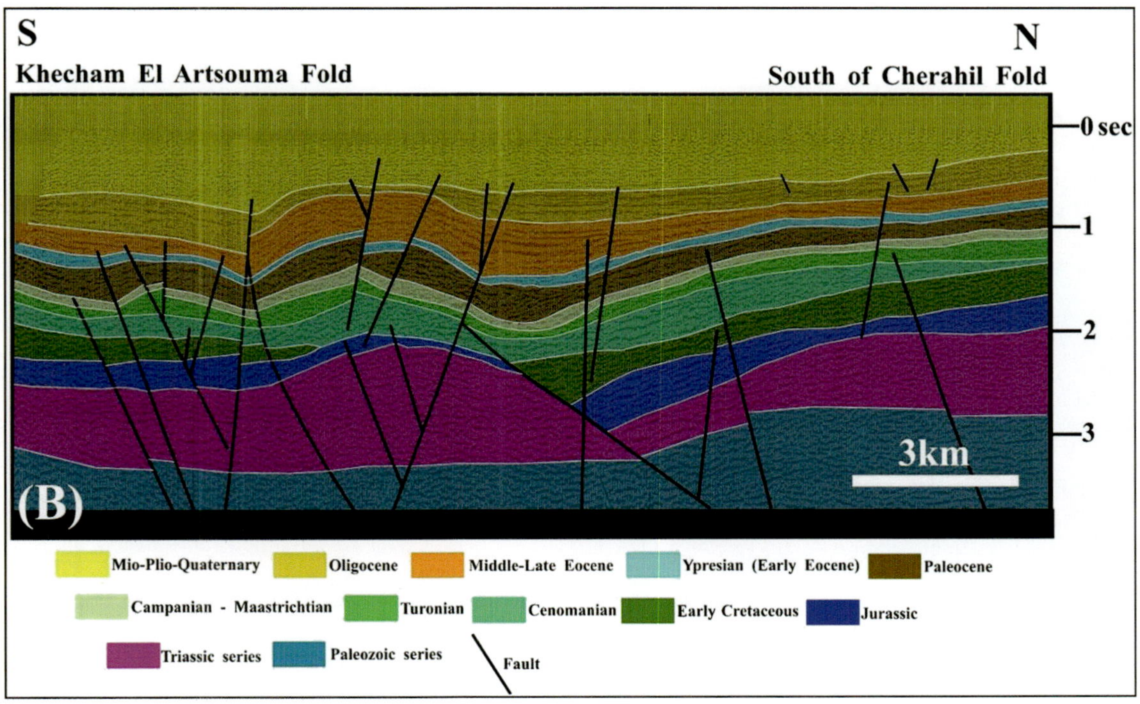

Fig. 8.44 Seismic section L3 oriented SN running from Khecham El Artsouma to the South of Cherahil showing a positive flower structure related to a strike-slip fault set the solicitation of these latter has guided the halokinetic activities since the Jurassic (modified from Yaich 1984). See location in Fig. 8.35

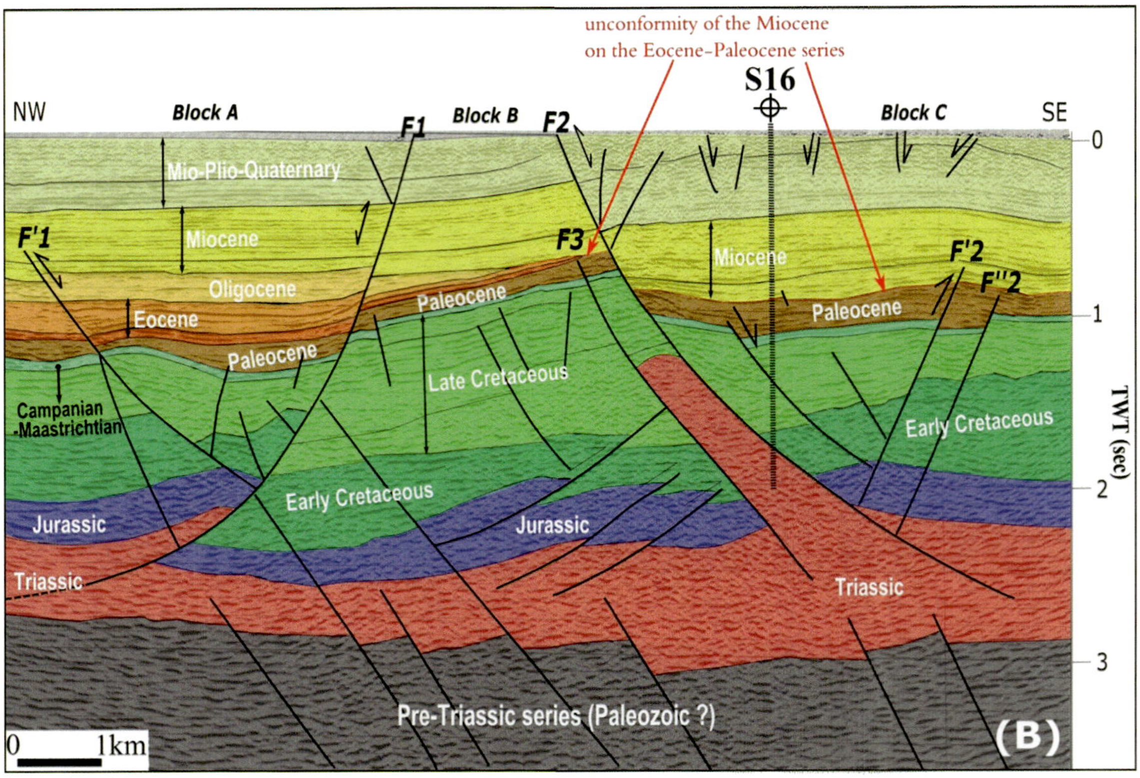

Fig. 8.45 Seismic section L5 showing an inverted structure bounded by the Chorbane fault and the vertical/lateral rise of Triassic salt along faults. Notice also the inverted structures (Mezni et al. 2019). See location in Fig. 8.35. Reprinted by permission from (Springer-Nature) [Arabian Journal of Geosciences] Mezni et al. (2019). Copyright (2019)

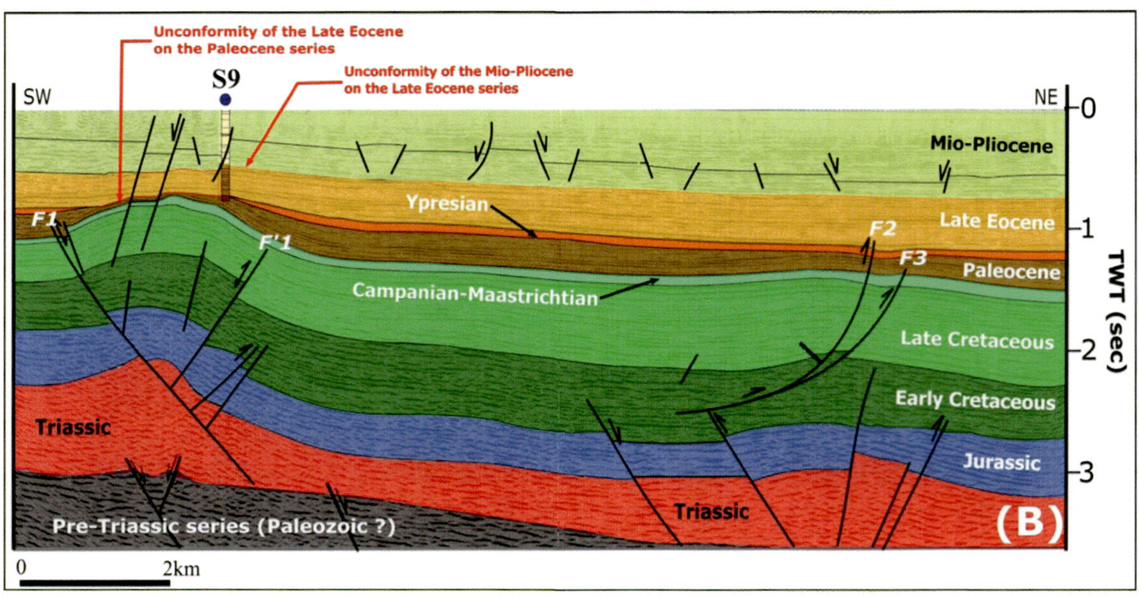

Fig. 8.46 Seismic section L6 showing the symmetric Mechertate fold, which is detached within the Triassic salt décollement level (Mezni et al. 2019). See location in Fig. 8.35. Reprinted by permission from (Springer-Nature) [Arabian Journal of Geosciences] Mezni et al. (2019). Copyright (2019)

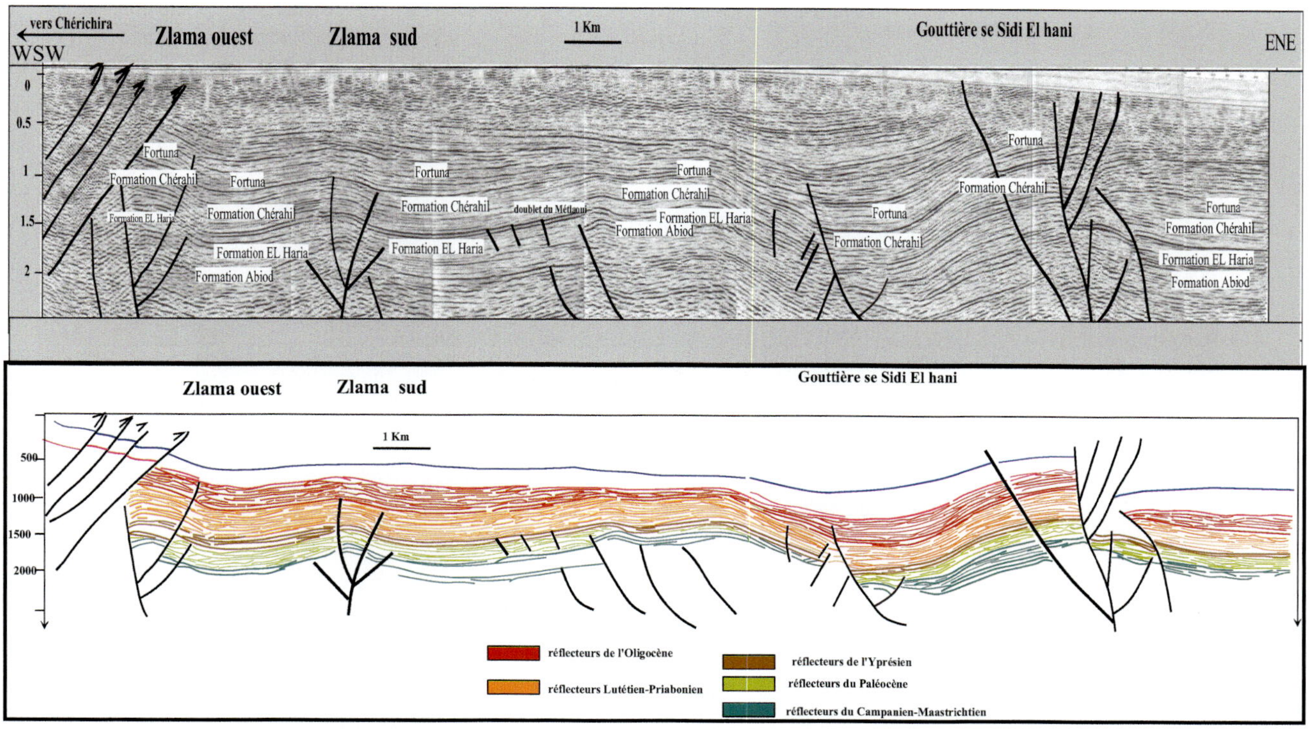

Fig. 8.47 Seismic section interpretation at the East of the N–S axis along Zlama–Sidi El Hani structures. Notice the control of deformation by deep-seated faults (Khomsi 2005)

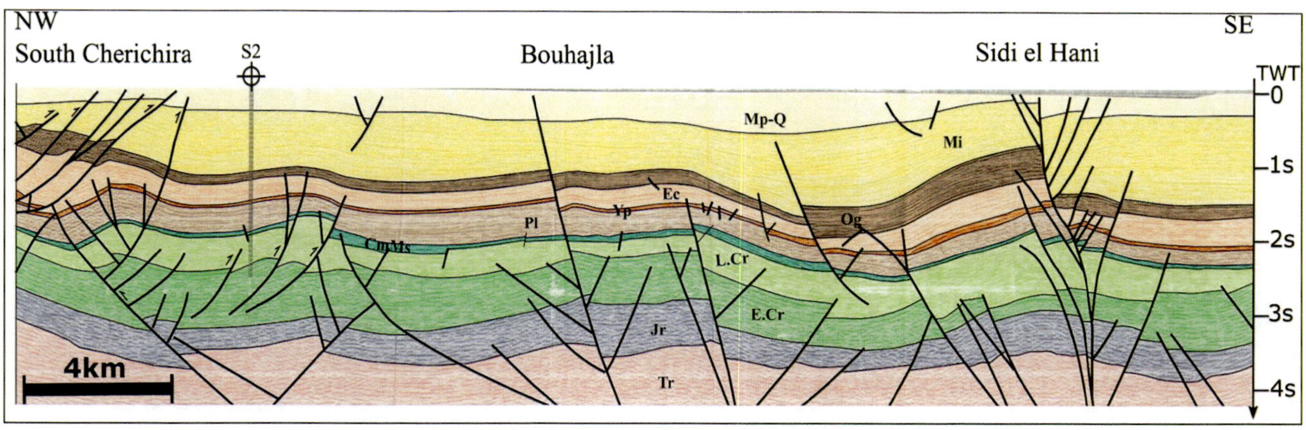

Fig. 8.48 Geoseismic cross-section L9 in the north area of the Sahel showing the changing in the structural style and the nature of the fault and the related fracture system between the south of Cherichira fold (northern extension of the North–South axis thrust) and the Sidi El Hani (foreland basin of the North–South axis front) (modified from Khomsi 2005). With Jr: Jurassic. Cr 1: Early Cretaceous. Cr 2: Late Cretaceous. CM: Campanian–Maastrichtian. Pl: Paleocene. Yp: Ypresian. Ec 2: Middle–Late Eocene. Og: Oligocene. M–Q: Miocene and Mio-Plio-Quaternary. See location in Fig. 8.35

In the Sahel domain, the foreland of the Atlasic belt, the Ypresian horizon is crosscut by multiple directed faults. In fact, at the East of the Cherahil structure (Nasrallah area), the high E–W structure is bounded by two E–W faults and bordered to its north by a NE–SW 7–2 elongated basin (Fig. 8.53). Then, we notice a migration of the high zone from the East of Chorbane toward Bouthadi (Fig. 8.53). However, we notice a deepening of the Ypresian horizon at Sidi El Hani and a northward migration of the El Jem depocenter (Fig. 8.53).

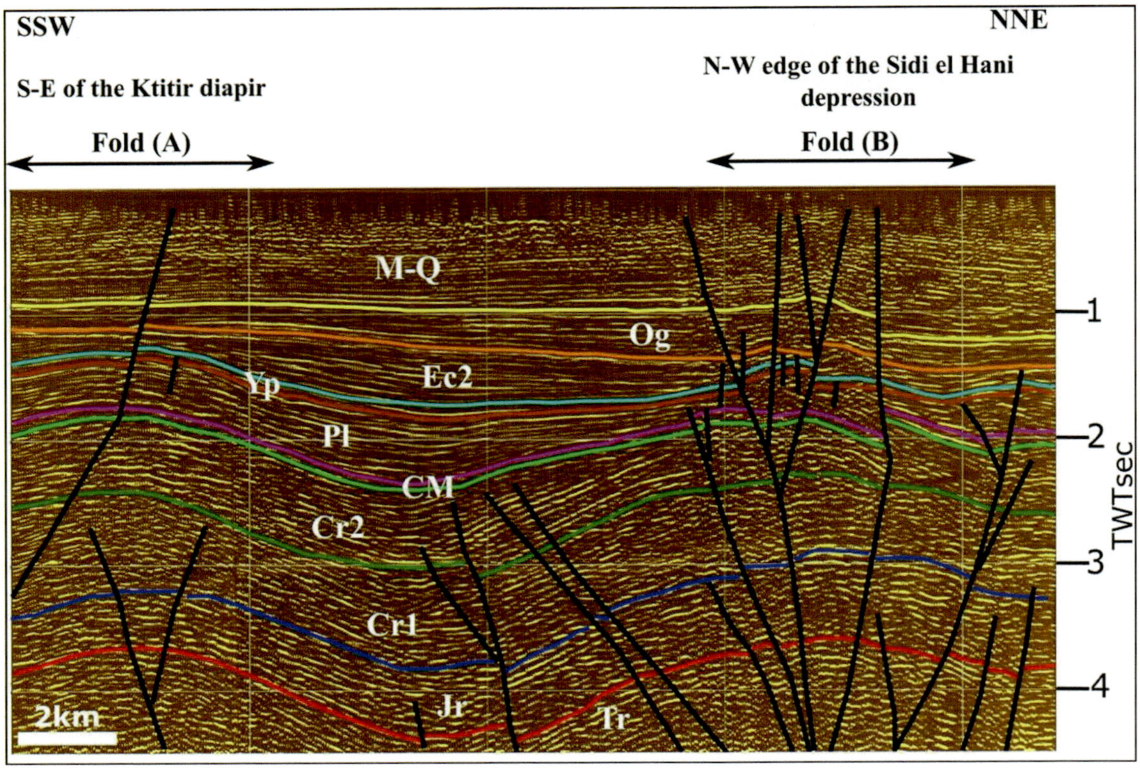

Fig. 8.49 Seismic cross-section L10 showing the fault system associated to fold A and fold B. The fault system is characterized by deep-seated oblique faults. With Tr: Triassic. Jr: Jurassic. Cr 1: Early Cretaceous. Cr 2: Late Cretaceous. CM: Campanian–Maastrichtian. Pl: Paleocene. Yp: Ypresian. Ec 2: Middle–Late Eocene. Og: Oligocene. M–Q: Miocene and Mio-Plio-Quaternary. See location in Fig. 8.35

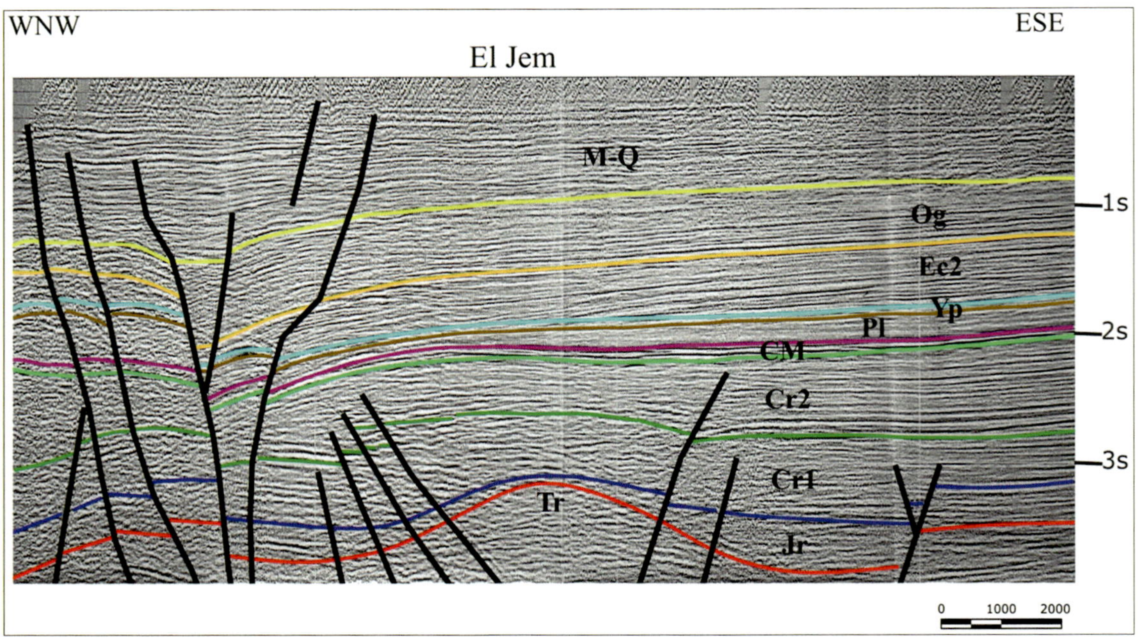

Fig. 8.50 Seismic cross-section L11 in the El Jem area showing a rollover structure limited by oblique deep-seated fault. With Tr: Triassic. Jr: Jurassic. Cr 1: Early Cretaceous. Cr 2: Late Cretaceous. CM: Campanian–Maastrichtian. Pl: Paleocene. Yp: Ypresian. Ec 2: Middle–Late Eocene. Og: Oligocene. M–Q: Miocene and Mio-Plio-Quaternary. (See location in Fig. 8.35)

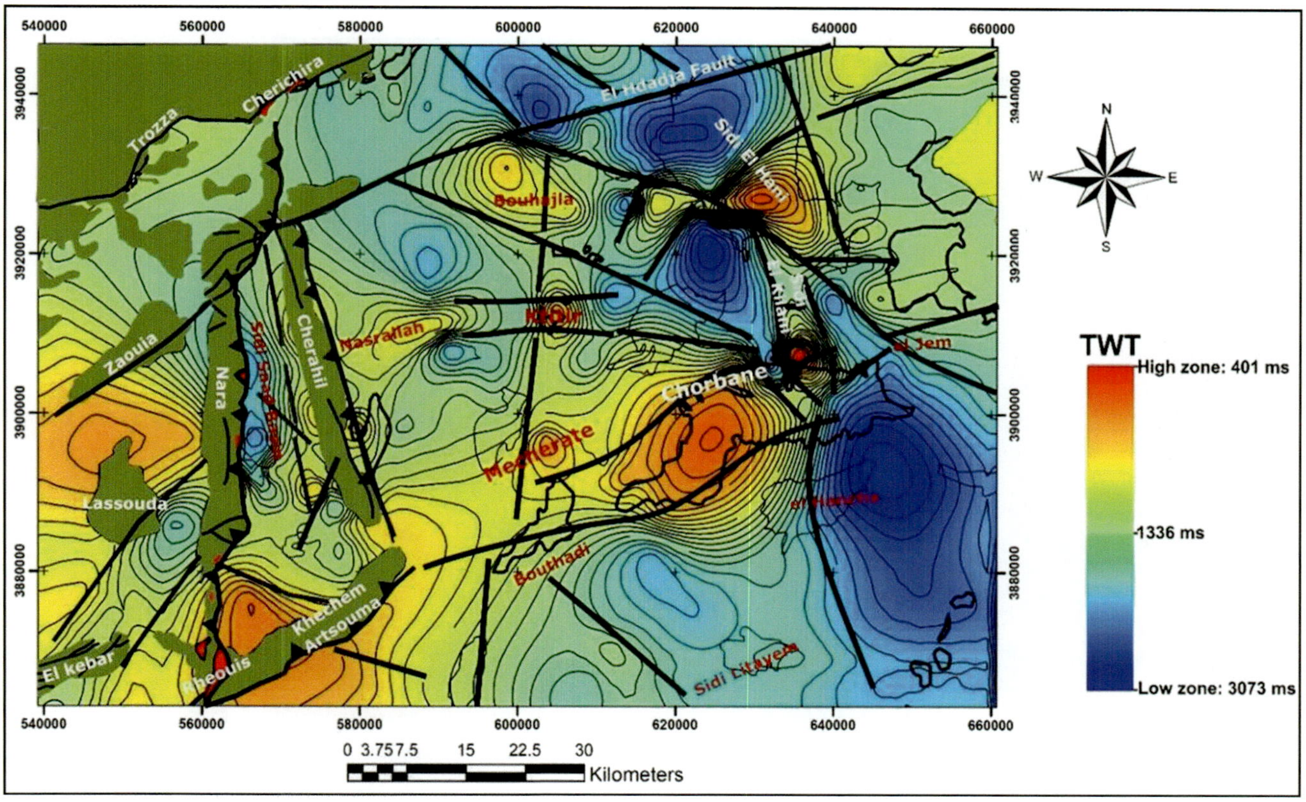

Fig. 8.51 Isochron map of the top of Turonian–Coniacian (Bireno, Douleb carbonates members). The iso-contour in TWT (ms) with equidistance of 100 ms

8.7 Discussion: Structural Style and Integration in the Structural Evolution of the Atlas

8.7.1 Major Structural Styles

The structural interpretations of seismic sections and wells correlations allow deciphering of the main tectonic evens and structural styles in Eastern Tunisia foreland basin. All these sections highlight an Oligocene–Early Miocene or quiescence time span which is general in the eastern Maghreb (Khomsi et al. 2006, 2016; Frizon de Lamotte et al. 2009).

Section L1 outlines the structural styles of the N–S Axis and its adjacent structures along the Ktitir platform. The N–S axis belt corresponds to a thrust fault system that propagates toward the East over a major décollement level situated along the Triassic salt. It is remarkable in Sect. 8.1 that the platform of Ktitir is slightly deformed and affected by a set of vertical deep-seated faults.

Section L2 illustrates principally the Diapiric salt activities. The two salt diapirs are deeply related to the N–S axis fault and the Cherahil bounding faults. The activity of this transtensive\tranpressive fault trigged the upward movement of salt material. This halokinetic activity has guided the sedimentary process corroborated by the thinning of the Meso-Cenozoic series and specially the Late Cretaceous–Eocene horizons toward the respective diapir flanks that bound the Sidi Saad basin.

Section L3 highlights the structures near Jebel Khechem–Artsouma fold. It is noticeable that the maximum deformation is present near the Khechem–Artsouma fold where the different stratigraphic series are affected by a set of vertical branched faults and flower structures. On other hand, this section shows a Triassic salt pillow bounded by these deep-seated oblique faults. This suggests that the halokinetic movements are controlled by these master faults. The thinning of the Jurassic horizon and the unconformity between the Late Cretaceous and Jurassic indicates that the occurrence of the early salt vertical movements started during Jurassic as indicated in the Northern Sahel foreland in Enfidha–Fadheloun (Fig. 8.54) at the footwall of the Zaghouan thrust system (Khomsi et al. 2016, 2019).

The migration of salt material toward this salt pillow is coupled with the collapsed structure identified, in this cross-section, as a rollover structure which is characterized by a thickening of Jurassic–Early cretaceous and Paleocene–late Eocene. The evolution of this rollover structure is coeval

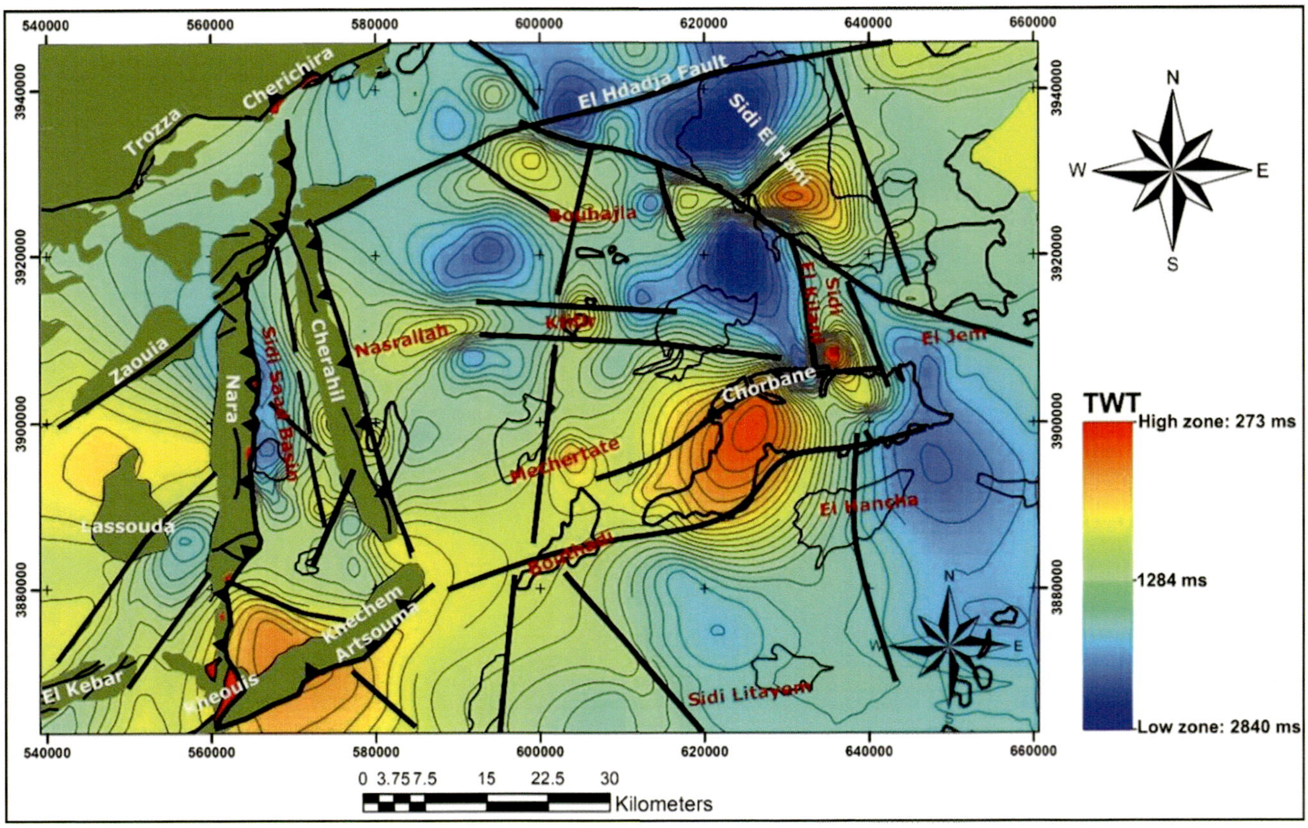

Fig. 8.52 Isochron map of the top of Abiod Horizon (Campanian–Maastrichtian). The iso-contour in TWT (ms) with equidistance of 100 ms

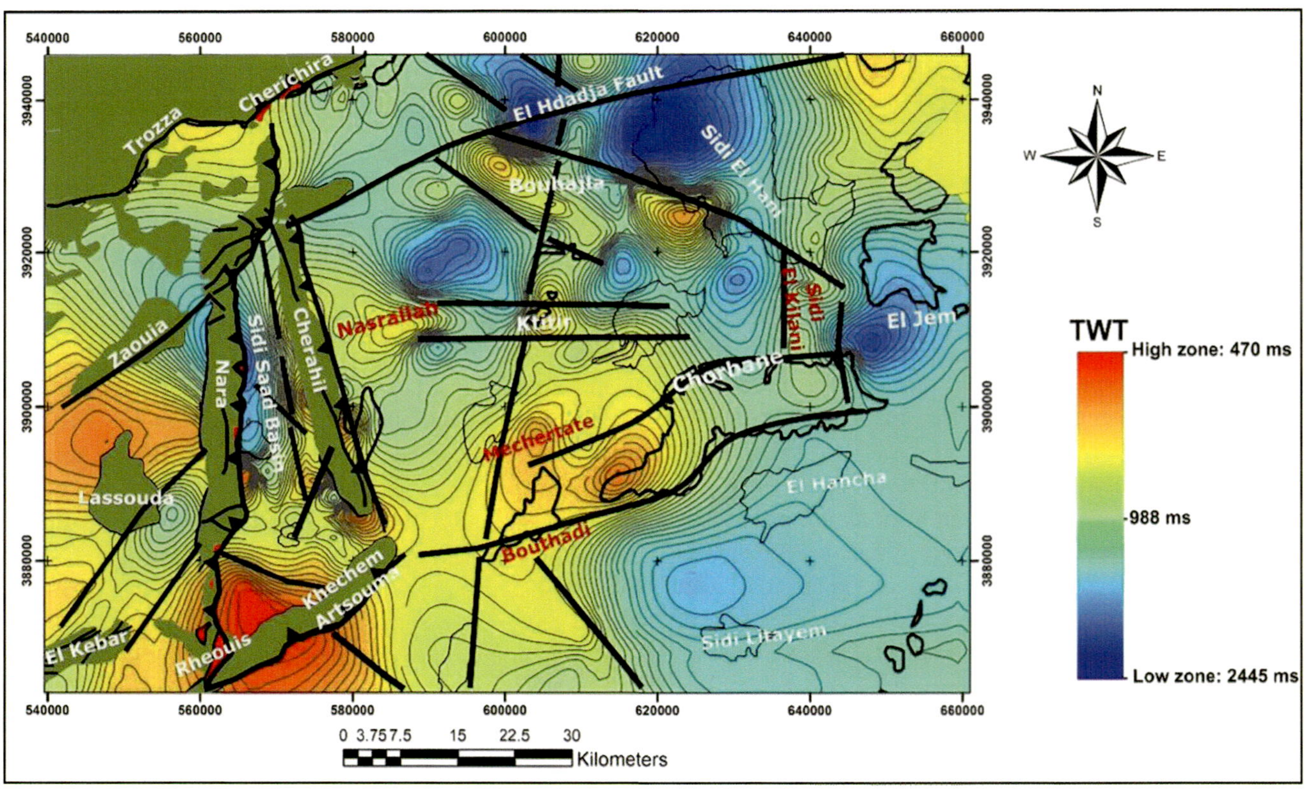

Fig. 8.53 Isochron map of the top of Ypresian Horizon (El Gueria/Bou Dabbous Formation)

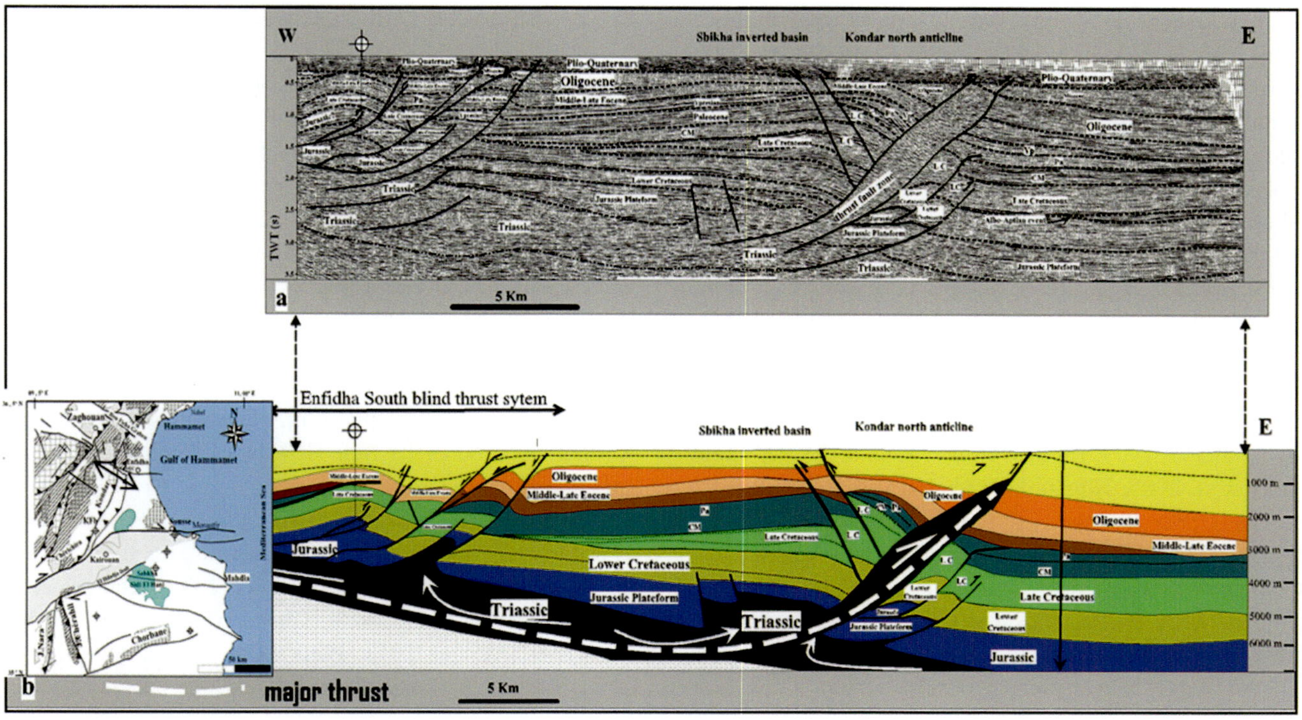

Fig. 8.54 Geoseismic section in the Northern Sahel foreland. This section shows thinning of Jurassic above a salt dome underneath Sbikha. It shows also balanced structural blocs in hemi-graben configurations related to the Tethyan rifting. These tilted blocs with thickness variations are inverted and the major bounding faults serve as decoupling weakness brittle surfaces along which Triassic salt are injected. Notice also the flexural Oligo–Miocene packages resulting from successively Late Eocene and Late Miocene thrusting events

to the lateral migration of salt materials toward the salt domes.

Section L4 characterizes thrust tectonics with a major décollement level situated within the Early Cretaceous shales. It shows a graben filled by thick Late Cretaceous deposits and bounded by the Bouthadi–Chorbane thrust fault. The rejuvenation of this main fault has induced the subsequent uplift/positive inversion of the former grabens.

Section L5 shows the thick Late Cretaceous series as well as the Formation of an extensional graben structure controlled by the reactivation of a deeper normal fault, i.e., the Bouthadi–Chorbane Fault. Since the Campanian–Maastrichtian, we recognize active tectonic inversion processes which are expressed by the positive reactivation of older faults (F1 and F′1). On the other hand, to the north of fault F2, we can also identify a progressive onlap unconformity of Early Miocene strata on top of the Paleogene series, with overall thinned intervening Eocene layers. This progressive onlapping of the Miocene series indicates a major compressional event occurring between the Late Cretaceous and Miocene times. In contrast, we observe a gap of the Eocene, Oligocene, and Burdigalian series in the southern part of fault F2, associated with an unconformity of the Aquitanian (Early Miocene) strata on top of the Paleocene beds. Thus, these discrepancies in sedimentary thickness account for differential subsidence as a consequence of active compression during Eocene.

Section L6 illustrates a Middle–Late Eocene compressional phase (Khomsi et al. 2006), which allowed the growth of the Mechertate anticline, interpreted here as a detachment fold limited by a thrust fault. It is a positive inverted structure characterized on its axial zone by a thinning of the Paleocene and Late Eocene series, with a lack of Ypresian and Oligocene series.

On the other hand, the interpretations of the seismic sections confirm the importance of Triassic evaporites and coeval salt tectonics presumably related to deep-seated faults (Truillet et al. 1981; Buness et al. 1992; Bédir et al. 1992, 2018; Bédir 1995; Khomsi et al. 2012, 2016). In fact, the reactivation of pre-Triassic faults has guided the evolution of the overlying diapiric structures. Thus, the structural interpretation of the seismic sections calibrated by well logs and well correlations across the major Bouthadi–Chorbane anticline in the southern part of the Sahel underlines the overall regional structural style, which is expressed by inverted grabens, half grabens, and detachment folds detached above the Upper Triassic salt. These compressional structures mostly result from positive inversions and uplift of former, inherited Cretaceous extensional structures. The seismic interpretations show also that the paroxysm of these positive inversions occurred during Middle and Late Eocene times (Frizon de Lamotte et al. 2009; Khomsi et al. 2009a). During the Oligocene, these structures underwent an extensional phase followed by the major compressional folding/thrusting event during Late Miocene–Quaternary times. These compressional structures are controlled by inherited faults cutting through the Jurassic–Cretaceous sedimentary cover of the eastern margin of Tunisia (Bédir 1995; Bouaziz et al. 2002; Khomsi et al. 2009a; Frizon de Lamotte et al. 2009; Raulin et al. 2011). Obviously, these inherited faults guided also the vertical migration of Upper Triassic salt, which could have started very early, i.e., from Jurassic times onward (Bédir 1995; Boukadi and Bédir 1996; Hlaiem 1999; Khomsi et al. 2004a, b, 2009a; Tanfous Amri et al. 2005; Houatmia et al. 2015).

The Upper Triassic salt constitutes a major regional detachment level for the overlying Mesozoic and Cenozoic sedimentary cover, thus allowing a partial or total decoupling between the basement and the sedimentary cover in many inversion structures.

8.7.2 Balanced and Restored Transects

Regional balanced structural cross-sections were tentatively presented and published by Mezni et al. (2019) and Khomsi et al. (2019), respectively, in the Southern Sahel and Northern Sahel (Fig. 8.55). These structural transects are issued from the combination of subsurface data and structural analyses of the main outcropping structures of the Eastern Atlas Fold and Thrust belt. The first runs from the N–S axis in the west to the Sahel of Kairouan the second from the N–S Axis ss anticline of Cherahil in the west (Houatmia et al. 2015, 2016; Khomsi et al. 2016) (Fig. 8.55) throughout the Ktitir platform. The second runs from Northern Kairouan To Sidi El Hani throughout Chorbane anticline.

The transects show that the N–S axis corresponds clearly to a thrust domain associated to Diapiric structures such as the well-known Ktitir Diapir (Haller 1983; Khomsi et al. 2004a, b; Bédir et al. 2018), Chaker (Haller 1983; Bobier et al. 1991; Azaiez 2011; Bédir et al. 2018), and Bir Ben

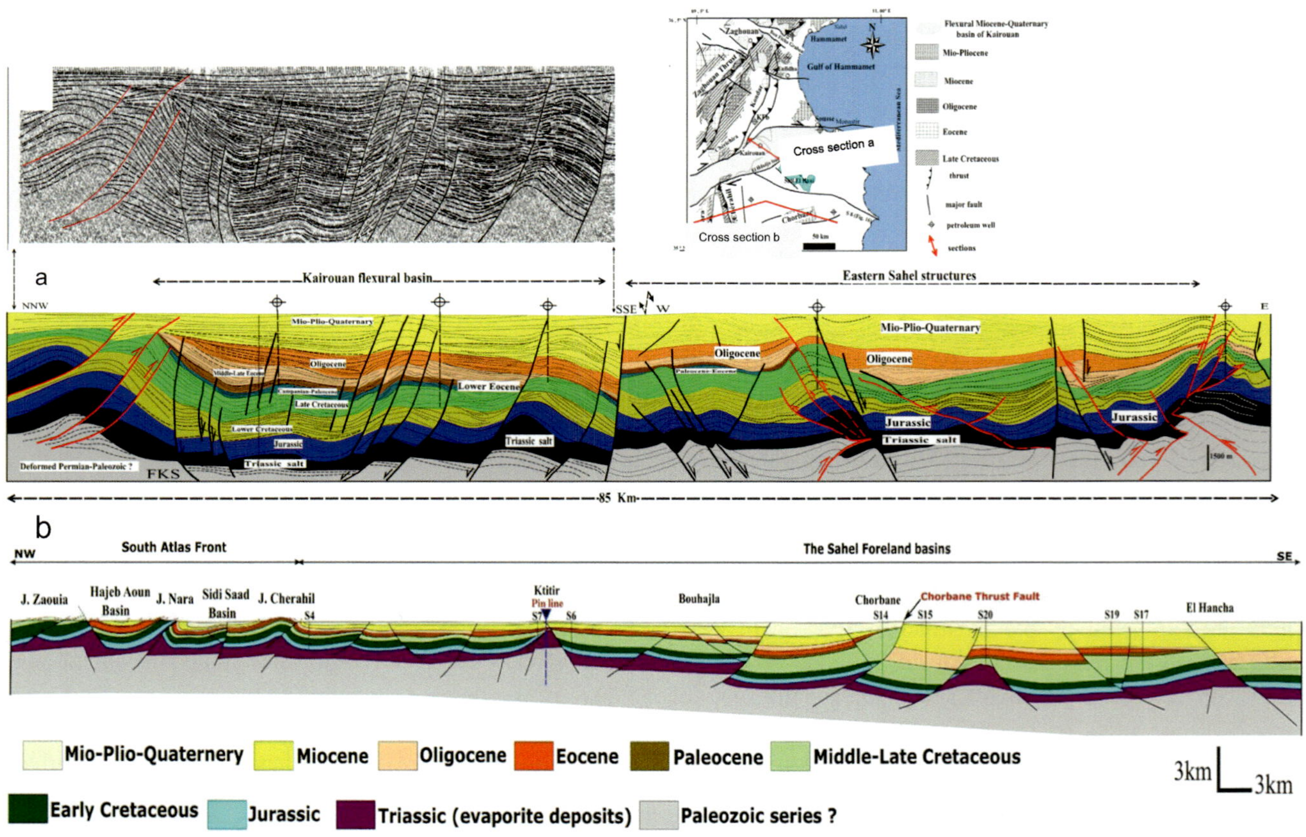

Fig. 8.55 Regional balanced cross-section crossing the South Atlas Front in the west and the Sahel foreland basin in the east. **a** In the Northern Sahel. **b** In the Southern Sahel. It shows the structural styles characterized by thick-skinned tectonic and locally thin skinned along the N–S axis. Notice the lateral thickness variations of the Miocene series on both sides of the Chorbane thrust. Notice also the lateral flow and rise of the Triassic salt, which is due to sedimentary loading, but is also controlled by the activity of major faults. **a** Reprinted from Khomsi et al. (2019). Copyright (2019), with permission from Elsevier. **b** Reprinted by permission from (Springer-Nature) [ArabianJournal of Geosciences] Mezni et al. (2019). Copyright (2019)

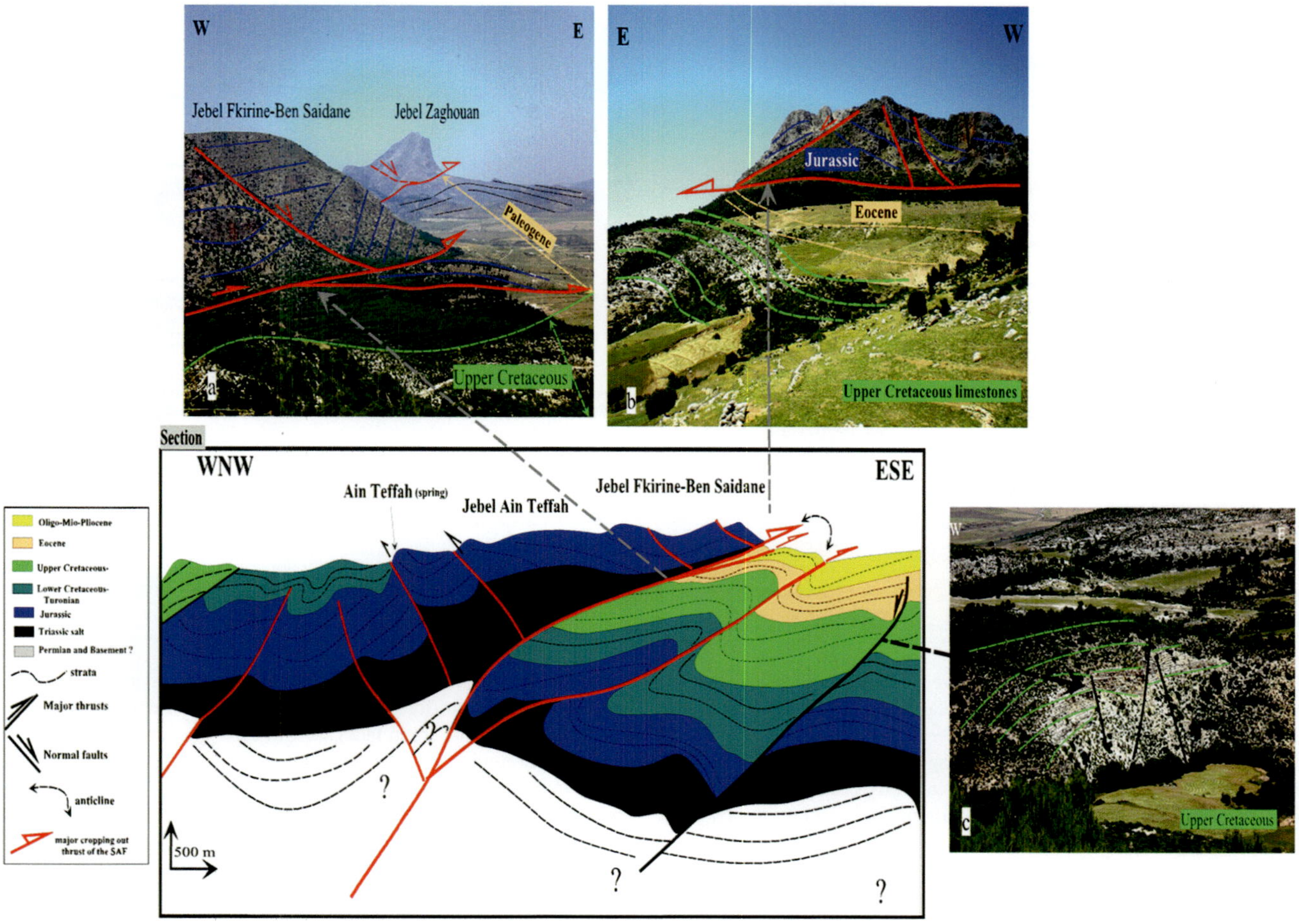

Fig. 8.56 The Alpine thrust of Zaghouan, overhangs tectonically the Northern Foreland domain. **a** Panoramic view looking North showing the Jurassic overhanging tectonically Upper Eocene strata. **b** Jurassic dolomites overhanging Eocene strata. **c** Synsedimentary tectonics affecting the Upper Cretaceous with tilted blocs in the footwall of Zaghouan thrust. Khomsi et al. (2016) Cross-section, modified from Turki (1985)

Jenale (Troudi et al. 2017) (Figs. 8.55 and 8.56). The Nara–Cherahil belts of the N–S axis constitutes the outcropping part of the South Atlasic Front (Khomsi et al. 2016) where the platform series of the Atlas belt are thinner compared to the foreland domain and Kairouan flexural basin. It corresponds also to an area of thrusting/buttressing associated to Triassic salt remobilization along major faults. Along this frontal thrust system, the Jurassic–Cretaceous platform deposits from the hinterland are thrust over the overturned Eocene–Oligocene beds (Fig. 8.55).

The balanced transects underline the importance of the Triassic salt in the control of the overall structures. The two transects (Fig. 8.55) show the control of the deep-seated faults on the sedimentation. Especially listric normal faults which limit inherited hemi-grabens issued form the Tetyhan rifting. Anyway, if we exclude the relatively small differences, it appears that both the northeastern foreland domain in the Sahel and the southern foreland domain have recorded the same events and that their structural styles are quite similar, with detached anticlines, salt-cored anticlines, and fault-propagation folds issued from the inversion of the former Tethyan Jurassic–Cretaceous extensional structures. The Eastern foreland Domain and the N–S Axis recorded two compression events: the Late Eocene and the Late Miocene–Quaternary.

The Late Eocene have similarly in the two domains quite equal shortening around 2 km on both transects, but the Late Miocene event is more stronger in the Northern Sahel Domain than along the Southern Sahel in Chorbane. This is presumably because the Northern Sahel foreland is closer to the thrust system of Zaghouan with Alpine plurikilometric thrusting displacement (Fig. 8.56).

The restored sections through time from Jurassic to Quaternary highlight (Figs. 8.57 and 8.58) the control of the rifting system by deep-seated faults controlling subsidence and so thicknesses variations. Some areas are quite

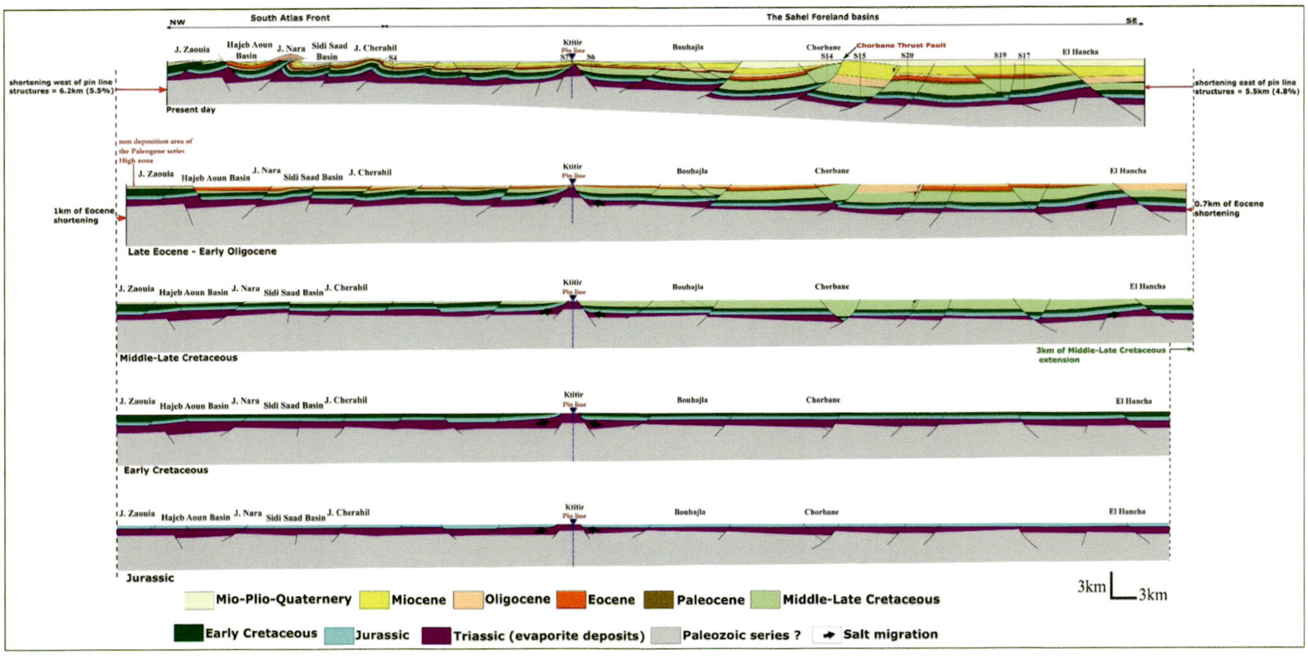

Fig. 8.57 Restored intermediate stages of the regional balanced cross-section of the Southern Sahel foreland for the Jurassic, the Early Cretaceous, the Middle–Late Cretaceous, end of Eocene–Early Oligocene, and Miocene–Quaternary stages, respectively. See text for discussions. Notice the Middle–Late Cretaceous stretching, amounting to 3 km, and the total Eocene shortening, amounting to 1.7 km. Altogether, the Late Eocene and Late Miocene–Quaternary compression events account for up to 11.7 km of shortening. The main compressional events are in fact Late Miocene–Quaternary in age, with an overall shortening value of 10 km along the entire section. Reprinted by permission from (Springer-Nature) [Arabian Journal of Geosciences] (Mezni et al. 2019) Copyright (2019)

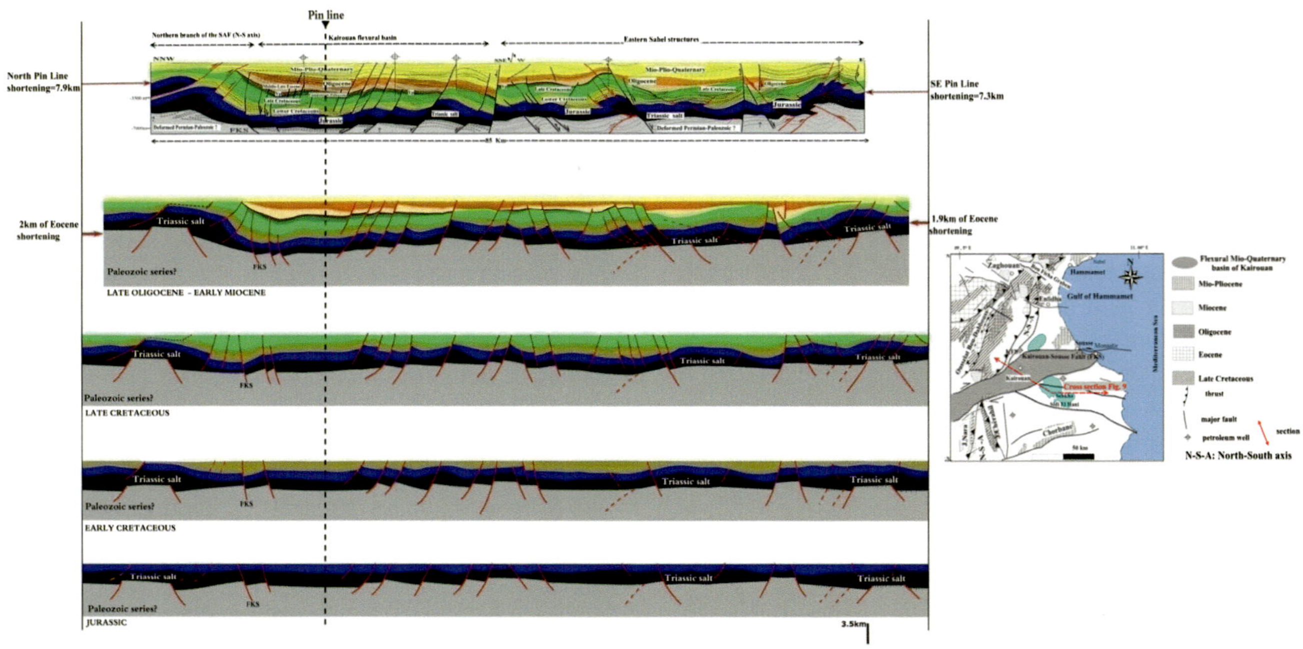

Fig. 8.58 Balanced and restored sections, respectively, in the Jurassic, Lower Cretaceous, Upper Cretaceous, and Lower Oligocene–Miocene, crossing the N–S axis toward the Kairouan flexural basin in the Atlas foreland, it shows the main stages and tectonic structures explaining the evolution from Jurassic to Miocene as well as Paleozoic structures. Note the rifting structures during the Jurassic–Lower Cretaceous and the beginning of the inversion from the Upper Cretaceous (Khomsi et al. 2009a, b). Reprinted from Khomsi et al. (2019). Copyright (2019), with permission from Elsevier

submitted since Upper Jurassic to early salt vertical movements along some deep faults. The tow restored transects underline also a huge Cretaceous subsidence with rifting rejuvenation.

8.7.3 The Triassic Salt Tectonics Related to the Major Faults Activities

In the North–South axis domain, the Triassic evaporitic material outcrops locally along the N–S thrust fault. This ductile material plays the main décollement level of the post-Triassic sedimentary cover detached over its pre-Triassic substratum (Paleozoic series). In the field outcrops, this thrusting tectonic style is corroborated by the overturned beds of the Eocene series and the clay of the Upper Eocene Cherahil formation playing a second décollement level locally. At Nara, the limestones beds of the Jurassic thrust the limestones beds of the Ypresian El Gueria and the shale series of the Late Eocene Cherahil formation (Fig. 8.58c). The discontinuity is characterized by the intrusion of the Triassic evaporites (Rheouis Formation) along the thrust fault (Fig. 8.58d) (Fig. 8.59).

Two fault sets crosscut the North–South Axis: The first set is an N160–180 sinistral strike-slip fault (Fig. 8.60), while the second fault set is an N80–90 dextral strike-slip fault (Fig. 8.60) (Abbés 2004). The intersection zone of these two fault sets is characterized by tectonic weakness, and it is highly fractured; consequently, the salt material finds its way up and reaches the surface (Abbés and Boukadi 1988; Boukadi 1994). The most known one is the diapir of Rheouis (Burollet 1973; Abbés and Boukadi 1988; Ouali 1985; Boukadi 1994; Abbés 2004; Ouali 2007). This diapir is installed in a tectonic node where the fault of the North–South axis intersects the NE–SW fault of Boudinar–Khechem El Artsouma.

In the subsurface, Diapric structures are recognized, such as the diapir of Ktitir (Haller 1983; Bédir 1995; Khomsi 2005) and the Ktifa dome (Khomsi 2005; Houatmia et al. 2015; Khomsi et al. 2016; Bédir 1995), and the diapir of Bir Ben Janele in the area of Bouhajla (Khomsi 2005; Troudi et al. 2017).

Section L8 (Fig. 8.61), running across the Sahel domain from Ktitir in the South to the E–W fault of El Hdadja allows to focus on the main mechanisms of Diapric structures set up. In fact, it shows three salt diapirs from south to north: the diapir of Ktitir, the diapir of Bouhajla, and in the southern part of this section, there is the diapir of El Hdadja.

The section shows that the Meso-Cenozoic series are thinning toward the diapirs (Fig. 8.61). The Ypresian series become thinner in the diapirs' vicinity until they disappear at the top of these salt features. Simultaneously, we notice a thinning of the Upper Eocene, which indicates that this salt feature has recorded a period of growth during the Eocene time-span that corresponds to the Atlas compression phase. During this uplift event, the reactivation/rejuvenation of the major fault N–S and NW–SE has trigged the lateral flow of the salt material toward the diapirs of Ktitir, Bouhajla, and El Hdadja. By consequence, they grow and rise upward. This upward movement of the salt diapir is coupled with flanking s rim synclines.

The Jurassic shows thinning toward the diapirs until it disappears at the top of these structures (Fig. 8.61). These indicate that the mobilization of the salt material began in the Early to Middle Jurassic (Bédir 1995; Tanfous Amri 2007; Khomsi 2005), guided by the activity of the E–W normal faults controlling horst and graben configuration during the Tethyan rifting.

Thus, the interpretation of subsurface maps of the N–S axis and Sahel domains shows structural and paleotectonic changes expressed by thickness variations, depocenters migration, and subsidence shifting of the Upper Cretaceous–Eocene sequences.

The Turonian–Coniacian time span corresponds to the Bireno–Douleb limestones deposits. During this period, the carbonates platform progressively deepens in the Pelagian domain's direction. The genesis of this subsiding area is tied to the Sirt opening, which is Late Cretaceous in age (Khomsi et al. 2016, 2019; Mezni et al. 2019). The major subsiding basins are located along fault corridors: The Bouthadi–Chorbane corridor oriented NE–SW to E–W, the Ktitir corridor oriented WNW–ESE, and the Sidi El Hani corridor directed NW–SE. The normal movement of this fault system during the Sirt rifting allows the lateral flow of the Triassic salt in the direction of the principal salt diapir as the diapir of Rheouis and Ktitir (Khomsi et al. 2016; Mezni et al. 2019).

During the Campanian–Maastrichtian, a compressive phase occurred in Tunisia (Haller 1983; Bédir 1995), corresponding to the beginning of the Atlasic orogenic event (Khomsi et al. 2009a, b, 2016). During this period, there is a reactivation of the major faults in the transpression regime (Bédir et al. 2001). The transition from a rifting mechanism during the Turonian–Coniacian to a transpression/compression regime during the Campanian–Maastrichtian is coupled to the space migration of the subsidence and subsidence inversions by places.

The Bouthadi–Chorbane E–W directed fault corridor represents the boundary of a set of E–W elongated depocenters. In addition, this E–W lineament constitutes a structural limit between a platform area with thin Abiod deposits located at north, and on the contrary at south, we have a more subsidente area characterized by thick Abiod series. Then as seen in the Turonian–Coniacian, during the Campanian–Maastrichtian, the Triassic salt movement, guided by the solicitation of the major faults, continues its lateral flow toward the different diapir structures.

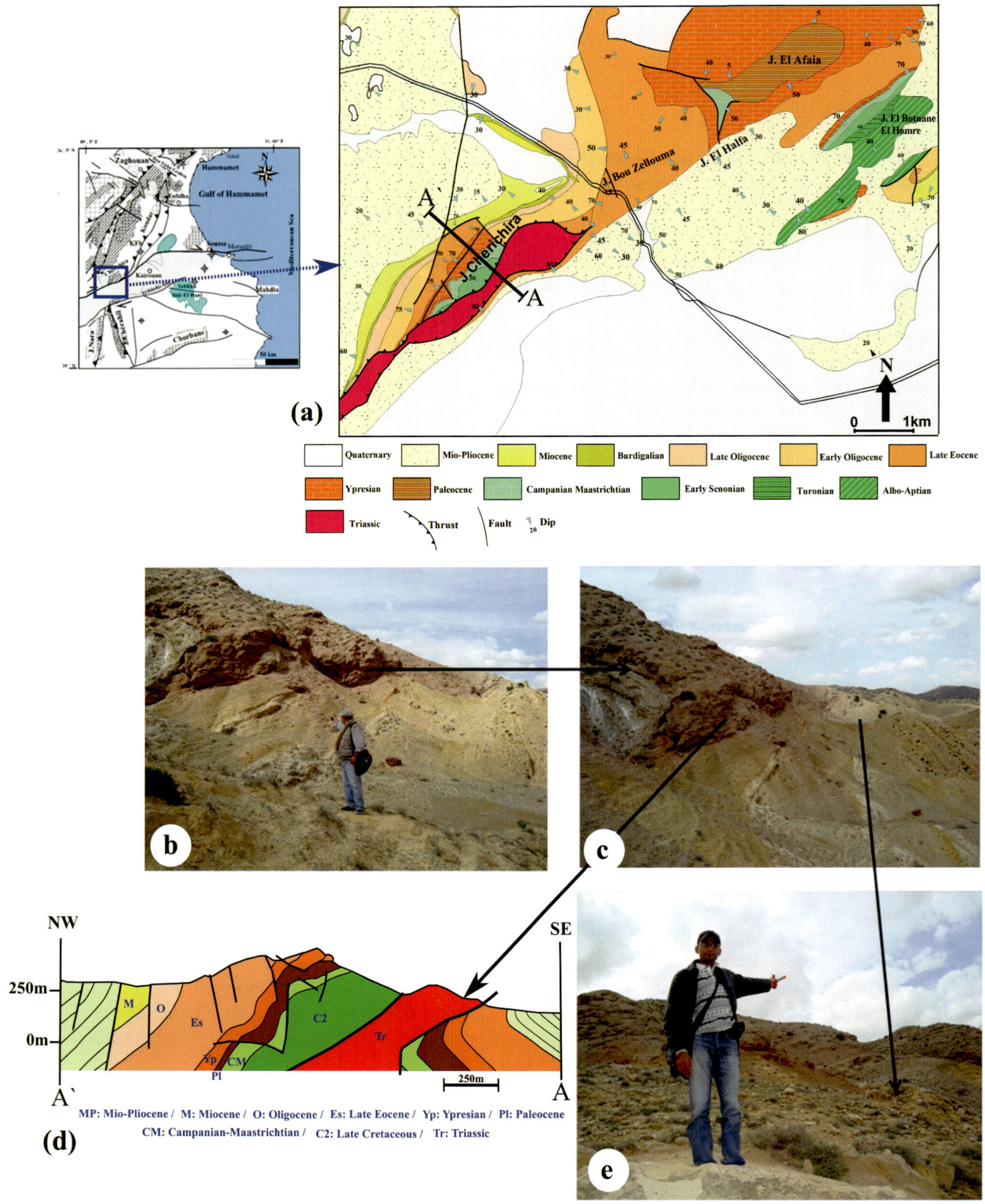

Fig. 8.59 Some aspects of the outcropping Triassic salt associated to the N–S Axis. **a** geological map of Cherichira outcrops **b** Triassic evaporates overhanging the Upper Eocene deposits in J. Cherichira. **c** and **e** details of the contact. **d** NW–SE structural cross section showing the Triassic-related structures (modified from Ouali, 2007)

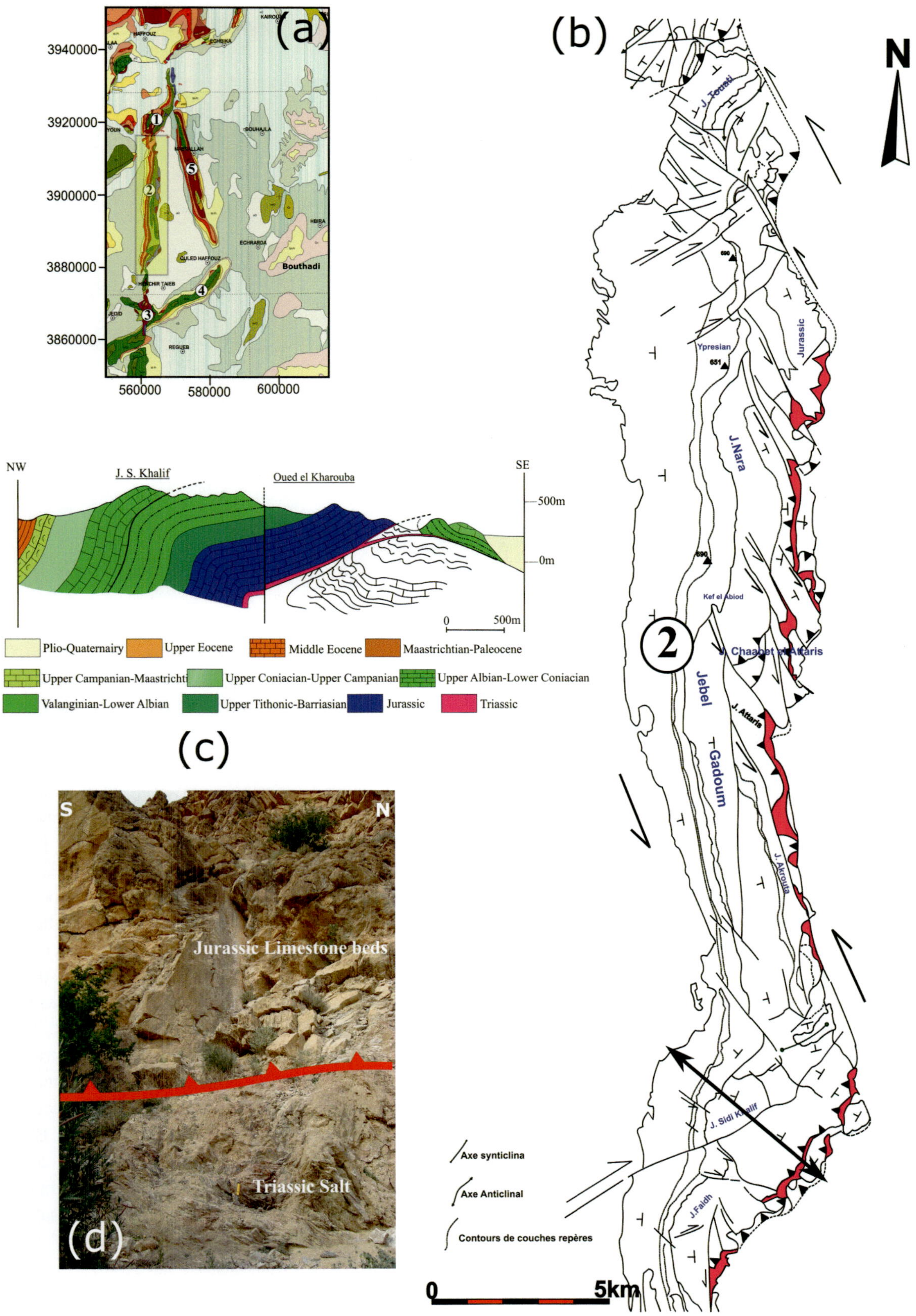

Fig. 8.60 The thrust tectonic of the North–South axis thrust belt. **a** A section of the geological map of Tunisia showing the position of the North–South axis structure. **b** Structural map of the N–S axis (Touila–Faidh segment) showing the main faults directions that crosscut the different structures of the North–South axis (Abbés 2004). **c** Cross-section of the Faid outcrop illustrating the thrust tectonics style of the N–S axis (after Delteil and Truillet (1983), Delteil et al. 1979, 1980). **d** A photo illustrating the thrust fault and the Triassic salt that outcrops under the Jurassic limestones beds

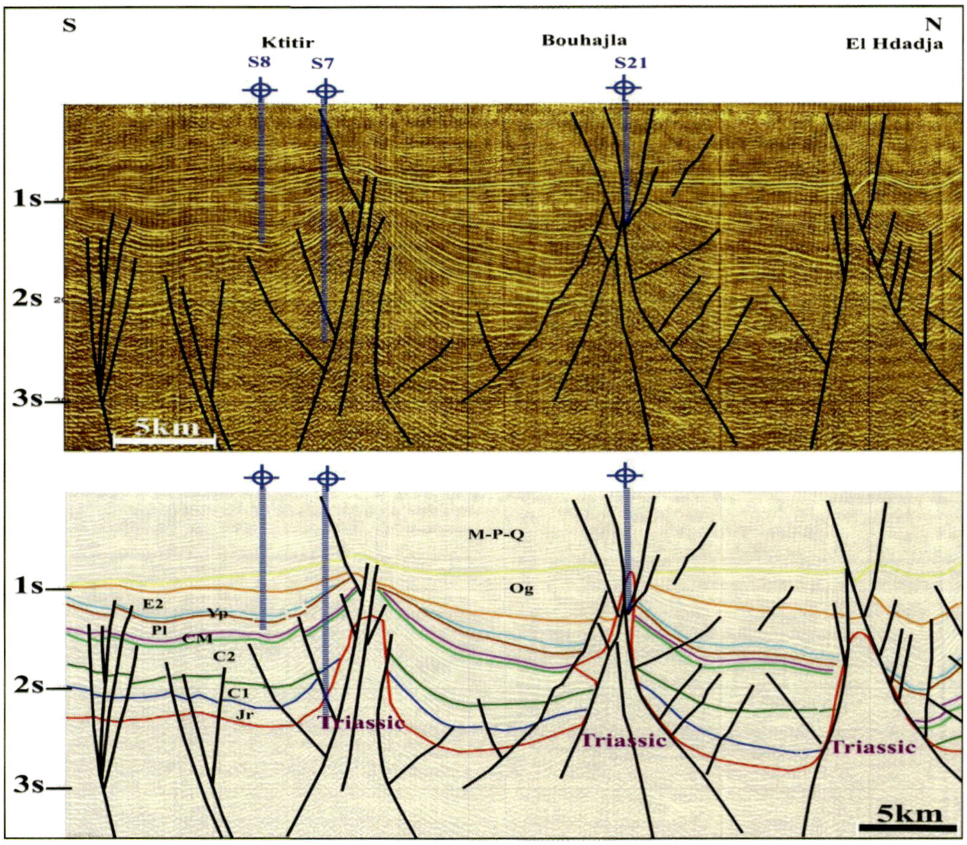

Fig. 8.61 Seismic section L8 showing the diapir of Ktitir, Bouhajla, and El Hdadja. The growth of these salt features is guided by the activity of the N–S and NW–SE fault systems in the Sahel domain. (Jr) Jurassic, (C1) Early Cretaceous, (C2) Late Cretaceous, (CM) Campanian–Maastrichtian, (Pl) Paleocene, (Yp) Ypresian, (E2) Late Eocene, (Og) Oligocene, and (M-P-Q) Miocene and Mio-Plio-Quaternary. (Location see Fig. 8.35). Modified from Khomsi (2005), see Fig. 8.40

During the Ypresian, the Bouhajla area still evolves in a subsidente structure. It corresponds to an NW–SE elongated graben limited by NW–SE oriented faults. However, During the Ypresian, the El Gueria–Bou Dabbous platform is crosscut by various oriented faults. This fault limited a set of subsiding basins. However, it is noticeable that the Triassic diapir did not pierce the Ypresian sediment cover during this period.

The subsurface investigations use the isochron maps of the top of these main carbonate formations and the isopach maps issued from seismic interpretations. These TWT structural maps help decipher the structural setting of these carbonate units and their relation with the different tectonic events that occurred during this geological time. In fact, during the Upper Cretaceous, the Sahel domain was dominated by an extension regime correlated to the Sirt rifting Cenomanian–Turonian in age (Khomsi et al. 2016; Mezni et al. 2019).

The Bireno–Douleb forms a carbonate platform where the limestones sequences do not show notable variations, but the thickness variations are important in the eastern region of the Sahel domain in relation with post-rift phase. This area is affected by NW–SE and E–W faults. It is important to indicate the role of the N–S Ktitir lineament in the subsurface structural architecture of the Sahel domain during the Sirt rifting. In fact, as mentioned in the previous section, through the restoration of the balanced cross-sections together with the subsurface maps presented (see isopach map of the Turonian–Coniacian), this lineament represents an important structural anomaly between a high domain localized at its west and corresponds to the Ktitir platform–NS axis domain versus a subsiding zone at it is East that corresponds to the eastern Sahel and Pelagian block.

By the Upper Cretaceous (Campanian–Maastrichtian), the domain is submitted to the first Atlasic compressional phase (Khomsi et al. 2006, 2016). During this uplift period, the corridor of Chorbane and the corridor of Sidi El Hani are reactivated partly in transpressional movement. This change in the tectonic regime is coupled with a structural inversion. In fact, the grabens are inverted as a high structures.

The Triassic material upward movement occurred earlier in the Jurassic–Early Cretaceous (Bédir 1995; Boukadi and Bédir 1996; Hlaiem 1999; Tanfous Amri et al. 2005; Azaeiz 2011; Khomsi et al. 2009a, b, 2016, 2019; Mezni et al. 2019) in the N–S axis and the Sahel domain. The salt flows laterally beneath subsiding rim syncline basins of the Jurassic–lower Cretaceous (Khomsi et al. 2016). The continuous reactivation of the deep-seated faults has trigged the lateral flow of the Triassic evaporites toward the salt diapirs. Furthermore, the salt material has played a role of a general

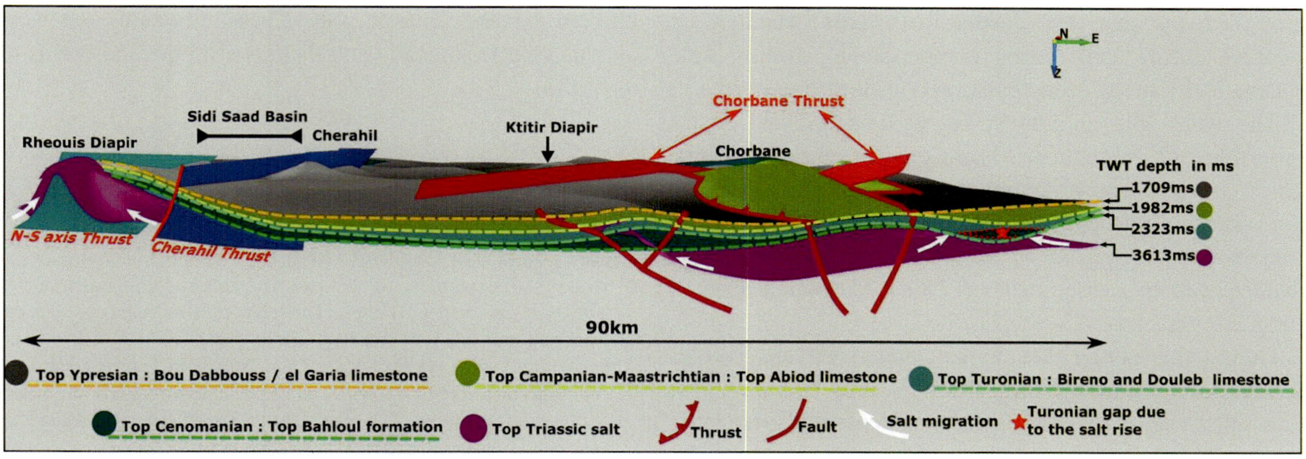

Fig. 8.62 WSW-ENE 3D cross-section model of the NS axis–Chorbane (Sahel domain). This cross-section shows the thrust and fault system-related structure and the Triassic salt that play the role of general décollement level and in another hand, the activity of the major faults that guided the migration and the rise of the salt material toward the diapir structure as the diapir of Rheouis and Ktitir

décollement of the post-Triassic cover during the shortening phases (Burollet 1973; Truillet et al. 1981; Khomsi 2005; Khomsi et al. 2009a, b, 2016, 2019; Mezni et al. 2019).

The cross-section presented in Fig. 8.62 corresponds to a time-depth cross-section extracted from the seismic data. This 3D model of the N–S axis–Chorbane domain reveals its structural configuration. In addition to that, it shows the geometry and the lateral extension of the three main carbonate formations presented above in subsurface maps: the Bireno and Douleb member (Turonian–Coniacian), the Abiod Formation (Campanian–Maastrichtian), and the El Gueria–Bou Dabbouss Formation (Ypresian).

By the end of the Cretaceous times and during Maastrichtian–Paleocene period began the first positive inversions of the former extensional structures inherited from the previous rifting period. These inversion movements were limited and preferentially situated on major pre-existing faults. These inversions were not strong enough to ensure the total reactivation of the previous structures which still locally preserve a normal offset, but they were more likely responsible for differential uplift and erosion of many structures in a marine environment with open marine conditions. During this time span occurred the first uplift of the N–S axis with the emersion of some paleogeographic domains as the resistant shoal of Jebel Zaouia (Figs. 8.58 and 8.59). The compressions continued through time in the Eocene with the major Atlas compressions leading to the uplift of many structures (Frizon de Lamotte et al. 2009; Khomsi et al. 2009a). The restored section at the end of Eocene–Oligocene times shows inversions of structures with fault-propagation folds and reverse offset of previous normal faults along the N–S axis, where the Eocene turned from marine to continental. By contrast, the eastern sub-domain of Chorbane which was also partly uplifted remained within a marine environment. After the Late Eocene compressions, the Oligocene deposits lie unconformably on the Eocene structures during a period of tectonic quiescence materialized by extensional tectonics which occurred during Oligocene to Middle Miocene times.

The Late Miocene–Quaternary period coincides with the major compressions related to the Alpine cycle with major inversions and thrusting/shortening and individualization of the present-day thrust system of the N–S axis (Figs. 8.57 and 8.58). At the same time, it is noticeable that the uplift and inversion of the Chorbane detachment anticline were strong enough. In fact, this period of important shortening and thrusting led to erosion of the emergent apical part of many structures along the N–S axis, with the deposition of erosional products in the subsiding footwall domains and depocenters adjacent to many active thrusts. The Chorbane area accommodated the deposition of important sedimentary thicknesses of Late Miocene–Quaternary deposits in the footwall of the major thrusts with an overall thickness of Late Miocene–Quaternary continental deposits exceeding 2000 m at the emplacement of the previous Cretaceous rift. At the eastern edge of the section, the El Hancha structure continued to be drowned, subsiding along an east-dipping listric normal fault associated with a rollover anticline.

8.8 Conclusion

The Eastern foreland basin of the Atlas shows a set of major structures related to the Meso-Cenozoic structural evolution of the North African Atlas belt: Diapirs, thrust structures, fault-propagation faults. These structures are mainly controlled by inherited deep-seated faults which were inverted during two major compressional events: Late Eocene and Late Miocene compressions/thrusting events.

The Atlas fold and thrust bet and its foreland basins underwent the subsequent Meso-Cenozoic tectonic events, which controlled its structural style. In fact during the Triassic–Early Cretaceous Tethys opening, this structural domain is characterized by an extensional structural style characterized by synsedimentary faults that bound a set of horst and grabens.

During this period, the domain underwent an NW–SE extension regime controlling NW–SE faults bounding grabens filled with thick Late Cretaceous deposits. These latter reaches, in some areas more than 1500 m. However, to the west, the Atlas hinterland domain and the N–S Axis correspond to a high resistant platform with thin Late Cretaceous deposits compared to the Eastern area.

By the end of the Late Cretaceous, the Sahel domain underwent the first pulses of the Atlasic compression phase. It is followed by the inversion of the Late Cretaceous basins and rejuvenating of the inherited normal faults. This shortening regime continues during the Paleocene–Eocene time span with the Atlas compressions phase which attains its paroxysmal level at Middle–Late Eocene time. It is mainly underlined by the thinning of the Paleocene–Eocene series and the Oligocene angular unconformities recognizable along the seismic cross-sections.

During the Oligocene–Early Miocene, the Sahel foreland basins underwent a period of quiescence characterized by the development of normal fault sets and extensional structures well expressed by important extensional features as listric normal faults.

The Late Miocene–Quaternary period coincides with the onset of the Alpine compressions and the building of the major structural elements of the Atlas front and its foreland basins. In fact, beneath the thick Mio-Quaternary deposits, the seismic sections recorded in this foreland domain show structural complexity characterized by thrust folds, pop-ups, inverted grabens, and fault-propagation folds.

During these major Alpine compressions and vertical uplift, the Upper Triassic salt plays the role of a general décollement level of the Meso-Cenozoic sedimentary pile on its substratum (Permian series). At the same time the rejuvenation of the deep-seated fault has guided the lateral flow of the Triassic salt material toward salt diapirs causing their growth which is underlined by the intra-Paleogene unconformities and the thinning of the Paleocene–Eocene series toward the salt features such as Ktitir, Bir Ben Zina, Bir Ben Janel, and Chorbane. Along the N–S axis, constituting the northern branch of the South Atlas Front of the Maghrebian Atlas, this Alpine compressional phase, solicited the inherited N–S faults which were reactivated as thrust faults injected by the Triassic salt. These structures are now overhanging tectonically and the Eastern Foreland basins of the Atlas.

References

Abbés C, Turki MM, Truillet R (1981) Un élément structural nouveau dans l'Atlas tunisien: le contact tangentiel decakilométrique a vergence ouest des Djebels Ousselat et Bou dabbous («axe» nord-sud tunisie). C R Acad Sci, Paris, Sér ii 292:173–176

Abbes C, Boukadi N (1988) Sur la chaîne N-S de la Tunisie centrale et ses infléchissements d'axe. Analyse des déformations du nœud tectonique du Jebel Ech Chérichira. Coulissement blocage et rotation. C R Acad Sci Paris 307:1277–1288

Abbés C (2004) Structuration et évolutions tectono-sédimentaires mésozoïques et cénozoïques associées aux accidents reghmatiques a la jonction des marges téthysienne et nord-africaine (chaine nord-sud, Tunisie centrale). Thèse es Sciences, Université de Tunis, p 437

Amari A, Bédir M (1989) Les bassins quaternaires du Sahel central de la Tunisie, genèse et évolution des sebkhas en contexte décrochant compressif et distensif. Rev Géodyn 4:49–65

Anderson JE (1996) The Neogene structural evolution of the western margin of the Pelagian Platform. Central Tunisia. J Struct Geol 18:819–833

Azaeiz H (2011) Structuration et sismo-stratigraphie des séries du crétacé inférieur et leurs réservoirs associés en Tunisie centrale (régions de Gafsa, Sidi Bouzid et Bir Ali Ben Khalifa). Thèse de Doctorat, Faculté des sciences de Tunis, Université de Tunis II, p 256

Bédir M, Bobier C (1987) Les grabens de Mahdia et Sidi Chérif (Tunisie orientale). Dynamique des fossés oligo-miocènes induits au toit d'anticlinaux crétacés-éocènes par les jeux, au Néogène, de décrochements est-ouest et nord-sud. Bull Soc Géol France 3 (6):1143–1151

Bédir M (1988) Géodynamique des bassins sédimentaires du Sahel de Mahdia (Tunisie orientale) de l'Aptien à l'Actuel. Sismo-stratigraphie, sismo-structurale et structurale. Répercussion pétrolière, hydrologique et sismique. Thèse 3ième Cycle, Tunis, p 242

Bédir M, Zargouni F, Tlig S, Bobier C (1992) Subsurface geodynamics and petroleum geology of transform margin basins in the Sahel of Mahdia and El Jem (eastern Tunisia). AAPG 76(9):1417–1442

Bédir M (1995) Mécanismes géodynamiques des bassins associés aux couloirs de coulissements de la marge atlasique de la Tunisie, seismo-stratigraphie, sismo-tectonique et implications pétrolières. Thèse ès Sciences, Université de Tunis II, Tunisie, p 412

Bédir M, Boukadi N, Tlig S, Ben Timzal F, Zitouni L, Alouani R, Slimane F, Bobier C, Zargouni F (2001) Subsurface Mesozoic basins in the central Atlas of Tunisia: tectonics, sequence deposit distribution and hydrocarbon potential. AAPG Bull 85(5):885–907

Bédir M, Soussi M, Gabtni H, Azaiez H, Zouaghi T, Khomsi S, Tanfous-Amri D, Arbil A, Houatmia F, Soltani A (2016) New petroleum systems related to the structuring of Meso-Cenozoic basins in North African plate Tunisian margin. Poster presentation given at AAPG annual convention and exhibition, Calgary, Alberta, Canada, June, pp 19–22

Bédir M, Gabtni H, Azaiez H, Arbi A (2018) Meso-Cenozoic strike slip salt tectonics, platform to basin structuring subsidence migrations, platform to basin structuring and new petroleum implications by seismic reflection analysis in the Eastern Tunisian margin. In: 7th Maghrebian geophysic congress, Algiers, Algeria, 20–22 Feb 2018, p 41

Bedir M, Soltani A, Belhaj Mohamed A, Arbi A, Saidi M (2020) Cretaceous petroleum system modeling of Kairouan Basin in eastern Tunisia. Arab J Geosci 13:683–713

Ben Brahim G, Ahmadi R, Brahim N, Turki F (2014) Neogene tectonic and fault-related fold in the Gulf of Hammamet area, Tunisian offshore. J Afr Earth Sci 97:78–86

Ben Ismail-Latrache K (1981) Etude micropaléontologique et biostratigraphique des séries paléogènes de l'anticlinal du Jebel Abderrahmane (Cap-Bon, Tunisie nord orientale). Thèse 3eme cycle, Fac. Sci. 'Tunis, p 224

Ben Ismail Latrache K (1994) Mise au point sur l âge des formations Metlaoui et Souar en Tunisie. Note Serv. Geol. de Tunisie. N 60, pp 59–85

Ben Ismail-Lattrache K, Bobier C (1996) Etude biostratigraphique, paléo-écologique et paléo-biogéographique des séries éocènes (Ypresien-Lutétien basal) de Tunisie centrale. In: Géologie de l'Afrique et de l'Atlantique Nord. Elf Aquitaine Editions, Pau, France, pp 563–575

Ben Ismail-Lattrache K (2000) Précision sur le passage Lutétien-Bartonien dans les dépôts éocènes moyens en Tunisie centrale et Nord orientale. Rev Micropaléontol 43(1–2):3–16

Ben Ferjani A, Burollet PF, Mejri F (1990) Petroleum geology of Tunisia. ETAP Memoir, p 194

Ben Jemia MG (1986) Evolution tectonique de la zone de failles de Trozza–Labeid (Tunisie centrale). Thèse 3ieme cycle, Univ. Paris Sud, p 158

Bishop WP (1988) Petroleum geology of East central Tunisia. AAPG Bull 72:1033–1058

Bismuth H, Gourmelen CL, Lamboy M, Philip J (1985) Existence d'un biostrome d'algues et rudistes d'âge Coniacien en Tunisie centrale (Jebel Bou Zer). In: Actes 1er Congrès N1. Sc. de la Terre, Tunis 1981, vol 1, pp 85–96, 4 fig

Blanpied C, Bellaiche G (1983) The Jeffara trough (Pelagian Sea); structural evolution and tectonic significance. Mar Geol 51:1–10

Bobier CL, Viguier C, Chaari A, Chine A (1991) The post-Triassic sedimentary cover of Tunisia: seismic sequences and structure. Tectonophysics 195:371–410

Bouaziz S, Barrier E, Soussi M, Turki MM, Zouari H (2002) Tectonic evolution of the northern African margin in Tunisia from paleostress data sedimentary record. Tectonophysics 357:227–253

Boukadi N (1994) Structuration de l'Atlas de Tunisie: signification géométrique et cinématique des nœuds et des zones d'interférences structurales au contact des grands couloirs tectoniques. Thèse es Sciences, Universite de Tunis, p 249

Boukadi N, Bédir M (1996) L'halocinèse en Tunisie: contexte tectonique et chronologie des événements. C R Geosci 322 (7):587–594

Bracène R, Frizon de Lamotte D (2002) The origin of intraplate deformation in the Atlas system of western and central Algeria: from Jurassic rifting to Cenozoic-Quaternary inversion. Tectonophysics 357:207–226

Bramaud M, Bismuth H, Khessibi M, Richert JP (1976) Mise en évidence d'une discordance de ravinement intra-Crétacé dans la région de Maknassy–Mezzouna (Tunisie Centrale). Notes Serv Géol Tunis, Tunis, 42:7–19

Bramaud M, Burollet PF (1981) Notice explicative de la carte géologique de la Tunisie, échelle 1/50.000, feuille N° 80: Chorbane. p 23, 3 fig, 1 carte

Buness H, Giese P, Bobier C, Eva C, Merlani F, Pedone R, Jenatton L, Nguyen DT, Thouvenot F, Egloff F, Makris Y, Lozej A, Maistrello M, Scarasga S, Tabacco I, Burollet PF, Morelli C, Nicolich R, Zaghouani T, Egger A, Freeman R, Mueller ST (1992) The EGT'85 seismic experiment in Tunisia: a reconnaissance of the deep structures. Tectonophysics, 207:245–267

Burollet PF (1956) Contribution à l'étude stratigraphique de la Tunisie centrale. Thèse Sci, Ann Mines Géol Tunis 18:350

Burollet PF (1973) Importance des facteurs salifères dans la tectonique tunisienne. Ann Mines Géol Tunis 13:11–30

Burollet PF, Mejri F, Ben Kilani H (Etap) (2000) What about the deep structures of Tunisia. In: Proceedings of the seventh Tunisian petroleum and production conference. ETAP Memoir no 16, pp 167–178

Busson G (1971) Le Mésozoïque Saharien, Essai de synthèse des données de sondages algéro-tunisiens, vol 10, issue 1–2. CNRS, Centre de recherche sur les zones arides, p 810

Casero P, Roure F (1994) Neogene deformations at the Sicilian–North African plate boundary. Peri-Tethyan Platforms. Editions Technic, Paris, pp 27–50

Castany G (1951) Etude géologique de l'Atlas Tunisien oriental. Ann Mines Géol Tunis 8:632

Comte D, Dufaure P (1973) Stratigraphie et paléogéographie tertiaire en Tunisie centrale. Ann Mines Géol Tunis 26:241–256

Creusot G, Mercier E, Ouali J, Tricart P (1993) La tectogenèse atlasique en Tunisie centrale: Apport de la modélisation géométrique. Eclogae Geol Helv 86(2):609–627

Delteil J, Truiellet R, Zargouni F (1979) Extension et ampleur de la tectonique tangentielle dans l'axe N-S. In: 7ième réunion. Ann. Sc. Terre, Lyon, p 158

Delteil J, Truiellet R, Zargouni F (1980) L'axe N-S: un élément structural original et complexe de l'orogenèse alpine en Tunisie centrale. In: Congrès Géol international, section 5, tectonique, p 331

Delteil J, Truillet R (1983) le style structural de lAtlas tunisien alpin. Rapp Comm Int Mer Medit 28:4

Etap (1998) Tunisian exploration review. ETAP and Schlumberger, p 107

Etap, Tunisian exploration review, 1991 p 107 Editors ETAP and Shclumberger coordinators: Z. Hmidi for Etap and W. Sadra for Schlumberger

El Euchi H, Saidi M, Fourati L, Ghenima R, Friha J, Hammouda F, Messaoudi F (2002) Northern Tunisia thrust belt: deformation models and hydrocarbon systems. In: Proceedings of the Eighth Tunisian petroleum and production conference. Etap Memoir no 19, pp 143–189

El Euchi H, Saidi M, Fourati, Marhessi C (2004) Northern Tunisia thrust belt: deformation models and hydrocarbon systems. In: Swennen R, Roure F, Granath JW (eds) Deformation, fluid flow and reservoir appraisal in foreland fold-and-thrust belts: AAPG Hedberg Series, no 1, pp 371–380

Ellouz N (1984) Etude de la subsidence de Tunisie atlasique, Orientale et de la mer pélagienne. Thèse de Doctorat 3ème cycle, Université Paris VI, p 139

Fakhfakh Ben Jemia H (1991) Les calcaires de l'Eocène inferieur en Tunisie Centro-septentrionale: sédimentologie, paléogéographie. Thèse, Uni, Franche Comte, p 161

Fournie D (1978) Nomenclature lithostratigraphique des séries du Crétacé supérieur au Tertiaire de Tunisie. Bull Cent Rech Explor Prod Elf Aquitaine, Pau 2(1):97–148

Frizon de Lamotte D, Mercier E, Outtani F, Addoum B, Ghandriche H, Ouali J, Bouaziz Andrieux J (1998) Structural inheritance and kinematics of folding and thrusting along the front of the Eastern Atlas mountains (Algeria and Tunisia). In: Crasquin-Soleau S, Barrier E (eds) Péri-Tethys memoir 3 (1998). Mém Mus Natl Hist 177:237–252

Frizon de Lamotte D, Saint Bézard B, Barcène R (2000) The two main steps of the Atlas building and geodynamics of the western Mediterranean. Tectonics 19(4):740–761

Frizon de Lamotte D, Leturmy P, Missenard Y, Khomsi S, Ruiz G, Saddiqi O, Guillocheau F, Michard A (2009) Mesozoic and

Cenozoic vertical movements in the Atlas system (Algeria, Morocco, Tunisia): an overview. Tectonophysics 475:9–28

Frizon de Lamotte D, Raulin C, Mouchot N, Wrobel Daveau JC, Blanpied C, Ringenbach JC (2011) The southernmost margin of the Tethys realm during the Mesozoic and Cenozoic: initial geometry and timing of the inversion processes. Tectonics 30:TC3002. https://doi.org/10.1029/2010TC002691

Gabtni H (2005) Apport de la gravimétrie à l'étude des structures profondes du Sahel de Tunisie (cas de la région de Kairouan–Sousse–Monastir). Gravity contribution on the deep structure study of the Tunisian Sahel domain (a Kairouan-Sousse Monastir area case). CRAS Géosci 337:1409–1414

Gharsalli R, Zouaghi T, Soussi M, Chebbi R, Khomsi S, Bedir M (2013) Seismic sequence stratigraphy of the Miocene deposits related to eustatic, tectonic and climatic events, Cap Bon Peninsula, northern Tunisa. C R Geosci 345:401–417

Gourmelen C, Rigane A, Broquet P, Truillet R, Aite MR (2000) Caractères structuraux et dynamiques d un bassin en transtension: la plate-forme tunisienne a l Yprésien terminal. Bull Soc Géol France 171(5):559–568

Haller P (1983) Structure profonde du Sahel tunisien. Interprétation géodynamique. Thèse de Doctorat 3ème cycle, Université de Franche Comté, Besançon, p 163

Hlaiem A (1999) Halokinesis and structural evolution of the major features in eastern and southern Tunisian Atlas. Tectonophysics 306:79–95

Houatmia F, Khomsi S, Bedir M (2015) Oligo-Miocene reservoir sequence characterization and structuring in the Sisseb El Alem – Kalaa Kebira Regions (Northeastern Tunisia). J Afr Earth Sci. https://doi.org/10.1016/j.jafrearsci.2015.08.019

Houatmia F, Khomsi S, Mlayah A, Bédir M (2016) Basin structure and water quality of the Oligocene aquifer in the Sebkhet El Behira basin (central Tunisia). Hydrol Sci J 61(5):868–880. https://doi.org/10.1080/02626667.2015.1008481

Jauzein A, Perthuisot V (1981) Accidents de socle et plissement de couverture. In: Premier Congr. Nat. Sa. de 'la terre. Tunis, pp 39–48

Khelil M, Khomsi S, Frizon de Lamotte D, Souloumiac P, Maillot B (2020) Reply to comment on "how to build an extensional basin in a contractional setting? Numerical and physical modeling applied to the Mejerda basin at the front of the eastern Tell of Tunisia. J Struct Geol 103936. https://doi.org/10.1016/j.jsg.2019.103936

Khelil M, Khomsi S, Roure F, Arfaoui MS, Echihi O, Zargouni F (2021) Late Miocene-Quaternary thrusting in the Utique-Kechabta foreland basin of the Tell, northern Tunisia. Arab J Geosci 14. https://doi.org/10.1007/s12517-020-06388-2

Khessibi M (1978) Etudes géologiques du secteur de Maknassy-Mezzouna et du Djebel Kebar (Tunisie centrale). Thèse, Univ. Claude Bernard, Lyon, p 175

Khomsi S, Bédir M, Ben Jemia GM (2004a) Highlight of a new thrust front in the oriental Atlas of Tunisia using seismic reflexion data. Structural Context and role of salt intrusions. C R Géosci 336:1401–1408

Khomsi S, Bédir M, Ben Jemia MG (2004b) Highlight and analysis of an Atlasic structure beneath the plain of Kairouan–Sahel. North eastern Tunisia. C R Géosci 336:1293–1300

Khomsi S (2005) Géodynamique des bassins du Paléogène et des réservoirs associés du Sahel et de Kairouan (Tunisie orientale): structuration, sismo-tectonique et organisation séquentielle. Implications pétrolières. PhD, Thèse d'université, Université Tunis El Manar, Faculté des sciences de Tunis, p 365

Khomsi S, Bédir M, Soussi M, Ben Jemia MG, Ben Ismail-Lattrache K (2006) Highlight of Middle-Late Eocene compressional events in the subsurface of eastern Tunisia (Sahel): generality of the Atlasic phase in North Africa. C R Geosci 338(1–2):41–49

Khomsi S, Bédir M, Soussi M, Ben Jemia MG, Ben Ismail-Lattrache K (2007) Reply to comment on the paper Mise en évidence en subsurface d'événements compressifs Eocène moyen-supérieur en Tunisie orientale (Sahel): Généralité de la phase atlasique en Afrique du Nord. C R Géosci 338(1–2):41–49, C R Géosci 339:173–177

Khomsi S, Ben Jemia MG, Frizon de Lamotte D, Maherssi C, Echihi O, Mezni R (2009a) An overview of the Late Cretaceous–Eocene positive inversions and Oligo-Miocene subsidence events in the foreland of the Tunisian Atlas: structural style and implications for the tectonic agenda of the Maghrebian Atlas system. Tectonophysics 475:38–582. https://doi.org/10.1016/j.tecto.2009.02.027

Khomsi S, Mahersi C, Fakfakh-Ben Jemia H, Riahi S, Bou Khalfa K (2009b) New insights on the structural style of the subsurface of the Tell units in north-western Tunisia issued from seismic imaging: geodynamic implications. C R Geosciences 341:347–356

Khomsi S, Echihi O, Slimani N (2012) Structural control on the deep hydrogeological and geothermal aquifers related to the fractured Campanian-Miocene reservoirs of north-eastern Tunisia foreland constrained by subsurface data. C R Géosci 344:247–265

Khomsi S, Frizon de Lamotte D, Bédir M, Echihi O (2016) The Late Eocene and Late Miocene fronts of the Atlas Belt in eastern Maghreb: integration in the geodynamic evolution of the Mediterranean Domain. Arab J Geosci 9:650–670. https://doi.org/10.1007/s12517-016-2609-1

Khomsi S, Roure F, Khelil M, Mezni R, Echihi O (2019) A review of the crustal architecture and related pre-salt oil/gas objectives of the eastern Maghreb Atlas and Tell: need for deep seismic reflection profiling. Tectonophysics 766:232–248. https://doi.org/10.1016/j.tecto.2019.06.009

Khomsi S, Khelil M, Roure F, Zargouni F (2021) Surface and subsurface architecture of the Kasseb structures: implications for petroleum exploration beneath the Tellian allochthon, the easternmost portion of the Maghrebides. Arab J Geosci 14. https://doi.org/10.1007/s12517-020-06401-8

Khomsi S, Roure F, Verges J (2022) Hinterland and foreland structures of the eastern Maghreb Tell and Atlas thrust belts: tectonic controlling fractors, pending question, and oil/gas exploration potential of the Pre-Trissic traps. Arab J Geosci 15:462–473. https://doi.org/10.1007/s12517-022-09707-x

Khouni R, Arfaoui MS, Dridi S et al (2018) Polyphasic evolution of the Jeffara basin in southern Tunisia, influence of halokinesis on the passive margin structuration in the Mesozoic and the Cenozoic. Arab J Geosci 11:68. https://doi.org/10.1007/s12517-017-3363-8

Klett TR (2001) Total petroleum systems of the Pelagian Province, Tunisia, Libya, Italy, and Malta—the Bou Dabbous–Tertiary and Jurassic-Cretaceous composite. US Geol Surv Bull 2202-D

Laffitte R (1939) Etude géologique de l'Aurès. Thèse ès Sciences, Paris, Pub. Serv. Carte géol. Algérie, nouv. Série, no 46, vol I, p 217, and vol II, p 281

Laridhi Ouazaa N, Bédir M (2004) Les migrations tectono-magmatiques du Trias au Miocène sur la marge orientale de la Tunisie. Afr Geosci Rev 11:177–194

Mejri M, Burollet PF, Ben Ferjani A (2006) Petroleum geology of Tunisia: a renewed synthesis. ETAP Memoir no 22, Tunis, p 233

Messaoudi F, Hammouda F (1994) Evènement structuraux et types de pièges dans l'offshore Nord-Est de la Tunisie. In: Proceedings of the 4th Tunisian petroleum exploration conference, Tunis, May 1994, vol 64, p 55

Mezni R, Khomsi S, Bédir M (2019) Structural styles, tectonic events, and deFormation features along a surface–subsurface structural transect from the South Atlas Front (N–S axis) to the Eastern Sahel foreland basin of Tunisia. Arab J Geosci 12:18. https://doi.org/10.1007/s12517-019-4366-4

Morgan MA, Grocott J, Moody RTJ (1998) The structural evolution of the Zaghouan–Ressas Structural Belt, northern Tunisia. In: Macgreggor DS, Moody RTJ, Clark-Lowes DD (eds) Petroleum geology of North Africa. Spec. Publ., No. 132. Geological Society, London, pp 405–422

Negra MH (1994) Les dépôts de plate-forme a bassin du Crétacé Supérieur en Tunisie centro-septentrionale (formation Abiod et facies associes). Stratigraphie, sédimentation, diagenèse et intérêts pétroliers. Thèses. Doct. d'Etat, Universite de Tunis, p 649

Negra MH, Mrabbet A, Troudi H, El Asmik, Saidi F (1996) Lithofacies and paleogeographic evolution of the Upper Cretaceous reservoir rock in Central Tunisia. In: Proceedings of fifth Tunisian petroleum exploration conference l'ETAP, Tunis, Tunisia, 22–24 Oct 15th–18th, 1996, pp 174–194

Ouali J (1984) Structure et evolution geodynamique du chainon Nara-Sidi Khalif (Tunisie centrale). Thèse 3eme cycle, Rennes, p 119

Ouali J (1985) Structure et evolution geodynamique du chainon Nara-Sidi Khalif (Tunisie centrale). Bull Cent Rech Expl Prod Elf Aquitaine 9(1):155–182

Ouali J, Tricart P, Deltail J (1987) Ampleurs et significations des recouvrements anormaux dans l'Axe Nord-Sud (Tunisie centrale), données nouvelles dans le chaînon Nara-Sidi Khalif. Eclogae Geol Helv 80:685–696

Ouali J (2007) Importance du réseau réghmatique dans la tectogenèse de la Tunisie atlasique à travers l'étude de l'Axe nord-sud. Thèse, Doctorat d'Etat, Université Tunis El Manar (Tunisia), p 399

Patriat M, Ellouze N, Dey Z, Gaulieraj M, Ben Kilani H (2003) The Hammamet, Gabes and Chotts basins (Tunisia): a review of the subsidence history. Sed Geol 156:241–262

Rabhi M (1999) Contribution a l'étude stratigraphique et analyse de l'évolution géodynamique de l'Axe N-S et des structures avoisinantes (Tunisie centrale). Thèse. Uni. Tunis. II. F S T. p 206 et planches photos

Raulin C, Frizon de Lamotte D, Bouaziz S, Khomsi S, Mouchot N, Ruiz G, Guillocheau F (2011) Late Triassic–early Jurassic block tilting along E-W faults, in southern Tunisia: new interpretation of the Tebaga of J Medenine. J Afr Earth Sci 61(1):94–104

Rigane A (1991) Les calcaires de l'Ypresien en Tunisie Centro Septentrionale: Cartographie, Cinématique et Dynamique des structures. Thèse d'Universite, Universite Franche-Comte, p 214

Rigane A, Gourmelen C (2011) Inverted intracontinental basin and vertical tectonic: the Saharan atlas in Tunisia. J Afr Earth Sci 61:109–128

Rigo L, Garde S, El Euchi H, Bandt K, Tiffert J (1996) Mesozoic fractured reservoirs in a compressional structural model for the North Eastern Tunisian Atlasic Zone. In: Proceedings of the fifth Tunisian petroleum and production conference. ETAP Memoir no 10, pp 233–256

Roure F, Casero P, Addoum B (2012) Alpine inversion of the North African margin and delamination 2 of its continental lithosphere. Tectonics 31(TC3006):2012. https://doi.org/10.1029/2011TC002989

Rouvier H (1977) Géologie de l'extrême Nord Tunisien: tectonique et paléogéographie superposées à l'extrémité orientale de la chaîne nord-maghrébine. Ann Mines Geol, Tunis 29:427

Soyer C, Tricart P (1989) Tectonique d'inversion en Tunisie centrale; le chainon atlasique Segdal-Boudinar. Bull Soc Géol Fr 4:829–836

Taktak F, Bouaziz S, Tlig S (2012) Depositional and tectonic constraints for hydrocarbon target of the Lutetian–Langhian sequences from the Gulf of Gabes—Tunisia. J Petrol Sci Eng 82–83:50–65

Tanfous Amri D, Bédir M, Soussi M, Azaiez H, Zitouni L, Inoubli MH, Ben Boubaker K (2005) Early halokinesis associated to the Jurassic rift faulting in Central Tunisia (Majoura-El Hfay area). C R Geosci 337(7):703–711

Tanfous Amri D (2007) Sismostratigraphie et sismotectonique du Jurassique dans l'Atlas centro-méridional de Tunisie. Thèse Troisième Cycle, Université de Tunis El Manar II, Tunisie, p 247

Troudi H, Tari G, Alouani W, Cantarella G (2017) Chapter 25—Styles of salt tectonics in central Tunisia: an overview. In: Permo-Triassic Salt Provinces of Europe, North Africa and the Atlantic Margins, p 543. https://doi.org/10.1016/B978-0-12-809417-4.00026-4

Truillet R, Turki MM (1980) La tectonique tangentielle dans la zone des diapirs. L'exemple du Dj. Amar de l'Ariana (Tunisie septentrionale). C R Acad Sci Paris, Sér D 291:325–327

Truillet R, Zargouni F, Delteil J (1981) La tectonique tangentielle dans l'axe Nord-Sud (Tunisie centrale). C R Acad Sci Paris Ser II 23:50–54

Turki MM (1981) Importance de la tectonique tangentielle dans la structure du Jebel Bou Kornine (Dorsale tunisienne). In: Resumes du 1er Congr. Nat. Sc. Terre, Tunis, p 53

Turki MM (1985) Polycinématique et contrôle sédimentaire associé sur la cicatrice Zaghouan Nebhana. Thèse Sci., Tunis, p 252

Turki MM, Delteil J, Truillet R, Yaich C (1988) Les inversions tectoniques de la Tunisie centro-septentrionale. BSGF 8(3):399–406

Yaich C (1984) Etude geologique des chainons du Cherahil et du Krechem El Artsouma (Tunisie centrale). Liaison avec les structures profondes des plaines adjacentes. Thèse 3emc cycle, Besançon, p 165

Zaïer A, Beji-Sassi A, Sassi S, Moody RTJ (1998) Basin evolution and deposition during the Early Paleogene in Tunisia. Geol Soc, Lond, Spec Publ 132(1):375–393

Zargouni F, Delteil J, Truillet R (1979) Interprétation des éléments structuraux alpins de l axe Nord-Sud dans le cadre d une genèse polyphasée (Tunisie centrale). In: 7eme RAST, Lyon, p 469

Zargouni F, Abbès C (1987) Zonation structurale de la Tunisie. Rev Sci Terre INRST 6:63–69

Zouaghi T (2008) Distribution des Séquences de Dépôt du Crétacé (Aptien–Maastrichtien) en Subsurface: Rôle de la Déformation Tectonique, l'Halocinèse et Evolution Géodynamique (Atlas central de Tunisie). PhD thesis, Université de Tunis El Manar, Tunis, Tunisie

Part IV

Fourth Thematic: Geology and Water Resources in Northern Africa: Cases Studies

9 Aquifer Structuring and Hydrogeological Investigation in North African Regions Using Geophysical Methods: Case Study of the Aquifer System in the Kairouan Plain (Central Tunisia)

Fethi Lachaal, Hajeur Azaiez, Rahma Bruni, Hakim Gabtni, and Mourad Bedir

Abstract

This study presents an integrated methodology to characterize the aquifer structuring, and geometry in the case of geological complexity and lack of hydrogeological data. Gravity methods analysis, seismic reflection, wireline logging, and electrical resistivity tomography techniques are used to identify the lithological and geometrical knowledge of the Mio-Plio-Quaternary aquifer in the Kairouan plain. In the Kairouan region, the Mio-Plio-Quaternary series form a complex aquifer system formed by siliciclastic fluvio–deltaic deposits. It consists of sandy and sandy clay horizons interbedded with clay levels. The geometry of the Mio-Plio-Quaternary systems is not yet precisely defined. The present study is dealing with the hydrogeological properties of these deeper sand layer reservoirs in the Kairouan basin using an integrated geophysical approach. Gravity is used in order to define regional structuring, which is characterized by the presence of a deep fault corridor. The interpretation of wireline logging of 15 petroleum wells shows that the Mio-Pliocene deposits (Ségui Formation) contain five reservoirs. The interpretation of fourteen 2D seismic reflection profiles is used to characterize the aquifer structuring and to identify the major aquifer boundaries. The Kairouan plain is marked in the subsurface by the presence of platforms, anticlines, synclines, and grabens: the Sidi El Hani and Chorbane blocks are separated by master deep-seated faults corridor directed N90 to N120. The groundwater flow and the water resources evolution are deduced from the water level in 26 piezometres in 2013 and monthly water level measurements in 4 monitoring wells during the 1968–2013 period.

Keywords

Hydrogeology structuring • Gravity • Seismic reflection • Well logging • ERT • Groundwater

9.1 Introduction

The aquifer structure characterization and geometric identification are of primary importance in hydrogeology, especially, to implement the geometric 3D model in order to apply a strategy of sustainable groundwater management aquifer and to provide prospective scenarios of groundwater evolution (Lachaal et al. 2011). Previous studies have proved the important contribution of geophysical methods in 3D geometric modeling (Lachaal et al. 2012a). Gravity and seismic reflection methods are widely used for oil and gas exploration. In hydrogeology Gravity and seismic reflection are applied to characterize the aquifer structuring, differentiate the aquifer bounders, and identify hydrogeological basin extension. In addition, aquifer bedrock depth can be deduced from seismic reflection, especially in the case of deep and fossil aquifers (Lachaal et al. 2011, 2012b; Gabtni et al. 2012; Houatmia et al. 2015; Poormirzaee et al. 2015). The borehole logging methods are used to identify the aquifer position, study the water salinity and aquifer productivity, and define the wells equipment (Lachaal et al. 2011; Thilagavathi et al. 2014; Chandra et al. 2015; Kumar and Yadav 2015). Time-domain electromagnetic exploration (TDEM) and electrical resistivity (ER) are widely used in hydrogeology for groundwater exploration (Mohamed et al. 2011; Salah 2013; Karunanidhi et al. 2014; Chekirbane et al. 2014; Aizebeokhai and Oyeyemi 2015; Mahmoud et al. 2015), particularly where the aquifer depth is about several

F. Lachaal (✉) · H. Azaiez · H. Gabtni · M. Bedir
Georesources Laboratory, Water Research and Technology Centre, Borj Cedria Ecopark, PO Box 273 8020 Soliman, Tunisia
e-mail: fethi.lachaal@certe.rnrt.tn

R. Bruni
Faculty of Sciences of Tunis, Department of Geology, University of Tunis El Manar, 2092 El Manar, Tunis, Tunisia

S. Khomsi and F. Roure (eds.), *Geology of North Africa and the Mediterranean: Sedimentary Basins and Georesources*, Regional Geology Reviews, https://doi.org/10.1007/978-3-031-18747-6_9

hundred meters. Electrical resistivity tomography (ERT) is used to study groundwater geometry, infiltration, pollution, and salinization in the case of seawater intrusion (Tamma Rao et al. 2014; Metwaly et al. 2014; Adhikary et al. 2015; Lachaal et al. 2022).

In the case of fluvio–deltaic sediment that is characterized by high geological and geometric complexity related to the high lateral and vertical variability, the aquifer characterization needs a multi-technique approach integrating all surface and subsurface methods. To better define aquifer geometry, all available hydrogeologic data and possible investigation methods must be integrated in order to identify the multilayer aquifer, especially, the reservoirs substratum, boundaries, and their lateral and vertical extension (Lachaal et al. 2011; Gabtni et al. 2012; Ayolabi et al. 2015; Araffa et al. 2015).

In this context, the Mio-Plio-Quaternary aquifer system in the Kairouan region is selected as a study case (Fig. 9.1), according to the recommendation of the General Direction of Water Resources, Ministry of Agriculture in Tunisia (DGRE). Especially, this area is a water stressed region (Leduc et al. 2007). A national project of groundwater resources characterization in the Sahel and Kairouan Basins was initiated in cooperation between the Water Research and Technology Centre, Ministry of Higher Education and Scientific Research in Tunisia (CERTE), and the DGRE, during the 2010–2014 period.

The studied area covers the Kairouan plain which is situated in east-central Tunisia. It is characterized by a semi-arid to arid climate and stressed water resources. The mean annual rainfall is about 300 mm·year^{-1} (Leduc et al. 2007). Groundwater is the principal origin of water demand. The hydrogeology of the Kairouan region is formed by a complex aquifer system, attributed to the Ségui formation (Mio-Plio-Quaternary) and formed by fluvio–deltaic sediments. The aquifer system is mainly formed of sandy and sandy clay horizons with clay intercalation. Previous studies have shown the Kairouan hydrogeology complexity (Besbes et al. 1978; Ben Ammar et al. 2006; Leduc et al. 2007). The geometry of the Mio-Plio-Quaternary aquifer system is not yet precisely defined. The aquifer is generally considered as two layers (Besbes et al. 1978). The aquifer substratum and the limit between shallow and deep aquifer are not precisely defined. In addition, the discharge zone that is considered as Sebkhat El Kelbia and Sebkhat Sidi Elhani is poorly unproven, because of the lack of hydrological data in the Eastern part of the aquifer. In this situation, the use of hydrogeophysical methods can give new knowledge to aquifer hydrogeology characterization.

The present study aims to deal with the aquifer geometry properties, in particular, to identify the aquifer structure, aquifer substratum, and boundaries using the combination of geophysical methods, especially Gravity, seismic reflection,

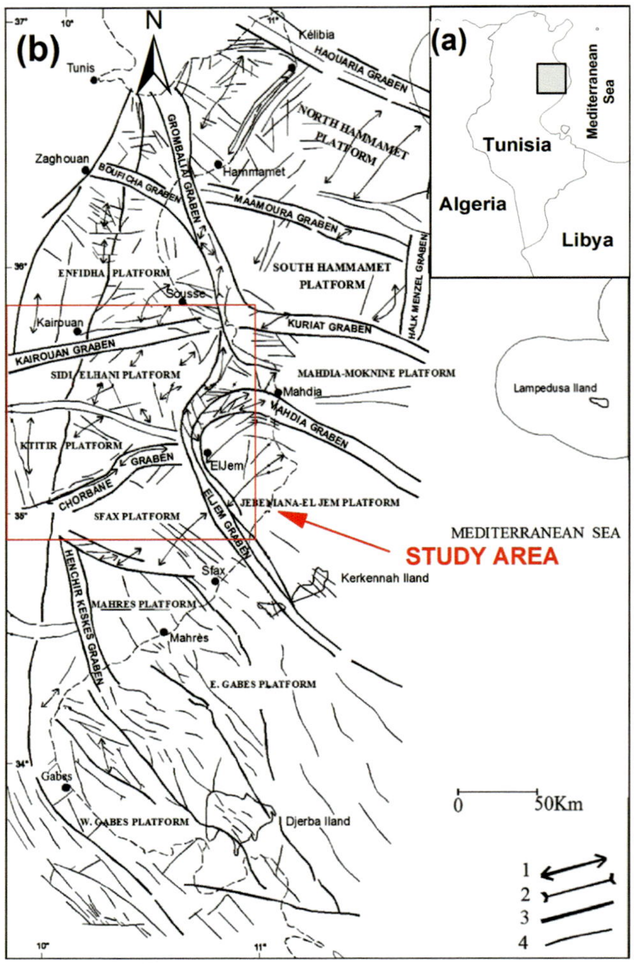

Fig. 9.1 Location map of the study area: **a** Tunisia location and **b** major structural elements in Eastern Tunisia (Bédir 1995). 1 anticlines, 2 synclines, 3 first order faults, 4 second order faults

borehole logging, and ERT methods. Due to the significant aquifer depth and the scarcity of drilling wells in the Kairouan basin, the Gravity and the seismic reflection profiles were adopted.

9.2 Geographical and Geological Setting

The study area is formed by the Kairouan plain. That is located in east-central Tunisia, between 3,870,000 and 3,970,000 north parallels and the 570,000 and 670,000 east meridians (UTM), with a total area of 10,000 km^2 (Figs. 9.1 and 9.2). The study area is limited to the North by Jebal Cherichira and Merguellil basin, to the South by Sebkhat Cherita Mechertat, to the West by Jebal Chérahil, and to the East by Sebkhat Sidi El Hani and El Kelbia.

The Kairouan plain is separated from NE–SW Atlasic structures by the NS string (Khomsi et al. 2004). The Kairouan platform is marked by collapsed outcrops of Triassic to Mio-Plio-Quaternary and characterized by extended

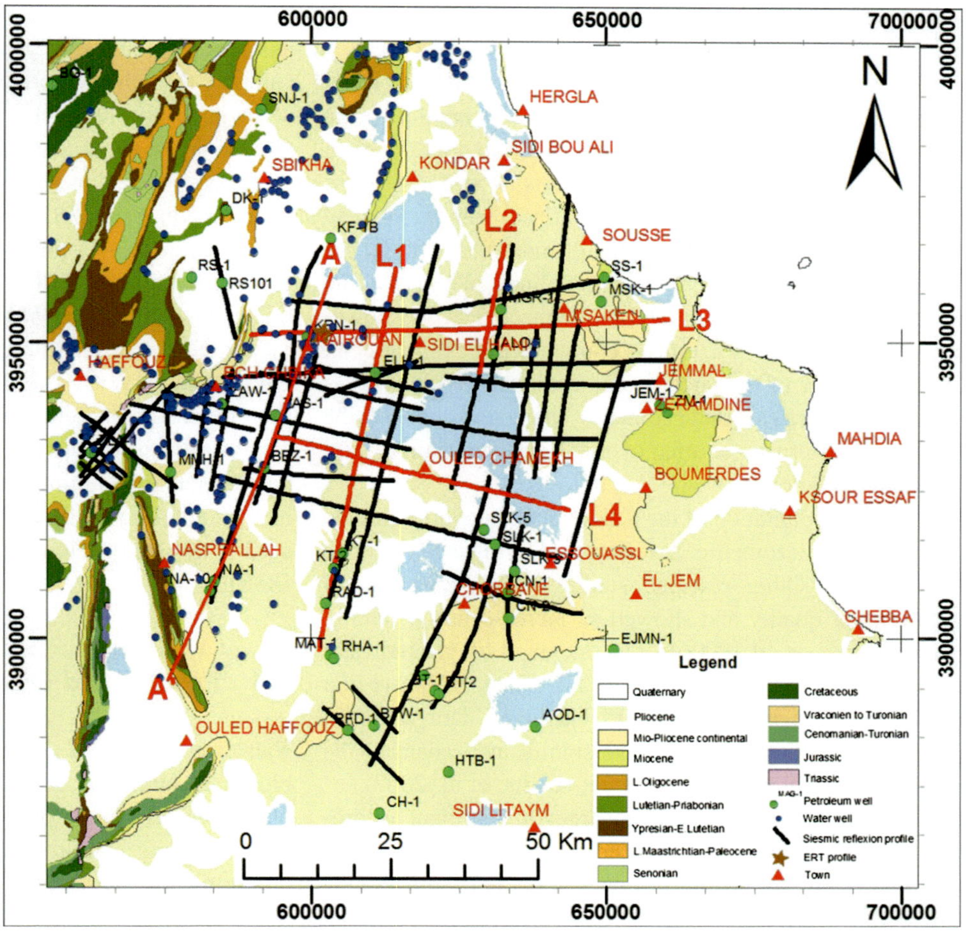

Fig. 9.2 Simplified geological map (Ben Haj Ali et al. (1985) modified), and location of water and petroleum wells, seismic reflection profiles, and hydrogeological cross sections

detrital continental Quaternary over several hundred meters thick formed eventually by sand, sand clay, and clay (Rabhi 1999). The Triassic outcrop (Rhéouis Formation) (Burrolet 1956) is considered the oldest sediment outcropping in the region. It is formed by a complex chaotic aspect of gypsum, clay, and marl, often containing various elements such as limestone, dolomite, and quartz black bipyramid.

The Ségui Formation is the most important outcrop in the region, which covers the essential of the Kairouan plain. The Mio-Pliocene series has a continental origin. It outcrops extensively at the Draa Affene anticline, Jebel Cherichira, Jebel Baten, and northern accident Trozza-Labaiedh (Fig. 9.1).

The Pliocene thickness of several hundred meters consists of gravel and sand intercalated with clay lenses in the western part of the plain, while in the eastern part there is a dominance of clay and silty sand with evaporates.

Quaternary deposits have a continental origin, mainly formed by alternating silts and sands with clays passage levels (Rabhi 1999). These deposits cover large areas and occupy the Kairouan plain.

9.3 Hydrogeology Context

The Mio-Plio-Quaternary aquifer system in the Kairouan plain, represents the most important and largest aquifer system in central Tunisia (Besbes et al. 1978; Nazoumou and Besbes 2001). The region is characterized by strong rainfall variability (Chargui et al. 2013, 2018, 2022) and most of the rainfall comes from violent and highly variable storms during the spring (February–May) and autumn (September, October) seasons (Kingumbi 2006). The region is characterized by the presence of Sebkhats (Salt Lake) as Sebkhat El Kalbia in the North, Sebkhat Sidi El Hani in the center, and Sebkhat Cherita in the South, which represents the outlets of rivers in the region.

The Kairouan plain constitutes a central collapse basin formed by continental clastic deposits of Mio-Plio-Quaternary (Besbes et al. 1978). These deposits are organized in lens layers with different sizes and extensions. The permeable sediments (sand and gravel) are concentrated in the rivers area in the upstream and central part of the plain, while medium and fine textures are situated downstream near the Sebkhat El Kelbia.

The Mio-Plio-Quaternary aquifer system consists of two hydrogeological unities: shallow and deep aquifers. The shallow system is relatively continuous throughout the plain and composed to Quaternary sediments, with depths less than 50 m. The deep aquifer is confined with more than 50 m of thickness and with depths of more than 100 m, the substratum of deep aquifer is not yet defined. Whereas to the western part, the two aquifers are connected, while downstream of a plain deep aquifer is confined (Ben Ammar et al. 2006). Because of the tectonic and sedimentary complexity affecting the region, the geometric characterization of the Mio-Plio-Quaternary aquifer of the Kairouan plain is still incompletely known.

The overexploitation of the Kairouan plain aquifer has led to a general piezometric drawdown of 1 m·year^{-1} (Leduc et al. 2007). Other research studies have addressed the groundwater quality and showed an increase in the salinity (Ben Ammar et al. 2006). The increase of groundwater pumping and the construction of two dams in the region (Sidi Saad dam in 1982, and El Haouareb dam in 1989) mainly for flood mitigation, have significantly changed the natural hydrodynamic comportment (Leduc et al. 2007).

The natural groundwater recharge of Kairouan is provided by infiltration from the Zeroud and Merguellil rivers and direct infiltration from behavior aquifers of Ain Beidha and Haffouz aquifers (Besbes et al. 1978). Since 1988, artificial groundwater recharge begins through the construction of hydraulic structures and small dams and principally from Sidi Saad and El Haouareb dams.

9.4 Materials and Methods

The used Gravity data cover a grid of 2 km spacing, and they were obtained from the Tunisian Company of Petroleum Activities (ETAP). The data were merged and reduced using the 1967 International Gravity formula (Morelli 1976). Free Air and Bouguer Gravity corrections were applied using 2.67 gcm^{-3} as a reduction density and sea level as a datum. Terrain corrections were done using the method of Plouff (1977), a 5-min topography grid, and a density of 2.67 gcm^{-3} (Row et al. 1995).

In addition to the Gravity data, we used seismic reflection data. It consists of a grid of 14 2D seismic reflection sections oriented roughly North–South and West–East, and 15 petroleum wells provided by ETAP (Fig. 9.2). Mio-Plio-Quaternary seismic reflectors has been calibrated and correlated along the seismic reflection grid. In addition, the top Ain Grab Formation reflector (Langhian, Middle Miocene) that is considered a regional repair (Bédir et al. 1996; Khomsi et al. 2004; Lachaal et al. 2011) is also drowned. The outcrops around the region and the time-depth conversion curve of the available petroleum wells were used to calibrate the seismic horizons.

In order to characterize the geometry of the Mio-Plio-Quaternary shallow aquifer in which the depth is less than 50 m, geophysical measurements of ERT were carried out in the region in April 2013. The ERT profile was located about 800 m from the Kairouan city and oriented from West to East coinciding with the western part of the L3 seismic reflection profile (Fig. 9.2).

The used piezometric data were collected by the DGRE during two surveys. The first one is realized in Mars 2013 (humid season) and the second is conducted in August 2013 (dry season). It consists of water level measurements in 26 piezometers. In addition, monthly water level measurements in 4 monitoring wells covering the 1968–2013 period are used in this study. These piezometric data were given by the DGRE.

9.5 Results and Discussion

Kairouan region is characterized by structuring in a block because of deep fault corridors. The Kairouan plain consists of two platforms (Bédir 1995): in the North Sidi El Hani platform and in the South the Chorbane platform called also Ktitir platform. These blocks are separated by fault corridors N45 to N45 (Bédir 1995) which coincides with the transverse fault of El Hdedja (Khomsi et al. 2004).

9.5.1 Gravity Interpretation

The Bouguer Gravity anomaly map was produced from the interpolation of a 2 km grid of Gravity data. The Bouguer Gravity anomaly values range from −50 to 25 mGal. The lower value is located in the western parts (Fig. 9.3). These data are affected by a regional gradient increasing eastward. This gradient represents the regional Gravity field that is probably related to crustal thickness variations (Buness et al. 1989; Gabtni 2005). Previous work shows the minor influence of the regional Gravity field on the local gravity anomalies (Gabtni et al. 2010).

Three trending anomalies were observed in the Bouguer Gravity anomaly map:

- In the NE, the Bouguer Gravity anomaly map shows, in the Zéramdine region, an N30–N45 trending anomaly which is associated to the Zéramdine fault corridor (Fig. 9.3) (Lachaal et al. 2012b).
- In the SE, the Bouguer Gravity anomaly map shows in the Mahdia and Jebeniana regions an N135 to N45

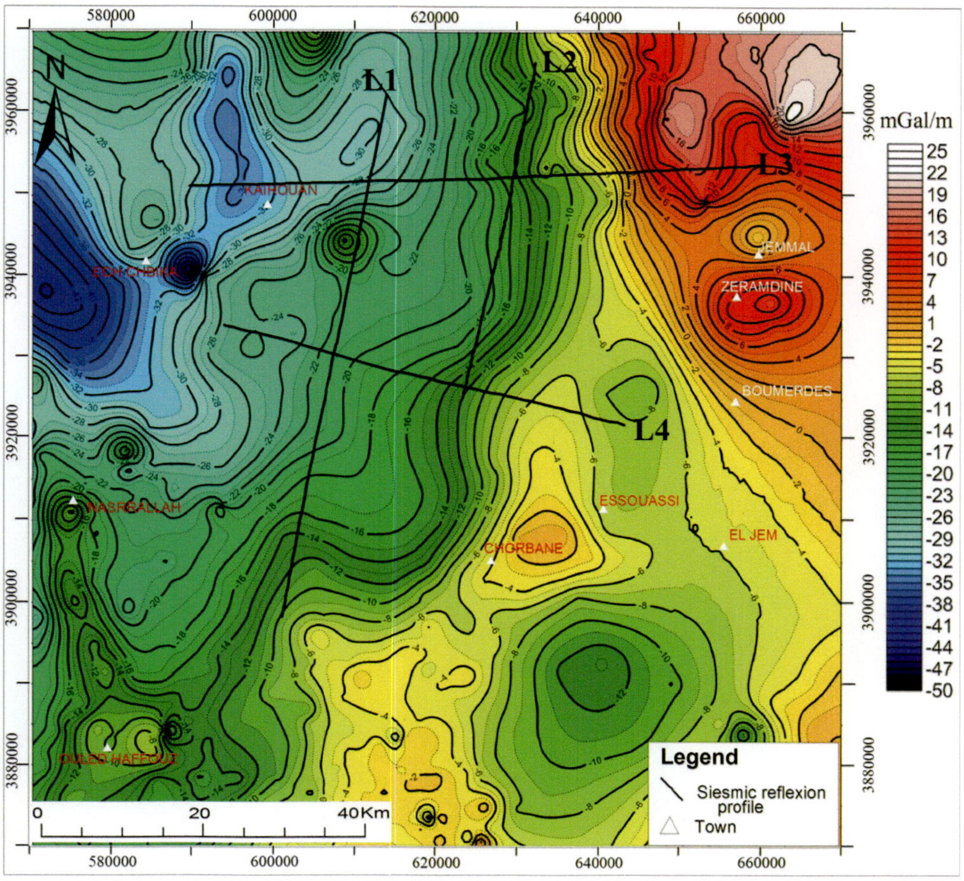

Fig. 9.3 Gravity Bouguer map of the Kairouan sector

trending anomaly with is correlated to the El Jem Half graben (Bédir et al. 1992; Lachaal et al. 2012b).
- In the Kairouan region, the Bouguer Gravity anomaly map shows an N45 trending anomaly which is correlated with the thrust fault of Kairouan-Sousse (Khomsi et al. 2004) and Kairouan Graben (Bédir et al. 1996).

In order to study the Gravity anomalies, and generally to delineate the vertical and lateral locations of the subsurface causative sources, a horizontal Gravity gradient map (HGG) was created. The interpretation of HGG anomaly map can contribute to the understanding of the structural and tectonic framework of the Kairouan region (Fig. 9.4). The superposition of the known faults of the HGG map shows HGG maxima (Fig. 9.1). These HGG anomalies are related to tectonic structures (S1 to S6) forming a deep fracturing corridor (Fig. 9.3):

- S1: the Zéramdine faults corridor, located in the NE of the study area with N0 to N40 direction;
- S2: the El Jem half graben situated in the SE of the study area with N135 to N45 direction;
- S3: Kairaoun-Sousse fault with N0 to N45 direction;
- S4: El Hdadja Fault with N45 to N45 direction;
- S5: Chorbane graben located in the SW with N30 to N45;
- S6: A probable graben with N90 to N45 direction;
- S6: Jebel Chérahil.

The HGG map analysis shows several blocs: Sidi El Hani, Chorbane (Ktitir), Mahdia-Jébéniana, and Zéramdine platforms.

9.5.2 Wireligne Logging Data Analysis and Aquifer Lithology

9.5.2.1 Sidi El Hani Platform

The Sidi El Hani block is limited to the North by Enfidha block through the Kairouan graben which coincides with the transverse fault of Kairouan-Sousse (Khomsi et al. 2004), to the Eastern side by the Zéramdine fault corridor, to the Southern side by Chorbane block and to the West by J. El Haouareb through the north–south axis. Several oil wells (ELH-1, KRN-1, GR-1, MSK-1, SAR-1, ALO-1, and ZAW-1) were drilled in this block. According to the El Hdadja-1 logging well, the Mio-Plio-Quaternary geological series presents 703 m in depth. The Mio-Plio-Quaternary series is composed of a multilayer aquifer system of sand and sand clay deposits. They are formed by five reservoir levels. Each level is characterized by a low gamma ray

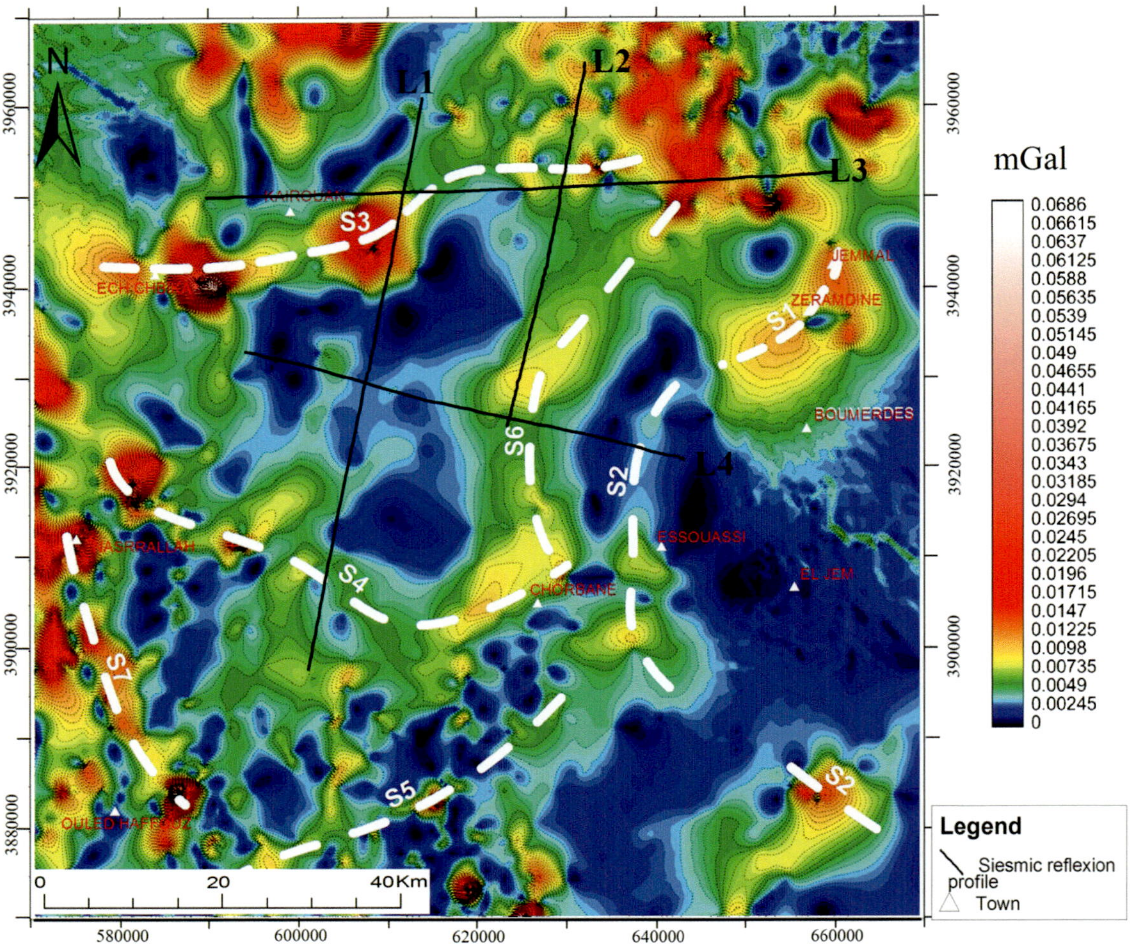

Fig. 9.4 Horizontal gravity gradient map of the Kairouan sector. S1: the Zéramdine fault corridor, S2: the El Jem half graben situated in the SE of the study area, S3: Kairaoun-Sousse fault, S4: El Hdadja fault, S5: Chorbane graben, and S6: Jebel Cherahil

registration and a negative spontaneous potential across permeable sandy series. The aquifer levels are distributed upwardly as follows (Fig. 9.5):

- The first reservoir layer (650–570 m) is formed of translucent sand with quartz.
- The second reservoir layer (560–530 m) is composed of translucent sand with quartz associated with some levels of clays.
- The third reservoir layer (520–440 m) is formed by sand with quartz, characterized by the presence of gravel.
- The fourth reservoir layer (410–390 m) is sand, clear, and translucent with quartz.
- The fifth reservoir layer (80–60 m): is formed by sand with quartz.

9.5.2.2 Chorbane Platform

Chorbane block is limited to the North by Sidi El Hani platform, from the South by the Chorbane graben, from the East by the Mahdia platform, through the El Jem Half graben, and from the West by Jebel Cherahil. Five oil wells are located in the Chorbane block (BBZ-1, NA-1, MH-1, SSS-1, and SLK-1). The Mio-Plio-Quaternary series are crossed by the Nasrallah-1 well. Indeed, the Ségui Formation reaches a depth of 525 m.

According to the NA-1 well logging, the Mio-Plio-Quaternary series contains a multilayer aquifer system composed of five levels of reservoirs. The sandy nature is deduced by low gamma ray readings, slight separation between the macro-resistivity and micro-resistivity recording, and through the records of scoring Neutron porosity levels. The Mio-Plio-Quaternary reservoir system is composed also of five layers, from bottom to top (Fig. 9.5):

- The first reservoir layer (450–300 m) is formed by sand white, ocher, yellow, medium to very coarse with small siliceous pebbles, sometimes weakly consolidated.
- The second reservoir layer (270–240 m) is formed by sand white and medium size.

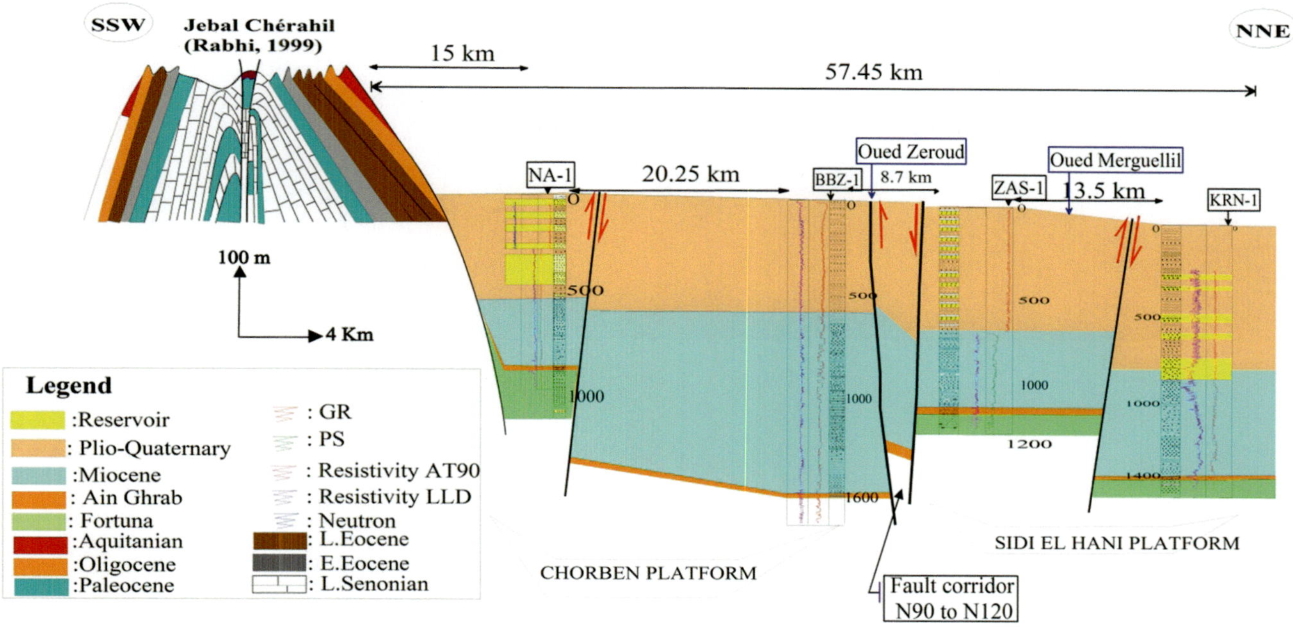

Fig. 9.5 A-A' hydrogeological correlation showing the different compartments, the distribution, the lateral and vertical evolution of Mio-Plio-Quaternary aquifers in the Kairouan basin, and the contribution of the wire line logging wells in the determination of the hydrogeological aquifer structure

- The third reservoir layer (180–150 m) is made with sand white and medium to very coarse, sometimes weakly consolidated.
- The fourth reservoir layer (120–90 m) is composed of sand white, medium to very coarse.
- The fifth reservoir layer (50–20 m) is formed by sand yellow, medium to very coarse, and sometimes weakly consolidated.

The NNE–SSW correlation was drowning using the interpretation of well logging (Fig. 9.5). It shows two structural blocks (Sidi El Hani and Chorbane blocks) which are separated by fault corridor N45 to N45 (Bédir 1995). Thinning of Mio-Plio-Quaternary deposits was observed from NE to SW. From the SW, the Mio-Plio-Quaternary series are limited by Jebal Chérahil which shows the ancient outcrops formed by Aquitanian to Upper Senonian deposits (Rabhi 1999).

9.5.3 Seismic Reflection Profiles Interpretation and Seismic Stratigraphy Study

The wireline logging of petroleum wells was used to identify the different aquifer layers in each geological block. The interpretation of 2D seismic reflection profiles shows the boundaries of each aquifer system and their lateral changes of facies. In order to identify the aquifer structure, limits, and reservoir evolution we use the interpretation of seismic profiles.

9.5.3.1 L1 Seismic Reflection Profile

The L1 seismic reflection profile extends 62.55 km from NNE to SSW (Fig. 9.6). It is calibrated with the time-depth conversion curve registered in the ELH-1 petroleum well where the Mio-Plio-Quaternary deposits reach 703 m in depth. The profile shows a tectonic structuring marked by the individualization of two tectonic blocks: Sidi El Hani and Chorbane blocks, bounded by deep faults corridors (Bédir 1995).

L1 profile shows the presence of several tectonic structures. To the NNE, the Kairouan-Sousse fault resulted from compression structures materialized by folds and syncline gutters. To the SSW, the El Hdadja fault, generates a positive flower structure (Khomsi et al. 2004). In addition, we note the presence of the Sidi El Kilani fault corridor with an N120 direction that separated the Sidi El Hani and Chorbane blocks.

The seismic-stratigraphic analysis highlights the presence of downlap configurations at the base of the Mio-Plio-Quaternary Ségui Formation. This seismic pattern is characteristic of channelized sedimentation marking the Mio-Plio-Quaternary horizons. Toplap features end the reservoirs horizons at the top.

From SSW to NNE, the profile shows the presence of channel sedimentation, especially, in the upper part of the

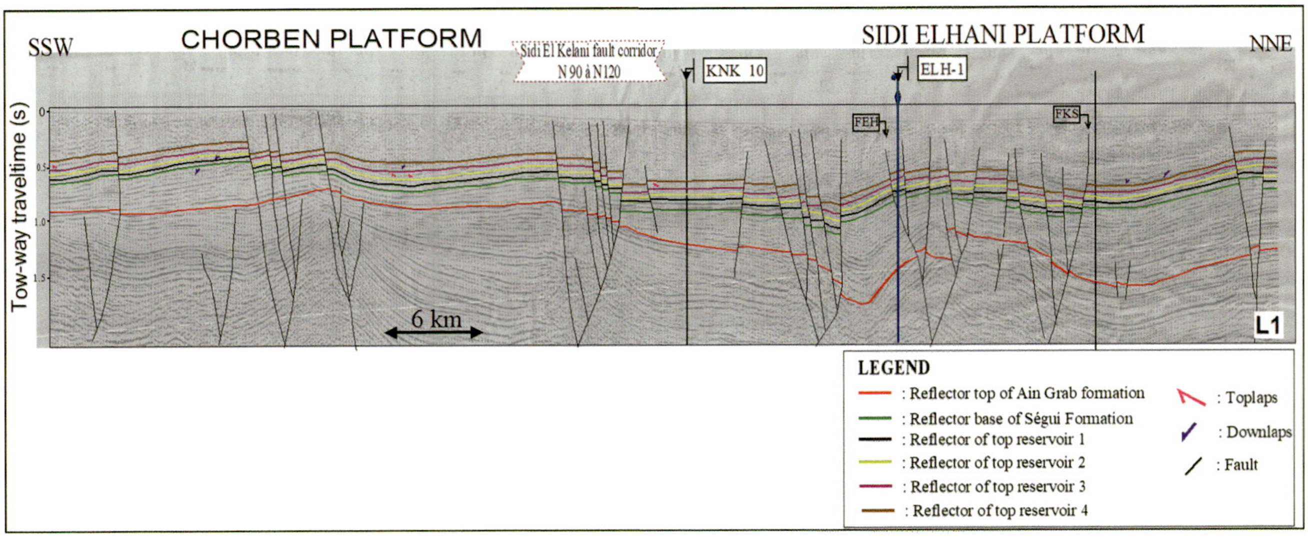

Fig. 9.6 L1 seismic reflection profile showing the NNE–SSW structuring of the Plio-Mio-Quaternary reservoir horizons

profile at the depth of 100 ms in TWT (two-way travel time). These deposits are characterized by reflections of low continuity and medium to high amplitude. Channel deposits are from fluvial plain deltaic environment consisting of pebbles and gravel mixed with locally clayey sands. They are characterized by high porosity and high permeability.

9.5.3.2 L2 Seismic Reflection Profile

The L2 profile is localized in Sidi El Hani plate form with NNE–SSW direction and is 21.75 km long (Fig. 9.7). In the southwestern part, the reflectors are continuous, parallel, and have the same amplitude.

The Mio-Plio-Quaternary series maintain the same thickness. That reflects uniform sedimentation with low hydrodynamic energy.

In the NNE direction, the profile shows the presence of faults zone related to deep faults corridor, formed by sub-vertical faults with low magnitudes. At the depth of 500 ms in TWT, the L2 profile shows the same channel sedimentation observed in the L1 profile. In the NNE part,

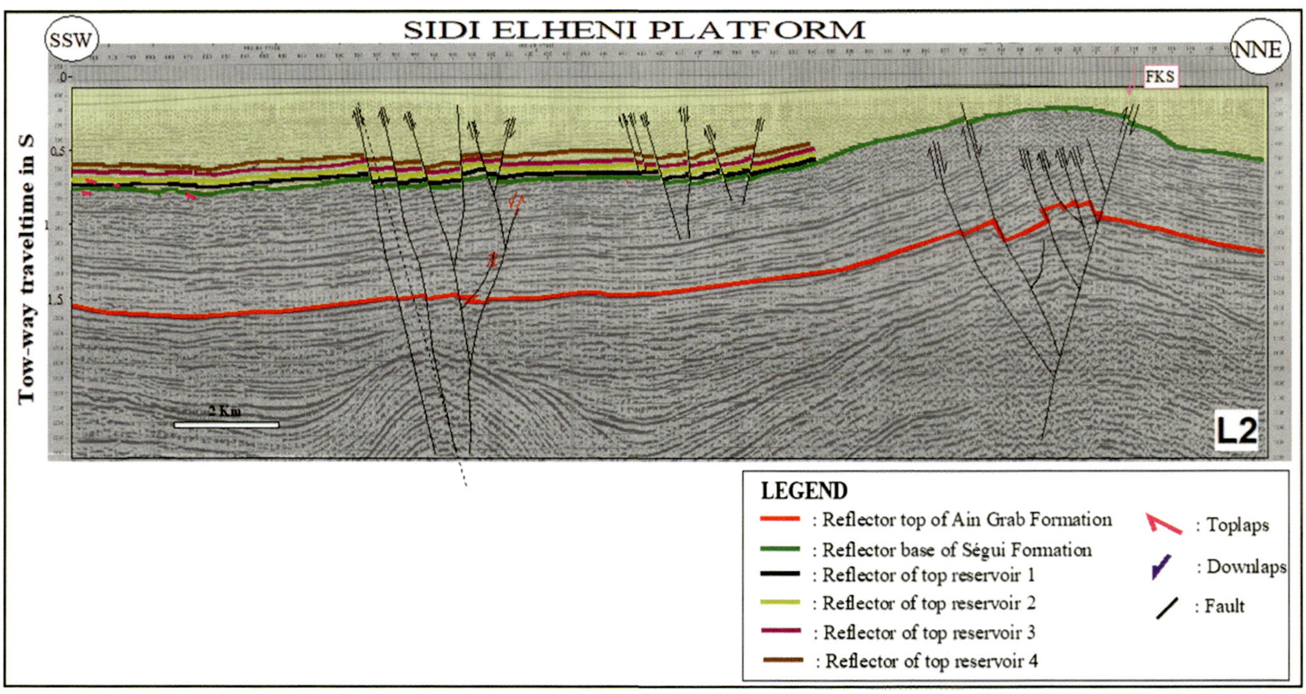

Fig. 9.7 L2 seismic profile reflection showing the NNE/SSW variation of the Plio-Mio-Quaternary reservoir layers and the NNE aquifer boundary

the profile highlights the effect of the Kairouan-Sousse fault which forms a geological limit to the Mio-Plio-Quaternary aquifer systems. It controls the spatial distribution and thicknesses of different series (Khomsi et al. 2004). In addition, the chaotic configuration associated with channel sedimentation was observed in the profile.

9.5.3.3 L3 Seismic Reflection Profile

This L3 profile covers Sidi El Hani and Zeramdine blocks with E–W direction (Fig. 9.8a). At the intersection with L1 geo-seismic line and in 100 ms in TWT, the profile shows a chaotic seismic facies configuration. This configuration has been associated with channel sedimentation that is formed by sands.

In the Eastern part of the seismic reflection line L2, fold structures called Bir El Taib has an NNE–SSW direction (Ghribi 2010). It separates the Sidi El Hani block from Zeramdine one. Several works suppose that El Kelbia and Sidi El Hani Sebkhats represent the Kairouane aquifer Eastern boundary. According to the L1 profile, the Mio-Plio-quaternary reservoir is continued under the El Kelbia Sebkhat. That concludes that the Kairouan basin is continued to the East side and the Eastern aquifer boundary is formed by the Bir El Taib fold and not by the El Kelbia and Sidi El Haria Sebkhats. This hypothesis needs a supplementary hydrogeological and geophysical investigation in the region.

The seismic-stratigraphic interpretation shows the presence of downlap horizon strata termed marking lens sedimentation. In addition, a toplap strata termed was observed characterizing the top of the first and second reservoir levels.

9.5.3.4 L4 Seismic Reflection Profile

The L4 profile is localized in the Sidi El Heni platform with 51 km from WNW to ESE (Fig. 9.9). In the West side, the L4 profile shows the continuity of the Mio-Plio-Quaternary reservoir under the Sidi El Heni Sebkhat. In fact, the Kairouan basin continued in the East direction. The hypothesis of Sidi El Heni Sebkhat forms the aquifer outlet, and the West boundary is rejected.

9.5.4 Electric Resistivity Tomography Study

9.5.4.1 ERT Data Acquisition

The ERT profile localization has been chosen in the Eastern side of the L2 profile. This area is characterized by the presence of detrital deposits of Mio-Plio-Quaternary, and the presence of channel deposits. The use of ERT method aims to confirm the sedimentation type (Fig. 9.2).

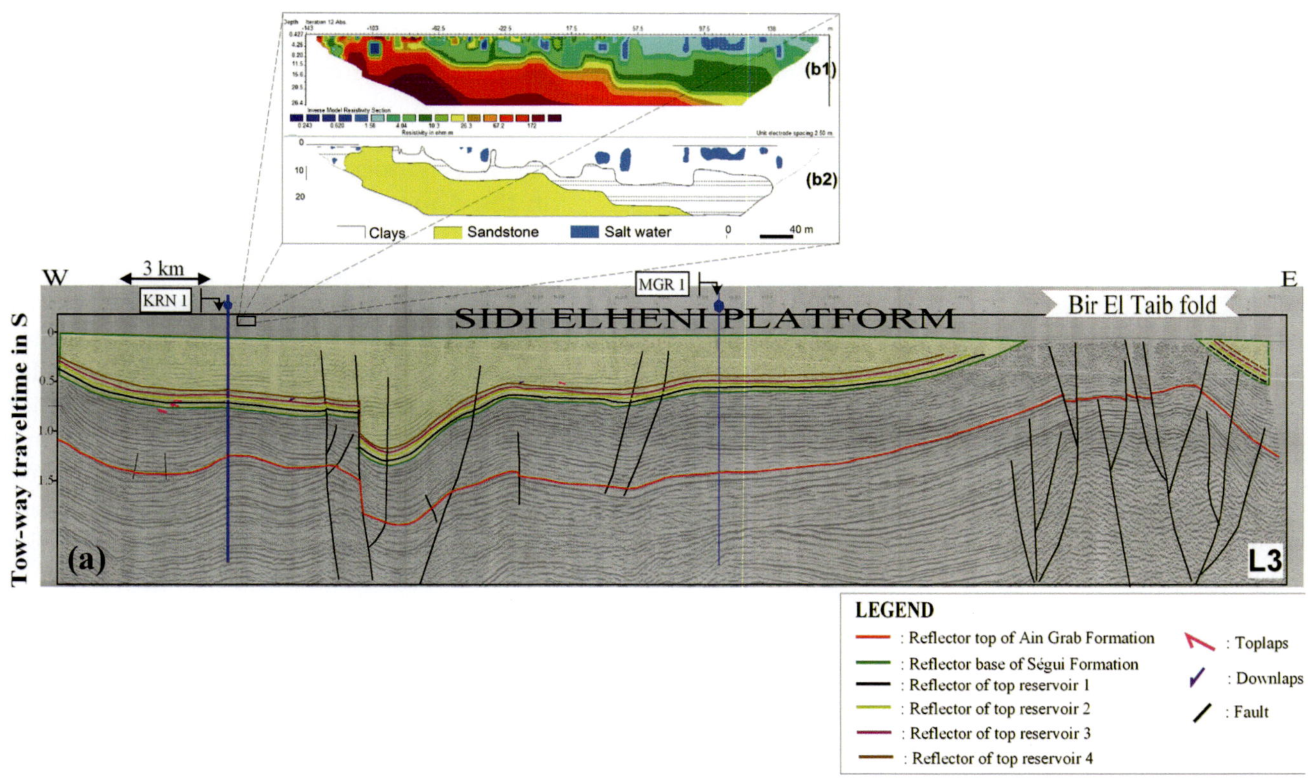

Fig. 9.8 **a** L3 seismic reflection profile showing the W–E variation of the Plio-Mio-Quaternary reservoir layers, and the aquifer East boundary. **b** 2D interpreted resistivity-depth sections along the ERT profile lines: $\mathbf{b_1}$ The inversion model resistivity section. $\mathbf{b_2}$ 2D interpreted lithological section

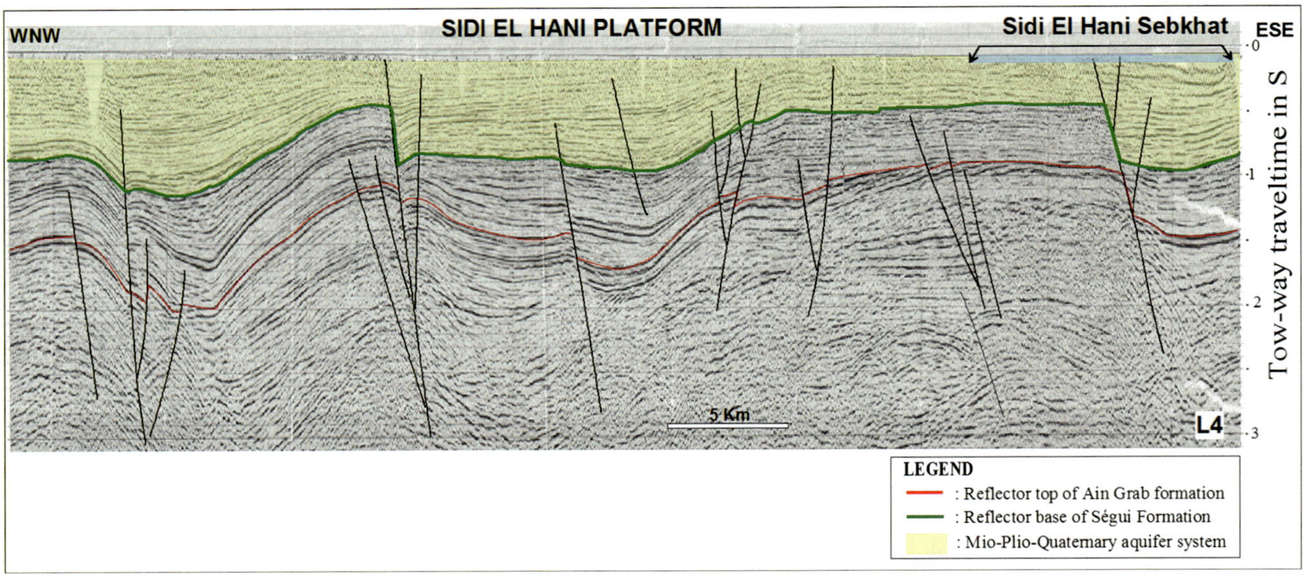

Fig. 9.9 L4 seismic reflection profile showing the WNW–ESE variation of the Plio-Mio-Quaternary reservoir layers

The used system is Terrameter SAS 4000 produced by ABEM Instruments, it is a multi-electrode system that can provide automatic management of the electrodes. This system is composed of 64 electrodes and products geoelectric profile along 315 m with a spacing of 5 m. The apparent resistivities are measured along the DDP method (device Dipole–Dipole). This device is very sensitive to horizontal changes in resistivity and therefore ideal for detecting vertical structures.

9.5.4.2 ERT Data Inversion and Interpretation

Thus, ERT data recorded in the field are apparent resistivities. To delineate the resistivity-depth image along the profile line using the ERT data, modeling of geo-electrical is performed. The ERT model consists of producing pseudo-meter depth profile (2D resistivity section) using an inversion modeling software: Res2Dinv. The iterative method gave us a geological model corresponding to the field data. 2D resistivity-depth images along the profile lines obtained by the inversion of the observed data are shown in Fig. 9.8b_1.

The resistivity of the Mio-Plio-Quaternary deposits varies within a wide range from 0.1 to 270 Ω·m. The variation of the resistivity shows a strong spatial variability of the Mio-Plio-Quaternary lithology. The lens structure of the Kairouan plain is very complex with successions of clays, marls, and sands. Figure 9.8b_2 shows the presence of lenticular structure that is characterized by resistivity varying from 70 to 200 Ω·m. This lens forms the shallow aquifer.

Referring to the location of this electrical profile (the eastern part of L3 seismic reflection profile which show the channels and taking into account the sand and gravel deposit), we conclude that the area highlights part of a channel deposit. This result confirms those obtained by seismic reflection and allows us to fully appreciate the presence of channel sedimentation located in the upper groundwater of Mio-Plio-Quaternary in the Kairouan region. This reservoir is covered by silt to clay layers marked by resistivity varying between 7 and 70 Ω·m (ABEM 2010). The soil is characterized by a saltwater zone with lower resistivity varying between 0.3 and 1 Ω·m (ABEM 2010), characterizing the lower permeability.

9.5.5 Hydrodynamics of the Shallow Aquifer in the Kairouan Plain

The superposition of the piezometric map of the humid season (Mars 2013) and the geological outcrops in the region are illustrated in Fig. 9.10. The piezometric map is drawn using the interpolation of 26 observation wells. The groundwater level varies from 120 in upstream to 25 m in downstream around Sebkhat El Kelbia. The piezometric map shows major sub-vertical water flow from SE to NW; from the El Houareb dam to Sebkhat El Kelbia and Sidi El Hani. The hydraulic gradient is ranging between 3.83‰ and 1.8‰, respectively, in the upstream and downstream.

Two sectors may be considered as recharge zones for the studied groundwater. The first one is the El Haouareb dam situated in the upstream region. The second one is assured by the infiltration from the Ain El Beidha aquifer. The two-water recharge origins are mixed in subsurface in the Karst reservoir with different proportion (Ben Ammar et al. 2006). The discharge fields are located in the Western direction.

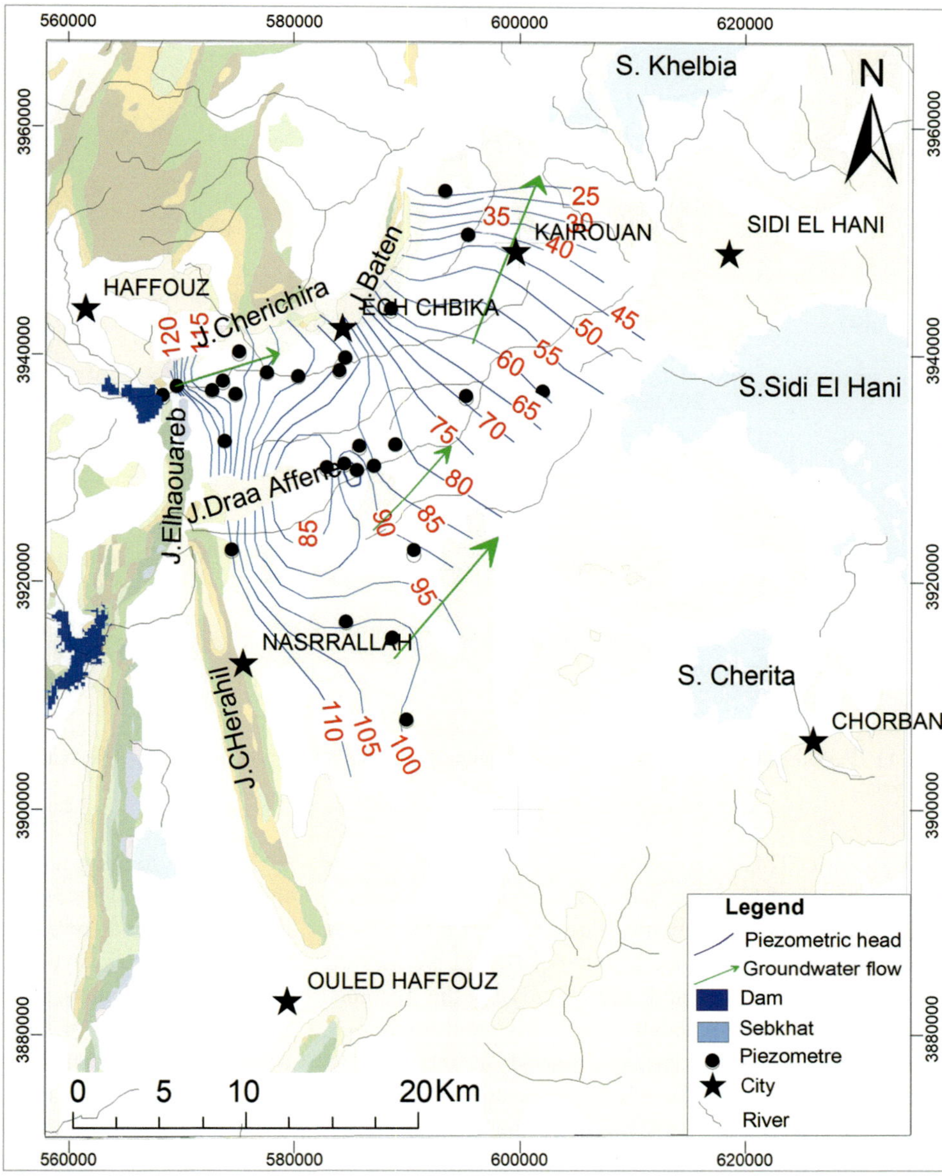

Fig. 9.10 Piezometric head map of Mio-Plio-Quaternary aquifer in the Kairouan plain (Mars 2013)

The recorded piezometric level over the 1968–2014 period revealed three piezometric behavior types (Fig. 9.11):

- A continuous piezometric drawdown: that was unregistered in all the monitoring wells (Fig. 9.11). The piezometric decrease varied from 0.44 m year^{-1} (12737/4 piezometer) to 1.45 m year^{-1} (20315/4 piezometer). The middle and the downstream areas are characterized by high piezometric drawdown. However, in upstream, the groundwater level was lower, due to the recharge influence from the El Hawareb dam and Ain El Bitha aquifer (Leduc et al. 2007). In the downstream region (12737/4 piezometer), the groundwater level has decreased by 19.4 m in 12737/4 piezometer and 24.61 m in 13264/4 piezometer during the last 44 years (from 1969 to 2013) because of strong pumping and a limited recharge. The piezometric drawdown is increasing during the last few years. In fact, and according to the 12727/4 piezometer the piezometric drawdown is about 0.17 m year^{-1} during the 1969–1996 period, 0.63 m year^{-1} during the 1996–2012 period and it growing to 4 m year^{-1} during the 2012–2013 period. The significant increase of the piezometric drawdown in the last period resulted from the important increase of illegal and unauthorized wells after the Tunisian revolution of September 2011–January 2012.
- The seasonal piezometric fluctuation was very important on the downstream, which represents the pumping area of the aquifer. It exceeds 1.91 m in the 20302/4 piezometer (Fig. 9.11a). During the humid season (September–March) the piezometric is higher, however, the dry season (April–August) is characterized by lower

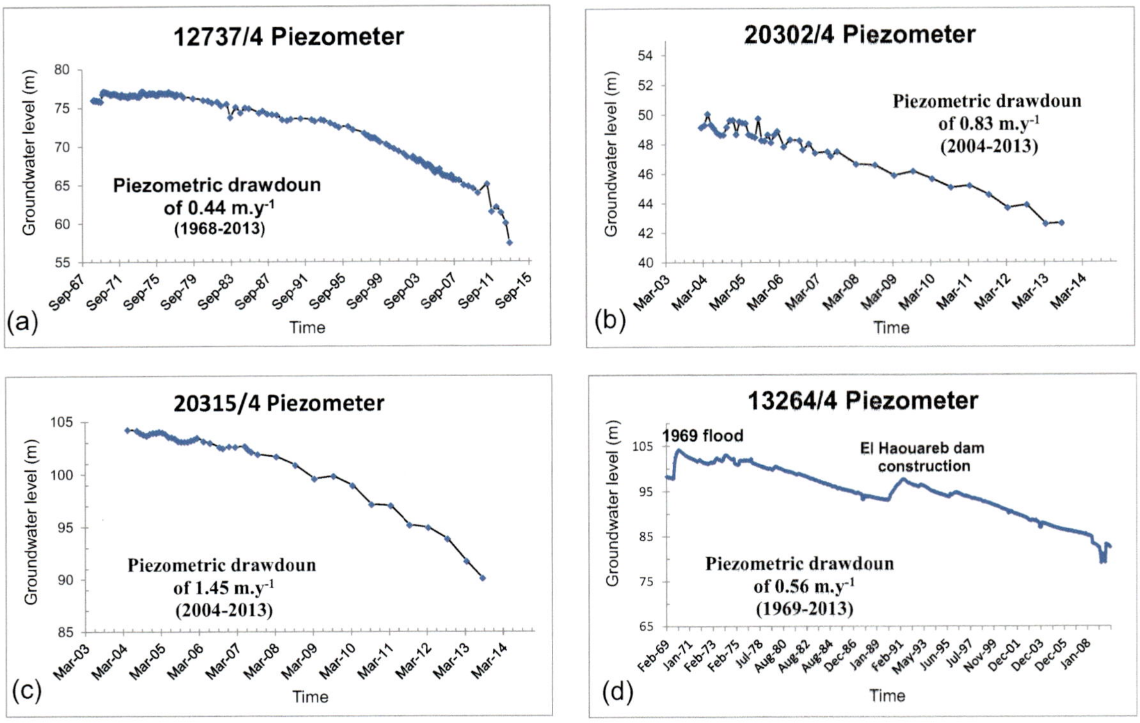

Fig. 9.11 Piezometric fluctuations of Kairouan aquifer: **a** 12737/4 piezometer, **b** 20302/4 piezometer, **c** 20315/4 piezometer, and **d** 13264/4 piezometer

piezometric. This piezometric fluctuation is related to the pumping effect for irrigation and rainwater infiltration.

- An occasional piezometric fluctuation: Leduc et al. (2007) show the presence of two occasional piezometric fluctuations in 1969 and 1989 studying the recorded groundwater level recorded in 13147/4 monitoring well. The same ascertainment is observed in the 13264/4 piezometer (Fig. 9.11d), which is located in upstream. The first one is caused by the 1969 flood, which represents typical exceptional infiltration, reflecting a strong inter-annual variability of precipitation (Chargui et al. 2013). The second occasional piezometric fluctuation is caused by the El Haouareb dam construction which increases the aquifer recharge (Leduc et al. 2007).

9.6 Conclusions and Perspectives

In this paper, different geophysical methods have been integrated (Gravity, seismic reflection, well logs, and ERT) in order to characterize reservoirs and water resources of the Mio-Plio-Quaternary aquifer system in the Kairouan plain. A geophysical study was undertaken to determine lateral and vertical reservoir extension, as well as to characterize the Mio-Plio-Quaternary aquifer.

Wireline logging analysis highlights the identification of the Mio-Plio-Quaternary multilayer aquifer system that consists of five reservoir aquifers levels separated by impermeable clay layers.

The interpretation of Gravity and seismic profiles permitted to establish of a consistent subsurface representation of reservoir architecture. In fact, sedimentary clastic deposits in the Kairouan plain are controlled and influenced by major deep faults such as the Kairouan-Sousse and El Hdadja faults. These structures have contributed to the reservoir levels distribution of Mio-Plio-Quaternary and their modes of arrangement. Mio-Plio-Quaternary series are compartmentalized into Sidi El Hani, Chorben, and Zeramdine blocks under the influence of the tectonic structuring. In addition to the upper Mio-Plio-Quaternary sandy level reservoir, other reservoir horizons were identified and correlated for the first time. And the base of the aquifer reservoirs was identified. In addition, the Eastern aquifer boundary was highlighted for the first time, which is the Bir El Taib fold.

Seismic-stratigraphic analysis of the Mio-Plio-Quaternary series shows the presence of channel sedimentation characterizing the Kairouan multilayer aquifers. This result was confirmed by ERT profile that shows sand and clay lenses corresponding to channel deposits. Downalp prograding horizons marked the base of the Ségui Formation which coincides to the base of the Mio-Plio-Quaternary aquifer system. The geometric characterization of the Mio-Plio-Quaternary aquifer system can be refined by an advanced study based on a dense seismic reflection grid and more wireline data.

The piezometric study in 26 piezometres shows that the principal groundwater flow is from west to east. In addition, a general drop in water table was defined using the groundwater water level measurements during the 1967–2013 period.

References

ABEM (2010) Instrument manual terrameter SAS 400/SAS 1000. ABEM Product 33 0020 26, ABEM printed Matter No 93109 (2010-06-04). ABEM Instrument AB, Sundbyberg, Sweden

Adhikary PP, Chandrasekharan H, Dubey SK, Trivedi SM, Dash ChJ (2015) Electrical resistivity tomography for assessment of groundwater salinity in west Delhi, India. Arab J Geosci 8(5):2687–2698

Aizebeokhai AP, Oyeyemi KD (2015) Application of geoelectrical resistivity imaging and VLF-EM for subsurface characterization in a sedimentary terrain, Southwestern Nigeria. Arab J Geosci 8 (6):4083–4099

Araffa SAS, El Shayeb HM, Hashesh MFA, Hassan NM (2015) Delineating subsurface structures and assessment of groundwater aquifer using integrated geophysical interpretation at the central part of Sinai, Egypt. Arab J Geosci 8:7993–8007. https://doi.org/10.1007/s12517-015-1824-5

Ayolabi EA, Oluwatosin LB, Ifekwuna CD (2015) Integrated geophysical and physicochemical assessment of Olushosun sanitary landfill site, southwest Nigeria. Arab J Geosci 8:4101–4115. https://doi.org/10.1007/s12517-014-1486-8

Bédir M, Zargouni F, Tlig S, Bobier C (1992) Subsurface geodynamics and petroleum geology of transform margin basins in the Sahel of Mahdia and El Jem (eastern Tunisia). Am Assoc Pet Geol Bull 76 (9):1417–1442

Bédir M (1995) Mécanismes géodynamiques des bassins associés aux couloirs de coulissement de la marge atlasique de la Tunisie: sismo-stratigraphie, sismo-tectonique et implications pétrolières. Thèse de Doctorat Es Sciences (unpublished), Université de Tunis, p 412

Bédir M, Tlig S, Bobier C, Zargouni F, Aïssaoui N (1996) Sequence stratigraphy, basin dynamics and petroleum geology of Miocene from the eastern Tunisia. Am Assoc Pet Geol Bull 80(1):63–81

Ben Ammar S, Zouari K, Leduc C, M'barek J (2006) Isotopic characterization of the dam–aquifer water transfer in the Merguellil catchment (Kairouan Plain, central Tunisia). Hydrol Sci J 51 (2):272–284. https://doi.org/10.1623/hysj.51.2.272

Ben Haj Ali M, Jedoui Y, Dali T, Ben Salem H, Memmi L (1985) Carte géologique de la Tunisie au 1/500.000. Office National des Mines, Service Géologique National, Tunis

Besbes M, Delhomme JP, De Marsily G (1978) Estimating recharge from ephemeral streams in arid regions: a case study at Kairouan, Tunisia. Water Resour Res 14(2):281–290

Buness H et al (1989) The EGT'85 seismic experiment in Tunisia: a reconnaissance of the deep structures. In: Freeman R, Muller S (eds) Sixth workshop on the European Geotraverse project, data compilations and synoptic interpretation. European Science Foundation, Strasbourg, pp 197–210

Burrolet PF (1956) Contribution à l'étude stratigraphique de la Tunisie centrale. Ann Mines Géol Tunis 18:350

Chandra S, Boisson A, Ahmed S (2015) Quantitative characterization to construct hard rock lithological model using dual resistivity borehole logging. Arab J Geosci 8(6):3685–3696

Chargui S, Gharbi H, Slimani M (2013) Runoff responses at different watershed scales in semi arid region: exploration of a developed rainfall runoff model (Merguellil and Skhira watershed, Central Tunisia). Earth Sci Inf 6:127–136

Chargui S, Jaberi A, Cudennec C, Lachaal F, Calvez R, Slimani M (2018) Statistical detection and no-detection of rainfall change trends and breaks in semiarid Tunisia-50+ years over the Merguellil agro-hydro-climatic reference basin. Arab J Geosci 11(675). https://doi.org/10.1007/s12517-018-4001-9

Chargui S, Lachaal F, Ben Khelifa W, Slimani M (2022) Trends in seasonal and monthly rainfall for semi-arid Merguellil basin, central Tunisia. Meteorol Atmos Phys 134:21. https://doi.org/10.1007/s00703-022-00859-9

Chekirbane A, Tsujimura M, Kawachi A, Lachaal F, Isoda H, Tarhouni J (2014) Use of time domain electromagnetic method (TDEM) with geochemical tracers to explore the salinity anomalies in a small coastal aquifer in north-east of Tunisia. Hydrogeol J 22 (8):1777–1794. https://doi.org/10.1007/s10040-014-1180-7

Gabtni H (2005) Apport de la gravimétrie à l'étude des structures profondes du Sahel de Tunisie (cas de la région de Kairouan-Sousse-Monastir). Gravity contribution on the deep structure study of the Tunisian Sahel domain (a Kairouan-Sousse-Monastir area case). C.R.A.S. Géosci 337:1409–1414

Gabtni H, Chulli Zenatti B, Jallouli C, Mickus KL, Bédir M (2010) The crustal structure of the Sahel Basin (eastern Tunisia) determined from gravity and geothermal gradients: implications for petroleum exploration. Arab J Geosci 4(3–4):507–516. https://doi.org/10.1007/s12517-010-0151-0

Gabtni H, Alyahyaoui S, Jallouli C, Hasni W, Lee Mickus K (2012) Gravity and seismic reflection imaging of a deep aquifer in an arid region: case history from the Jeffara basin, southeastern Tunisia. J Afr Earth Sci 66–67:85–97

Ghribi R (2010) Etude morpho-structurale et évolution des paléochamps de contraintes du Sahel tunisien: implications géodynamiques. Thèse en sciences géologiques. Thèse Doc. Es. Sc. Univ. Sfax, Tunisie, p 255

Houatmia F, Khomsi S, Malayah A, Andolssi M, Bedir M (2015) Neogene aquifer: geochemistry and structuring in the Sidi Saad basin, central Tunisia. Arab J Geosci 8(6):4221–4238

Karunanidhi D, Vennila G, Suresh M, Karthikeyan P (2014) Geoelectrical Schlumberger investigation for characterizing the hydrogeological conditions using GIS in Omalur Taluk, Salem District, Tamil Nadu, India. Arab J Geosci 7(5):1791–1798

Khomsi S, Bédir M, Ben Jemia GM, Zouari H (2004) Mise en évidence d'un nouveau front de chevauchement dans l'Atlas tunisien oriental de Tunisie par sismique réflexion. Contexte structural régional et rôle du Trias salifère. C R Geosci 336:1401–1408

Kingumbi A (2006) Modélisation hydrologique d'un bassin affecte par des changements d'occupation. Cas du Merguellil en Tunisie centrale. Thèse Doc. Génie Hydraul. ENIT. Univ. Tunis, Tunisie, p 218

Kumar R, Yadav GS (2015) Geohydrological investigation using Schlumberger sounding in part of hard rock and alluvial area of Ahraura region, Mirzapur district, Uttar Pradesh, India. Arab J Geosci 8(6):3645–3654

Lachaal F, Bédir M, Tarouni J, Ben Gacha A, Leduc C (2011) Characterizing a complex aquifer system using geophysics, hydrodynamics and geochemistry: a new distribution of Miocene aquifers in the Zéramdine and Mahdia-Jébéniana blocks (east-central Tunisia). J Afr Earth Sci 60:222–236. https://doi.org/10.1016/j.jafrearsci.2011.03.003

Lachaal F, Mlayah A, Bédir M, Tarhouni J, Leduc C (2012a) Development and application of three–dimensional groundwater flow numerical model to complex aquifer system in arid and semi-arid regions using MODFLOW and GIS tools: Zéramdine--Béni Hassen Miocene aquifer system (east–central Tunisia). Comput Geosci 48:187–198. https://doi.org/10.1016/j.cageo.2012.05.007

Lachaal F, Gabtni H, Bédir M, Tarhouni J (2012b) Seismic, gravity, and wire line logging characterization of the Zéramdine fault corridor and its influence in the deep Miocene aquifers distribution (east-central Tunisia). Arab J Geosci 5:1391–1398

Lachaal F, Chargui S, Jeballa N, Ayari K, Triki L, Gabtni H (2022) Adapting groundwater artificial recharge to global and climate change in water-stressed coastal region: the case of Ras Jebel aquifer (North Tunisia). Arab J Geosci 15(1202). https://doi.org/10.1007/s12517-022-10453-3

Leduc C, Ben Ammar S, Favreau G, Beji R, Virrion R, Lacombe G, Tarhouni J, Aouadi C, Zenati Chelli B, Jebnoun N, Oi M, Michelot JL, Zouari K (2007) Impacts of hydrological changes in the Mediterranean zone: environmental modifications and rural development in the Merguellil catchment, central Tunisia. Hydrol Sci J 52(6):1162–1178

Mahmoud HH, Barseem MSM, Youssef AMA (2015) Application of the two dimensional geoelectric imaging technique to explore shallow groundwater in wadi El Gerafi basin, Eastern Central Sinai – Egypt. Arab J Geosci 8(6):3589–3601

Metwaly M, Elawadi E, Moustafa SSR, Al Arifi N, El Alfy M, Al Zaharani E (2014) Groundwater contamination assessment in Al-Quwy'yia area of central Saudi Arabia using transient electromagnetic and 2D electrical resistivity tomography. Environ Earth Sci 71(2):827–835

Mohamed NE, Yaramanci U, Kheiralla KM, Abdelgalil MY (2011) Assessment of integrated electrical resistivity data on complex aquifer structures in NE Nuba Mountains – Sudan. J Afr Earth Sci 60:337–345

Morelli (1976) Modern standards for gravity surveys. Geophysics 41:1051

Nazoumou Y, Besbes M (2001) Estimation de la recharge et modélisation de nappe en zone aride: cas de la nappe de Kairouan, Tunisie. IAHS Publ 26:75–88

Plouff D (1977) Preliminary documentation for a FORTRAN program to compute gravity terrain corrections based on topography digitized on a geographic grid. US Geological Survey Open-File Report 77-535, Denver, Colorado, USA, p 111

Poormirzaee R, Moghadam RH, Zarean A (2015) Inversion seismic refraction data using particle swarm optimization: a case study of Tabriz, Iran. Arab J Geosci 8:5981–5989. https://doi.org/10.1007/s12517-014-1662-x

Row LW, Hastings DA, Dunbar PK (1995) TerrainBase: worldwide digital terrain data. US Texas, US Department of Commerce, National Geophysical Data Center, Boulder, p 163

Rabhi M (1999) Contribution à l'étude stratigraphique et analyse de l'évolution géodynamique de l'axe Nord-Sud et des structures avoisinantes (Tunisie centrale). Thèse Doc. Es. Sc. Univ. Tunis, Tunisie, p 217

Salah H (2013) Application of VES and TDEM techniques to investigate sea water intrusion in Sidi Abdel Rahman area, northwestern coast of Egypt. Arab J Geosci 6(8):3093–3101

Tamma Rao G, Gurunadha Rao VVS, Padalu G, Dhakate R, Subrahmanya Sarma V (2014) Application of electrical resistivity tomography methods for delineation of groundwater contamination and potential zones. Arab J Geosci 7(4):1373–1384

Thilagavathi R, Chidambaram S, Prasanna MV, Pethaperumal S (2014) A study on the interpretation of spontaneous potential and resistivity logs in layered aquifer sequence of Pondicherry Region, South India. Arab J Geosci 7(9):3715–3729

10 Use of Geochemical Tracers for the Characterization and Quantification of Water Leakage at the Joumine Dam Site, Tunisia

Mohamed Fethi Ben Hamouda, Souha Sari, and Luis Araguas-Araguas

Abstract

The Joumine reservoir, located in the northwest of Tunisia has an upstream watershed area of 418 km^2 and the reservoir capacity is 130 Mm^3. Shortly after the first filling of the reservoir in 1987, an important water leak was detected at the dam toe immediately after its construction. The emerging flow rate at the maximum level in the reservoir was close to 500 L/s. An important sinkhole was detected in a limestone block outcropping in the left abutment. The emergency work aimed at impermeabilizing the sinkhole by the injection of mine tailings lowered the flow rate of the leakage to about 120 L/s. The flow rate of the leakage was monitored in two drains D1 and D2, located at the dam toe. The decrease in the flow rate of the leakage due to emergency work was accompanied by an increasing salinity in drain D2, which was interpreted as a risk to the integrity of the dam. A combined hydrogeological, geochemical and isotopic investigation was carried out to determine the origin, flow path and other characteristics of leakage, integrating water level data from monitoring piezometers, water temperature and electrical conductivity, as well as hydrochemical and environmental isotopes. The study is a good example illustrating the usefulness of geochemical tracer methods for obtaining more precise and rapid information of the main features of water leakages, which if ignored may result in great repair costs or even affect the stability of a dam.

M. F. Ben Hamouda (✉) · S. Sari
CNSTN, Isotope Hydrology and Geochemistry Unit, Technopark of Sidi Thabet, 2020 Sidi Thabet, Tunisia
e-mail: f_benhamouda@yahoo.fr

M. F. Ben Hamouda
LRSTE, Institut National Agronomique de Tunisie, Tunis, Tunisia

L. Araguas-Araguas
IAEA, Isotope Hydrology Section, Vienna International Centre, P.O. Box 100 1400 Vienna, Austria

Keywords

Environmental tracers • Dam • Joumine reservoir • Drains • Isotopes • Leakage • Tunisia

10.1 Introduction

In arid and semi-arid regions, the growing water demand for irrigation, industrial uses and domestic water supply is often satisfied by the combined use of a network of irrigation systems fed by reservoirs and local groundwater. Millions of people depend on dams and reservoirs for domestic, industrial and agricultural water supply for flood protection and electricity. In the case of Tunisia, water transfer systems consist of pipelines and canals constructed to transfer surface water from dams in the northern part of the country to the coastal urban and farming areas, where the demand for water is high (Boiten 1992). The ratio of water supply from surface and groundwater is about 50/50 (Irie et al. 2009). About 97% of available surface runoff in Tunisia has already been used through the construction of reservoirs or planned to exploit (Louati and Bucknall 2010). Due to the difficulties of developing new surface water resources, it is necessary to optimize existing infrastructures to minimize water losses through leaks in dams and reservoirs. Therefore, sedimentation and leak control are important issues in sustainable surface water resource management.

One of the most efficient and least costly methods of detecting and characterizing leaks from dams and canals is the use of geochemical and isotope tracers to delineate the source and movement of water (Contreras and Hernández 2010). These methods will often reveal the flow path of the leak and will help in adopting engineering approaches to reduce the safety risks and water losses, instead of the costly method of repair. Such tests may also reveal that sudden changes in flow-rates downstream are not necessarily due to

S. Khomsi and F. Roure (eds.), *Geology of North Africa and the Mediterranean: Sedimentary Basins and Georesources*, Regional Geology Reviews, https://doi.org/10.1007/978-3-031-18747-6_10

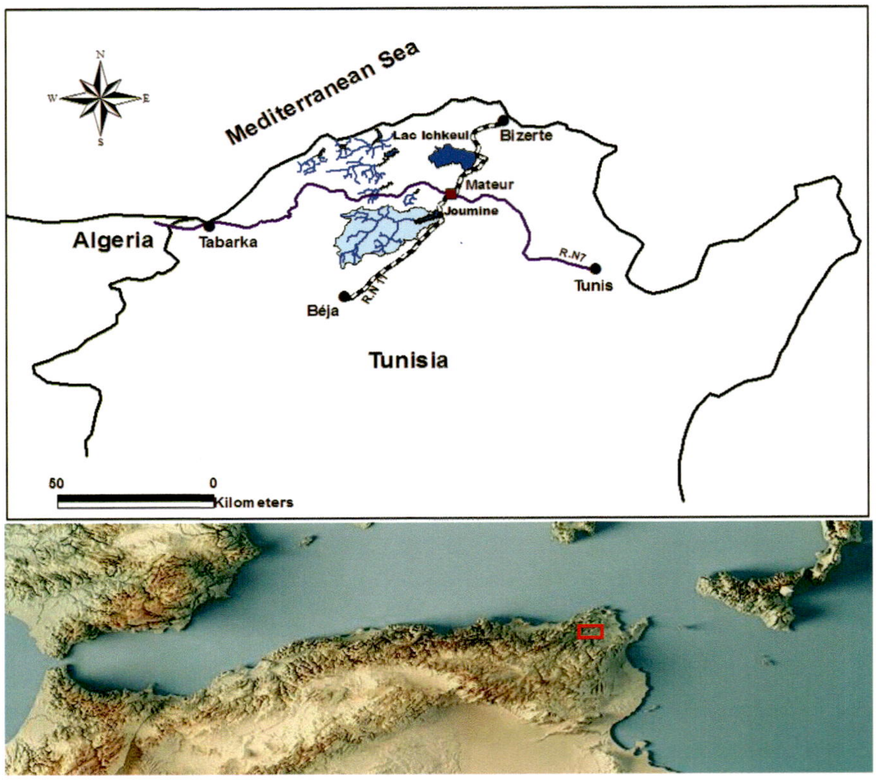

Fig. 10.1 Location of the study area

a leak from the reservoir, but may actually be the result of natural groundwater discharges unconnected with the dam.

The Joumine reservoir, located in the northwest of Tunisia (Fig. 10.1), constitutes an important water resource for the region and the sedimentation and leaks in the dam site have been the subject of previous studies (Plata and Araguas 2002; Irie et al. 2009, 2011).

The objective of the study is to illustrate the crucial role of isotope hydrology and related geochemical tools in characterizing the origin and flow paths of leaks from dams, contributing to the adoption of corrective measures to reduce water losses and/or minimize safety risks. The study presented in this paper is a good example of the combined use of hydrological and geochemical tools for the identification of infiltration sites and preferential pathways of water escaping from the reservoir (Ben Hamouda 2000; Plata and Araguas 2002).

10.2 Main Features of the Study Area

The climate of the study area is sub-humid; the average monthly temperature varies from 8 °C for the coldest period (January) to 23 °C for the hottest month (August). The wettest month is December and the driest is August. Average annual rainfall in the catchment area is about 730 mm but varies greatly: the extremes observed range from 402 mm/a in 1993 and 1131 mm/a in 2003. Potential evapotranspiration, calculated using the Riou method (Riou 1980), exceeds 1700 mm/a. The reservoir is located at about 60 km west of the capital city of Tunis (Fig. 10.1). The reservoir has an upstream watershed area of 418 km^2 and the capacity is 130 Mm^3 which ranks 5th in water storage capacity among all the water reservoirs in Tunisia (Fig. 10.2).

Water salinity in the reservoir is lower than most reservoirs in Tunisia, and due to this property, water stored in the Joumine is utilized not only for drinking purposes and irrigation (about 1800 ha) but also is mixed with water from other reservoirs to produce water of lower salinity. The Joumine reservoir has been in use since 1987. Water from the Joumine reservoir is mainly used for irrigation purposes as well as for water supply for the city of Tunis. The dam is an earth-filled dam with impervious clay nucleus (Fig. 10.3a).

The length of the dam is about 600 m, and its maximum height reaches 90 m. The two abutments of the dam are embedded in outcrops of Cretaceous limestone, separated by a thin marl layer. The Cretaceous limestone is underlain by Triassic formations (Fig. 10.3b).

Shortly after the first filling of the reservoir in 1987, an important leak was detected at the dam toe. The emerging flow rate reached a value close to 500 L/s at the maximum filled level of the reservoir. An important sinkhole was detected in a limestone block outcropping in the left abutment at a short distance from the uptake tower (Fig. 10.4). Most of the water emerging downstream infiltrated through

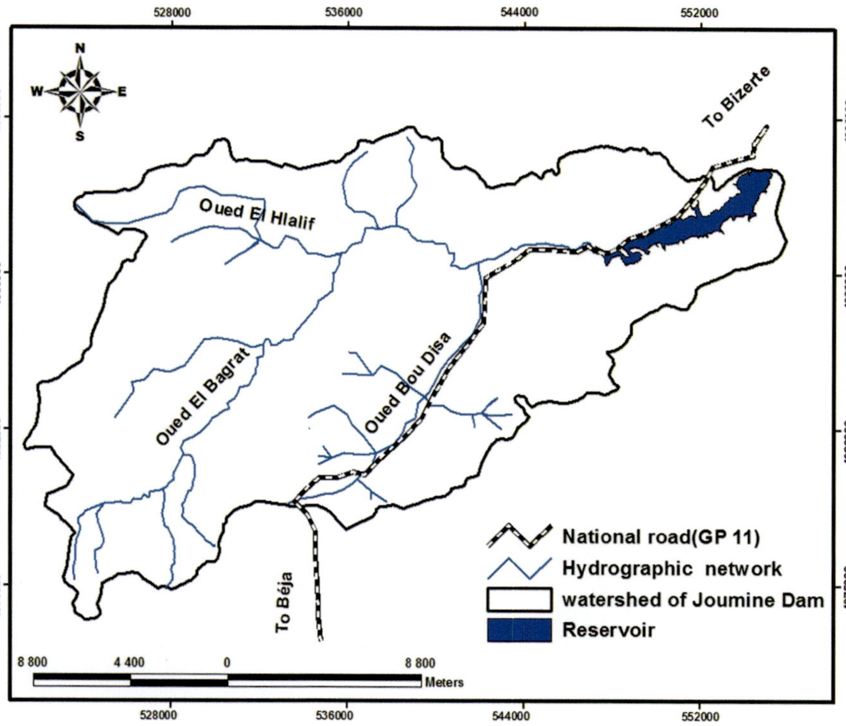

Fig. 10.2 Watershed Joumine dam

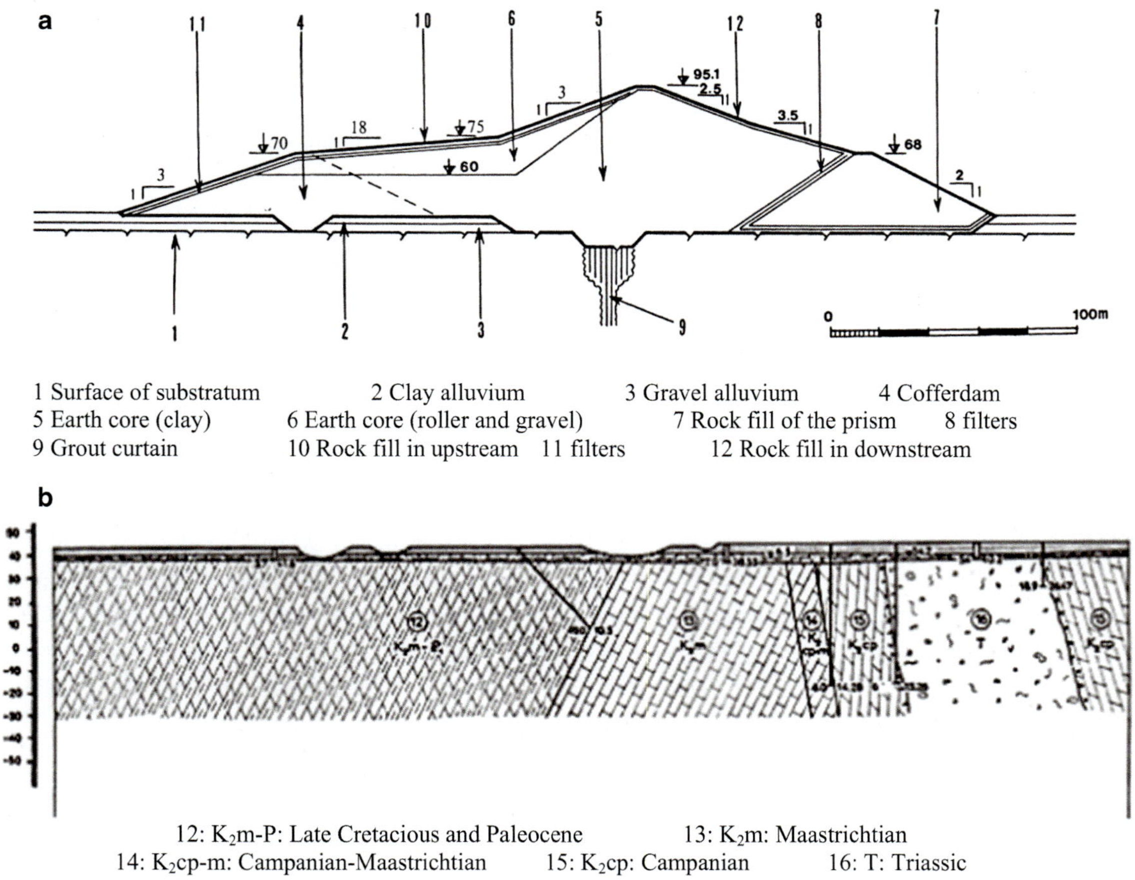

Fig. 10.3 **a** Cross section of the Joumine dam, **b** Cross section through the line IV-IV in the Joumine dam

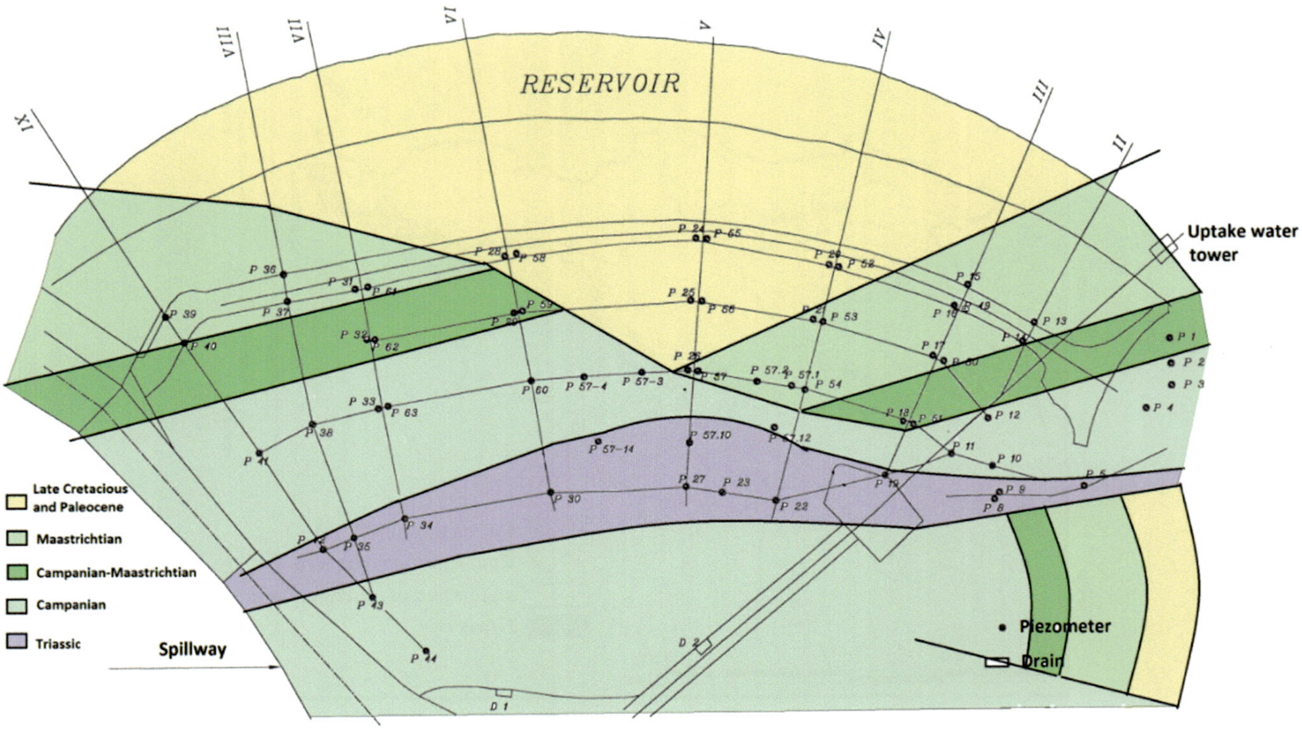

Fig. 10.4 Geologic map of the dam and groundwater monitoring and sampling well location

this visible sinkhole. The technicians involved in the leakage study in the 1980s decided to undertake corrective measures aiming at impermeabilizing the sinkhole by filling it with about 2000 tons of mine tailings from a blast furnace. The result of the mine tailings injection was that the escaping flow rate was lowered to about 120L/s at the maximum level of the reservoir (reduction of 75% of leakage). After this corrective measure, the water leakage emerged in two drains located at the dam toe (drains D1 and D2) (Fig. 10.4). Forty deep piezometers (numbered 1 to 48) were installed in the downstream slope to control the aquifer and another twenty piezometers (numbered 49 to 63) to control water leaks and hydraulic grade line in the dike (Ben Hamouda 2000, 2013; Sari 2013).

10.3 Material and Methods

The first part of this investigation of the leakage problem in the Joumine reservoir was carried out in 1998 within the framework of an international workshop on dam leakage studies organized in Tunis by the International Atomic Energy Agency (IAEA) and the National Centre of Nuclear Science and Technology of Tunisia. For simplicity, the individual values of these parameters obtained at each borehole have been omitted. The flow rates measured in drains D1 and D2 by means of the tracer dilution technique were, respectively, 45 and 58 L/s.

Previous hydrochemical data of surface and groundwater in the area was obtained as part of a reconnaissance study in 1964. For this study, the geographical position, depth to the water table, pH, electrical conductivity and temperature were measured in 62 sampling points in the study area (Fig. 10.4) during field campaigns carried out in 1998 and 2013 (Table 10.1). Samples were also collected for analysis of major ions, stable isotopes of water (^{2}H and ^{18}O), carbon-13 in total dissolved inorganic carbon (TDIC) and tritium (Table 10.1). Figures 10.5, 10.6 and 10.7 present, respectively, the results obtained from the measurements of potentiometric levels, conductivity and temperature profiles in all accessible piezometers at the dam site (Fig. 10.4).

The chemical analyses of the water samples were carried out at the Isotope Hydrology and Geochemistry Unit at the CNSTN (Centre National des Sciences et Technologies Nucléaires, Tunisia). Major cations (Ca, Mg, Na and K) were analyzed in filtered samples using an atomic absorption spectrometer with a furnace. Major anions (Cl, SO_4 and NO_3) were analyzed in filtered samples using a Dionex DX 120 ion chromatograph equipped with AG14 and AS14 Ion Pac columns and an AS-40 auto-sampler. The electrical charge balance between major anions and cations is in most samples better than $\pm$ 5%.

Isotope analysis of the water samples (^{2}H and ^{18}O) was conducted at the Isotope Hydrology Laboratory (IAEA), Vienna, Austria, through the technical cooperation project RAF/8/028. Stable isotopes of oxygen and hydrogen were

Table 10.1 Analysis of major ions, stable isotopes of water using samples

Sample name	Water level	Piezometry level	pH	T (°C)	EC mS/cm	TDS	Ca	Mg	Na	K	SO_4	Cl	NO_3	HCO_3	$\delta^{18}O$ %	$\delta^{18}O$ %	^3H TU	$\delta^{18}O$ %
						(Mg/L)												
R	–	–	8,1	17,1	0,682	437	74	22,32	29,9	2,34	95,69	63,71	6	200,88	−3.78	−23.3	5.01	−10.35
D1	–	–	7,8	14	0,716	472	72,40	34,56	32,2	1,95	102,95	70,15	4	281,48	−3.89	−23.8	5.36	−12.3
D2	–	–	7,7	13,6	1,3	951	71,20	32,64	135,7	1,56	154,26	205,69	4,8	258,54	−4.04	−24.5	4.65	−11.25
P1	9.87	107.12	8.4	21.6	3.01	1957	311	98.52	259.9	22.62	388	530.47	749.72	68.2				
P2	18.1	100.90	8.4	21.1	4.04	2626	367	121.68	418.6	20.28	692	690.32	759.46	77.5	−5.64	−30.6	1.27	
SJD	–	–	7.9	22.3	0.64	416	59.2	33.6	25.3	2.34	100.95	69.31	34.62	99.2				
S. Krips	–	–	7.1	22.7	1.55	1008	94.4	44.16	154.1	4.29	128.56	233.58	12	347.82				
P4	32.12	87.78	–	–	3.59	2334	249.2	115.44	478.4	12.48	648	716.13	125.34	409.2				
P5	–	–	–	–	–	–	63,20	32,4	128,8	4,29	143,7	153,1	13	323,02				
P8	16,18	61,45	7,3	19,9	9,22	6755	204	267,12	1138,5	21,06	1038,59	1518,37	12	698,74				
P9	31,82	45,76	–	20.3	5,86	–	–	–	–	–	–	–	–	–				
P10	–	–	–	–	–	–	–	–	–	–	–	–	–	–				
P11	22,21	51,43		19,1	7,5	–	–	–	–	–	–	–	–	–				
P12	18,93	63,93	8,1	19,3		3856	292,80	71,28	692,3	15,99	1009,48	802,41	12	330,46	−5.17	−30.4	1.29	−2.63
P13	–	–	–	–	–	–	–	–	–	–	–	–	–	–				
P14	9.63	86.57	–	–	–	–	–	–	–	–	–	–	–	–				
P15	–	–	–	–	–	–	–	–	–	–	–	–	–	–				
P16	43.88	52.48	7.4	19.4	0.83	540	90.4	35.76	46	3.51	155.10	100.86	10	195.92				
P17	25.75	55.37	8.4	19.5	0.424	276	50	26.4	50.6	11.31	10.35	86.69	29.42	257.3				
P18	20.91	48.07	7.9	21,8	1.33	865	117.6	50.76	163.3	5.85	269.02	212.90	15	350.92				
P19	8,87	42,17	8.1	15,8	0,832	585	50	32,88	71,3	20,67	129,26	101,11	7,5	200,88				
P20	26.39	70.06	8,1	19.1	0.43	280	51	29.76	57.5	7.41	7.08	106.69	14	264.74				
P21	23.66	58.06	8.3	17.6	0.422	274	47.8	22.56	59.8	7.02	3.76	110.63	24.98	196.54	−4.01	−24	5.18	
P22	7,72	41,31	6,7	17,9	1,165	842	304	176,16	372,6	13,26	153,05	1534,18	49,49	257,3	−4.46	−27.1	4.88	
P23	7,47	41,18	–	14,2	1,06	–	–	–	–	–	–	–	–	–				
P24	48.77	47.80	8.3	20.1	1.418	922	102.2	33	246.1	8.19	266.11	296.39	50	225.68	−5.28	−27		
P25	38.51	43.25	–	–	–	–	–	–	–	–	–	–	–	–				
P26	+ 6.5	75	7.9	22.6	0.6	390	84.4	38.52	34.5	2.73	123.57	85.25	14	288.92				
P27	8,16	40,94	7,6	14,9	1,188	792	304	176,16	372,6	13,26	153,05	1534,18	49,49	257,3				

(continued)

Table 10.1 (continued)

Sample name	Water level	Piezometry level	pH	T (°C)	EC mS/cm	TDS (Mg/L)	Ca	Mg	Na	K	SO_4	Cl	NO_3	HCO_3	$\delta^{18}O$ %	$\delta^{18}O$ %	^3H TU	$\delta^{18}O$ %
P28	16.56	79.98	8.3	19.8	3.95	2568	247.4	107.64	641.7	8.19	855	915.27	9	116.56				
P29	39.3	42.57	–	–	–	–	–	–	–	–	–	–	–	–				
P30	10.27	40.24	7,3	14,8	1,997	1465	124	70,08	223,1	4,29	363,92	231,29	6	628,68				
P31	21.23	75.02	8.3	19.5	1.069	695	73.2	37.2	138	2.34	226	206.19	12	83.7				
P32	31.69	49.77	7.4	23.2	1.45	943	145	54.84	158.7	3.12	276.66	307.87	10	350.92				
P33	28.73	39.86	7.8	18.4	7.48	4862	337	101.28	1359.3	40.56	892	1804.39	10	661.54				
P34	10,26	39,04	7,5	16,7	1,173	810	108,80	34,8	112,7	3,51	152,93	152,75	51,35	360,84				
P35	8	43,40	7,1	18,9	2,67	1988	339,20	112,8	135,7	4,29	975,55	175,27	9	254,82				
P36	–	–	–	–	—	–	–	–	–	–	–	–	–	–				
P37	10.43	85.58	8.3	19.4	0.953	619	64.8	25.32	163.3	4.29	216	195.98	10	84.94				
P38	18.65	49.89	8.2	19.2	2.76	1794	192.6	53.88	473.8	23.4	468.78	628.01	124.97	356.5				
P39	6,15	89,04	7,3	20,5	1,364	810	36	27,36	193,2	8,97	127,82	278,29	50,15	88,04				
P40	10,4	85,61	7,2	20,6	0,661	365	28,80	21,36	89,7	3,51	66,62	144,45	24,98	103,54				
P41	12	56,38	–	–	–	–	–	–	–	–	–	–	–	–	−5.41	−31.3		
P42	12,1	43,87	7,7	20,2	5,78	3975	244,80	160,32	655,5	44,46	685,18	992,84	280,45	251,1	−5.42	−29.1	1.24	−8.99
P43	6	43,06	6.9	20,3	16.59	10784	514.4	222.6	1688.2	437.58	2886.25	2145.00	13	739.04				
P44	6,5	38,75	6.6	19,3	5.5	3575	297	134.64	632.5	3.12	683.73	1161.63	12	336.04				
P49	34,59	14,87	7,4	19,1	1,580	1212	98,40	42,72	121,9	7,41	26,37	246	5	394,94	−4.97	−29.4		
P50	16	65,16	7,2	18,7	0,940	562	92	53,28	27,6	0,78	32,02	81,8	25,11	441,44				
P51	14,88	54,11	7,2	18,5	0,645	433	77,20	47,52	18,4	1,17	145,44	52,88	25,38	253,58				
P52	17,72	78,72	7,6	18,9	1,110	622	64	28,08	32,2	6,63	69,04	147,98	2	124				
P53	35,48	46,25	7,1	16,1	1,538	984	72,80	29,28	48,3	19,11	28,33	146,79	50,19	174,84				
P54	25,67	43,13		16,7	5,3	–	–	–	–	–	–	–	–	–				
P55	33,39	63,21	7,9	19,8	7,37	4783	406,40	133,44	837,2	5,07	544,03	1646,78	250,78	67,58	−4.25	−21.4	5.83	
P56	–	–	–	–	–	–	–	–	–	–	–	–	–	–				
P57	6,27	62,40	–	–	–	–	–	–	–	–	–	–	–	–	−4.32	−26.8	4.84	−13.95
57-1	26,5	41,00	6,6	13,3	0,764	519	64,40	40,32	39,1	3,51	129,75	91,31	9	175,46	−3.89	−23.6		
57-2	22,92	44,77	6,7	13,2	0,382	352	58	29,76	29,9	3,9	116,17	86,83	11	145,7	−3.87	−21.6	4.81	
57-3	27,05	40,85		14,4	0,723	–	–	–	–	–	–	–	–	–				
57-4	27,45	40,50		13,3	0,747	–	–	–	–	–	–	–	–	–	−3.84	−23.7		

(continued)

Table 10.1 (continued)

Sample name	Water level	Piezometry level	pH	T (°C)	EC mS/cm	TDS	Ca	Mg	Na	K	SO_4	Cl	NO_3	HCO_3	$\delta^{18}O$ %	$\delta^{18}O$ %	^3H TU	$\delta^{18}O$ %
						(Mg/L)												
57-10	9	40,51	7.8	14	0.536	348	51	32.16	32.2	1.95	133.42	83.58	14	55.18				
57-12	8	41,1	7.6	16	2.19	1424	119.2	44.88	285.2	2.73	199.55	385.08	12	327.36				
57-14	–	–	–	–	–	–	–	–	–	–	–	–	–	–	−3.84	−22.9		
P58	28,1	68,54	7,7	20,2	8,18	5807	410,80	139,92	1016,6	6,63	523,2	1864,76	250,28	217				
P59	33,96	47,95	7,9	17,5	3,67	2767	226	106,8	202,4	5,07	17,65	785,61	124,94	203,98				
P60	28,4	40,55		13,8	0,712	–	–	–	–	–	–	–	–	–				
P61	44,4	51,92	7,7	20,2	4,4	3130	284	106,32	489,9	3,9	508,8	968,19	124,98	217				
P62	21,98	59,41	7,7	18,7	0,204	216	28	30	20,7	5,07	49,38	72,96	25	69,44				
P63	27,55	41,04	–	15,8	1,192	–	–	–	–	–	–	–	–	–				

determined by isotope ratio mass spectrometry. The $\delta^{18}O$ values in samples were analysed via equilibration with CO_2 at 25 °C for 24 h (Epstein and Mayeda 1953) and for the δ^2H values via reaction with Zn at 450 °C (Coleman et al. 1982). The results are reported as $\delta^{18}O$ and δ^2H, where $\delta = ((R_{sample}/R_{standard})-1) \times 1000$. Both $\delta^{18}O$ and δ^2H values were determined relative to internal standards that were calibrated using IAEA SMOW standards. Data were normalized following Coplen (1988) and are expressed in the V-SMOW-SLAP scale. Samples were measured at least in duplicates and the precision of the analytical measures is ± 0.1‰ for $\delta^{18}O$ and ± 1‰ for δ^2H. Tritium (^3H) was also measured at the IAEA's Isotope Hydrology Laboratory in Vienna, Austria. Tritium contents were determined by liquid scintillation counting after electrolytic enrichment of the samples (Taylor and Schwarz 1977). Tritium concentrations are reported in Tritium Units (TU). One TU is defined as the isotope ratio $^3H/^1H = 10^{-18}$.

10.4 Results and Discussion

10.4.1 Water Levels in the Dam Site

Water level measurements at piezometers provide valuable information on the local groundwater flow and hydraulic connections in the dam and the surrounding areas. The potentiometric contour lines shown in Fig. 10.5 indicate that the subsurface waters emerging at both drains are related to the infiltration occurring at a zone of the reservoir close to the uptake tower, that is, from the zone where the sinkhole was initially found. The piezometric high observed in the zone of piezometer P57 is explained by the infiltration of local precipitation into the two affected piezometers. Local precipitation remained stored inside the casing of these piezometers due to the partial clogging of their screens, which are located at the bottom. The piezometric level outside this casing is normal. The low conductivity of the

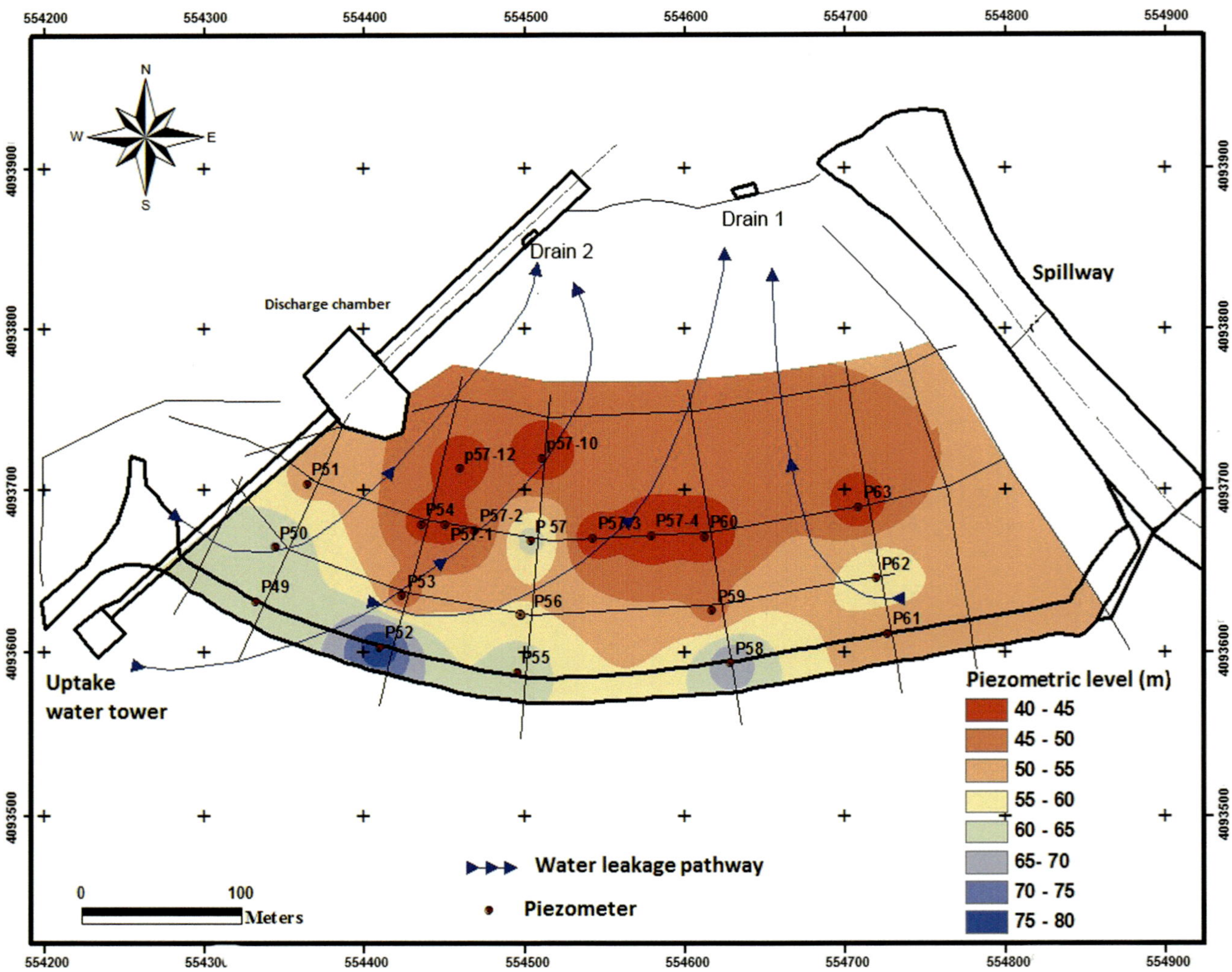

Fig. 10.5 Piezometric map at Joumine dam

waters of these piezometers confirms the previous statement. The arrows marked in the figure show the direction of the preferential pathway of the water escaping from the reservoir (Sari 2013).

10.4.2 Electric Conductivity

The distribution map of the EC values in the subsurface waters is shown in Fig. 10.6. The EC of the reservoir water was 0.73 mS/cm. The EC value of the water emerging at drain D1 was 0.76 mS/cm, that is, very similar to the electrical conductivity value found in the reservoir. On the contrary, the EC of drain D2 was 2.34 mS/cm. Initially, it was not clear whether the higher salinity found in D2 was due to the dissolution of material used for the dam construction or if it was a consequence of groundwater discharging from the underlying Triassic formations. The various geochemical tracers demonstrated that the latter hypothesis was the valid one. Important groundwater flows were determined in some piezometers where the electrical conductivity of water was higher than 4 mS/cm, using the single-well tracer technique described later in paragraph 4.3. Moreover, the analysis of historic hydrogeological data of the study area from the 1960s showed that before the dam construction, springs with high salinity water were present at this site.

The distribution map of EC (Fig. 10.6) clearly reflects the preferential flow path from the reservoir toward the dam toe. This flow path coincides satisfactorily with the flow line defined by the potentiometric data. Moreover, this preferential flow path coincides also with the former bed of the river Joumine before the dam construction. The lower EC values found at piezometers P19 and P51, which are located close to the discharge chamber, are derived from leaks of the neighboring discharge pipeline.

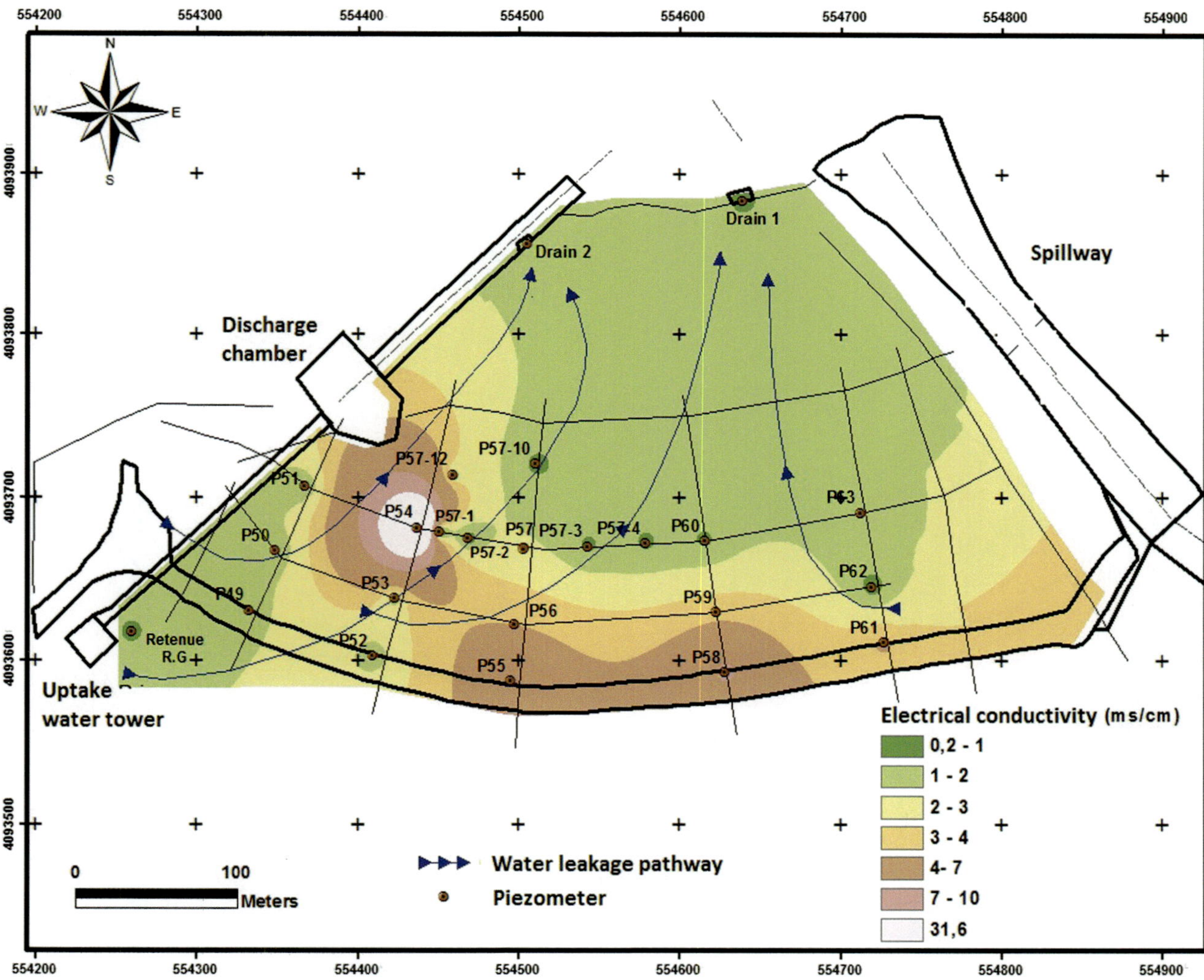

Fig. 10.6 Distribution of EC values at the Joumine dam expressed in mS/cm at 25 °C

10.4.3 Determination of Magnitude of Groundwater Contribution

The magnitude of groundwater flow inside the piezometers was measured using tracers to label the whole water column (Plata and Araguas 2002; Bolève et al. 2011). As expected, a rapid groundwater flow velocity was found in all piezometers located along the preferential pathway. The tracer injected in these boreholes (saturated solution of common salt) escaped almost instantaneously. The interpreted results showed that groundwater at the dam site related to the Triassic formations had EC values ranging from 4 to 5 mS/cm. Therefore, it was concluded that subsurface water emerging at drain D2 was a mixture of this deep groundwater from the Triassic formations with reservoir water having an EC equal to 0.73 mS/cm. Assuming for the Triassic water an average value of 4.5 mS/cm, the contribution (expressed as a percentage) of this component to the flow rate emerging at drain D2 is given by the expression: $(2.34-0.73)/(4.5-0.73) = 0.43$, that is to say, 43%. This percentage corresponds to a flow rate of $0.43 \times 45 = 19.4$ l/s (Drost and Neumaier 1974; Plata and Iraguen 1992).

10.4.4 Temperature

Water temperature is one of the best tracers for leak investigations at reservoir. The distribution of water temperature in the study area is shown in Fig. 10.7. The temperature of the reservoir water was 14.5 °C, measured at the surface. Due to the fact that the study was conducted during the cold season (December), no significant thermal stratification existed in the reservoir. Therefore, this temperature can be considered representative for the whole water column of the reservoir. On the contrary, the temperature values measured at the piezometers ranged from 15.8 to 21.6 °C. The piezometers located along the preferential pathway of the water escaping from the reservoir showed temperature

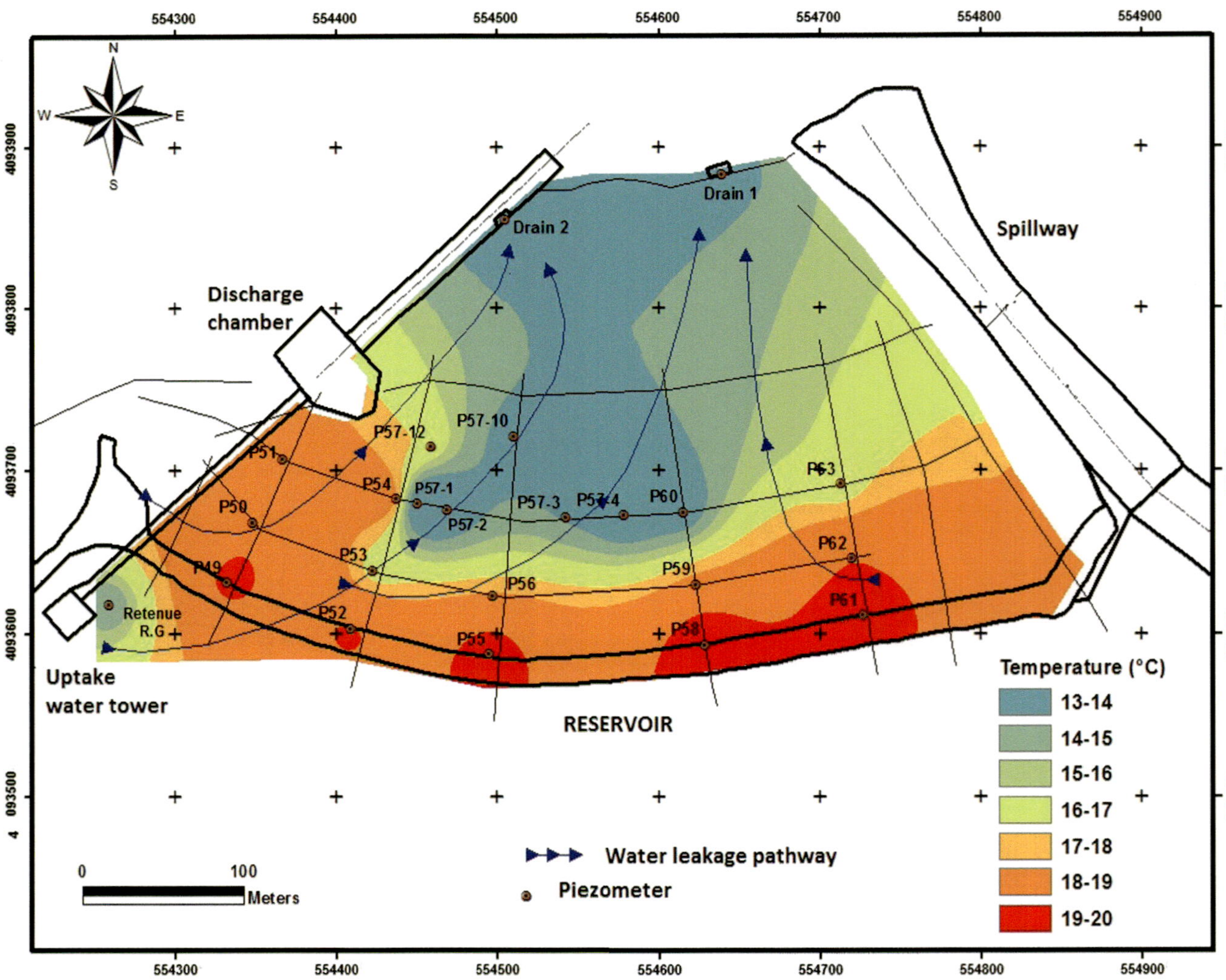

Fig. 10.7 Distribution of temperature in °C values at the Joumine dam

values ranging from 15.8 to 16.7 °C. An increasing gradient of temperature in the direction of the flow is also observed at these piezometers. Outside this preferential pathway, the temperature values were higher due to the presence of stagnant water or of water not related to the reservoir. The lowest temperature value was obtained for piezometer P52, indicating that this borehole was preferentially connected to the reservoir (Fig. 10.7). The difference in temperature between the reservoir and the piezometers located at the preferential flow area can be explained by heat exchange with soil materials. These materials could still show a higher temperature due to the heat stored during the previous months, when the temperature of the reservoir water was much higher (Plata 1983).

10.4.5 Water Quality

The study of water quality changes in the piezometers can help in assessing the integrity of the embankment (Lee et al. 2005). The salinity content of groundwater samples ranges from 360 to 6600 mg/L (the highest values are measured in D2 after injection of mine tailings). The chemical composition of groundwater plotted in a Piper diagram (Fig. 10.8) shows waters belonging to various chemical types depending on their origin. The computation of saturation indices (SI) using WATEQF (Plummer et al. 1976) shows a progressive increase in the saturation index in gypsum, with an increase in SO_4 concentrations (Fig. 10.9), contributing to an increase in the salinity of water (Turkmen 2010).

An abnormal leakage observed in February 1985 at D1 and D2 was accompanied by a decrease of total dissolved salts (TDS) in D1 and D2 (Fig. 10.10). Groundwater discharge with high salinity in D1 and D2 was reported at this site before the dam construction. These relative saline groundwaters present TDS contents above 6000 mg/L. After the impermeabilization work in the left abutment, waters measured at D1 and D2 showed lower TDS values (<1000 mg/L) due to the contribution of water from the reservoir ($\sim$600 mg/L) (Fig. 10.10). After the mine tailings injection, the decrease of leakage was accompanied by an increase of TDS, which stabilized at about 1800 mg/L. This increase in salinity is expected, and the TDS value will depend on the mixing ratio (dam leakage/water from aquifer).

10.4.6 Tracer Injection

Two interconnection experiments were carried out in the study area for the estimation of flow rates in both drains D1 and D2. Small quantities of uranine and rhodamine WT were injected in piezometers P57.2 and P57.3, respectively, and located a few tens of meters downstream of the dam at a distance of only 3 m from each other. The rock formation consists of karstified limestone. Tracer tests in piezometers using the labeling technique of the entire water column described in Sect. 10.4.3 were conducted in both boreholes (Adar 1994; Benischke 1987; Mozafari et al. 2012). A high flow velocity was detected at the bottom of both boreholes,

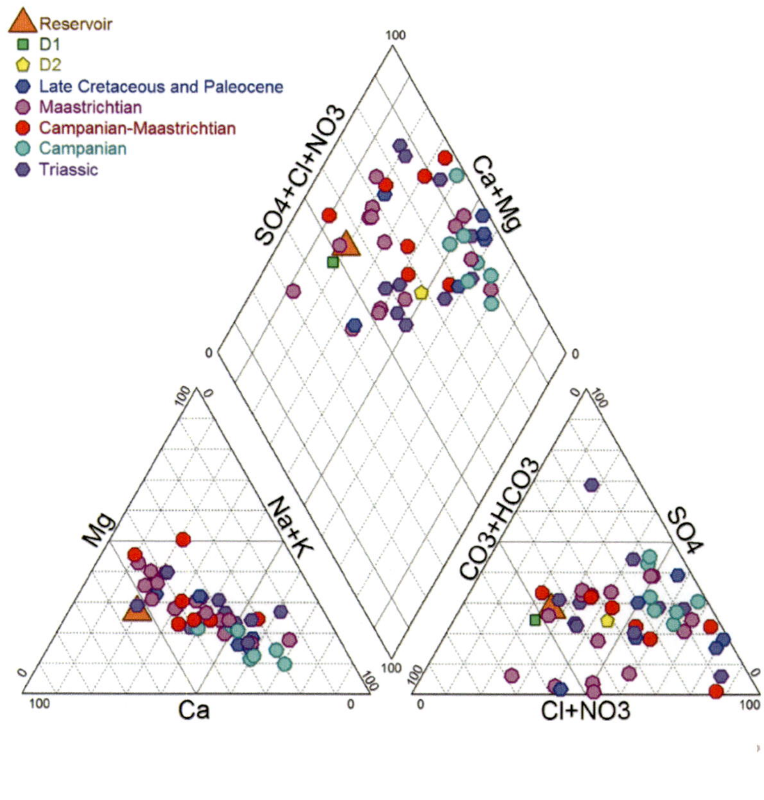

Fig. 10.8 Piper diagram

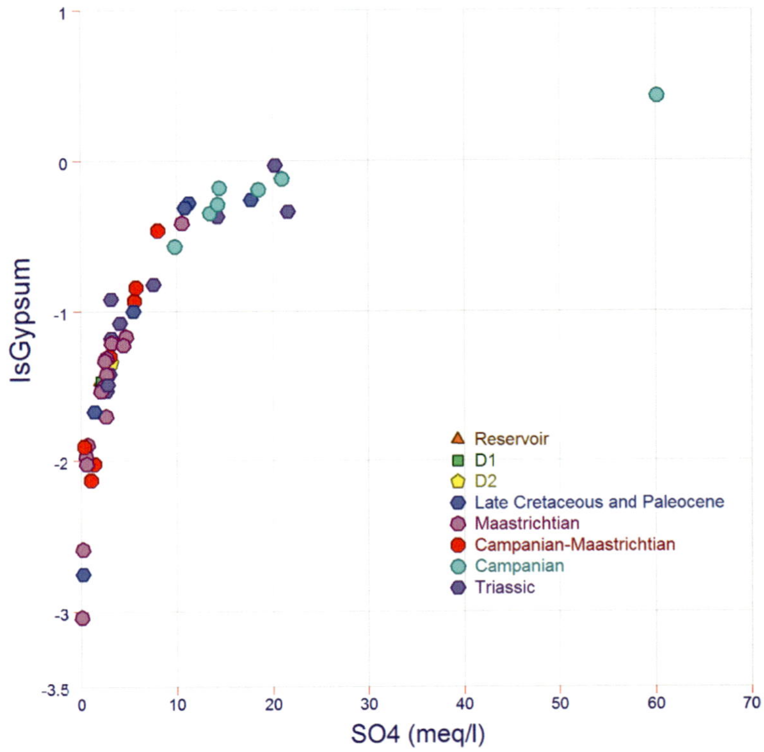

Fig. 10.9 Plot of saturation indices versus SO4

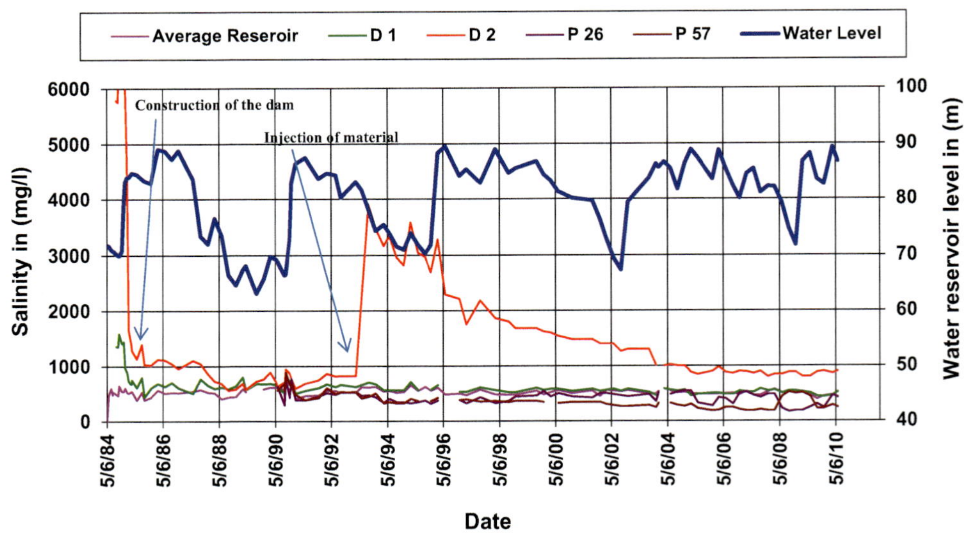

Fig. 10.10 Evolution of salinity and water level in reservoir from 1985 to 2010

where the tracers were injected. Two springs located at a distance of about 200 m from these boreholes were used to obtain the breakthrough curves shown in Fig. 10.11. Approximately 95% of the injected uranine was recovered in drain D2 and about 5% in drain D1. Therefore, the uranine recovery was about 100%. The transit time of uranine from P57.3 to drain D2 was about 4 h and about 20 h for drain D1 (Fig. 10.11). On the contrary, most of the rhodamine WT was recovered at drain D1 (only about 65%) with a transit time of about 10 h. The percentage of recovery was much lower than in the case of uranine, and the tracer breakthrough curve showed a long tail (Fig. 10.11). This was interpreted as a consequence of the strong interaction of rhodamine WT with the rock formations and soil material. This interaction was much lower in the case of uranine.

These results indicate, as expected, a more conservative behavior of uranine with respect to rhodamine WT under the existing experimental conditions. Rhodamine B and WT are not recommended for quantitative groundwater tracing experiments in the presence of fine-grain materials.

Fig. 10.11 Breakthrough curves obtained in drains D1 and D2

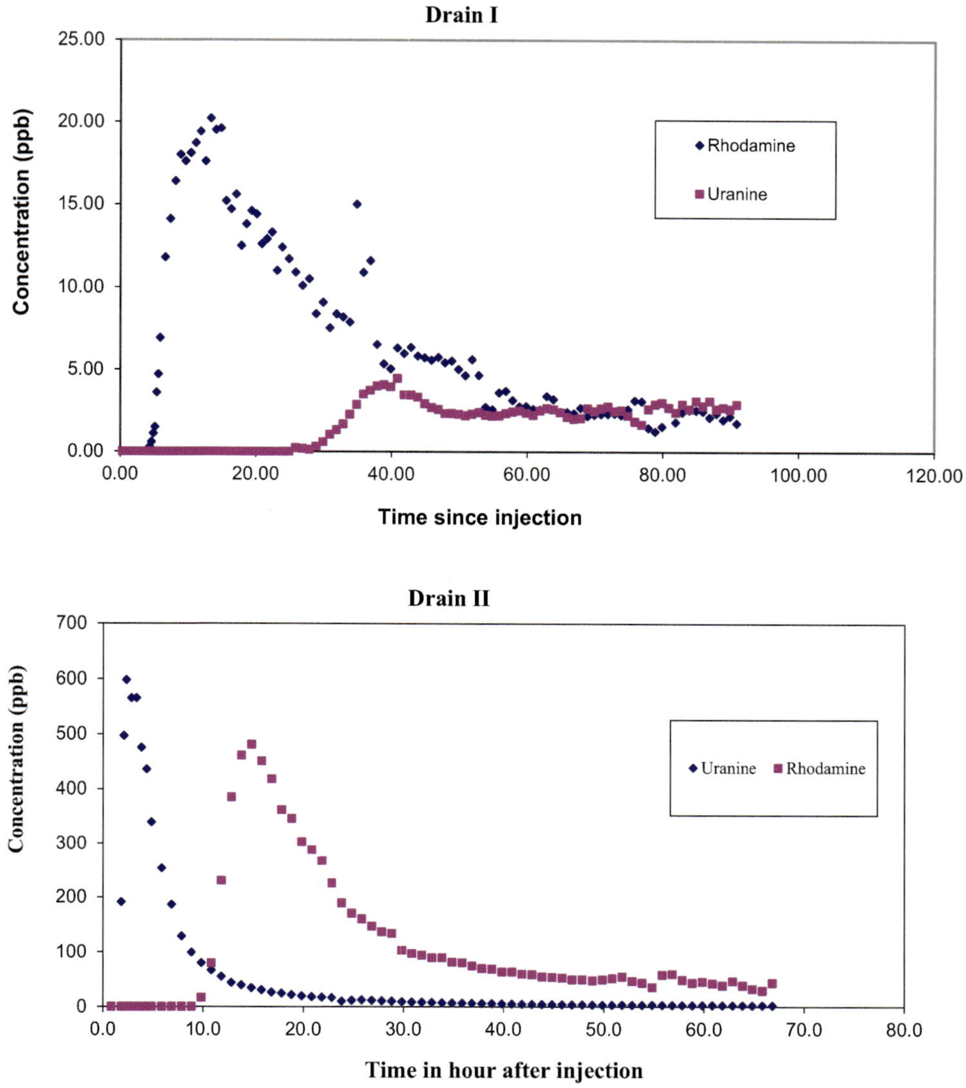

10.4.7 Stables Isotopes of Water (^2H, ^{18}O)

Stable isotopes of hydrogen and oxygen are very suitable tracers for hydrogeological investigations because (1) they are part of the water molecule and follow their behavior through the hydrological cycle; (2) they are conservative in most low-temperature environments (Aggarwal et al 2010, Ben Hamouda 2013; Clark and Fritz 1997; Plata and Araguas 2002; Peng and Wang 2008). The measured isotopic composition of groundwater in the study area is compiled in Table 10.1. The plot of $\delta^{18}O$ versus δ^2H (Fig. 10.12) reveals that groundwater close to the reservoir has a stable isotope composition which is indicative of an appreciable contribution of water coming from reservoir (P57, P17, P21, P22 and P55). Moving away from the reservoir the δ values become more depleted, owing to the contribution of infiltrating precipitation which also recharges the local groundwater system of piezometers P42, P41, P12 and P49. Leakage water in D1 and D2, in addition to reservoir water R with piezometers: P57, P17, P57-2 and P55 show a similar isotopic composition and hence reflecting the same origin. Regression line that best fit the aforementioned group has a slope of 4.36, which reflect the effect of evaporation and/ or mixing on δ values. This cluster P57, P17, P57-2 and P55 represents the first end member which is the surface water suffering from partial evaporation. The second end-member P42, P41, P24 and P49 (which is more depleted in stable isotope contents) and whose $\delta^{18}O$ contents vary between −5.41 and −4.96 ‰, thus less than −4.41 ‰, the weighted average of rain in Tunis-Carthage (Celle-Jeanton et al. 2001), is mainly composed by the local groundwater system which shows a different isotopic composition that has ruled out of the theory that the reservoir water is the only source of water in the piezometers (Fig. 10.12). This separate cluster plot to the left of meteoric water line is indicative of non-local recharge. The points located between the two end

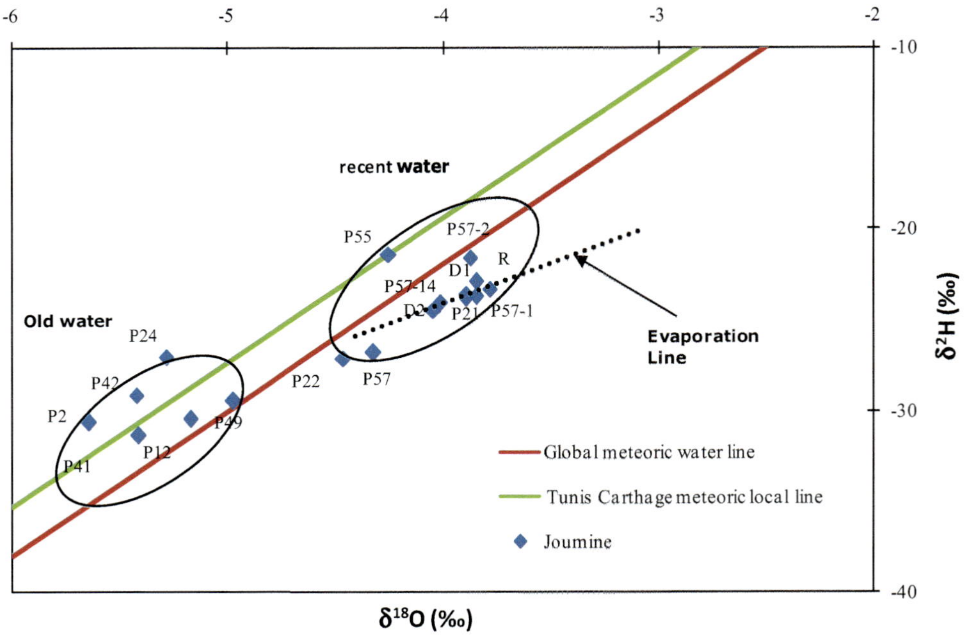

Fig. 10.12 Plot of δ^2H versus δ^{18}O

members in Fig. 10.12 (P22, P57) represent a mixture of the reservoir water and the local groundwater with different mixing proportions.

10.4.8 Carbon 13

The abundance of the stable isotope of carbon (δ^{13}C) in TDIC shows important variations in natural waters, depending on its origin. The δ^{13}C values are used to differentiate the origin of some waters discharges. More depleted δ^{13}C values (−20 to −15 ‰) are associated with groundwater whose alkalinity mainly derives from biogenic CO_2 (plant decay). Surface waters are generally fully exchange with atmospheric CO_2 and therefore show values around (−6 to −8 ‰) (Plata 1991, Plata and Araguas 2002). At the Joumine study area, δ^{13}C in TDIC was determined in six water samples collected during the field campaign in December 1998 (Table 10.1). The δ^{13}C value of the surface water collected in the reservoir was −10.35 ‰, and a bit more depleted value was obtained for the sample collected in P57 (−13.95 ‰ vs PDB) located on the pathway of the leakage. The δ^{13}C values obtained from the samples collected in D1 and D2 were −12.3 and −11.25 ‰, values slightly more depleted than the reservoir. The content of ^{13}C in P42 shows a value (−8.99 ‰) in good agreement with the equilibrium process with atmospheric CO_2. The higher value (−2.63 ‰ vs PDB) more positive was found in P12 probably due to the exchange with the aquifer carbonate.

10.4.9 Tritium

Environmental tritium was used as a natural tracer to determine the pathways of water leaks and estimate any recent recharge to the aquifer from the reservoir (Plata 1991). The spatial distribution of tritium for the local aquifer in 1998 is shown in Fig. 10.13. The concentration of tritium in the reservoir was at that time ~5 TU at a depth of 20 m (Fig. 10.13). Tritium in the aquifer and the water leakage varies between 1.24 and 4.8 TU., following the same direction of karst and confirming the above fact that the reservoir water is the main source of water leakage. These waters when they emerge from the reservoir are mixed with water without tritium (local groundwater) and gives lower concentrations resulting from the mixing of these two types of water.

The samples analyzed for tritium can be divided into two groups: the first group contains water with low tritium concentrations (about 1.0 TU); these waters are present in the piezometers P2, P12 and P42. These tritium levels indicate the existence of old waters originating from the deep Triassic aquifers.

The second group contains waters with higher levels of tritium (about 5 TU), similar to those observed in the reservoir. This observation is explained by the fact that water leakage from the reservoir recharges piezometers P21, P22, P55, P57 and P57-2. We can consider that all the waters of this group are recent groundwater and do not contain tritium from thermonuclear tests in the 1960s.

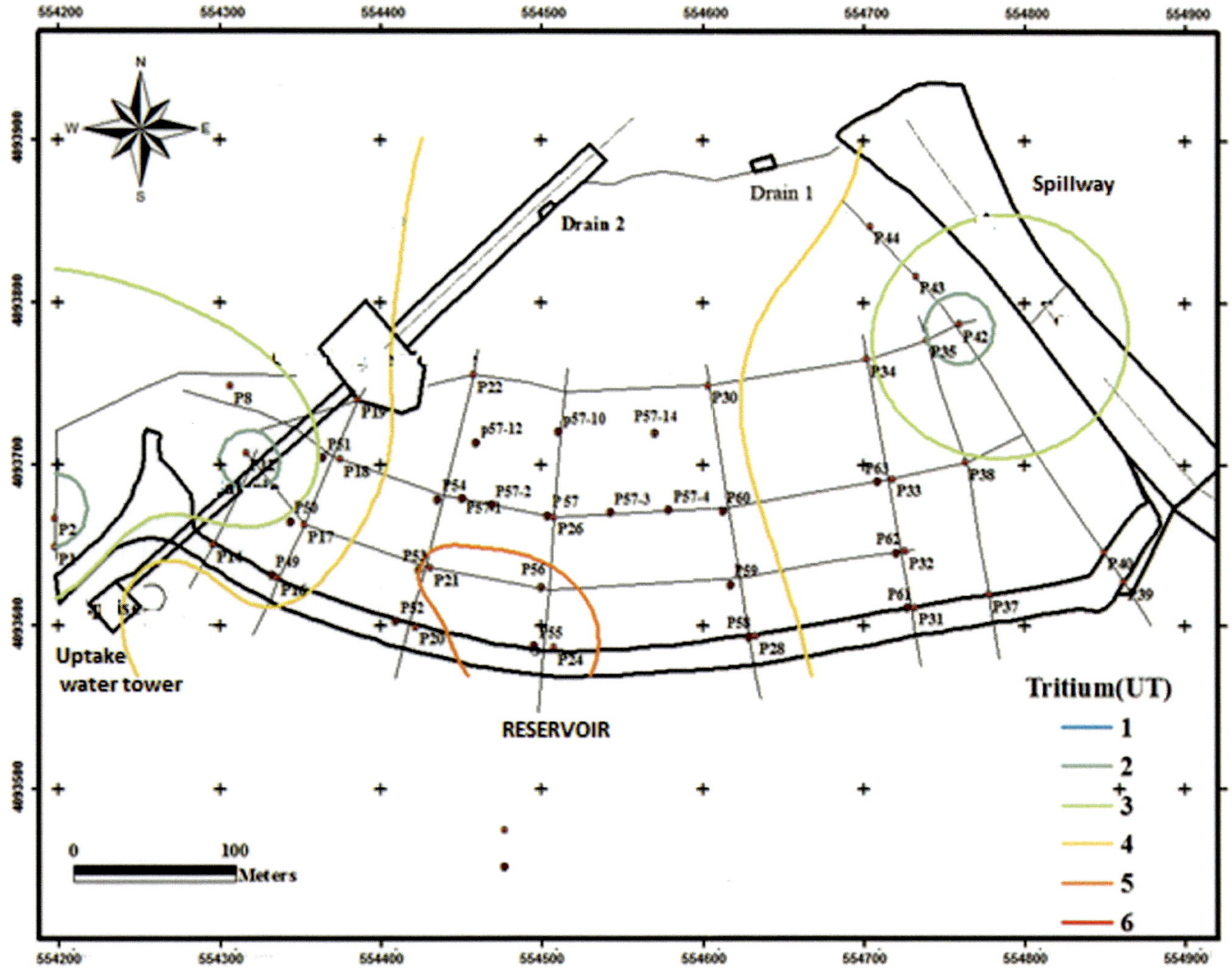

Fig. 10.13 Map of distribution of tritium

10.5 Conclusions and Suggestions

The results presented in this paper have provided a deeper insight into the nature and characteristic of the leaks and discharge of local groundwater in the vicinity of the Joumine dam. The main results are:

(a) Water leakage from the reservoir mainly occurs in the karstic area situated in the left bank foundation and abutment and moves toward the middle part of the downstream toe of the dam, and then dissipates into two main flows that join through the drainage system drains D1 and D2.
(b) The spatial distribution of temperature, electrical conductivity and the chemical composition of water in the piezometers and in D1 and D2 shows that the origin of leakage occurs through a perched aquifer situated in the left abutment and has a direct impact on the water quality in D2 by increasing the salinity and has no influence about the stability of the dam.
(c) Finally, the change in the salinity of water observed in D2 is due to the contribution of brackish water from the mound on the Triassic formations in the left bank. Hydrological maps were produced showing the water leakage paths at the dam site. Many of the conclusions presented in this paper could not have been obtained using exclusively classical hydrological methods.

Acknowledgements This work greatly benefited from financial and logistical support provided by several organizations to which we address our thanks: Ministry of Agriculture: Barrage de Joumine, Direction des Etudes et Grands Travaux Hydrauliques and the Centre National des Sciences et Technologies Nucléaires, Tunisia. These initiatives have been supported by a technical cooperation project RAF/8/028 of the International Atomic Energy Agency (IAEA), Vienna, Austria.

References

Adar ME (1994) Application of tracers in arid zone hydrology. In: Proceedings of Symposium. IAHS Publication No.232, pp 451

Aggarwal PK, Froehlich K, Gonfiantini R (2010) Contributions of the international atomic energy agency to the development and practice of isotope hydrology. Hydrogeol J 19:5–8. https://doi.org/10.1007/s10040-010-0648-3

Ben Hamouda MF (2000) The use of nuclear techniques for the detection of dam leakages: case of joumine dam in the North of Tunisia. In: Proceedings of fifth Arab conference on the peaceful uses of atomic energy, 13–17 Nov, Beirut, Lebanon

Ben Hamouda MF (2013) Use of geochemical tracers for characterization and quantification of water leakage at the Joumine dam site, Tunisia. In: 10th applied isotope geochemistry AIG 10, 22–27 Sept, Budapest, Hungary

Benischke R (1987) Fluorescent tracers in hydrology, post graduate training course on groundwater tracing techniques. Institute for Geothermics and Hydrogeology, Research Centre Joanneum, Graz, Austria

Boiten W (1992) Northern Tunisia water resource management project. Report of a preappraisal mission for the World Bank, April 26–May 4. ISSN 0926-230X

Bolève A, Janod F, Revil A, Lafon A, Fry JJ (2011) Localization and quantification of leakages in dams using time-lapse self-potential measurements associated with salt tracer injection. J Hydrol 403:242–252. https://doi.org/10.1016/j.jhydrol.2011.04.008

Celle-Jeanton H, Zouari K, Travi Y, Daoud A (2001) Caractérisation isotopique des pluies en Tunisie. Essai de typologie dans la région de Sfax. Isotopic characterisation of the precipitation in Tunisia. [Variations of the stable isotope compositions of rainfall events related to the origin of air masses]. Earth Planet Sci Lett 333:625–631. https://doi.org/10.1016/S1251-8050(01)01671-8

Clark ID, Fritz P (1997) Environmental isotopes in hydrogeology. CRC Lewis Publishers, New York, p 328

Coleman ML, Shepherd TJ, Durham JJ, Rouse JE, Moore GR (1982) Reduction of water with zinc for hydrogen isotope analysis. Anal Chem 54:993–995. https://doi.org/10.1021/ac00243a035

Contreras IA, Hernández SH (2010) Techniques for prevention and detection of leakage in dams and reservoirs. In: Collaborative management of integrated watersheds, 30th annual USSD conference sacramento, California, April 12–16, pp 785–814

Coplen TB (1988) Normalization of oxygen and hydrogen isotope data. Chem Geol 72:293–297. https://doi.org/10.1016/0168-9622(88)90042-5

Drost W, Neumaier F (1974) Application of single borehole methods in groundwater research, isotope techniques in groundwater hydrology. In: Proceedings of symposium, 11–15 March, Vienna, IAEA, p 241

Epstein S, Mayeda TK (1953) Variations of ^{18}O content of waters from natural sources. Geochim Cosmochim Acta 4:213–224. https://doi.org/10.1016/0016-7037(53)90051-9

Irie M, Abe Y, Isoda H (2009) Water resource and reuse in Tunisia. J Arid Land Stud 18:171–175

Irie M, Kawachi A, Tarhouni J, Ghrabi A (2011) Development of sedimentation and characteristics of sediment on the reservoir in Tunisia. Annu J Hydrosci Hydraul Eng 55:163–168

Lee JY, Choi YK, Kim HS, Yun ST (2005) Hydrologic characteristics of a large rockfill dam: implications for water leakage. Eng Geol 80:43–59

Louati MH, Bucknall J (2010) Tunisia's experience in water resource mobilization and management. Word development record, WDR, background note 25 p

Mozafari M, Raeisi E, ZareWater M (2012) Leakage paths in the Doosti Dam, Turkmenistan and Iran. Environ Earth Sci 65 (103):117. https://doi.org/10.1007/s12665-011-1069-x

Riou C (1980) A simple empiric formula for estimating potential evapotranspiration in Tunisia. Cah. O.R.S.T.O.M., sér Hydrol XVII (2):129–137 [in French]

Peng TR, Wang CH (2008) Identification of sources and causes of leakage on a zoned earth dam in northern Taiwan: hydrological and isotopic evidence. Appl Geochem 23:2438–2451. https://doi.org/10.1016/j.apgeochem.2008.05.015

Plata BA, Araguás AL (2002) Detection and prevention of leaks from dams. A.A. Balkema Publishers, p 419

Plata A, Iragüen E (1992) Leakage study at lake Laja Chile, In: Hötzl and Werner (eds) Tracer hydrology. Balkema, pp 378–396

Plata A (1983) Single well techniques using radioactive tracers, tracer methods in isotope hydrology. In: Proceedings of advisory group meeting, 27 Sept–1 Oct, IAEA, Vienna, IAEA-TECDOC-291, pp 17–33

Plata A (1991) Detection of leaks from reservoirs and lakes, IAEA, TECDOC-601

Plummer LN, Jones BF, Truesdell AH (1976) WATEQF, a Fortan IV version of WATEQ, a computer program for calculating chemical equilibrium of natural waters. (revised and reprinted, January, 1984): U.S. Geological Survey Water-Resources Investigations pp 76–13

Sari S (2013) Detection of water leak in the dam Joumine and study of sedimentation in the dam Ghezala by nuclear method. Memoire de projet de fin d'etude, ESIER, University of Jendouba, 130 p

Taylor CB, Schwarz P (1977) Deionization procedure for large quantities of tritium-free water using an ion-exchange column. Int J Appl Radiat Isot 28:605–605

Turkmen S (2010) Water leakage from the power tunnel of Gezende Dam, southern Turkey: a case study. Environ Earth Sci 61:419–427. https://doi.org/10.1007/s12665-009-0354-4

Determination of Suspended Sediments Using Nuclear Probe in the Medjerda River, Tunisia

11

Mohamed Fethi Ben Hamouda, Mohamed Mondher Rejeb, Noura Azizi, and Mohamed Hedi Trabelsi

Abstract

The suspended sediment concentration in Medjerda river is detected by the attenuation of radioactivity emitted from a radioactive source of americium (^{241}Am) dipped in the water. This attenuation of Gamma rays emitted from the ^{241}Am, due to the presence of the sediments transported in great amounts during events of flood is measured using a scintillation detector made up of a crystal of NaI. The suspended particles are measured in Slouguia hydrometric station (Northwest) located on the Medjerda river which is the largest river in Tunisia. The acquisition takes place in an automatic station that permits also continuous measurements of the hydrological and meteorological parameters such as temperature, rainfall, height of water and suspended solid concentration. The quantity of particles that was transported by the river during the period of observation in 2005 is estimated at 778×10^3 tons, which is equivalent to a volume of 294×10^3 m^3.

Keywords

Medjerda river • Nuclear probe • Sedimentation • Suspended elements • Solid transport • Tunisia

11.1 Introduction

Research work in Tunisia on the impact of sediment studies on hydraulic infrastructure was operated using bathymetric and topographic methods (Baccouche 1982; Ben Mamou 1998; Abdelhadi 2001). This research has been unsuccessful since the equipment that allows sampling of sediment-laden water has not been satisfactory, as well as faulty manual sampling and handling errors. As a result, recent nuclear methods constitute a great contribution to research on the measurement of the sediment concentration in runoff water (Tazioli 1980, 1981; Tazioli et al. 1987; Becchi and Tazioli 1987; Ben Hamouda and Trabelsi 2002; Ben Hamouda 2004, Tazioli 2009).

These methods not only quantify erosion but also spatially determine the sensitivity to erosion. In fact, at least 25–30 million m^3 of sediments are deposited annually in the reservoirs of dams which lose on average 0.5% of their storage capacities (Saadaoui 1996, Saadaoui and Ben Mansour 2003, Ben Mammou and Louati 2007; Ben Moussa et al. 2019).

In addition, sediment transport in surface water contributes to the premature clogging of the hydraulic infrastructures installed on the rivers as well as the enlargement of the beds of the wadis. In this scientific chapter, we propose to determine the suspended sediment transport in the Medjerda river at the level of the Slouguia station located in the northwest of Tunisia, 70 km from the city of Tunis.

The Medjerda, one of the rivers located in the north of Tunisia, is the main perennial river of the country both in terms of its length (484 km), the area of its watershed (7600 km^2) and the volume of water that it carries (Fig. 11.1). The Medjerda valley is characterized by a sub-humid to semi-arid climate with a mean annual rainfall on the watershed of 480 mm/year (Zahar et al. 2008).

The concentration of suspended elements transported in Medjerda river is very high. It can reach 100 g l^{-1} which results in continuous enlargement of its bed. It is therefore imperative to assess more precisely the quantities of

M. F. B. Hamouda (✉) · N. Azizi · M. H. Trabelsi
Isotope Hydrology and Geochemistry Unit, CNSTN, Tunis, Tunisia
e-mail: f_benhamouda@yahoo.fr

M. F. B. Hamouda
LRSTE, Institut National Agronomique de Tunisie, Tunis, Tunisia

M. M. Rejeb
Direction Générale Des Ressources en Eau, Tunis, Tunisia

S. Khomsi and F. Roure (eds.), *Geology of North Africa and the Mediterranean: Sedimentary Basins and Georesources*, Regional Geology Reviews, https://doi.org/10.1007/978-3-031-18747-6_11

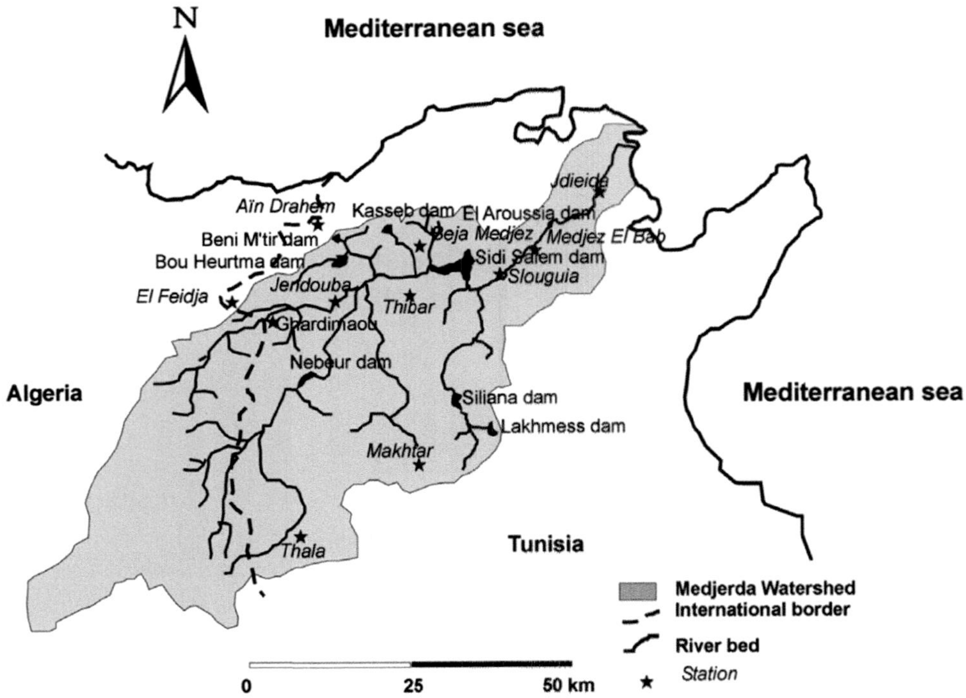

Fig. 11.1 Watershed of the Medjerda river, Tunisia (Zahar et al. 2008)

sediment transported. The continuous in situ measurement of the suspended element in Medjerda during flood provide essential information for geoscientist and water resources managers (Ben Hamouda et al. 2011; Ben Hamouda 2004, Louati et al. 2014).

11.2 Method

11.2.1 Principle of a Nuclear Gauge

The measurement of the concentration of suspended sediments with a nuclear gauge is based on the absorption of gamma radiation from an artificial radioactive source or alternatively on the measurement of the natural gamma radioactivity emitted by the particles in suspension (Caillot et al. 1984; Crikmore et al. 1990; Florkowski 1970; Magloire 1972; Tazioli and Caillot 1983).

The intensity I of monoenergetic electromagnetic radiation transmitted through x cm of water containing a concentration C, by weight of sediment is related to the source of intensity I_o, in the absence of water and sediment, by the equation:

$$\frac{I}{I_0} = exp(-\rho_m \times [\mu_s C + \mu_w (1 - C)])$$

where μ_s is the coefficient mass attenuation of sediment; μ_w is the coefficient of mass attenuation of water and ρ_m is the density of the water-sediment mixture (Florkowski 1970), the density ρ_m is given by the equation

$$\rho_m = \frac{\rho_s \rho_m}{\rho_s - C(\rho_s - \rho_w)}$$

with ρ_s and ρ_w are the densities of sediment and water, respectively.

The sensitivity and accuracy of the measurements increase with the increasing difference in mass attenuation coefficients of sediment and water, therefore low energy radiation electromagnetic is preferable since $\mu_s > \mu_W$ and the error due to the variation in water density with temperature is small. It is also necessary to know the temperature and the salinity of the suspended elements in water in the medium in order to make the necessary corrections of measurements (Tazioli and Caillot 1983). The suspended elements are measured by an NaI scintillation detector from the detection of gamma radiation emitted by a radioactive source of ^{241}Am americium with 100 mCi activity (Fig. 11.2). The detector and the source are 31.5 cm apart. This radiation will be attenuated by the amount of sediment that will cross the section of the river. The detector transforms the physical quantity into an electrical quantity in the form of pulses. These are sent to the Data-Logger to be counted and recorded in the form of data expressed in counts. Since the water level in the Medjerda is variable, the Slouguia station was designed with the possibility of making vertical translational movements of the column that holds the probe in order to be able to measure the particles in suspension according to the level of the water in Medjerda river (Fig. 11.3a, b).

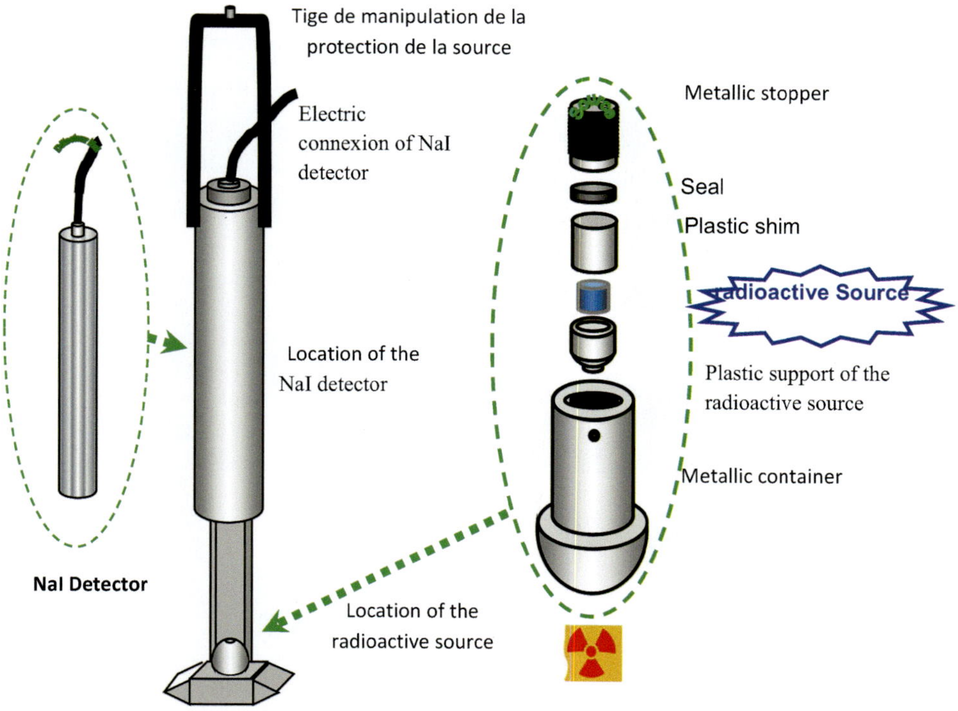

Fig. 11.2 Implementation of the radioactive source and the NaI detector in the probe

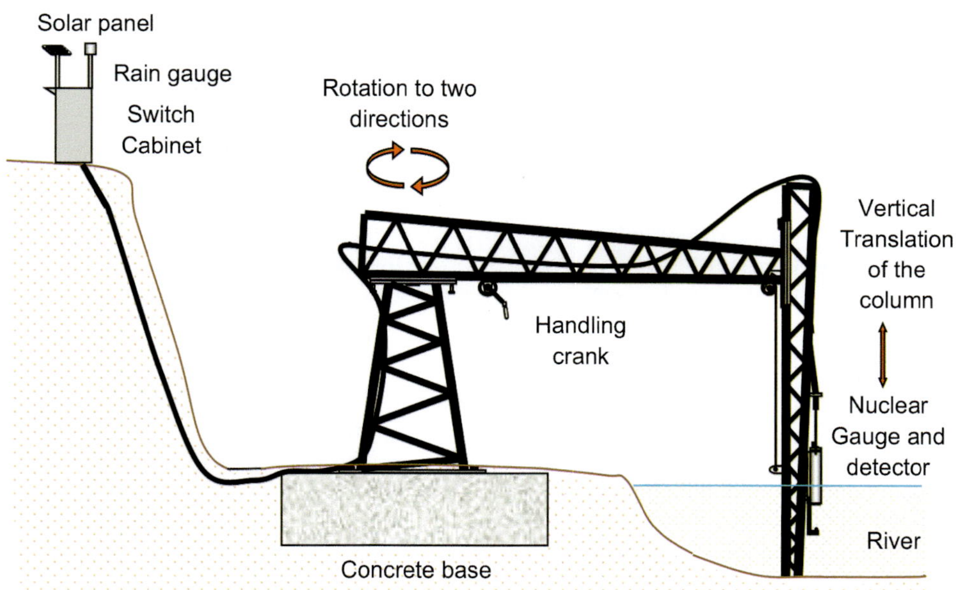

Fig. 11.3 **a** General outline of the measuring station with the metallic structures. **b** Photos showing the nuclear probe at Slouguia hydrologic station

11.2.2 Determination of the Water Depth in the River

The stage in the wadi (water level) is recorded, by the hydraulic pressure sensor expressed in the intensity of the electric current. This current is expressed in milli-ampere (mA) and it is converted into stage expressed in meters (m). This requires the establishment of the calibration curve of the pressure switch in the wadi (Fig. 11.4). The water level is read directly on a graduated scale placed at the level of the

b

Fig. 11.3 (continued)

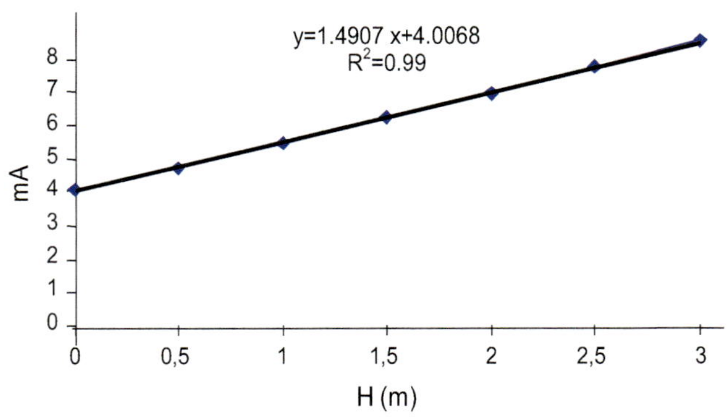

Fig. 11.4 Level of water in river versus pressure

wadi to make the necessary corrections to the pressure switch measurements. The equation that allows the transformation of electric current data in a stage is (11.1):

$$H(m) = \frac{mA - 4}{1.49} \tag{11.1}$$

with H: water stage and mA: water pressure calculated in amperes.

To this stage, we add the distance between the bottom of the wadi with the location of the pressure sensor.

11.2.3 Determination of the Flow Rate of the Wadi

From the water depth (stage) measured in meters, the flow rate of the wadi is determined after having previously established the gauging curve (stream measurement) for the section of the wadi (Fig. 11.5).

11.2.4 Determination of the Concentration of Suspended Elements

The correct operation of the nuclear gauge depends on the calibrations undertaken both in the laboratory and in situ. The low-energy radiation instrument becomes more sensitive to changes in the chemical composition of the sediment.

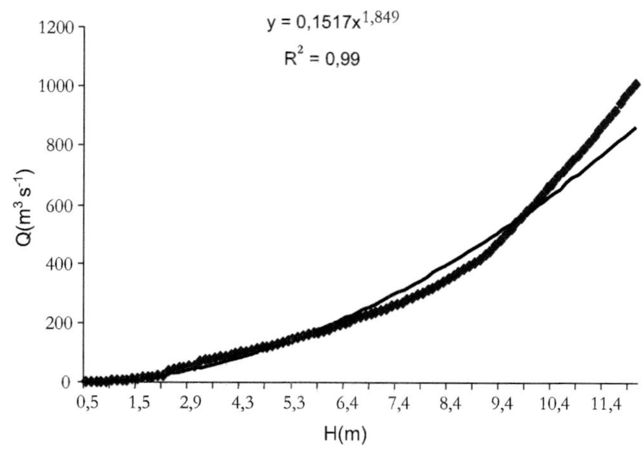

Fig. 11.5 Gauging curve at Slouguia station

In practice, it is necessary to calibrate the instruments with the material to be measured or with the simulated means. This calibration of the gauge, in the laboratory with the materials simulated in the sediments (sodium sulfates or sodium silicate solutions), provides the typical measurement curve of the sediment composition (Tazioli and Caillot 1983).

In the calibration of the nuclear probe, sodium chloride (NaCl) solution was used at various concentrations. The equation relating the concentration to the counting is obtained from the calibration curve of the NaI detector established in the laboratory with pre-determined concentrations (Fig. 11.6).

The validity of the calibration curve is further verified in situ with sampling during the flood event. It is determined by the relationship between the continuous count detected from the gamma radiation (γ) and the values of varying concentrations of suspended sediments (Crikmore et al. 1990). Thus the concentration is related to the counting by Eq. (11.2)

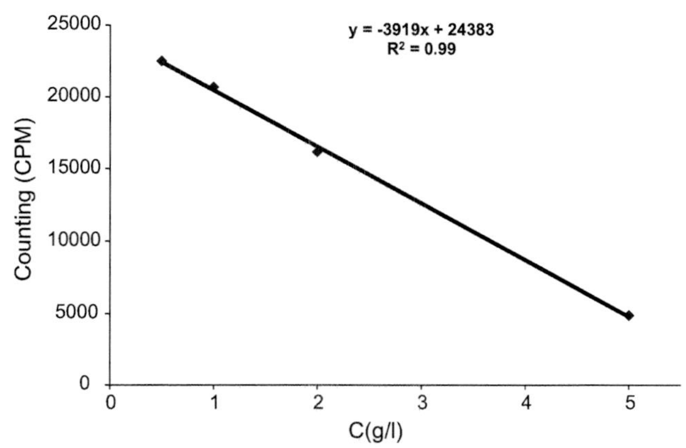

Fig. 11.6 Counting and concentration of SE relationship

$$C(g/l) = \frac{24383 - CPM}{3919} \tag{11.2}$$

with CPM: counting per minute and C: concentration of suspended elements.

11.2.5 Determination of the Solid Flow Gs of the Elements in Suspension

The flow of suspended particles (Gs) crossing a section of a river is given by the equation:

$$Gs(Kg/s) = Q \times C \tag{11.3}$$

with Q: the liquid flow ($m^3\,s^{-1}$) and C: the concentration of suspended elements ($kg\,m^{-3}$).

11.2.6 Determination of the Amount of Sediment Crossing the Section of the River

The determination of the amount of the sediments (weight) crossing the section of the river is established from the calculation of the area of the curve Gs = f (t) and therefore the integral of the solid flow Gs between two instants of measurement.

$$m = \int G_s dt$$

The total mass is determined by the sum of the masses calculated during the measurement period.

$$m_{tot} = \sum_{i=1}^{n} m_i$$

11.3 Results and Discussion

The network de of surface water mobilized by dams in Tunisia constitutes an important part of the water resources of this country. Given That the Tunisian wadis are characterized by a fairly strong sediment flow rate, the continuous control of the sedimentation flow in upstream and downstream of the large water reservoirs is very important for the management of the country's hydraulic infrastructure.

Indeed, the installation of the Sidi Salem dam caused a consequent decrease in the flow rate of the wadi Medjerda in downstream of this hydraulic structure whose vocation (irrigation, drinking water, hydroelectricity and protection against floods) has oriented managers to frequent and low-speed water releases from the Sidi Salem dam. These

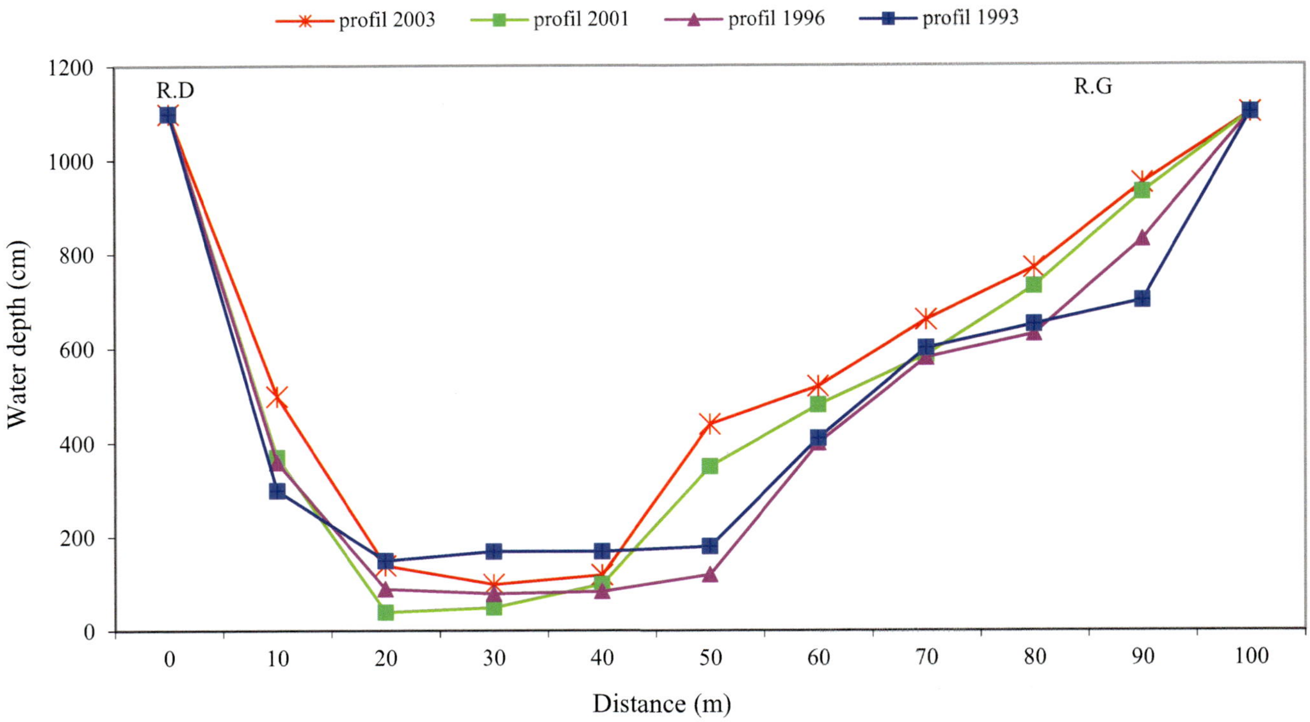

Fig. 11.7 Profile across the wadi Medjerda at Slouguia station

low water speed releases are sometimes combined with impoundment clearance dam maneuvers.

The Medjerda has therefore experienced continuous enlargement since 1981 when Sidi Salem dam was put into service. In places, the flow causing the overflow has fallen to less than 200 m^3s^{-1}, a value much lower than 700 m^3s^{-1}, the overflow flow existing before the construction of the dam which was. To illustrate the enlargement of the river, we studied the evolution of the cross section at the level of the hydrometric station of Slouguia. The evolution of the section at this station is seen in the plotted three cross sections in Fig. 11.7.

By examining the evolution of the cross sections in the Slouguia hydrometric station, we can postulate an enlargement of the bed of the Medjerda with some digging which coincides with periods of more frequent water releases and at higher flows. This phenomenon of enlargement of Medjerda shows that the river bed is out of equilibrium, that is to say, the deposit and the erosion are not equivalents (Gafrej 1990; Ghorbal et al. 2000, 2001).

Radioactive methods for turbidity measurement allow the exploration of areas with high concentrations of suspended elements. As a result, the first station for measuring suspended elements by nuclear technique in Tunisia is installed downstream of the Sidi Salem dam to ensure precise measurement of suspended sediments in the water and continuous control of sedimentation at the level of Wadi Medjerda at Slouguia station. Figure 11.8 shows an example of measurements taken in February and March 2005.

The measurements of the hydrological parameters at the Slouguia station on the Medjerda river show a large fluctuation in the concentration of solid elements over time. It goes from 0.01 to 6.22 g l^{-1}. The average concentration is 3.72 g l^{-1}. The water releases from Sidi Salem dam upstream of the Slouguia station have a great influence on the quality of the water that crosses through the measurement wadi section, particularly following the operations to clear the water reservoir. The quantity of sediment having crossed the section of the river during the observation period is evaluated at a mass of 778 10^3 tons of sediment equivalent to a volume equal to 293.6 × 10^3 m^3 for an average density of 2.65.

The determination of the suspended sediment was also estimated by the filtration method during the same measurement period. The concentration of suspended elements has an average value of 5 g.l^{-1}. The difference between the results of the two methods used is due, on the one hand, to the fact that the turbidity of a river is a phenomenon presenting in the same medium and at the same time large random variations. On the other hand, it is the difference in the location of the measurement point and the choice of the location of the nuclear probe. Indeed, the filtration method is carried out on a surface sample. While the nuclear method takes a measurement of 30 cm from the bottom of the river.

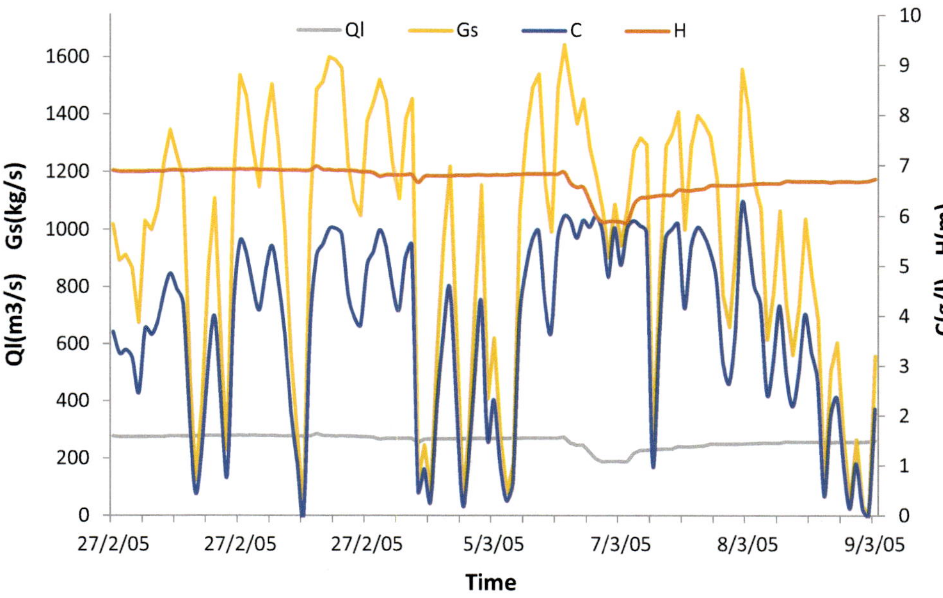

Fig. 11.8 Example of hydrological parameters over 2005 at Slouguia Station

Indeed, analysis of the Rouse-Vanoni equation (Rouse 1939) has shown that for sediments with a diameter of <20 μm, the concentration profile, at equilibrium, is approximately uniform. For sediments with a diameter of d > 20 μm (the case of Medjerda river d_{75} > 20 μm), the concentration is representative unless the probe is located at mid-height of the water column. However, surface sampling for flow rates Q > 70 m^3s^{-1} corresponding to a water column H > 2 m, requires that the probe be positioned at a level H = 1 m. But, for safety conditions, the current installation of the probe only allows the suspended matter to be measured at a height of H = 30 cm. Also, the physical phenomenon of vertical and horizontal diffusion of suspended elements requires that it can only be measured with vertical two-dimensional analysis.

11.4 Conclusion

Except under special conditions, nuclear probes do not allow the measurement of turbidity lower than 1–2 $g.l^{-1}$. Their use is recommended for high turbidity 3–10 $g.l^{-1}$ which is frequently found in watersheds in Tunisia and they are able to measure the suspended sediment concentration during the floods.

It is difficult to interpret a measurement of suspended sediments with punctual and instantaneous measurements over a long period of time with a hope of reproducible results. This is the case with measurement by a nuclear technique which incorporates the phenomenon of turbidity over 30 cm and during a chosen time interval. In addition, the filtration method opposes its precision to its slowness. This is a good method for research. While the nuclear method combines speed with precision in measurements.

The sediment concentrations recorded by the nuclear technique showed the importance of the deposits downstream of the Sidi Salem dam, which was estimated to the enhancement of a layer of 15 cm after the 2003 floods.

The results obtained in this study on the assessment of erosion and sedimentation in the Medjerda Basin contribute to the advancement of knowledge on the transport mechanisms of suspended materials, a crucial parameter for the assessment of sedimentation in the management of solid transport in Tunisia, in particular in the northern regions of Tunisia which have recently been subjected to severe climatic constraints (floods) and to growing demographic, agricultural and industrial development.

Acknowledgements This study was funded by the IAEA Technical Cooperation Project TUN/8/014 "Evaluation of erosion and Sedimentation in the Medjerda watershed, Tunisia ".

References

Abdelhedi T (DGBGTH) (2001) Acte des 18ième journées des ressources en eau: Tunis le 26 avril 2001 à l'INAT. Le transport solide des oueds en Tunisie

Baccouche S (1982) Contribution à l'étude du transport solide par l'oued Zeroud à sidi Saâd. Mémoire de fin d'études du cycle de spécialisation à l'Institut National d'Agronomie de Tunis

Becchi I, Tazioli GS (1987) On the river processes methodological approach: experience with the river Musone, Italy. In: Proceedings of an international symposium on the use of isotopes techniques in water resources development organised by the IAEA, Vienne 30/3-3/04/1987

Ben Hamouda MF, Tarhouni J, Leduc C et al (2011) Understanding the origin of salinization of the Plio-quaternary eastern coastal aquifer of Cap Bon (Tunisia) using geochemical and isotope investigations. Environ Earth Sci 63:889–901. https://doi.org/10.1007/s12665-010-0758-1

Ben Hamouda MF (2004) Rapport final du projet AIEA, TUN8/014, Evaluation de l'érosion et de la sédimentation dans le bassin de la Medjerda, Tunisie

Ben Hamouda MF, Trabelsi MH (2002) Rapport de stage AIEA effectué à l'universita delgi studi di Ancona, Italie du 14 Mai au 14 Juillet 2002

Ben Mammou A (1998) Barrages Nebeur, Sidi Salem, Sidi Saâd et Sidi Bouraker

Quantification, Etude sédimentontologique et géotechnique des sédiments piégés. Apports des images satellitaires. Thèse de doctorat Es-sciences géologiques, 345 p

Ben Mammou A, Louati MH (2007) Temporal silting evolution in Tunisian réservoirs. Revue des sciences de l'eau. J Water Sci 20:201–210. https://doi.org/10.7202/015813ar

Ben Moussa T, Amrouni O, Hzami A, Dezileau L, Mahé G, Abdeljaouad S (2019) Progradation and retrogradation of the Medjerda delta during the 20th century (Tunisia, western Mediterranean). CR Geosci 351(4):340–350. https://doi.org/10.1016/j.crte.2018.10.004

Caillot A, Meyer G, Chambellan D, Tanguy JC (1984) A new nuclear gauge to measure high turbidities in muddy areas. Coast Eng Conf Houston. https://doi.org/10.1061/9780872624382.213

Crikmore et al (1990) The use of nuclear techniques in sediment transport and sedimentation problems. IHP-III Project 5.2 UNESCO International Hydrological Programme, Paris, pp 35–36. https://hydrologie.org/BIB/Publ_UNESCO/TD_SC90_WS049.pdf

Florkowski (1970) Portable radioisotope gauges for suspended sediments Symposium on the use of isotopes in hydrology; Vienna, Austria; 9 Mar 1970, 8 p, 2 tables, 6 figs. 1001728.pdf (iaea.org)

Gafrej R (1990) Modélisation statique des matières en suspension sur un bassin versant expérimental représentatif : l'Orgeval. Mémoire de DEA de Mécanique Appliquée. Faculté des Sciences de Tunis

Ghorbel A, Magri A, Zitouni M, Ben Belgacem M, Balti N, Fouzai M, Gadri B (2000) Crue des 26 et 27 Mai 2000. Rapport interne de la DGRE

Ghorbel A, Réjeb MM, Balti N (2001) Engravement du lit de la Medjerda. Acte des 18ième journées des ressources en eau: Tunis le 26 avril 2001 à l'INAT. Le transport solide des oueds en Tunisie

Louati M, Saïdi H (2014) Zargouni F (2014) Shoreline change assessment using remote sensing and GIS techniques: a case study of the Medjerda delta coast. Tunisia Arab J Geosci 8:4239. https://doi.org/10.1007/s12517-014-1472-1

Magloire M (1972) Mesure de turbidité par des méthodes radioactives. Thèse de Docteur Ingénieur. Université Paris VI, 107 p. 4074223.pdf (iaea.org)

Rouse H (1939) An analysis of sediment transportation in the light of fluid turbulence. United States Department of Agriculture Soil Conservation Service, 1–25. An Analysis of sediment transportation in the light of-1999.pdf (caltech.edu)

Saadaoui M (1996) Erosion et transport solide en Tunisie, Mesure et prévision du transport solide dans les bassins versants et de l'envasement dans les retenues des barrages. Bulletin de la direction des sols, no 17

Saadaoui M, Ben Mansour H (2003) Crues et inondations dans la haute vallée de la Medjerda. Cures de janvier - Février 2003

Tazioli A (2009) Evaluation of erosion in equipped basins: preliminary results of a comparison between the Gavrilovic model and direct measurements of sediment transport. Environ Geol 56:825–831. https://doi.org/10.1007/s00254-007-1183-y

Tazioli GS (1980) The measurement of suspended sediment transport in natural streams using automatic radioisotope gauges. J Hydrol 47:173–185. https://doi.org/10.1016/0022-1694(80)90055-4

Tazioli GS (1981) Nuclear techniques for measuring sediment transport in natural streams. Examples from instrumented basins. In: Proceedings of the Florence symposium on erosion and sediment transport measurement. IAHS Publ no 133. iahs_133_0063.pdf (http://www.hydrologie.org)

Tazioli GS, Caillot A (1983) Suspended sediment. Guidebook on nuclear techniques in hydrology 1983 Edition. Technical reports series no 91. International Atomic Energy Agency, Vienna, pp 93–95

Tazioli GS, Billi P, Tacconi P (1987) On fluvial dynamics in Italy. The lithosphere in Italy, Advances in Earth Science Research. Accademia Nationale dei Lincei, Roma, May 1987

Zahar Y, Ghorbel A , Albergel J (2008) Impacts of large dams on downstream flow conditions of rivers: Aggradation and reduction of the Medjerda channel capacity downstream of the Sidi Salem dam (Tunisia) J Hydrol 351, 318–330. https://doi.org/10.1016/j.jhydrol.2007.12.019

GPSR Compliance

The European Union's (EU) General Product Safety Regulation (GPSR) is a set of rules that requires consumer products to be safe and our obligations to ensure this.

If you have any concerns about our products, you can contact us on ProductSafety@springernature.com

In case Publisher is established outside the EU, the EU authorized representative is:

Springer Nature Customer Service Center GmbH
Europaplatz 3
69115 Heidelberg, Germany

Batch number: 10756390

Printed by Printforce, the Netherlands